CAMBRIDGE STUDIES IN
ADVANCED MATHEMATICS : 56

COHOMOLOGY OF DRINFELD MODULAR VARIETIES, PART II

Already published

1 W.M.L. Holcombe *Algebraic automata theory*
2 K. Petersen *Ergodic theory*
3 P.T. Johnstone *Stone spaces*
4 W.H. Schikhof *Ultrametric calculus*
5 J.-P. Kahane *Some random series of functions, 2nd edition*
6 H. Cohn *Introduction to the construction of class fields*
7 J. Lambek & P.J. Scott *Introduction to higher-order categorical logic*
8 H. Matsumura *Commutative ring theory*
9 C.B. Thomas *Characteristic classes and the cohomology of finite groups*
10 M. Aschbacher *Finite group theory*
11 J.L. Alperin *Local representation theory*
12 P. Koosis *The logarithmic integral I*
13 A. Pietsch *Eigenvalues and s-numbers*
14 S.J. Patterson *An introduction to the theory of the Riemann zeta-function*
15 H.J. Baues *Algebraic homotopy*
16 V.S. Varadarajan *Introduction to harmonic analysis on semisimple Lie groups*
17 W. Dicks & M. Dunwoody *Groups acting on graphs*
18 L.J. Corwin & F.P. Greenleaf *Representations of nilpotent Lie groups and their applications*
19 R. Fritsch & R. Piccinini *Cellular structures in topology*
20 H Klingen *Introductory lectures on Siegel modular forms*
22 M.J. Collins *Representations and characters of finite groups*
24 H. Kunita *Stochastic flows and stochastic differential equations*
25 P. Wojtaszczyk *Banach spaces for analysts*
26 J.E. Gilbert & M.A.M. Murray *Clifford algebras and Dirac operators in harmonic analysis*
27 A. Frohlich & M.J. Taylor *Algebraic number theory*
28 K. Goebel & W.A. Kirk *Topics in metric fixed point theory*
29 J.F. Humphreys *Reflection groups and Coxeter groups*
30 D.J. Benson *Representations and cohomology I*
31 D.J. Benson *Representations and cohomology II*
32 C. Allday & V. Puppe *Cohomological methods in transformation groups*
33 C. Soulé et al *Lectures on Arakelov geometry*
34 A. Ambrosetti & G. Prodi *A primer of nonlinear analysis*
35 J. Palis & F. Takens *Hyperbolicity, stability and chaos at homoclinic bifurcations*
37 Y. Meyer *Wavelets and operators I*
38 C. Weibel *An introduction to homological algebra*
39 W. Bruns & J. Herzog *Cohen-Macaulay rings*
40 V. Snaith *Explicit Brauer induction*
41 G. Laumon *Cohomology of Drinfield modular varieties I*
42 E.B. Davies *Spectral theory and differential operators*
43 J. Diestel, H. Jarchow & A. Tonge *Absolutely summing operators*
44 P. Mattila *Geometry of sets and measures in Euclidean spaces*
45 R. Pinsky *Positive harmonic functions and diffusion*
46 G. Tenenbaum *Introduction to analytic and probabilistic number theory*
47 C. Peskine *An algebraic introduction to complex projective geometry I*
50 I. Porteous *Clifford algebras and the classical groups*
51 M. Audin *Spinning Tops*
54 J. Le Potier *Lectures on Vector bundles*

Cohomology of Drinfeld Modular Varieties, Part II

Automorphic forms, trace formulas and Langlands correspondence

Gérard Laumon
Université Paris–Sud,
CNRS

with an appendix by
Jean-Loup Waldspurger

CAMBRIDGE UNIVERSITY PRESS
Cambridge, New York, Melbourne, Madrid, Cape Town, Singapore, São Paulo, Delhi

Cambridge University Press
The Edinburgh Building, Cambridge CB2 8RU, UK

Published in the United States of America by Cambridge University Press, New York

www.cambridge.org
Information on this title: www.cambridge.org/9780521109901

First published 1997
This digitally printed version 2009

A catalogue record for this publication is available from the British Library

ISBN 978-0-521-47061-2 hardback
ISBN 978-0-521-10990-1 paperback

Contents

Appendices

Preface

The second volume of *Drinfeld modular varieties* is devoted to the Arthur–Selberg trace formula and to the proof in some cases of the Ramanujan–Petersson conjecture and the global Langlands correspondence for function fields.

As in the first volume we fix a function field F together with a place ∞ of F and a positive integer d.

The group $F_\infty^\times\backslash GL_d(\mathbb{A})$ acts by right translation on the Hilbert space

$$L^2_{GL_d,1_\infty} = L^2(F_\infty^\times GL_d(F)\backslash GL_d(\mathbb{A}))$$

and any locally constant function f with compact support on $F_\infty^\times\backslash GL_d(\mathbb{A})$ induces by convolution an operator $R_{GL_d,1_\infty}(f)$ on $L^2_{GL_d,1_\infty}$. This operator admits a kernel

$$K(h,g) = \sum_{\gamma\in GL_d(F)} f(h^{-1}\gamma g)$$

and, at least formally, its trace is the integral

$$J(f) = \int_{F_\infty^\times GL_d(F)\backslash GL_d(\mathbb{A})} K(g,g)\frac{dg}{dz_\infty d\gamma}.$$

Unfortunately, for an arbitrary function f the operator $R_{GL_d,1_\infty}(f)$ is not of trace class and the integral $J(f)$ is not absolutely convergent. To tide over this difficulty Arthur has introduced a truncated version $J^T(f)$ of the above integral which is absolutely convergent. It depends on some truncation parameter T in the positive Weyl chamber $\mathfrak{a}_\emptyset^+$.

Let us fix some level I and some place o which is prime to I. If $f = f_\infty f^{\infty,o} f_o$, where f_∞ is the very cuspidal Euler–Poincaré function introduced in chapter 5 of the first volume, $f^{\infty,o}$ is an arbitrary element of the Hecke algebra of level I and f_o is the Drinfeld function of level r (for some positive integer r) introduced in chapter 4 of the first volume, we will see that $J(f)$ is convergent and that we have

$$J^T(f) = J(f)$$

for any value of the truncation parameter T which is far enough from the walls of $\mathfrak{a}_\emptyset^+$. Moreover, if we take r large enough with respect to $f^{\infty,o}$ (Kazhdan's trick) we will see that $J(f)$ is equal to the number $\mathrm{Lef}_r(f^{\infty,o})$ of fixed points of $\mathrm{Frob}_o^r \times f^{\infty,o}$ acting on the reduction in "characteristic" o of the Drinfeld modular variety M_I^d.

But, on the one hand, if r is large enough with respect to $f^{\infty,o}$ we will see that the Lefschetz number $\mathrm{Lef}_r(f^{\infty,o})$ is equal to the trace of $\mathrm{Frob}_o^r \times f^{\infty,o}$ acting on the ℓ-adic cohomology with compact supports of $\overline{F} \otimes_F M_I^d$. This is due to Grothendieck's theorem if $f^{\infty,o}$ is the trivial Hecke operator and it is due to Deligne's conjecture, proved by Fujiwara and Pink, for a general $f^{\infty,o}$. On the other hand, following Arthur the truncated integral $J^T(f)$ admits a spectral expression and, using some residue computations, which have been done by Waldspurger and which are included as an appendix at the end of this volume, we will make explicit this spectral expression.

Putting together all these results we will obtain an explicit expression for the alternating sum of traces,

$$\sum_n (-1)^n \operatorname{tr}(\mathrm{Frob}_o^r \times f^{\infty,o}, H_c^n(\overline{F} \otimes_F M_I^d, \mathbb{Q}_\ell)),$$

in terms of cuspidal automorphic representations of $F_\infty^\times \backslash GL_{d'}(\mathbb{A})$ ($d' = 1, \ldots, d$).

Finally, by a standard procedure we will deduce the Petersson conjecture and the Langlands correspondence for cuspidal automorphic representations of $F_\infty^\times \backslash GL_d(\mathbb{A})$ the local component at ∞ of which is isomorphic to the Steinberg representation.

The numbering of this volume is the continuation of the numbering of the first one. Here is a brief description of its contents.

In chapter 9 we review some basic definitions and results about the cuspidal spectrum of $L^2_{GL_d,1_\infty}$.

In chapter 10 we study the geometric side of Arthur's non-invariant trace formula for our function $f = f_\infty f^{\infty,o} f_o$ and we prove that, if r is large enough with respect to $f^{\infty,o}$ (and f_∞), it has a simple form. In fact f_∞ may be any very cuspidal function and, as a special case, we obtain the Flicker–Kazhdan simple trace formula. The arguments are adapted from those used by Arthur in the number field case.

In chapter 11 we study the spectral side. Again we adapt Arthur's arguments. But here we have not been courageous enough to transpose all of his arguments to the function field case. Actually $J^T(f)$ is a sum over the cuspidal data of expressions $J_c^T(f)$. Using Waldspurger's residue computations we obtain an explicit formula for $J_c^T(f)$ when c is a regular cuspidal datum. For the other cuspidal data we only state a conjectural formula. This formula has been recently proved by Lafforgue.

In chapter 12 we deduce the Ramanujan–Petersson conjecture and the global Langlands correspondence (for cuspidal automorphic representations of $F_\infty^\times \backslash GL_d(\mathbb{A})$ the local component at ∞ of which is isomorphic to the

Steinberg representation) from the results of the previous chapters. We also give a complete description of the virtual $(\mathrm{Gal}(\overline{F}/F) \times GL_d(\mathbb{A}^\infty))$-module

$$\sum_n (-1)^n H_c^n(\overline{F} \otimes_F M_I^d, \overline{\mathbb{Q}}_\ell).$$

Up to this point we have only considered the cohomology with compact supports of Drinfeld modular varieties. We may also consider the intersection cohomology of the Satake compactification of M_I^d. In chapter 13 we give a conjectural description of the intersection complex of this Satake compactification. We have discovered this conjectural description by transposing to the function field case a formula for the L^2-Lefschetz number of a Hecke operator which has been proved by Arthur in the number field case.

There are four appendices. An addendum to appendix D contains some rationality results and a definition of the Grothendieck group of admissible representations.

The main results of reduction theory are reviewed in appendix E. Our proofs differ from Harder's original ones in the way that we systematically use the Harder–Narasimhan filtration.

In appendix F we give the proof of Harish-Chandra's results on orbital integrals which are needed in chapter 10.

In appendix G we present some of the basic results concerning the spectral decomposition of Langlands and Morris. In particular we explain the first step in Langlands' computation of the scalar product of two pseudo-Eisenstein series associated with cuspidal automorphic forms of Levi subgroups.

I would like to thank J.-L. Waldspurger once more for his help during the elaboration of this project. His residue computations are fundamental for the results of the second volume. I would also like to thank R. Pink for his comments on chapter 13. During the preparation of the manuscript I visited the University of Toronto (Winter 1993). My thanks go to J. Arthur for his kind hospitality and for the numerous discussions that I had with him. Special thanks go to the editors who again did a beautiful job for this second volume.

9

Trace of $f_{\mathbb{A}}$ on the discrete spectrum

(9.0) Introduction

In this chapter we will use again the notations of chapters 1, 2, 3 and 6 of the first volume. So F is a function field of positive characteristic p, ∞ and o are two distinct places of F, I is a proper, non-zero ideal of the ring

$$A = \{a \in F \mid x(a) \geq 0,\ \forall x \in |X| - \{\infty\}\}$$

such that $o \notin V(I)$ and $\mathbb{A}$ is the ring of adeles of F. In fact, in this volume we will use the notation $\mathcal{I}$ for the ideal I in order to avoid any confusion with the subsets I of Δ.

The purpose of this chapter is to compute the trace of the compactly supported, locally constant function

$$f_{\mathbb{A}} = f_\infty f^{\infty,o} f_o$$

acting on L^2-automorphic irreducible representations of $F_\infty^\times\backslash GL_d(\mathbb{A})$. Here

$$f_\infty \in \mathcal{C}_c^\infty(F_\infty^\times\backslash GL_d(F_\infty)/\!\!/\mathcal{B}_\infty^\circ)$$

is our very cuspidal Euler–Poincaré function (see (5.2.1)),

$$f^{\infty,o} \in \mathcal{C}_c^\infty(GL_d(\mathbb{A}^{\infty,o})/\!\!/K_{\mathcal{I}}^{\infty,o})$$

is an arbitrary Hecke operator and

$$f_o \in \mathcal{C}_c^\infty(GL_d(F_o)/\!\!/K_o)$$

is the Drinfeld function with Satake transform

$$f_o^\vee(z) = p^{\deg(o)r(d-1)/2}(z_1^r + \cdots + z_d^r)$$

for some fixed positive integer r.

(9.1) Automorphic representations

Let M be a standard Levi subgroup of GL_d. We have $M = M_I$ for some $I \subset \Delta$ and, if $d_I = (d_1, \ldots, d_s)$ is the corresponding partition of d, M is canonically isomorphic to $GL_{d_1} \times \cdots \times GL_{d_s}$. We denote by

$$\mathcal{C}_M^\infty = \mathcal{C}^\infty(M(F)\backslash M(\mathbb{A}), \mathbb{C}) \tag{9.1.1}$$

the $\mathbb{C}$-vector space of the complex functions φ on $M(\mathbb{A})$ which are invariant under left translation by $M(F)$ and which are invariant under right translation by some compact open subgroup of $M(\mathbb{A})$ (depending on φ). The unimodular, locally compact, totally discontinuous, separated topological group $M(\mathbb{A})$ acts by right translation on $M(F)\backslash M(\mathbb{A})$ and therefore acts smoothly on $\mathcal{C}_M^\infty$. We denote this action by R_M.

An **automorphic form** for M is a function $\varphi \in \mathcal{C}_M^\infty$ such that the subrepresentation of $(\mathcal{C}_M^\infty, R_M)$ generated by φ, i.e. the $\mathbb{C}$-linear span of

$$R_M(M(\mathbb{A}))(\varphi) \subset \mathcal{C}_M^\infty,$$

is admissible. We denote by

$$\mathcal{A}_M = \mathcal{A}(M(F)\backslash M(\mathbb{A}), \mathbb{C}) \subset \mathcal{C}_M^\infty \tag{9.1.2}$$

the subset of the automorphic forms for M. Obviously, $\mathcal{A}_M$ is a $\mathbb{C}$-vector subspace and is stable under $R_M(M(\mathbb{A}))$ (the subquotients and the finite sums of admissible representations are admissible). We again denote by R_M the restriction of R_M to $\mathcal{A}_M$; $(\mathcal{A}_M, R_M)$ is a smooth representation of $M(\mathbb{A})$.

An **automorphic irreducible representation** of $M(\mathbb{A})$ is an irreducible representation of $M(\mathbb{A})$ which is isomorphic to a subquotient of $(\mathcal{A}_M, R_M)$.

LEMMA (9.1.3). — *Any automorphic irreducible representation of $M(\mathbb{A})$ is admissible and is isomorphic to*

$$(\mathcal{W}_2, \rho_2)/(\mathcal{W}_1, \rho_1)$$

where

$$(\mathcal{W}_1, \rho_1) \subsetneqq (\mathcal{W}_2, \rho_2)$$

are admissible subrepresentations of $(\mathcal{A}_M, R_M)$.

Proof : Up to isomorphism the automorphic irreducible representation $(\mathcal{V},\pi)$ of $M(\mathbb{A})$ is equal to

$$(\mathcal{W}_2,\rho_2)/(\mathcal{W}_1,\rho_1)$$

for some subrepresentations

$$(\mathcal{W}_1,\rho_1)\subsetneqq(\mathcal{W}_2,\rho_2)\subset(\mathcal{A}_M,R_M).$$

Let us arbitrarly fix $\varphi\in\mathcal{W}_2-\mathcal{W}_1$. As $(\mathcal{V},\pi)$ is irreducible the subrepresentation of $(\mathcal{W}_2,\rho_2)$ generated by φ maps onto $(\mathcal{V},\pi)$. Therefore we may assume that $(\mathcal{W}_2,\rho_2)$ is generated by φ. Then, as φ is an automorphic form, $(\mathcal{W}_2,\rho_2)$ is admissible and the lemma follows (see (D.2)). □

Let Z_M be the center of M. If we identify M with $GL_{d_1}\times\cdots\times GL_{d_s}$ ($M=M_I$ and $d_I=(d_1,\ldots,d_s)$), Z_M is identified with $(GL_1)^s$.

Let us define

$$\deg:\mathbb{A}^\times\longrightarrow\mathbb{Z}$$

by

$$\deg(a_{\mathbb{A}})=\frac{1}{f}\sum_{x\in|X|}\deg(x)x(a_{\mathbb{A}})\qquad(\forall a\in\mathbb{A}^\times),$$

where f is the degree over $\mathbb{F}_p$ of the field of constants in F (for each $x\in|X|$, f divides $\deg(x)$). Then deg is a group homomorphism and is trivial on $F^\times\subset\mathbb{A}^\times$ and $\mathcal{O}^\times\subset\mathbb{A}^\times$. If we denote by $(\mathbb{A}^\times)^1$ the kernel of deg,

$$F^\times\backslash(\mathbb{A}^\times)^1/\mathcal{O}^\times\cong\mathrm{Pic}^0_{X/\mathbb{F}_p}(\mathbb{F}_p)$$

is finite and $F^\times\backslash(\mathbb{A}^\times)^1$ is compact and we have an exact sequence of abelian groups,

$$1\longrightarrow F^\times\backslash(\mathbb{A}^\times)^1\longrightarrow F^\times\backslash\mathbb{A}^\times\xrightarrow{\deg}\mathbb{Z}\,.$$

Moreover we have

LEMMA (9.1.4). — *The homomorphism* $\deg:F^\times\backslash\mathbb{A}^\times\to\mathbb{Z}$ *is onto.*

Proof : The zeta function of X is a rational function (see [We 4] Ch. 4, 22). More precisely we have

$$\prod_x\left(1-T^{\frac{\deg(x)}{f}}\right)^{-1}=\frac{P(T)}{(1-T)(1-p^fT)}$$

for some polynomial $P(T)\in\mathbb{Q}[T]$. Therefore, if e is the greatest common divisor of the integers $\frac{\deg(x)}{f}$ $(x\in|X|)$ the rational function

$$\frac{P(T)}{(1-T)(1-p^fT)}$$

is invariant under $T\mapsto\zeta T$ for any e-th root of unity ζ. But, by the Riemann hypothesis for curves (see loc. cit.), $P(T)$ is prime to $(1-T)(1-p^fT)$. Thus we should have $e=1$. □

Therefore, if we define

$$\deg_{Z_M} : Z_M(\mathbb{A}) \longrightarrow \mathbb{Z}^s$$

by

$$\deg_{Z_M}(z_{\mathbb{A},1}, \ldots, z_{\mathbb{A},s}) = (\deg(z_{\mathbb{A},1}), \ldots, \deg(z_{\mathbb{A},s}))$$

and if we set

$$Z_M(\mathbb{A})^1 = \mathrm{Ker}(\deg_{Z_M})$$

we have $Z_M(F) \subset Z_M(\mathbb{A})^1$ and $Z_M(\mathcal{O}) \subset Z_M(\mathbb{A})^1$. The group $Z_M(F)\backslash Z_M(\mathbb{A})^1/Z_M(F)$ is finite and the group $Z_M(F)\backslash Z_M(\mathbb{A})^1$ is compact. We have an exact sequence of abelian groups,

$$1 \longrightarrow Z_M(F)\backslash Z_M(\mathbb{A})^1 \longrightarrow Z_M(F)\backslash Z_M(\mathbb{A}) \xrightarrow{\deg_{Z_M}} \mathbb{Z}^s \longrightarrow 0.$$

Let $(\mathcal{V}, \pi)$ be a smooth representation of the locally compact, totally discontinuous, separated, abelian group $Z_M(F)\backslash Z_M(\mathbb{A})$. A vector $v \in \mathcal{V}$ is said to be $Z_M(\mathbb{A})$-**finite** if the $\mathbb{C}$-linear span of $\pi(Z_M(F)\backslash Z_M(\mathbb{A}))(v)$ in $\mathcal{V}$ is finite dimensional.

We denote by $\mathcal{X}_M$ the abelian group of the smooth complex characters of $Z_M(F)\backslash Z_M(\mathbb{A})$.

LEMMA (9.1.5). — *Let $v \in \mathcal{V}$. Then the following conditions are equivalent:*

(i) *v is $Z_M(\mathbb{A})$-finite,*

(ii) *there exists a function $\mu : \mathcal{X}_M \to \mathbb{Z}_{\geq 0}$ with finite support (i.e. with $\mu(\chi) = 0$ for almost all $\chi \in \mathcal{X}_M$) such that*

$$\Big(\prod_{\chi \in \mathcal{X}_M} \prod_{i=1}^{\mu(\chi)} (\pi(z_{\mathbb{A},\chi,i}) - \chi(z_{\mathbb{A},\chi,i}))\Big)(v) = 0$$

for every family $(z_{\mathbb{A},\chi,i})_{(i=1,\ldots,\mu(\chi);\chi \in \mathcal{X}_M)}$ of elements of $Z_M(\mathbb{A})$,

(iii) *there exist finitely many characters $\chi_1, \ldots, \chi_N$ in $\mathcal{X}_M$ and a positive integer m such that*

$$\Big(\prod_{n=1}^{N} (\pi(z_{\mathbb{A}}) - \chi_n(z_{\mathbb{A}}))^m\Big)(v) = 0$$

for every $z_{\mathbb{A}} \in Z_M(\mathbb{A})$.

Proof: Let $K' \subset Z_M(\mathbb{A})$ be a compact open subgroup such that $v \in \mathcal{V}^{K'}$ and let

$$\mathcal{Z} = Z_M(F)\backslash Z_M(\mathbb{A})/K'.$$

The abelian group $\mathcal{Z}$, and hence its group algebra $\mathbb{C}[\mathcal{Z}]$, acts on $\mathcal{V}$. Let $\mathcal{J}$ be the annihilator of v in $\mathbb{C}[\mathcal{Z}]$. Condition (i) is equivalent to

$$\dim_{\mathbb{C}}(\mathbb{C}[\mathcal{Z}]/\mathcal{J}) < +\infty.$$

If it is satisfied there are finitely many maximal ideals $\mathfrak{p}_1, \ldots, \mathfrak{p}_N$ in $\mathbb{C}[\mathcal{Z}]$ such that

$$\sqrt{\mathcal{J}} = \mathfrak{p}_1 \cap \cdots \cap \mathfrak{p}_N = \mathfrak{p}_1 \cdots \mathfrak{p}_N$$

and a positive integer m such that

$$\mathfrak{p}_1^m \cdots \mathfrak{p}_N^m = (\sqrt{\mathcal{J}})^m \subset \mathcal{J},$$

i.e. $(\sqrt{\mathcal{J}}/\mathcal{J})^m = (0)$ in $\mathbb{C}[\mathcal{Z}]/\mathcal{J}$. For each $n = 1, \ldots, N$ there exists a unique $\chi_n \in \mathcal{X}_M$ such that

$$\mathrm{Ker}(\chi_n) = \mathfrak{p}_n$$

(we also denote by χ_n the $\mathbb{C}$-linear extension of χ_n to $\mathbb{C}[\mathcal{Z}]$). Therefore condition (ii) is satisfied if we define μ by $\mu(\chi_n) = m$ for $n = 1, \ldots, N$ and by $\mu(\chi) = 0$ for $\chi \in \mathcal{X}_M - \{\chi_1, \ldots, \chi_N\}$.

Obviously condition (ii) implies condition (iii).

Finally, if condition (iii) is satisfied we may assume that $\chi_1, \ldots, \chi_N$ are trivial on $K' \subset Z_M(\mathbb{A})$. The abelian group $\mathcal{Z}$ is finitely generated as it is an extension of $\mathbb{Z}^s$ by the finite group $Z_M(F)\backslash Z_M(\mathbb{A})^1/K'$. Therefore the ideal $\mathcal{I}$ in $\mathbb{C}[\mathcal{Z}]$ generated by the elements

$$\prod_{n=1}^{N} (z - \chi_n(z))^m \qquad (z \in \mathcal{Z})$$

has finite codimension in $\mathbb{C}[\mathcal{Z}]$ (if $\{z_1, \ldots, z_r\}$ generates the abelian group $\mathcal{Z}$ then

$$\{z_1^{\alpha_1} \cdots z_r^{\alpha_r} \mid \alpha \in \{0, 1, \ldots, Nm-1\}^r\}$$

generates the $\mathbb{C}$-vector space $\mathbb{C}[\mathcal{Z}]/\mathcal{I}$). But $\mathcal{I} \subset \mathcal{J}$ and therefore condition (i) is satisfied. □

For each $\chi \in \mathcal{X}_M$ we set

$$\mathcal{V}_{\chi,\mathrm{gen}} = \bigcup_{m \geq 1} \mathcal{V}_{\chi,m}$$

where, for each positive integer m, we have put

$$\mathcal{V}_{\chi,m} = \{v \in \mathcal{V} \mid (\pi(z_{\mathbb{A}}) - \chi(z_{\mathbb{A}}))^m(v) = 0, \ \forall z_{\mathbb{A}} \in Z_M(\mathbb{A})\}.$$

Let

$$\mathcal{V}_f \subset \mathcal{V}$$

be the subset of all $Z_M(\mathbb{A})$-finite vectors in $\mathcal{V}$. Then $\mathcal{V}_{\chi,m}$, $\mathcal{V}_{\chi,\mathrm{gen}}$ and $\mathcal{V}_f$ are $\mathbb{C}$-vector subspaces of $\mathcal{V}$ and are stable under $\pi(Z_M(F)\backslash Z_M(\mathbb{A}))$. We have

$$\sum_{\chi\in\mathcal{X}_M} \mathcal{V}_{\chi,\mathrm{gen}} \subset \mathcal{V}_f$$

(see (9.1.5)).

COROLLARY (9.1.6). — *The $\mathbb{C}$-vector space $\mathcal{V}_f$ is the direct sum of its subspaces $\mathcal{V}_{\chi,\mathrm{gen}}$ $(\chi \in \mathcal{X}_M)$.*

Proof: Let χ', $\chi'' \in \mathcal{X}_M$ with $\chi' \neq \chi''$, let m', m'' be positive integers and let $v \in \mathcal{V}$. We assume that

$$(\pi(z_{\mathbb{A}}) - \chi'(z_{\mathbb{A}}))^{m'}(v) = 0 = (\pi(z_{\mathbb{A}}) - \chi''(z_{\mathbb{A}}))^{m''}(v)$$

for every $z_{\mathbb{A}} \in Z_M(\mathbb{A})$. Let us fix $z_{\mathbb{A}} \in Z_M(\mathbb{A})$ such that $\chi'(z_{\mathbb{A}}) \neq \chi''(z_{\mathbb{A}})$. Then by the Bezout theorem there exist $P'(T)$, $P''(T) \in \mathbb{C}[T]$ such that

$$P'(T)(T - \chi'(z_{\mathbb{A}}))^{m'} + P''(T)(T - \chi''(z_{\mathbb{A}}))^{m''} = 1.$$

Replacing T by $\pi(z_{\mathbb{A}})$ and applying the operator to v we get $0 = v$. Therefore we have proved that $\mathcal{V}_{\chi',\mathrm{gen}} \cap \mathcal{V}_{\chi'',\mathrm{gen}} = (0)$.

Let $v \in \mathcal{V}_f$. Thanks to (9.1.5) there exists $\mu : \mathcal{X}_M \to \mathbb{Z}_{\geq 0}$ such that

$$\Big(\prod_{\chi\in\mathcal{X}_M} (\pi(z_{\mathbb{A},\chi}) - \chi(z_{\mathbb{A},\chi}))^{\mu(\chi)}\Big)(v) = 0$$

for every family $(z_{\mathbb{A},\chi})_{\chi\in\mathcal{X}_M} \in (Z_M(\mathbb{A}))^{\mathcal{X}_M}$. Let us prove that

$$v \in \sum_{\chi\in\mathcal{X}_M} \mathcal{V}_{\chi,\mathrm{gen}}$$

by induction on the number of elements in the support of μ.

If $\mathrm{Supp}(\mu) = \{\chi\}$ we have $v \in \mathcal{V}_{\chi,\mathrm{gen}}$. If $\chi' \neq \chi''$ are in $\mathrm{Supp}(\mu)$ let us fix $z_{\mathbb{A}} \in Z_M(\mathbb{A})$ such that $\chi'(z_{\mathbb{A}}) \neq \chi''(z_{\mathbb{A}})$. Then by the Bezout theorem there exist polynomials $P'(T)$, $P''(T) \in \mathbb{C}[T]$ such that

$$P'(T)(T - \chi'(z_{\mathbb{A}}))^{\mu(\chi')} + P''(T)(T - \chi''(z_{\mathbb{A}}))^{\mu(\chi'')} = 1.$$

Let us set

$$v' = \big(P'(\pi(z_{\mathbb{A}}))(\pi(z_{\mathbb{A}}) - \chi'(z_{\mathbb{A}}))^{\mu(\chi')}\big)(v)$$

and

$$v'' = \big(P''(\pi(z_{\mathbb{A}}))(\pi(z_{\mathbb{A}}) - \chi''(z_{\mathbb{A}}))^{\mu(\chi'')}\big)(v).$$

We have

$$v = v' + v''$$

and

$$\Big(\prod_{\substack{\chi\in\mathcal{X}_M \\ \chi\neq\chi'}} (\pi(z_{\mathbb{A},\chi}) - \chi(z_{\mathbb{A},\chi}))^{\mu(\chi)}\Big)(v') = 0$$

and

$$\Big(\prod_{\substack{\chi\in\mathcal{X}_M \\ \chi\neq\chi''}} (\pi(z_{\mathbb{A},\chi}) - \chi(z_{\mathbb{A},\chi}))^{\mu(\chi)}\Big)(v'') = 0$$

for every family $(z_{\mathbb{A},\chi})_{\chi\in\mathcal{X}_M} \in Z_M(\mathbb{A})^{\mathcal{X}_M}$. By the induction hypothesis we have

$$v', v'' \in \sum_{\chi\in\mathcal{X}_M} \mathcal{V}_{\chi,\mathrm{gen}}$$

and the corollary follows. □

Let

$$\mathcal{C}^{\infty}_{Z_M} = \mathcal{C}^{\infty}(Z_M(F)\backslash Z_M(\mathbb{A}), \mathbb{C})$$

be the $\mathbb{C}$-vector space of the complex functions φ on $Z_M(\mathbb{A})$ which are invariant under $Z_M(F)$ and under some compact open subgroup of $Z_M(\mathbb{A})$ (depending on φ). The group $Z_M(F)\backslash Z_M(\mathbb{A})$ acts smoothly by translation on $\mathcal{C}^{\infty}_{Z_M}$.

For each $\chi \in \mathcal{X}_M$ let

$$\iota_\chi : \mathbb{C}[X_1, \ldots, X_s] \longrightarrow \mathcal{C}^{\infty}_{Z_M} \tag{9.1.7}$$

be the $\mathbb{C}$-linear map defined by

$$\iota_\chi(P(X))(z_{\mathbb{A}}) = P(\deg_{Z_M}(z_{\mathbb{A}}))\chi(z_{\mathbb{A}})$$

for every $P(X) \in \mathbb{C}[X_1, \ldots, X_s]$ and every $z_{\mathbb{A}} \in Z_M(\mathbb{A})$.

We denote by $\mathbb{C}[X_1, \ldots, X_s]_m$ the $\mathbb{C}$-vector space of the polynomials $P(X)$ in $\mathbb{C}[X_1, \ldots, X_s]$ such that the degree of $P(X)$ in each variable is strictly smaller than m.

LEMMA (9.1.8). — *The map ι_χ is injective and its image is exactly the subspace $(\mathcal{C}^{\infty}_{Z_M})_{\chi,\mathrm{gen}}$ of $\mathcal{C}^{\infty}_{Z_M}$. More precisely, for each positive integer m we have*

$$\iota_\chi(\mathbb{C}[X_1, \ldots, X_s]_m) = (\mathcal{C}^{\infty}_{Z_M})_{\chi,m}.$$

Proof : Let us fix a positive integer m. The $\mathbb{C}$-linear map

$$(\mathcal{C}_{Z_M}^\infty)_{1,m} \longrightarrow (\mathcal{C}_{Z_M}^\infty)_{\chi,m},\ \varphi \longmapsto \varphi\chi,$$

is an isomorphism. Therefore we may assume that $\chi = 1$.

Let $\mathcal{C}(\mathbb{Z}^s, \mathbb{C})$ be the $\mathbb{C}$-vector space of the complex functions on $\mathbb{Z}^s$ and let r be the action by translation of $\mathbb{Z}^s$ on this space. It is well known and easy to prove by induction on s and m that

$$\{\varphi \in \mathcal{C}(\mathbb{Z}^s, \mathbb{C}) \mid (r(n) - 1)^m(\varphi) = 0,\ \forall n \in \mathbb{Z}^s\}$$

is exactly

$$\{n \longmapsto P(n) \mid P(X) \in \mathbb{C}[X_1, \ldots, X_s]_m\}$$

(use the fact that

$$\{(X+1)^\ell - X^\ell \mid \ell = 1, \ldots, k\}$$

is a basis of $\mathbb{C}[X]_k$ for each positive integer k). Therefore we have

$$\iota_1(\mathbb{C}[X_1, \ldots, X_s]_m) \subset (\mathcal{C}_{Z_M}^\infty)_{1,m}$$

and the equality holds if we can prove that any $\varphi \in (\mathcal{C}_{Z_M}^\infty)_{1,m}$ is invariant under $Z_M(\mathbb{A})^1$.

Let $\varphi \in (\mathcal{C}_{Z_M}^\infty)_{1,m}$ and let $K' \subset Z_M(\mathbb{A})$ be a compact open subgroup such that φ is invariant under K'. We may view φ as a function on the abelian group

$$\mathcal{Z} = Z_M(F)\backslash Z_M(\mathbb{A})/K'$$

which is an extension of $\mathbb{Z}^s$ by the finite group

$$\mathcal{Y} = Z_M(F)\backslash Z_M(\mathbb{A})^1/K'.$$

We want to prove that φ is invariant under $\mathcal{Y}$. Let us fix $y_0 \in \mathcal{Y}$ and $z \in \mathcal{Z}$ and let us set

$$\psi = \sum_{y \in \mathcal{Y}} (\varphi(yz) \cdot y) \in \mathbb{C}[\mathcal{Y}].$$

As $\varphi \in (\mathcal{C}_{Z_M}^\infty)_{1,m}$ we have

$$(y_0 - 1)^m \psi = 0$$

in $\mathbb{C}[\mathcal{Y}]$, so that

$$y_0^n (y_0 - 1)^{m-1} \psi = (y_0 - 1)^{m-1} \psi$$

for every integer n. If $m = 1$ this implies that $\varphi(y_0 z) = \varphi(z)$. If $m \geq 2$ this implies that

$$0 = (y_0^N - 1)(y_0 - 1)^{m-2} \psi = N(y_0 - 1)^{m-1} \psi$$

where N is the order of y_0 in the finite group $\mathcal{Y}$ (we have

$$y_0^N - 1 = (y_0^{N-1} + \cdots + y_0 + 1)(y_0 - 1)$$

in $\mathbb{C}[\mathcal{Y}]$), so that

$$(y_0 - 1)^{m-1} \psi = 0.$$

Therefore, by induction on m we also get that $\varphi(y_0 z) = \varphi(z)$. This completes the proof of the lemma. □

Let us define a homomorphism

$$\deg_M : M(\mathbb{A}) \longrightarrow \bigoplus_{j=1}^{s} \frac{1}{d_j}\mathbb{Z} \subset \mathbb{Q}^s$$

by

$$\deg_M(m_{\mathbb{A}}) = \Big(\frac{\deg(\det(g_{\mathbb{A},1}))}{d_1}, \ldots, \frac{\deg(\det(g_{\mathbb{A},s}))}{d_s}\Big)$$

for all $m_{\mathbb{A}} = (g_{\mathbb{A},1}, \ldots, g_{\mathbb{A},s}) \in M(\mathbb{A}) \cong GL_{d_1}(\mathbb{A}) \times \cdots \times GL_{d_s}(\mathbb{A})$. We have

$$\deg_M |Z_M(\mathbb{A}) = \deg_{Z_M} .$$

Let $M(\mathbb{A})^1 \subset M(\mathbb{A})$ be the kernel of $\deg_M$. We have $M(F) \subset M(\mathbb{A})^1$ and $M(\mathcal{O}) \subset M(\mathbb{A})^1$ and we have

$$Z_M(\mathbb{A})^1 = Z_M(\mathbb{A}) \cap M(\mathbb{A})^1.$$

For each $\chi \in \mathcal{X}_M$ let us denote by

$$\mathcal{C}_{M,\chi}^{\infty} = \mathcal{C}_{\chi}^{\infty}(M(F)\backslash M(\mathbb{A}), \mathbb{C}) \subset \mathcal{C}_M^{\infty} \tag{9.1.9}$$

the $\mathbb{C}$-vector subspace of the functions $\varphi \in \mathcal{C}_M^{\infty}$ such that

$$\varphi(z_{\mathbb{A}} m_{\mathbb{A}}) = \chi(z_{\mathbb{A}})\varphi(m_{\mathbb{A}}) \qquad (\forall z_{\mathbb{A}} \in Z_M(\mathbb{A}),\ m_{\mathbb{A}} \in M(\mathbb{A})).$$

The subspace $\mathcal{C}_{M,\chi}^{\infty}$ is stable under $R_M(M(\mathbb{A}))$ and we will denote by $R_{M,\chi}$ the restriction of R_M to $\mathcal{C}_{M,\chi}^{\infty}$.

We consider the $\mathbb{C}$-linear map

$$\iota : \bigoplus_{\chi \in \mathcal{X}_M} \mathbb{C}[X_1, \ldots, X_s] \otimes_{\mathbb{C}} \mathcal{C}_{M,\chi}^{\infty} \longrightarrow \mathcal{C}_M^{\infty} \tag{9.1.10}$$

defined by

$$\iota(P(X) \otimes \varphi)(m_{\mathbb{A}}) = P(\deg_M(m_{\mathbb{A}}))\varphi(m_{\mathbb{A}})$$

for every $P(X) \in \mathbb{C}[X_1, \ldots, X_s]$, every $\varphi \in \mathcal{C}_{M,\chi}^{\infty}$ and every $\chi \in \mathcal{X}_M$. On $\mathbb{C}[X_1, \ldots, X_s]$ we have the action r of $\mathbb{Q}^s$ defined by

$$r(x)(P(X)) = P(X + x)$$

$(\forall x \in \mathbb{Q}^s,\ \forall P(X) \in \mathbb{C}[X_1, \ldots, X_s])$. Therefore $M(\mathbb{A})$ acts on $\mathbb{C}[X_1, \ldots, X_s]$ by $r \circ \deg_M$ and on the source of ι by

$$\bigoplus_{\chi \in \mathcal{X}_M} (r \circ \deg_M) \otimes R_{M,\chi}.$$

For this action on the source of ι and for the action R_M of $M(\mathbb{A})$ on its target ι is obviously $M(\mathbb{A})$-equivariant.

LEMMA (9.1.11). — *The $\mathbb{C}$-linear map (9.1.10) is injective and its image is equal to $(\mathcal{C}_M^\infty)_f \subset \mathcal{C}_M^\infty$. More precisely, for each $\chi \in \mathcal{X}_M$ and each positive integer m, ι induces an isomorphism ι_χ from $\mathbb{C}[X_1, \ldots, X_s]_m \otimes_{\mathbb{C}} \mathcal{C}_{M,\chi}^\infty$ onto $(\mathcal{C}_M^\infty)_{\chi,m}$.*

Here $(\mathcal{C}_M^\infty)_f$ and $(\mathcal{C}_M^\infty)_{\chi,m}$ are defined for the representation $R_M|(Z_M(F)\backslash Z_M(\mathbb{A}))$.

Proof: We have $\mathcal{C}_{M,\chi}^\infty = (\mathcal{C}_M^\infty)_{\chi,1}$ and it is clear that

$$\iota(\mathbb{C}[X_1, \ldots, X_s]_m \otimes_{\mathbb{C}} \mathcal{C}_{M,\chi}^\infty) \subset (\mathcal{C}_M^\infty)_{\chi,m}$$

(see the proof of (9.1.8)).

Conversely let $\varphi \in (\mathcal{C}_M^\infty)_{\chi,m}$. If we fix $m_{\mathbb{A}} \in M(\mathbb{A})$ the function $z_{\mathbb{A}} \mapsto \varphi(z_{\mathbb{A}} m_{\mathbb{A}})$ belongs to $(\mathcal{C}_{Z_M}^\infty)_{\chi,m}$. Therefore, applying (9.1.8) we obtain a unique family of complex numbers

$$(\varphi_\alpha(m_{\mathbb{A}}))_{\alpha \in \{0,1,\ldots,m-1\}^s}$$

such that, for each $z_{\mathbb{A}} \in Z_M(\mathbb{A})$, we have

$$\varphi(z_{\mathbb{A}} m_{\mathbb{A}}) = \sum_{\alpha \in \{0,1,\ldots,m-1\}^s} \varphi_\alpha(m_{\mathbb{A}}) \chi(z_{\mathbb{A}}) (\deg_{Z_M}(z_{\mathbb{A}}) + \deg_M(m_{\mathbb{A}}))^\alpha$$

(the polynomials $(X + \deg_M(m_{\mathbb{A}}))^\alpha$, $\alpha \in \{0, 1, \ldots, m-1\}^s$, form a basis of the $\mathbb{C}$-vector space $\mathbb{C}[X_1, \ldots, X_s]_m$). By uniqueness the functions

$$m_{\mathbb{A}} \mapsto \varphi_\alpha(m_{\mathbb{A}}) \qquad (\alpha \in \{0, 1, \ldots, m-1\}^s)$$

are all in $\mathcal{C}_{M,\chi}^\infty$. □

REMARK (9.1.12). — The functions φ_α of the proof of (9.1.11) can be computed in the following way. We have

$$(\deg_{Z_M}(z_{\mathbb{A}}) + \deg_M(m_{\mathbb{A}}))^\alpha = \sum_\beta \binom{\alpha}{\beta} \deg_{Z_M}(z_{\mathbb{A}})^\beta \deg_M(m_{\mathbb{A}})^\alpha$$

where β runs through the set $\prod_{j=1}^s \{0, 1, \ldots, \alpha_j\}$ and where

$$\binom{\alpha}{\beta} = \prod_{j=1}^s \binom{\alpha_j}{\beta_j}.$$

So, if we choose a family $(z_{\mathbb{A},n})_{n=1,\ldots,m^s}$ of elements of $Z_M(\mathbb{A})$ such that the matrix

$$(\chi(z_{\mathbb{A},n}) \deg_{Z_M}(z_{\mathbb{A},n})^\beta)_{(n=1,\ldots,m^s;\ \beta \in \{0,1,\ldots,m-1\}^s)}$$

is invertible and if we denote by $(c_{\beta,n})$ the inverse of this matrix, we have

$$\sum_{\alpha}\binom{\alpha}{\beta}\varphi_{\alpha}(m_{\mathbb{A}})\deg_M(m_{\mathbb{A}})^{\alpha-\beta}=\sum_{n=1}^{m^s}c_{\beta,n}\varphi(z_{\mathbb{A},n}m_{\mathbb{A}})$$

for every $m_{\mathbb{A}}\in M(\mathbb{A})$ and every $\beta\in\{0,1,\ldots,m-1\}^s$, where α runs through the set $\prod_{j=1}^{s}\{\beta_j,\beta_j+1,\ldots,m-1\}$. But, if we set

$$\psi_{\beta}=\sum_{\alpha}\binom{\alpha}{\beta}\varphi_{\alpha}(\deg_M)^{\alpha-\beta},$$

we have

$$\varphi_{\alpha}=\sum_{\beta}(-1)^{|\beta-\alpha|}\binom{\beta}{\alpha}\psi_{\beta}(\deg_M)^{\beta-\alpha}$$

where β runs through the set $\prod_{j=1}^{s}\{\alpha_j,\alpha_j+1,\ldots,m-1\}$ and where $|\beta-\alpha|=(\beta_1-\alpha_1)+\cdots+(\beta_s-\alpha_s)$. Therefore we obtain

$$\varphi_{\alpha}(m_{\mathbb{A}})=\sum_{\beta}(-1)^{|\beta-\alpha|}\binom{\beta}{\alpha}\sum_{n=1}^{m^s}c_{\beta,n}\varphi(z_{\mathbb{A},n}m_{\mathbb{A}})\deg_M(m_{\mathbb{A}})^{\beta-\alpha}.$$

□

LEMMA (9.1.13). — *Let $(\mathcal{V},\pi)$ be an admissible representation of $Z_M(F)\backslash M(\mathbb{A})$. Then we have*

$$\mathcal{V}=\mathcal{V}_f$$

(any vector in $\mathcal{V}$ is $Z_M(\mathbb{A})$-finite).

Moreover, if $(\mathcal{V},\pi)$ is irreducible there exists a unique $\chi\in\mathcal{X}_M$ such that

$$\mathcal{V}=\mathcal{V}_{\chi,1}$$

(existence of a central character).

Proof: Let $v\in\mathcal{V}$ and let K' be a compact open subgroup of $M(\mathbb{A})$ such that $v\in\mathcal{V}^{K'}$. Then $\pi(Z_M(\mathbb{A}))(v)$ is contained in $\mathcal{V}^{K'}$ and $\dim_{\mathbb{C}}(\mathcal{V}^{K'})<+\infty$. Therefore v is $Z_M(\mathbb{A})$-finite.

If $(\mathcal{V},\pi)$ is irreducible we have $\mathcal{V}=\mathcal{V}_{\chi,m}$ for any $\chi\in\mathcal{X}_M$ and any positive integer m such that $\mathcal{V}_{\chi,m}\neq(0)$ ($\mathcal{V}_{\chi,m}$ is stable under $\pi(M(\mathbb{A}))$). Therefore there exists a unique $\chi\in\mathcal{X}_M$ such that $\mathcal{V}_{\chi,m}\neq(0)$ for some m. Let m be the smallest positive integer such that $\mathcal{V}_{\chi,m}\neq(0)$. For any $z_{\mathbb{A}}\in Z_M(\mathbb{A})$ we have $(\pi(z_{\mathbb{A}})-\chi(z_{\mathbb{A}}))(\mathcal{V}_{\chi,m})\subset\mathcal{V}_{\chi,m-1}=(0)$ and therefore $m=1$. See also (D.1.12).

□

For each $\chi \in \mathcal{X}_M$ let us set

$$(9.1.14) \qquad \mathcal{A}_{M,\chi} = \mathcal{A}_\chi(M(F)\backslash M(\mathbb{A}),\mathbb{C}) = \mathcal{A}_M \cap \mathcal{C}^\infty_{M,\chi}.$$

PROPOSITION (9.1.15). — *The* $\mathbb{C}$*-linear map* (9.1.10) *induces an isomorphism from*

$$\bigoplus_{\chi\in\mathcal{X}_M} \mathbb{C}[X_1,\ldots,X_s]\otimes_{\mathbb{C}}\mathcal{A}_{M,\chi}$$

onto $\mathcal{A}_M$.

More precisely, for each $\chi \in \mathcal{X}_M$ *and each positive integer* m*,* ι_χ *maps* $\mathbb{C}[X_1,\ldots,X_s]_m\otimes_{\mathbb{C}}\mathcal{A}_{M,\chi}$ *isomorphically onto* $(\mathcal{A}_M)_{\chi,m}$ *and we have*

$$\mathcal{A}_M = \bigoplus_{\chi\in\mathcal{X}_M}(\mathcal{A}_M)_{\chi,\mathrm{gen}}.$$

Proof: First of all let us show that the image of ι is contained in $\mathcal{A}_M$. Let $P(X)\in\mathbb{C}[X_1,\ldots,X_s]_m$ and let $\varphi\in\mathcal{A}_{M,\chi}$ for some positive integer m and some $\chi\in\mathcal{X}_M$. It is sufficient to check that the subrepresentation of

$$(\mathbb{C}[X_1,\ldots,X_s]\otimes_{\mathbb{C}}\mathcal{A}_{M,\chi},(r\circ\deg_M)\otimes R_M)$$

which is generated by $P(X)\otimes\varphi$ is admissible. But this subrepresentation is contained in

$$\mathbb{C}[X_1,\ldots,X_s]_m\otimes_{\mathbb{C}}\mathcal{V},$$

where $\mathcal{V}$ is the $\mathbb{C}$-linear span of $R_{M,\chi}(M(\mathbb{A}))(\varphi)$ in $\mathcal{A}_{M,\chi}$, we have $\dim_{\mathbb{C}}(\mathbb{C}[X_1,\ldots,X_s]_m) < +\infty$ and $(\mathcal{V},R_{M,\chi}|\mathcal{V})$ is admissible. As any compact open subgroup of $M(\mathbb{A})$ is contained in $M(\mathbb{A})^1$ and acts trivially on $\mathbb{C}[X_1,\ldots,X_s]$ our assertion follows.

The same argument proves that

$$P(\deg_M)\psi\in\mathcal{A}_M$$

for every $P(X)\in\mathbb{C}[X_1,\ldots,X_s]$ and every $\psi\in\mathcal{A}_M$.

Now if $\varphi\in(\mathcal{A}_M)_{\chi,m}$ we may decompose φ into

$$\varphi=\sum_\alpha\varphi_\alpha(\deg_M)^\alpha$$

with $\varphi_\alpha\in\mathcal{C}^\infty_{M,\chi}$ for every $\alpha\in\{0,1,\ldots,m-1\}^s$ (see the proof of (9.1.11)). From remark (9.1.12) it follows that $\varphi_\alpha\in\mathcal{A}_{M,\chi}$ for every α and we have proved that

$$\varphi\in\iota_\chi(\mathbb{C}[X_1,\ldots,X_s]_m\otimes_{\mathbb{C}}\mathcal{A}_{M,\chi}).$$

Finally lemma (9.1.13) implies that

$$\mathcal{A}_M=(\mathcal{A}_M)_f.$$

□

COROLLARY (9.1.16). — *An irreducible representation of* $M(\mathbb{A})$ *is automorphic if and only if it is isomorphic to a subquotient of* $(\mathcal{A}_{M,\chi}, R_{M,\chi})$ *for some* $\chi \in \mathcal{X}_M$ *(this* χ*, if it exists, is the central character of the representation and is thus uniquely determined).*

Here we again denote by $R_{M,\chi}$ the restriction of $R_{M,\chi}$ to $\mathcal{A}_{M,\chi} \subset \mathcal{C}^{\infty}_{M,\chi}$.

Proof: The "if" part is obvious. Conversely, let $(\mathcal{V}, \pi)$ be an automorphic irreducible representation of $M(\mathbb{A})$. It is admissible (see (9.1.13)) and trivial on $Z_M(F)$. Therefore we have $\mathcal{V} = \mathcal{V}_{\chi,1}$ for some $\chi \in \mathcal{X}_M$ (see (9.1.13)).

Let us choose admissible subrepresentations

$$(\mathcal{W}_1, \rho_1) \subsetneqq (\mathcal{W}_2, \rho_2) \subset (\mathcal{A}_M, R_M)$$

such that $(\mathcal{V}, \pi)$ is isomorphic to

$$(\mathcal{W}_2, \rho_2)/(\mathcal{W}_1, \rho_1)$$

(see (9.1.3)). Replacing $\mathcal{W}_1$ and $\mathcal{W}_2$ by $(\mathcal{W}_1)_{\chi,m}$ and $(\mathcal{W}_2)_{\chi,m}$ respectively for some sufficie large enough positive integer m we may assume that

$$\mathcal{W}_1 \subsetneqq \mathcal{W}_2 \subset (\mathcal{A}_M)_{\chi,m}$$

for some m (see (9.1.15); choose m such that the image of $(\mathcal{W}_2)_{\chi,m}$ in $\mathcal{V}$ is non-zero).

Let K' be a compact open subgroup of $M(\mathbb{A})$ such that $\mathcal{V}^{K'} \neq (0)$. Then we have $\mathcal{W}_2^{K'} \neq (0)$ (the functor $(\cdot)^{K'}$ is exact, see (D.1.5)). Let $(\mathcal{W}_2', \rho_2')$ be a subrepresentation of $(\mathcal{W}_2, \rho_2)$ such that $\mathcal{W}_2'^{K'}$ maps onto $\mathcal{V}^{K'}$ and has the smallest possible dimension for this property. Let $(\mathcal{W}_2'', \rho_2'')$ be the intersection of all the subrepresentations of $(\mathcal{W}_2, \rho_2)$ containing $\mathcal{W}_2'^{K'}$. Then if $(\mathcal{W}_2''', \rho_2''')$ is a proper subrepresentation of $(\mathcal{W}_2'', \rho_2'')$ the map $(\mathcal{W}_2''', \rho_2''') \to (\mathcal{V}, \pi)$ is zero (otherwise it would be surjective and $(\mathcal{W}_2''')^{K'} \longrightarrow \mathcal{V}'^{K'}$ would be surjective too, so that the inclusion

$$(\mathcal{W}_2''')^{K'} \subset (\mathcal{W}_2'')^{K'} = (\mathcal{W}_2')^{K'}$$

would be an equality and we would have a contradiction). Therefore, replacing $(\mathcal{W}_2, \rho_2)$ by $(\mathcal{W}_2'', \rho_2'')$ and $(\mathcal{W}_1, \rho_1)$ by its intersection with $(\mathcal{W}_2'', \rho_2'')$ we may assume that any proper subrepresentation of $(\mathcal{W}_2, \rho_2)$ is contained in $(\mathcal{W}_1, \rho_1)$.

Now to prove the corollary it is sufficient to construct a non-zero homomorphism $(\mathcal{W}_2, \rho_2) \to (\mathcal{A}_{M,\chi}, R_{M,\chi})$. But applying proposition (9.1.15) we obtain a non-zero homomorphism

$$(\mathcal{W}_2, \rho_2) \longrightarrow (\mathbb{C}[X_1, \ldots, X_s]_m \otimes_{\mathbb{C}} \mathcal{A}_{M,\chi}, (r \circ \det_M) \otimes R_{M,\chi})$$

and we leave it to the reader to check that there is at least one filtration

$$(0) = (\mathcal{U}_0, \sigma_0) \subset (\mathcal{U}_1, \sigma_1) \subset \cdots \subset (\mathcal{U}_L, \sigma_L) = (\mathbb{C}[X_1, \cdots, X_s]_m, r \circ \det_M)$$

such that the successive subquotients

$$(\mathcal{U}_\ell, \sigma_\ell)/(\mathcal{U}_{\ell-1}, \sigma_{\ell-1}) \qquad (\ell = 1, \ldots, L)$$

are isomorphic to the trivial representation $(\mathbb{C}, 1)$ of $M(\mathbb{A})$. By the proof of the corollary is completed. □

(9.2) Cuspidal automorphic representations

Let $M = M_I$ be a standard Levi subgroup of GL_d as in (9.1). Let $P' \subset M$ be a standard parabolic subgroup of M with its standard Levi decomposition $P' = M'N'$, $P' = P_J^I$, $M' = M_J$ and $N' = N_J^I = N_J \cap M_I$ for some $J \subset I$.

LEMMA (9.2.1). — *The topological space $N'(F)\backslash N'(\mathbb{A})$ is compact.*

Proof: Let R_M^+ and $R_{M'}^+$ be the sets of positive roots for $(M, T, B \cap M)$ and $(M', T, B \cap M')$ and for each $\beta = \epsilon_i - \epsilon_j \in R_M^+ - R_{M'}^+$ let

$$x_\beta : \mathbb{G}_a \to N',\ t \mapsto 1 + tE_{ij},$$

be the corresponding 1-parameter subgroup (E_{ij} is the elementary matrix with all entries 0 except the entry on the i-th row and the j-th column which is equal to 1). Then we have

$$N' = \prod_{\beta \in R_M^+ - R_{M'}^+} x_\beta(\mathbb{G}_a).$$

There exists a total ordering $\beta_1 < \beta_2 < \cdots < \beta_L$ on $R_M^+ - R_{M'}^+$ such that, for each $\ell = 0, 1, \ldots, L$,

$$V_\ell = x_{\beta_1}(\mathbb{G}_a) \cdots x_{\beta_\ell}(\mathbb{G}_a)$$

is a normal closed algebraic subgroup of N'. As $V_\ell / V_{\ell-1}$ is isomorphic to $\mathbb{G}_a$ for $\ell = 1, \ldots, L$ and as $F\backslash\mathbb{A}$ is compact (the group

$$F\backslash\mathbb{A}/\mathcal{O} = H^1(X, \mathcal{O}_X)$$

is finite), $V_\ell(F)\backslash V_\ell(\mathbb{A})$ is compact for $\ell = 1, \ldots, L$ (induction on ℓ) and the lemma is proved ($V_L = N'$). □

For any $\varphi \in \mathcal{C}_M^\infty$ we set

(9.2.2) $\varphi_{P'}(m_{\mathbb{A}})$

$$= \frac{1}{\mathrm{vol}(N'(F)\backslash N'(\mathbb{A}), dv'\backslash dn'_{\mathbb{A}})} \int_{N'(F)\backslash N'(\mathbb{A})} \varphi(n'_{\mathbb{A}} m_{\mathbb{A}}) \frac{dn'_{\mathbb{A}}}{dv'} \quad (\forall m_{\mathbb{A}} \in M(\mathbb{A}))$$

where $dn'_{\mathbb{A}}$ is the Haar measure on $N'(\mathbb{A})$ which is normalized by $\mathrm{vol}(N'(\mathbb{A}) \cap K_{\mathcal{I}}, dn'_{\mathbb{A}}) = 1$ and where dv' is the counting measure on $N'(F)$. The function

$$\varphi_{P'} : M'(F)N'(\mathbb{A})\backslash M(\mathbb{A}) \longrightarrow \mathbb{C}$$

is called the **constant term** of φ along P'. It is invariant under right translation by some compact open subgroup of $M(\mathbb{A})$.

The function $\varphi \in \mathcal{C}_M^\infty$ is said to be **cuspidal** if

$$\varphi_{P'}(m_{\mathbb{A}}) = 0 \qquad (\forall m_{\mathbb{A}} \in M(\mathbb{A}))$$

for any proper standard parabolic subgroup P' of M. We denote by

$$\mathcal{C}_{M,\mathrm{cusp}}^\infty = \mathcal{C}_{\mathrm{cusp}}^\infty(M(F)\backslash M(\mathbb{A}), \mathbb{C}) \subset \mathcal{C}_M^\infty \tag{9.2.3}$$

the $\mathbb{C}$-vector subspace of the cuspidal $\varphi \in \mathcal{C}_M^\infty$. It is invariant under $R_M(M(\mathbb{A}))$. We denote by $R_{M,\mathrm{cusp}}$ the restriction of R_M to $\mathcal{C}_{M,\mathrm{cusp}}^\infty$.

A **cusp form** for M is an automorphic form for M which is cuspidal. We set

$$\mathcal{A}_{M,\mathrm{cusp}} = \mathcal{A}_{\mathrm{cusp}}(M(F)\backslash M(\mathbb{A}), \mathbb{C}) = \mathcal{A}_M \cap \mathcal{C}_{M,\mathrm{cusp}}^\infty. \tag{9.2.4}$$

This is a $\mathbb{C}$-vector subspace of $\mathcal{A}_M$ which is invariant under $R_M(M(\mathbb{A}))$. We again denote by $R_{M,\mathrm{cusp}}$ the restriction of R_M to $\mathcal{A}_{M,\mathrm{cusp}}$.

LEMMA (9.2.5). — *The isomorphism ι of* (9.1.11) (*resp.* (9.1.15)) *induces an isomorphism from*

$$\bigoplus_{\chi \in \mathcal{X}_M} \mathbb{C}[X_1, \ldots, X_s] \otimes_{\mathbb{C}} \mathcal{C}_{M,\chi,\mathrm{cusp}}^\infty$$

(*resp.*

$$\bigoplus_{\chi \in \mathcal{X}_M} \mathbb{C}[X_1, \ldots, X_s] \otimes_{\mathbb{C}} \mathcal{A}_{M,\chi,\mathrm{cusp}} \,)$$

onto $\mathcal{C}_{M,f}^\infty \cap \mathcal{C}_{M,\mathrm{cusp}}^\infty$ (*resp.* $\mathcal{A}_{M,\mathrm{cusp}}$) *where we have set*

$$\mathcal{C}_{M,\chi,\mathrm{cusp}}^\infty = \mathcal{C}_{M,\chi}^\infty \cap \mathcal{C}_{M,\mathrm{cusp}}^\infty$$

(*resp.*

$$\mathcal{A}_{M,\chi,\mathrm{cusp}} = \mathcal{A}_{M,\chi} \cap \mathcal{A}_{M,\mathrm{cusp}} \,).$$

Proof : The lemma follows from remark (9.1.12): for any $P(X) \in \mathbb{C}[X_1, \ldots, X_s]$, any $\psi \in \mathcal{C}_M^\infty$ and any standard parabolic subgroup P' of M we have

$$(P(\deg_M)\psi)_{P'} = P(\deg_M)\psi_{P'}.$$

□

THEOREM (9.2.6) (Harder). — *Let K' be a compact open subgroup of $M(\mathbb{A})$. Then there exists an open subset $C_{K'}$ in $M(\mathbb{A})$ such that $Z_M(\mathbb{A})M(F)C_{K'}K' = C_{K'}$, the quotient $Z_M(\mathbb{A})M(F)\backslash C_{K'}/K'$ is finite and*

$$\mathrm{Supp}(\varphi) \subset C_{K'} \qquad (\forall \varphi \in (\mathcal{C}_{M,\mathrm{cusp}}^\infty)^{K'}).$$

Before proving the theorem we need to recall some basic results from reduction theory.

Recall that $M = M_I$. For any integers $c_1 \leq c_2$ let

$$T(\mathbb{A})^I_{]-\infty,c_2]} \quad (\text{resp. } T(\mathbb{A})^I_{[c_1,c_2]})$$

be the open subset of $T(\mathbb{A})$ defined by the conditions

$$\deg(\alpha(t_\mathbb{A})) \leq c_2 \qquad (\forall \alpha \in I)$$

(resp.

$$c_1 \leq \deg(\alpha(t_\mathbb{A})) \leq c_2 \qquad (\forall \alpha \in I)).$$

Let g_X be the geometric genus of X, i.e. the genus of an arbitrary connected component of $X \otimes_{\mathbb{F}_p} k$ where k is an algebraic closure of $\mathbb{F}_p$. We have

$$\dim_{\mathbb{F}_p}(H^1(X, \mathcal{O}_X)) = fg_X.$$

LEMMA (9.2.7) (Harder). — *For any integer $c_2 \geq 2g_X$ we have*

$$M(\mathbb{A}) = M(F)U^I(\mathbb{A})T(\mathbb{A})^I_{]-\infty,c_2]}M(\mathcal{O})$$

(recall that $U^I = U \cap M_I$). □

A proof of (9.2.7) is given in (E.1.1).

LEMMA (9.2.8) (Harder). — *For any compact open subgroup K' of $M(\mathbb{A})$ and any integer c_2 there exists an integer $c_1 \leq c_2$ having the following property: if $\alpha \in I$ and $N' = N^I_{I-\{\alpha\}}$ we have*

$$N'(\mathbb{A}) = N'(F)(N'(\mathbb{A}) \cap b^I_\mathbb{A} K'(b^I_\mathbb{A})^{-1})$$

for any $b^I_\mathbb{A} = u^I_\mathbb{A} t_\mathbb{A} \in B^I(\mathbb{A}) = U^I(\mathbb{A})T(\mathbb{A})$ with $t_\mathbb{A} \in T(\mathbb{A})^I_{]-\infty,c_2]}$ and with

$$\deg(\alpha(t_\mathbb{A})) < c_1.$$

Proof : We use the same notations as in the proof of (9.2.1). For each $\beta \in R_M^+$, $x_\beta^{-1}(K') \subset \mathbb{A}$ is an open subgroup and for each $b_{\mathbb{A}}^I = u_{\mathbb{A}}^I t_{\mathbb{A}} \in B^I(\mathbb{A}) = U^I(\mathbb{A})T(\mathbb{A})$ we have

$$x_\beta^{-1}(b_{\mathbb{A}}^I K'(b_{\mathbb{A}}^I)^{-1}) = \beta(t_{\mathbb{A}})x_\beta^{-1}(K') \subset \mathbb{A}.$$

Let $\alpha \in I$ and let us set $M' = M_{I-\{\alpha\}}$. If $\beta \in R_M^+ - R_{M'}^+$, i.e. if β occurs in Lie (N'), $\beta - \alpha$ is a sum of simple roots

$$\beta - \alpha = \alpha_1 + \cdots + \alpha_r$$

with $\alpha_1, \cdots, \alpha_r \in I$ and $r \leq d-1$. Therefore we have

$$\deg(\beta(t_{\mathbb{A}})) \leq (d-1)\sup(c_2, 0) + \deg(\alpha(t_{\mathbb{A}})) \qquad (\forall t_{\mathbb{A}} \in T(\mathbb{A})^I_{]-\infty,c_2]}).$$

If $\mathcal{U} \subset \mathbb{A}$ is an open subgroup there exists $b_{\mathbb{A}} \in \mathbb{A}^\times$ such that $b_{\mathbb{A}}\mathcal{O} \subset \mathcal{U}$ and for any $a_{\mathbb{A}} \in \mathbb{A}^\times$ we have

$$F\backslash\mathbb{A}/a_{\mathbb{A}}b_{\mathbb{A}}\mathcal{O} = H^1(X, \mathcal{O}_X(-D))$$

where the divisor D on X is defined by

$$D = \sum_{x\in|X|} (x(a_{\mathbb{A}}b_{\mathbb{A}}) \cdot x).$$

By the Serre duality we have

$$H^1(X, \mathcal{O}_X(-D)) \cong H^0(X, \Omega^1_{X/\mathbb{F}_p}(D))^\vee = (0)$$

as long as

$$\deg(D) = \deg(a_{\mathbb{A}}b_{\mathbb{A}}) < 2g_X - 2.$$

Therefore we have

$$F + a_{\mathbb{A}}b_{\mathbb{A}}\mathcal{O} = F + a_{\mathbb{A}}\mathcal{U} = \mathbb{A}$$

as long as

$$\deg(a_{\mathbb{A}}) < 2g_X - 2 - \deg(b_{\mathbb{A}}).$$

Applying this to $\mathcal{U} = x_\beta^{-1}(K')$ with $\beta \in R_M^+ - R_{M'}^+$ we get that there exists an integer $c_1 \leq c_2$ which satisfies the property

$$F + x_\beta^{-1}(b_{\mathbb{A}}^I K'(b_{\mathbb{A}}^I)^{-1}) = \mathbb{A}$$

for any $\beta \in R_M^+ - R_{M'}^+$ and any $b_{\mathbb{A}}^I = u_{\mathbb{A}}^I t_{\mathbb{A}} \in B^I(\mathbb{A}) = U^I(\mathbb{A})T(\mathbb{A})$ with $t_{\mathbb{A}} \in T(\mathbb{A})^I_{]-\infty,c_2]}$ and with

$$\deg(\alpha(t_{\mathbb{A}})) < c_1.$$

Now, choosing a total ordering $\beta_1 < \beta_2 < \cdots < \beta_L$ on $R_M^+ - R_{M'}^+$ as in the proof of (9.2.1), we obtain a tower of normal closed algebraic subgroups of N',

$$(0) = V_0 \subset V_1 \subset \cdots \subset V_L = N'$$

with

$$V_\ell / V_{\ell-1} \cong \mathbb{G}_a \qquad (\forall \ell = 1, \ldots, L)$$

(set $V_\ell = x_{\beta_1}(\mathbb{G}_a) \cdots x_{\beta_\ell}(\mathbb{G}_a)$). Then by induction on ℓ we easily check that

$$V_\ell(\mathbb{A}) = V_\ell(F)(V_\ell(\mathbb{A}) \cap b_{\mathbb{A}}^I K'(b_{\mathbb{A}}^I)^{-1})$$

for any $\ell = 1, \ldots, L$ and any $b_{\mathbb{A}}^I = u_{\mathbb{A}}^I t_{\mathbb{A}} \in B^I(\mathbb{A}) = U^I(\mathbb{A})T(\mathbb{A})$ with $t_{\mathbb{A}} \in T(\mathbb{A})^I_{]-\infty, c_2]}$ and with

$$\deg(\alpha(t_{\mathbb{A}})) < c_1.$$

This completes the proof of the lemma. □

Corollary (9.2.9). — *For any compact open subgroup K' of $M(\mathbb{A})$ and any integer c_2 there exists an integer $c_1 \leq c_2$ having the following property: if $\alpha \in I$ ($M = M_I$) and if $P' = P^I_{I-\{\alpha\}}$ is the corresponding maximal proper standard parabolic subgroup of M we have*

$$\varphi(u_{\mathbb{A}}^I t_{\mathbb{A}} m_{\mathcal{O}}) = \varphi_{P'}(u_{\mathbb{A}}^I t_{\mathbb{A}} m_{\mathcal{O}})$$

for any $\varphi \in (C_M^\infty)^{K'}$, any $u_{\mathbb{A}}^I \in U_{\mathbb{A}}^I$, any $t_{\mathbb{A}} \in T(\mathbb{A})^I_{]-\infty, c_2]}$ with $\deg(\alpha(t_{\mathbb{A}})) < c_1$ and any $m_{\mathcal{O}} \in M(\mathcal{O})$.

Proof: Replacing K' by an open subgroup we may assume that K' is an open normal subgroup of $M(\mathcal{O})$. Let us fix an integer $c_1 \leq c_2$ satisfying the property of (9.2.8) and let $\alpha \in I$. Then we have

$$N'(\mathbb{A}) = N'(F)(N'(\mathbb{A}) \cap (b_{\mathbb{A}}^I m_{\mathcal{O}} K' m_{\mathcal{O}}^{-1} (b_{\mathbb{A}}^I)^{-1}))$$

for any $b_{\mathbb{A}}^I = u_{\mathbb{A}}^I t_{\mathbb{A}} \in U^I(\mathbb{A})T(\mathbb{A})^I_{]-\infty, c_2]}$ with $\deg(\alpha(t_{\mathbb{A}})) < c_1$ and any $m_{\mathcal{O}} \in M(\mathcal{O})$ (K' is normal in $M(\mathcal{O})$). Therefore, for any such $b_{\mathbb{A}}^I$ and $m_{\mathcal{O}}$ the function

$$N'(\mathbb{A}) \to \mathbb{C},\ n'_{\mathbb{A}} \mapsto \varphi(n'_{\mathbb{A}} b_{\mathbb{A}}^I m_{\mathcal{O}})$$

is constant and c_1 satisfies the property of the corollary. □

Proof of (9.2.6) : We may assume that K' is an open normal subgroup of $M(\mathcal{O})$. Let $m_{\mathcal{O},1},\ldots,m_{\mathcal{O},N}\in M(\mathcal{O})$ be a system of representatives of the classes in $M(\mathcal{O})/K'$. Let us fix an integer $c_2\geq 2g_X$. Then we have

$$M(\mathbb{A})=\bigcup_{n=1}^{N} M(F)U^I(\mathbb{A})T(\mathbb{A})^I_{]-\infty,c_2]}m_{\mathcal{O},n}K'$$

(see (9.2.7)).

Let us fix an integer $c_1\leq c_2$ having the property of (9.2.9) with respect to K' and let us set

$$C_{K'}=\bigcup_{n=1}^{N} M(F)U^I(\mathbb{A})T(\mathbb{A})^I_{[c_1,c_2]}m_{\mathcal{O},n}K'.$$

Then we have

$$\mathrm{Supp}(\varphi)\subset C_{K'}$$

for any cuspidal function $\varphi\in(\mathcal{C}_M^\infty)^{K'}$. Indeed, if

$$m_{\mathbb{A}}=\mu u_{\mathbb{A}}^I t_{\mathbb{A}} m_{\mathcal{O},n}k'\notin C_{K'}$$

($\mu\in M(F)$, $u_{\mathbb{A}}^I\in U^I(\mathbb{A})$, $t_{\mathbb{A}}\in T(\mathbb{A})^I_{]-\infty,c_2]}$, $n\in\{1,\ldots,N\}$, $k'\in K'$) there exists $\alpha\in I$ such that

$$\deg(\alpha(t_{\mathbb{A}}))<c_1$$

and we have

$$\varphi(m_{\mathbb{A}})=\varphi(u_{\mathbb{A}}^I t_{\mathbb{A}} m_{\mathcal{O},n})=\varphi_{P'}(u_{\mathbb{A}}^I t_{\mathbb{A}} m_{\mathcal{O},n})=0$$

where $P'=P^I_{I-\{\alpha\}}$.

As we have

$$Z_M(\mathbb{A})M(F)C_{K'}K'=C_{K'},$$

in order to finish the proof of the theorem it is sufficient to check that the topological space

$$Z_M(\mathbb{A})M(F)\backslash C_{K'}$$

is compact. But the quotient space $U^I(F)\backslash U^I(\mathbb{A})$ is compact (see (9.2.1)) and the quotient space

$$Z_M(\mathbb{A})T(F)\backslash T(\mathbb{A})^I_{[c_1,c_2]}$$

is compact too (if we put

$$(\mathbb{A}^\times)_{[c_1,c_2]}=\{a_{\mathbb{A}}\in\mathbb{A}^\times\mid c_1\leq\deg(a_{\mathbb{A}})\leq c_2\}$$

the set

$$F^\times\backslash(\mathbb{A}^\times)_{[c_1,c_2]}/\mathcal{O}^\times = \bigcup_{c_1\leq c\leq c_2} \mathrm{Pic}^c_{X/\mathbb{F}_p}(\mathbb{F}_p)$$

is finite). Therefore the quotient space

$$Z_M(\mathbb{A})B^I(F)\backslash U^I(\mathbb{A})T(\mathbb{A})^I_{[c_1,c_2]}$$

is compact and $Z_M(\mathbb{A})M(F)\backslash C_{K'}$ is a quotient of the compact topological space

$$Z_M(\mathbb{A})B^I(F)\backslash\Big(\bigcup_{n=1}^{N} U^I(\mathbb{A})T(\mathbb{A})^I_{[c_1,c_2]}m_{\mathcal{O},n}K'\Big).$$

□

Corollary (9.2.10). — *For any function $\mu : \mathcal{X}_M \longrightarrow \mathbb{Z}_{\geq 0}$ with finite support, the representation of $M(\mathbb{A})$ which is induced by R_M on the $\mathbb{C}$-vector space*

$$\Big(\bigoplus_{\chi\in\mathcal{X}_M}(\mathcal{C}^\infty_M)_{\chi,\mu(\chi)}\Big)\cap\mathcal{C}^\infty_{M,\mathrm{cusp}}$$

is admissible.

In particular a function $\varphi\in\mathcal{C}^\infty_M$ is a cusp form if and only if it is $Z_M(\mathbb{A})$-finite and cuspidal.

Proof: Thanks to (9.2.5) it is sufficient to check that for any $\chi\in\mathcal{X}_M$

$$(\mathcal{C}^\infty_{M,\chi,\mathrm{cusp}}, R_{M,\chi}|\mathcal{C}^\infty_{M,\chi,\mathrm{cusp}})$$

is admissible.

But, if K' is a compact open subgroup of $M(\mathbb{A})$, $(\mathcal{C}^\infty_{M,\chi,\mathrm{cusp}})^{K'}$ is contained in the $\mathbb{C}$-vector space (see (9.2.6))

$$\begin{aligned}\{\varphi : M(F)\backslash C_{K'}/K' \to \mathbb{C} \mid \varphi(z_{\mathbb{A}}m_{\mathbb{A}})\\ = \chi(z_{\mathbb{A}})\varphi(m_{\mathbb{A}}),\ \forall z_{\mathbb{A}}\in Z_M(\mathbb{A}),\ \forall m_{\mathbb{A}}\in C_{K'}\}\end{aligned}$$

which obviously is finite dimensional. □

Corollary (9.2.11). — *Let $\chi\in\mathcal{X}_M$ and let $\mathbb{Q}(\chi)\subset\mathbb{C}$ be the subfield generated by the values of χ. Then the $\mathbb{Q}(\chi)$-vector subspace*

$$\mathcal{A}^\circ_{M,\chi,\mathrm{cusp}} \overset{\mathrm{dfn}}{=\!=} \{\varphi\in\mathcal{A}_{M,\chi,\mathrm{cusp}} \mid \varphi(M(\mathbb{A}))\subset\mathbb{Q}(\chi)\}$$

is an $M(\mathbb{A})$-invariant $\mathbb{Q}(\chi)$-structure on $\mathcal{A}_{M,\chi,\mathrm{cusp}}$.

Proof: For any compact open subgroup K' of $M(\mathbb{A})$, $(\mathcal{A}_{M,\chi,\mathrm{cusp}})^{K'}$ is the $\mathbb{C}$-vector subspace of the finite dimensional $\mathbb{C}$-vector space

$$\{\varphi : M(F)\backslash M(\mathbb{A})/K' \to \mathbb{C} \mid \varphi(z_{\mathbb{A}}m_{\mathbb{A}}) = \chi(z_{\mathbb{A}})\varphi(m_{\mathbb{A}}),\ \forall z_{\mathbb{A}} \in Z_M(\mathbb{A}),\\ \forall m_{\mathbb{A}} \in M(\mathbb{A}),\ \text{and } \mathrm{Supp}(\varphi) \subset C_{K'}\}$$

defined by the vanishing of the $\mathbb{C}$-linear forms

$$\varphi \longmapsto \varphi_{P'}$$

for all the proper standard parabolic subgroups P' of M (see (9.2.10) and its proof). But this last $\mathbb{C}$-vector space and these $\mathbb{C}$-linear forms are obviously defined over $\mathbb{Q}(\chi)$. Therefore the natural map

$$\mathbb{C} \otimes_{\mathbb{Q}(\chi)} (\mathcal{A}^{\circ}_{M,\chi,\mathrm{cusp}})^{K'} \to (\mathcal{A}_{M,\chi,\mathrm{cusp}})^{K'}$$

is an isomorphism. □

It follows that

$$(\mathcal{A}_{M,\mathrm{cusp}})^{\circ}_{\chi,\mathrm{gen}} \overset{\mathrm{dfn}}{=\!=} \iota(\mathbb{Q}(\chi)[X_1,\ldots,X_s] \otimes_{\mathbb{Q}(\chi)} \mathcal{A}^{\circ}_{M,\chi,\mathrm{cusp}})$$

is an $M(\mathbb{A})$-invariant $\mathbb{Q}(\chi)$-structure on $(\mathcal{A}_{M,\mathrm{cusp}})_{\chi,\mathrm{gen}}$.

A **cuspidal automorphic irreducible representation** of $M(\mathbb{A})$ is an irreducible representation of $M(\mathbb{A})$ which is isomorphic to a subquotient of $(\mathcal{A}_{M,\mathrm{cusp}}, R_{M,\mathrm{cusp}})$. A cuspidal automorphic irreducible representation of $M(\mathbb{A})$ is obviously an automorphic irreducible representation of $M(\mathbb{A})$ and therefore is admissible (see (9.1.3)). The same arguments as in the proof of (9.1.16) show that a cuspidal automorphic irreducible representation of $M(\mathbb{A})$ is isomorphic to a subquotient of

$$(\mathcal{A}_{M,\chi,\mathrm{cusp}}, R_{M,\chi,\mathrm{cusp}} = R_{M,\chi}|\mathcal{A}_{M,\chi,\mathrm{cusp}})$$

for a unique $\chi \in \mathcal{X}_M$ (χ is its central character).

For any $\chi \in \mathcal{X}_M$ we denote by

$$\Pi_{M,\chi,\mathrm{cusp}} = \Pi_{\chi,\mathrm{cusp}}(M(F)\backslash M(\mathbb{A})) \tag{9.2.12}$$

a system of representatives of the isomorphism classes of cuspidal automorphic representations of $M(\mathbb{A})$ which admit χ as a central character, i.e. which are isomorphic to a subquotient of $(\mathcal{A}_{M,\chi,\mathrm{cusp}}, R_{M,\chi,\mathrm{cusp}})$.

For each smooth irreducible representation $(\mathcal{V},\pi)$ of $M(\mathbb{A})$ which admits χ as a central character we denote by

$$m_{\mathrm{cusp}}(\pi) \tag{9.2.13}$$

the dimension of the $\mathbb{C}$-vector space

$$\mathrm{Hom}_{\mathrm{Rep}_s(M(\mathbb{A}))}((\mathcal{V},\pi), (\mathcal{A}_{M,\chi,\mathrm{cusp}}, R_{M,\chi,\mathrm{cusp}}))$$

(if $m_{\mathrm{cusp}}(\pi) > 0$ the representation $(\mathcal{V},\pi)$ is a cuspidal automorphic one).

THEOREM (9.2.14) (Gelfand and Piatetski-Shapiro). — *Let* $\chi \in \mathcal{X}_M$. *Then the set* $\Pi_{M,\chi,\mathrm{cusp}}$ *is countable. For each* $(\mathcal{V},\pi)$ *in* $\Pi_{M,\chi,\mathrm{cusp}}$ *the multiplicity* $m_{\mathrm{cusp}}(\pi)$ *is finite and non-zero. The representation* $(\mathcal{A}_{M,\chi,\mathrm{cusp}}, R_{M,\chi,\mathrm{cusp}})$ *is (non-canonically) isomorphic to*

$$\bigoplus_{(\mathcal{V},\pi)\in\Pi_{M,\chi,\mathrm{cusp}}} (\mathcal{V},\pi)^{\oplus m_{\mathrm{cusp}}(\pi)}.$$

Moreover, for any compact open subgroup K' *of* $M(\mathbb{A})$ *there are only finitely many* $(\mathcal{V},\pi) \in \Pi_{M,\chi,\mathrm{cusp}}$ *such that* $\mathcal{V}^{K'} \neq (0)$.

Proof : The representation $(\mathcal{A}_{M,\chi,\mathrm{cusp}}, R_{M,\chi,\mathrm{cusp}})$ is admissible (see (9.2.10)).

If μ is a complex character of $\bigoplus_{j=1}^{s}\left(\frac{1}{d_j}\mathbb{Z}\right)$ and if we set

$$\chi' = ((\mu \mid \mathbb{Z}^d) \circ \deg_{Z_M})\chi$$

the map

$$\begin{aligned}(\mathcal{A}_{M,\chi,\mathrm{cusp}}, R_{M,\chi,\mathrm{cusp}} \otimes (\mu \circ \deg_M)) &\longrightarrow (\mathcal{A}_{M,\chi',\mathrm{cusp}}, R_{M,\chi',\mathrm{cusp}}),\\ \varphi &\longmapsto (\mu \circ \deg_M)\varphi,\end{aligned}$$

is an isomorphism. Therefore we may assume that χ is unitary (choose μ such that $\mu|\mathbb{Z}^d = |\chi|^{-1}$).

Now for any $\varphi_1, \varphi_2 \in \mathcal{A}_{M,\chi,\mathrm{cusp}}$ we set

$$(*) \qquad \langle \varphi_1, \varphi_2 \rangle_M = \int_{Z_M(\mathbb{A})M(F)\backslash M(\mathbb{A})} \overline{\varphi_1(m_{\mathbb{A}})}\varphi_2(m_{\mathbb{A}})\frac{dm_{\mathbb{A}}}{dz_{\mathbb{A}}d\mu}$$

($dm_{\mathbb{A}}$ and $dz_{\mathbb{A}}$ are arbitrary Haar measures on $M(\mathbb{A})$ and $Z_M(\mathbb{A})$ and $d\mu$ is the counting measure on $M(F)$; because χ is unitary, the integral is well-defined and, thanks to (9.2.6), it is absolutely convergent and in fact can be reduced to a finite sum). Then $(*)$ is an $M(\mathbb{A})$-invariant, positive definite, Hermitian scalar product on $\mathcal{A}_{M,\chi,\mathrm{cusp}}$ and the theorem follows from (D.6.7) and remarks (D.6.8.1) and (D.6.8.2) (we leave it to the reader to check that $M(\mathbb{A})$ admits a countable basis of neighborhoods of 1). □

REMARK (9.2.15). — It can be proved and we will use later that $m_{\mathrm{cusp}}(\pi) = 1$ for any $(\mathcal{V},\pi) \in \Pi_{M,\chi,\mathrm{cusp}}$ (see [Sha 1] Theorem 5.5). □

We will conclude this section by studying some rationality properties of cuspidal automorphic irreducible representations of $M(\mathbb{A})$.

Let us fix a character $\chi \in \mathcal{X}_M$ which is trivial on $Z_M(F_\infty)$. As $Z_M(F_\infty)Z_M(F)\backslash Z_M(\mathbb{A})$ is finite (see (9.1)) the subfield $\mathbb{Q}(\chi)$ of $\mathbb{C}$ is a number field (i.e. a finite extension of $\mathbb{Q}$).

We will simply denote by

$$\mathcal{H}_M^\circ \quad (\text{resp. } \mathcal{H}_M)$$

the $\mathbb{Q}(\chi)$-algebra $\mathcal{C}_c^\infty(Z_M(F_\infty)\backslash M(\mathbb{A}), \mathbb{Q}(\chi))$ (resp. $\mathbb{C}$-algebra $\mathcal{C}_c^\infty(Z_M(F_\infty)\backslash M(\mathbb{A}), \mathbb{C})$) of smooth functions with compact support on $Z_M(F_\infty)\backslash M(\mathbb{A})$ and with values in $\mathbb{Q}(\chi)$ (resp. $\mathbb{C}$). We have

$$\mathcal{H}_M = \mathbb{C} \otimes_{\mathbb{Q}(\chi)} \mathcal{H}_M^\circ.$$

Having fixed a Haar measure $dm_{\mathbb{A}}$ on $M(\mathbb{A})$ and the Haar measure $dz_{M,\infty}$ on $Z_M(F_\infty)$ which is normalized by $\mathrm{vol}(Z_M(\mathcal{O}_\infty), dz_{M,\infty}) = 1$, the $\mathbb{C}$-algebra $\mathcal{H}_M$ acts on the complex vector space $\mathcal{A}_{M,\chi,\mathrm{cusp}}$. We will assume that $dm_{\mathbb{A}}$ is rational, i.e. that $\mathrm{vol}(K_M, dm_{\mathbb{A}}) = 1$ for some (and therefore any) compact open subgroup K_M of $M(\mathbb{A})$ (see (D.1)). Then, by (9.2.11),

$$\mathcal{A}_{M,\chi,\mathrm{cusp}}^\circ = \{\varphi \in \mathcal{A}_{M,\chi,\mathrm{cusp}} \mid \varphi(M(\mathbb{A})) \subset \mathbb{Q}(\chi)\}$$

is an $\mathcal{H}_M^\circ$-invariant $\mathbb{Q}(\chi)$-structure on $\mathcal{A}_{M,\chi,\mathrm{cusp}}$.

Let $(\mathcal{V}, \pi)$ be a cuspidal automorphic irreducible representation of $M(\mathbb{A})$ with central character χ. Let us denote by $\mathbb{Q}(\pi)$ the subfield of $\mathbb{C}$ generated by the complex numbers $\mathrm{tr}\, \pi(f)$ for $f \in \mathcal{H}_M^\circ$. It is easy to see that $\mathbb{Q}(\pi) \supset \mathbb{Q}(\chi)$ (if $z_{\mathbb{A}}$ is a central element in $M(\mathbb{A})$ and if K_M is a compact open subgroup of $M(\mathbb{A})$ such that $\mathcal{V}^{K_M} \neq (0)$,

$$\mathrm{tr}\ \pi(1_{Z_M(F_\infty)z_{\mathbb{A}}K_M}) = \chi(z_{\mathbb{A}})\, \mathrm{vol}(Z_M(\mathcal{O}_\infty)K_M, dm_{\mathbb{A}}) \mathrm{dim}_{\mathbb{C}} \mathcal{V}^{K_M}$$

is a non-zero rational multiple of $\chi(z_{\mathbb{A}})$).

LEMMA (9.2.16). — *The field $\mathbb{Q}(\pi)$ is a number field and the isotypical component of $(\mathcal{V}, \pi)$ in $(\mathcal{A}_{M,\chi,\mathrm{cusp}}, R_{M,\chi,\mathrm{cusp}})$ (which is non-canonically isomorphic to $(\mathcal{V}, \pi)^{m_{\mathrm{cusp}}(\pi)}$ by (9.2.14)) has a natural $\mathbb{Q}(\pi)$-structure.*

Moreover $(\mathcal{V}, \pi)$ itself has a rational structure over a finite extension of $\mathbb{Q}(\pi)$.

Proof: As $(\mathcal{A}_{M,\chi,\mathrm{cusp}}, R_{M,\chi,\mathrm{cusp}})$ is admissible and admits a $\mathbb{Q}(\chi)$-structure the lemma follows from general considerations (see (D.10)). □

(9.3) L^2-automorphic representations

Let us fix a **unitary** character $\chi \in \mathcal{X}_M$. An automorphic form $\varphi \in \mathcal{A}_{M,\chi}$ is said to be **square-integrable** if

$$\int_{Z_M(\mathbb{A})M(F)\backslash M(\mathbb{A})} |\varphi(m_{\mathbb{A}})|^2 \frac{dm_{\mathbb{A}}}{dz_{\mathbb{A}}d\mu} < +\infty$$

($dm_{\mathbb{A}}, dz_{\mathbb{A}}$ and $d\mu$ are the same as in the proof of (9.2.14)). We denote by

$$\mathcal{A}^2_{M,\chi} = \mathcal{A}^2_\chi(M(F)\backslash M(\mathbb{A}), \mathbb{C}) \subset \mathcal{A}_{M,\chi} \tag{9.3.1}$$

the $\mathbb{C}$-vector subspace of the square-integrable automorphic forms. Obviously it is stable under $R_{M,\chi}(M(\mathbb{A}))$. We denote by $R^2_{M,\chi}$ the restriction of $R_{M,\chi}$ to $\mathcal{A}^2_{M,\chi}$.

The smooth representation $(\mathcal{A}^2_{M,\chi}, R^2_{M,\chi})$ of $M(\mathbb{A})$ is unitarizable: if we set

(9.3.2)
$$\langle\varphi_1, \varphi_2\rangle_M = \int_{Z_M(\mathbb{A})M(F)\backslash M(\mathbb{A})} \overline{\varphi_1(m_{\mathbb{A}})}\varphi_2(m_{\mathbb{A}})\frac{dm_{\mathbb{A}}}{dz_{\mathbb{A}}d\mu} \quad (\forall \varphi_1, \varphi_2 \in \mathcal{A}^2_{M,\chi})$$

$\langle\cdot,\cdot\rangle_M$ is an $M(\mathbb{A})$-invariant, positive definite, Hermitian scalar product on $\mathcal{A}^2_{M,\chi}$.

LEMMA (9.3.3). — *We have*

$$\mathcal{A}_{M,\chi,\mathrm{cusp}} \subset \mathcal{A}^2_{M,\chi}$$

and

$$\mathcal{A}^2_{M,\chi} = \mathcal{A}_{M,\chi,\mathrm{cusp}} \oplus \mathcal{A}^2_{M,\chi,\mathrm{Eis}}$$

where $\mathcal{A}^2_{M,\chi,\mathrm{Eis}}$ is the orthogonal subspace of $\mathcal{A}_{M,\chi,\mathrm{cusp}}$ in $\mathcal{A}^2_{M,\chi}$.

Proof : The first assertion has already been checked in the proof of (9.2.14). The second follows from the admissibility of $(\mathcal{A}_{M,\chi,\mathrm{cusp}}, R_{M,\chi,\mathrm{cusp}})$ (see (9.2.10)) and from (D.6.6). □

An **L^2-automorphic irreducible representation** of $M(\mathbb{A})$ is an irreducible representation of $M(\mathbb{A})$ which is isomorphic to a subquotient of $(\mathcal{A}^2_{M,\chi}, R^2_{M,\chi})$ (for some unitary $\chi \in \mathcal{X}_M$). It is automatically an automorphic irreducible representation of $M(\mathbb{A})$ and therefore it is admissible (see (9.1.3)). Then it follows from (9.1.3) and (D.6.6) that any L^2-automorphic irreducible representation of $M(\mathbb{A})$ is isomorphic to an orthogonal direct summand of $(\mathcal{A}^2_{M,\chi}, R^2_{M,\chi})$ with respect to $\langle\cdot,\cdot\rangle_M$ (for some unitary $\chi \in \mathcal{X}_M$) and therefore is unitarizable.

Any cuspidal automorphic irreducible representation of $M(\mathbb{A})$ with unitary central character is automatically L^2-automorphic.

The **discrete spectrum**

$$\mathcal{A}^2_{M,\chi,\mathrm{disc}} = \mathcal{A}^2_{\chi,\mathrm{disc}}(M(F)\backslash M(\mathbb{A}), \mathbb{C}) \subset \mathcal{A}^2_{M,\chi} \tag{9.3.4}$$

is the sum of all $M(\mathbb{A})$-invariant $\mathbb{C}$-vector subspaces $\mathcal{V}$ of $\mathcal{A}^2_{M,\chi}$ such that $(\mathcal{V}, R^2_{M,\chi}|\mathcal{V})$ is irreducible. It is invariant under $R^2_{M,\chi}(M(\mathbb{A}))$. We denote by $R^2_{M,\chi,\mathrm{disc}}$ the restriction of $R^2_{M,\chi}$ to $\mathcal{A}^2_{M,\chi,\mathrm{disc}}$.

It follows from (9.3.3) and (9.2.14) that

$$\mathcal{A}_{M,\chi,\mathrm{cusp}} \subset \mathcal{A}^2_{M,\chi,\mathrm{disc}}.$$

By the admissibility of $(\mathcal{A}_{M,\chi,\mathrm{cusp}}, R_{M,\chi,\mathrm{cusp}})$ (see (9.2.10)) we have

$$\mathcal{A}^2_{M,\chi,\mathrm{disc}} = \mathcal{A}_{M,\chi,\mathrm{cusp}} \oplus \mathcal{A}^2_{M,\chi,\mathrm{res}} \tag{9.3.5}$$

(orthogonal direct sum) where we have set

$$\mathcal{A}^2_{M,\chi,\mathrm{res}} = \mathcal{A}^2_{M,\chi,\mathrm{disc}} \cap \mathcal{A}^2_{M,\chi,\mathrm{Eis}}$$

(see (D.6.6)).

For each smooth irreducible representation $(\mathcal{V}, \pi)$ of $M(\mathbb{A})$ which admits χ as a central character we denote by

$$m^2(\pi) \tag{9.3.6}$$

the dimension of the $\mathbb{C}$-vector space

$$\mathrm{Hom}_{\mathrm{Rep_s}(M(\mathbb{A}))}((\mathcal{V}, \pi), (\mathcal{A}^2_{M,\chi}, R^2_{M,\chi}))$$

(we have $m^2(\pi) > 0$ if and only if $(\mathcal{V}, \pi)$ is L^2-automorphic). We also set

$$m^2_{\mathrm{res}}(\pi) = m^2(\pi) - m_{\mathrm{cusp}}(\pi).$$

We have $m^2_{\mathrm{res}}(\pi) \geq 0$ for any $(\mathcal{V}, \pi)$.

THEOREM (9.3.7) (Harish-Chandra; Borel and Jacquet). — *For any admissible irreducible representation $(\mathcal{V}, \pi)$ of $M(\mathbb{A})$ the dimension of the $\mathbb{C}$-vector space*

$$\mathrm{Hom}_{\mathrm{Rep_s}(M(\mathbb{A}))}((\mathcal{V}, \pi), (\mathcal{C}^\infty_M, R_M))$$

is finite.

COROLLARY (9.3.8). — (i) *For any smooth irreducible representation $(\mathcal{V},\pi)$ of $M(\mathbb{A})$ which admits χ as central character $m^2(\pi)$ is finite.*

(ii) *Let $\Pi^2_{M,\chi}$ be a system of representatives of the isomorphism classes of L^2-automorphic irreducible representations of $M(\mathbb{A})$ with central character χ. Then the smooth representation $(\mathcal{A}^2_{M,\chi,\mathrm{disc}}, R^2_{M,\chi,\mathrm{disc}})$ of $M(\mathbb{A})$ is (non-canonically) isomorphic to*

$$\bigoplus_{(\mathcal{V},\pi)\in\Pi^2_{M,\chi}} (\mathcal{V},\pi)^{\oplus m^2(\pi)}.$$

Moreover, if we choose $\Pi_{M,\chi,\mathrm{cusp}}$ and $\Pi^2_{M,\chi}$ in such a way that $\Pi_{M,\chi,\mathrm{cusp}}\subset \Pi^2_{M,\chi}$, we can find an isomorphism between $(\mathcal{A}^2_{M,\chi,\mathrm{disc}}, R^2_{M,\chi,\mathrm{disc}})$ and

$$\bigoplus_{(\mathcal{V},\pi)\in\Pi^2_{M,\chi}} (\mathcal{V},\pi)^{\oplus m^2(\pi)}$$

such that the decomposition (9.3.5) *corresponds to the decompositions*

$$m^2(\pi) = m_{\mathrm{cusp}}(\pi) + m^2_{\mathrm{res}}(\pi) \qquad (\forall(\mathcal{V},\pi)\in\Pi^2_{M,\chi}).$$

Proof of the corollary assuming the theorem : Part (i) is a direct consequence of (9.1.3) and (9.3.7). If we apply (D.6.9) to $(\mathcal{A}_{M,\chi,\mathrm{res}}, R_{M,\chi,\mathrm{res}})$ we get a (non-canonical) isomorphism

$$(\mathcal{A}_{M,\chi,\mathrm{res}}, R_{M,\chi,\mathrm{res}}) \cong \bigoplus_{(\mathcal{V},\pi)\in\Pi^2_{M,\chi}} (\mathcal{V},\pi)^{m^2_{\mathrm{res}}(\pi)}$$

(thanks to part (i), $m^2_{\mathrm{res}}(\pi)\leq m^2(\pi)$ is finite) and together with (9.2.14) this implies part (ii) of the corollary. □

We will deduce theorem (9.3.7) from the following stronger result.

Let us fix a place x of F. We consider the convolution algebra $\mathbb{C}\otimes \mathcal{C}^\infty_c(M(F_x))$ with respect to the Haar measure dm_x which is normalized by $\mathrm{vol}(M(\mathcal{O}_x), dm_x) = 1$. It acts on any smooth representation of $M(F_x)$ (see (D.1)).

The restrictions of R_M to the closed subgroups $M(F_x)$ and $M(\mathbb{A}^x)$ of $M(\mathbb{A})$ are smooth representations on $\mathcal{C}^\infty_M$. Therefore, if $\mathcal{J}_x$ is a left ideal of $\mathbb{C}\otimes\mathcal{C}^\infty_c(M(F_x))$ and if K'^x is a compact open subgroup of $M(\mathbb{A}^x)$ we may consider the $\mathbb{C}$-vector subspace

$$\mathcal{C}^\infty_M[\mathcal{J}_x, K'^x]\subset \mathcal{C}^\infty_M$$

of the functions $\varphi\in\mathcal{C}^\infty_M$ which are annihilated by $\mathcal{J}_x$ and which are invariant under right translation by K'^x.

A left ideal $\mathcal{J}_x$ of $\mathbb{C}\otimes\mathcal{C}^\infty_c(M(F_x))$ is said to be **admissible** if there exist an admissible representation $(\mathcal{V}_x,\pi_x)$ of $M(F_x)$ and a vector $v_x\in\mathcal{V}_x$ such that $\mathcal{J}_x$ is exactly the annihilator of v_x in $\mathbb{C}\otimes\mathcal{C}^\infty_c(M(F_x))$.

THEOREM (9.3.9) (Harish-Chandra; Borel and Jacquet). — *Let $\mathcal{J}_x$ be an admissible left ideal of $\mathbb{C} \otimes C_c^\infty(M(F_x))$ and let K'^x be a compact open subgroup of $M(\mathbb{A}^x)$. Then the $\mathbb{C}$-vector space*

$$\mathcal{C}_M^\infty[\mathcal{J}_x, K'^x]$$

is finite dimensional.

Proof of (9.3.7) *assuming* (9.3.9) : Thanks to (D.7.1)(ii) there exist admissible irreducible representations $(\mathcal{V}_x, \pi_x)$ and $(\mathcal{V}^x, \pi^x)$ of $M(F_x)$ and $M(\mathbb{A}^x)$ respectively and an isomorphism

$$\iota : (\mathcal{V}_x, \pi_x) \otimes (\mathcal{V}^x, \pi^x) \xrightarrow{\sim} (\mathcal{V}, \pi)$$

of smooth representations of $M(\mathbb{A}) = M(F_x) \times M(\mathbb{A}^x)$. Let v_x and v^x be non-zero vectors in $\mathcal{V}_x$ and $\mathcal{V}^x$ respectively, let $\mathcal{J}_x$ be the annihilator of v_x in $\mathbb{C} \otimes C_c^\infty(M(F_x))$ and let K'^x be a compact open subgroup of $M(\mathbb{A}^x)$ fixing v^x. The left ideal $\mathcal{J}_x$ is admissible. The $\mathbb{C}$-linear map

$$\mathrm{Hom}_{\mathrm{Rep}_s(M(\mathbb{A}))}((\mathcal{V}, \pi), (\mathcal{C}_M^\infty, R_M)) \longrightarrow \mathcal{C}_M^\infty, \ u \longmapsto u(\iota(v_x \otimes v^x))$$

is injective ($\mathcal{V}$ is the $\mathbb{C}$-linear span of $\pi(M(\mathbb{A}))(\iota(v_x \otimes v^x))$ by irreducibility of $(\mathcal{V}, \pi)$) and its image is contained in $\mathcal{C}_M^\infty[\mathcal{J}_x, K'^x]$. Therefore (9.3.7) follows from (9.3.9). □

Before proving (9.3.9) let us give some properties of admissible left ideals of $\mathbb{C} \otimes C_c^\infty(M(F_x))$.

LEMMA (9.3.10). — *Let $\mathcal{J}_x$ be an admissible left ideal of $\mathbb{C} \otimes C_c^\infty(M(F_x))$.*

(i) *There exist a compact open subgroup K'_x of $M(F_x)$ and an ideal $\mathcal{I}_x$ of the group algebra $\mathbb{C}[Z_M(F_x)]$ of finite codimension and having the following property: for any smooth representation $(\mathcal{V}_x, \pi_x)$ of $M(F_x)$ and any vector $v_x \in \mathcal{V}_x$ which is annihilated by $\mathcal{J}_x$ we have*

$$v_x \in \mathcal{V}_x^{K'_x}$$

and

$$\pi_x(\mathcal{I}_x)(v_x) = (0)$$

($\mathbb{C}[Z_M(F_x)] \subset \mathbb{C}[M(F_x)]$ acts on $\mathcal{V}_x$ by π_x).

(ii) *If P' is a standard parabolic subgroup of M with standard Levi decomposition $P' = M'N'$ there exists an admissible left ideal $\mathcal{J}'_x$ of $\mathbb{C} \otimes C_c^\infty(M'(F_x))$ having the following property: for any smooth representation $(\mathcal{V}_x, \pi_x)$ of $M(F_x)$ and any vector $v_x \in \mathcal{V}_x$ which is annihilated by $\mathcal{J}_x$ the canonical image v'_x of v_x in the Jacquet module*

$$\mathcal{V}'_x = \mathcal{V}_x / \mathcal{V}_x(N'(F_x))$$

is annihilated by $\pi_x(\mathcal{J}'_x)$.

Proof: Let us set

$$\mathcal{W}_x = \mathbb{C} \otimes \mathcal{C}_c^\infty(M(F_x))/\mathcal{J}_x$$

and let ρ_x be the natural representation of $M(F_x)$ on $\mathcal{W}_x$ ($M(F_x)$ acts by right translation on itself). Then $(\mathcal{W}_x, \rho_x)$ is admissible and there exists a compact open subgroup K_x' of $M(F_x)$ such that, if we set $e_{K_x'} = 1_{K_x'}/\operatorname{vol}(K_x', dg_x)$, we have

$$f_x * e_{K_x'} - f_x \in \mathcal{J}_x \qquad (\forall f_x \in \mathbb{C} \otimes \mathcal{C}_c^\infty(M(F_x)))$$

(pick an admissible representation $(\mathcal{V}_x^0, \pi_x^0)$ of $M(F_x)$ and a vector $v_x^0 \in \mathcal{V}_x^0$ such that $\mathcal{J}_x$ is the annihilator of v_x^0, choose for K_x' any compact open subgroup of $M(F_x)$ fixing v_x^0 and consider the monomorphism

$$(\mathcal{W}_x, \rho_x) \hookrightarrow (\mathcal{V}_x^0, \pi_x^0), \;\; f_x + \mathcal{J}_x \mapsto \pi_x^0(f_x)(v_x^0)).$$

Let

$$w_x = e_{K_x'} + \mathcal{J}_x \in \mathcal{W}_x.$$

Then w_x is fixed by K_x' and there exists an ideal $\mathcal{I}_x$ of $\mathbb{C}[Z_M(F_x)]$ of finite codimension such that

$$\rho_x(\mathcal{I}_x)(w_x) = (0)$$

(we have

$$\rho_x(\mathbb{C}[Z_M(F_x)])(w_x) \subset \mathcal{W}_x^{K_x'}\text{).}$$

If $(\mathcal{V}_x, \pi_x)$ is a smooth representation of $M(F_x)$ and if $v_x \in \mathcal{V}_x$ is annihilated by $\mathcal{J}_x$ then v_x is fixed by K_x'. Indeed v_x is fixed by some open subgroup K_x'' of K_x' and

$$e_{K_x'} - e_{K_x''} = e_{K_x''} * e_{K_x'} - e_{K_x''} \in \mathcal{J}_x.$$

Therefore the morphism

$$u_x : (\mathcal{W}_x, \rho_x) \to (\mathcal{V}_x, \pi_x), \;\; f_x + \mathcal{J}_x \mapsto \pi_x(f_x)(v_x),$$

maps w_x onto v_x and we have

$$\pi_x(\mathcal{I}_x)(v_x) = u_x(\rho_x(\mathcal{I}_x)(w_x)) = (0).$$

This completes the proof of part (i).

Let

$$\begin{cases} \mathcal{W}_x' = \mathcal{W}_x/\mathcal{W}_x(N'(F_x)), \\ \rho_x' = \rho_x | M'(F_x) \text{ modulo } \mathcal{W}_x(N'(F_x)) \end{cases}$$

be the Jacquet module of $(\mathcal{W}_x, \rho_x)$ and let $w'_x \in \mathcal{W}'_x$ be the canonical image of w_x. Then

$$(\mathcal{W}'_x, \rho'_x) = \left(r_{M(F_x)}^{M'(F_x),P'(F_x)}(\mathcal{W}_x, \rho_x)\right) \otimes (\mathbb{C}, \delta_{P'(F_x)}^{1/2})$$

is admissible (see (7.1.4)(i)) and the annihilator $\mathcal{J}'_x$ of w'_x in $\mathbb{C}\otimes C_c^\infty(M'(F_x))$ is therefore admissible. By functoriality we get a morphism

$$u'_x : (\mathcal{W}'_x, \rho'_x) \longrightarrow (\mathcal{V}'_x, \pi'_x)$$

which maps w'_x onto the canonical image v'_x of v_x in $\mathcal{V}'_x = \mathcal{V}_x/\mathcal{V}_x(N'(F_x))$. Therefore $\mathcal{J}'_x$ annihilates v'_x and part (ii) is also proved. □

Proof of (9.3.9) : First of all let us consider

$$\mathcal{C}^\infty_{M,\mathrm{cusp}}[\mathcal{J}_x, K'^x] = \mathcal{C}^\infty_M[\mathcal{J}_x, K'^x] \cap \mathcal{C}^\infty_{M,\mathrm{cusp}}.$$

Let us fix a compact open subgroup K'_x of $M(F'_x)$ and an ideal $\mathcal{I}_x \subset \mathbb{C}[Z_M(F_x)]$ of finite codimension as in (9.3.10)(i). Then there exists a function $\mu : \mathcal{X}_M \to \mathbb{Z}_{\geq 0}$ with finite support such that

$$\mathcal{C}^\infty_M[\mathcal{J}_x, K'_x] \subset \Big(\bigoplus_{\chi\in\mathcal{X}_M} (\mathcal{C}^\infty_M)_{\chi,\mu(\chi)}\Big) \cap (\mathcal{C}^\infty_M)^{K'}$$

where we have set

$$K' = K'_x K'^x.$$

Indeed the group

$$Z_M(F_x)Z_M(F)\backslash Z_M(\mathbb{A})/(Z_M(\mathbb{A}^x) \cap K'^x)$$

is finite (for any compact open subgroup $\mathcal{U}^x$ of $(\mathbb{A}^x)^\times$ the group

$$F_x^\times F^\times\backslash\mathbb{A}^\times/\mathcal{U}^x$$

is finite) ; therefore there exists an ideal of finite codimension,

$$\mathcal{I}_{\mathbb{A}} \subset \mathbb{C}\big[Z_M(F)\backslash Z_M(\mathbb{A})/(Z_M(\mathbb{A}) \cap K')\big],$$

which annihilates any φ in $\mathcal{C}^\infty_M[\mathcal{J}_x, K'^x]$ and we can apply (9.1.5) (or more precisely its proof). Hence it follows from (9.2.10) that $\mathcal{C}^\infty_{M,\mathrm{cusp}}[\mathcal{J}_x, K'^x]$ is finite dimensional over $\mathbb{C}$.

Next let us prove the theorem by descending induction on the integer s such that $d_I = (d_1, \ldots, d_s)$ if $M = M_I$. For $s = d$ we have $\mathcal{C}^\infty_M = \mathcal{C}^\infty_{M,\mathrm{cusp}}$ and the theorem is already proved. Let us assume the theorem for any $s' > s$ and let us prove it for s ($s < d$). For each proper standard parabolic subgroup P' of M let $C_{P'}$ be a system of representatives of the double cosets in $P'(\mathbb{A})\backslash M(\mathbb{A})/K'$. It is finite (we have the Iwasawa decomposition

$$M(\mathbb{A}) = P'(\mathbb{A})M(\mathcal{O}),$$

see (4.1)). Then $\mathcal{C}_{M,\mathrm{cusp}}^{\infty}[\mathcal{J}_x, K'^x]$ is the intersection of the kernels of the $\mathbb{C}$-linear maps

$$\mathcal{C}_M^{\infty}[\mathcal{J}_x, K'^x] \to (\mathcal{C}_{M'}^{\infty})^{c_{\mathbb{A}}(M'(\mathbb{A})\cap K')c_{\mathbb{A}}^{-1}},$$

$$\varphi \mapsto \left(m'_{\mathbb{A}} \mapsto \int_{N'(F)\backslash N'(\mathbb{A})} \varphi(n'_{\mathbb{A}} m'_{\mathbb{A}} c_{\mathbb{A}}) \frac{dn'_{\mathbb{A}}}{d\nu'}\right)$$

for all $c_{\mathbb{A}} \in C_{P'}$ and all proper standard parabolic subgroups P' of M with standard Levi decomposition $P' = M'N'$ ($dn'_{\mathbb{A}}$ and $d\nu'$ are as in (9.2.2)). It is therefore sufficient to show that, for any given P' and $c_{\mathbb{A}} \in C_{P'}$, the image of the corresponding $\mathbb{C}$-linear map is finite dimensional. Replacing $\mathcal{J}_x$ by

$$\{m_x \longmapsto f_x(c_x^{-1} m_x c_x) \mid f_x \in \mathcal{J}_x\}$$

and K'^x by $c^x K'^x (c^x)^{-1}$ we may assume that $c_{\mathbb{A}} (= c_x c^x) = 1$. But the morphism

$$(\mathcal{C}_M^{\infty}, R_M | M'(F_x)) \to (\mathcal{C}_{M'}^{\infty}, R_{M'} | M'(F_x)),\ \varphi \mapsto \varphi_{P'} | M'(F_x),$$

factors through the Jacquet module

$$(\mathcal{C}_M^{\infty}/\mathcal{C}_M^{\infty}(N'(F_x)),\ R_M | M'(F_x) \text{ modulo } \mathcal{C}_M^{\infty}(N'(F_x))).$$

Therefore, if $\mathcal{J}'_x$ is an admissible left ideal of $\mathbb{C}\otimes\mathcal{C}_c^{\infty}(M'(F_x))$ as in (9.3.10)(ii), $\mathcal{J}'_x$ annihilates $\varphi_{P'}|M'(F_x)$ for all $\varphi \in \mathcal{C}_M^{\infty}[\mathcal{J}_x, K'^x]$. It follows that $\varphi \mapsto \varphi_{P'}|M'(F_x)$ maps $\mathcal{C}_M^{\infty}[\mathcal{J}_x, K'^x]$ into $\mathcal{C}_{M'}^{\infty}[\mathcal{J}'_x, M'(\mathbb{A}^x) \cap K'^x]$ and, since this last space is finite dimensional over $\mathbb{C}$ by our induction hypothesis, the proof of the theorem is completed. □

(9.4) One dimensional automorphic representations

We denote by Ξ_M the abelian group of smooth complex characters

$$\xi : M(\mathbb{A}) \longrightarrow \mathbb{C}^{\times}$$

which are trivial on $M(F)$ and by

$$\det_M : M(\mathbb{A}) \longrightarrow (\mathbb{A}^{\times})^s$$

the group homomorphism defined by

$$\det_M(m_{\mathbb{A}}) = (\det(g_{\mathbb{A},1}), \dots, \det(g_{\mathbb{A},s}))$$

for all $m_{\mathbb{A}} = (g_{\mathbb{A},1}, \dots, g_{\mathbb{A},s}) \in M(\mathbb{A}) = GL_{d_1}(\mathbb{A}) \times \cdots \times GL_{d_s}(\mathbb{A})$ if $M = M_I$ and $d_I = (d_1, \dots, d_s)$.

LEMMA (9.4.1). — *The map*

$$\zeta \longmapsto \zeta \circ \det_M$$

induces a group isomorphism from the abelian group of smooth complex characters

$$\zeta : (F^\times \backslash \mathbb{A}^\times)^s \longrightarrow \mathbb{C}^\times$$

onto Ξ_M.

Proof: We may assume that $s = 1$. Then $\det_M = \det$ is open, continuous and surjective and maps $GL_d(F)$ onto $F^\times$ (it admits the continuous section

$$\mathbb{A}^\times \to GL_d(\mathbb{A}),\ a_{\mathbb{A}} \mapsto \begin{pmatrix} a_{\mathbb{A}} & & & \\ & 1 & & \\ & & \ddots & \\ & & & 1 \end{pmatrix}).$$

Therefore, if ζ is a complex character of $\mathbb{A}^\times$, $\zeta \circ \det$ is smooth (resp. trivial on $GL_d(F)$) if and only if ζ is smooth (resp. trivial on $F^\times$).

Now to finish the proof of the lemma it is sufficient to show that any $\xi \in \Xi_{GL_d}$ is trivial on $\mathrm{Ker}(\det) = SL_d(\mathbb{A})$. But $SL_d(\mathbb{A})$ is generated by $N_{\Delta-\{\alpha\}}(\mathbb{A}) \cup \widetilde{N}_{\Delta-\{\alpha\}}(\mathbb{A})$ where $\alpha = \epsilon_1 - \epsilon_2 \in \Delta$ (see the proof of (8.5.5)). Moreover, if we fix $a_{\mathbb{A}} \in \mathbb{A}^\times$ such that $a_{\mathbb{A}} - 1 \in \mathbb{A}^\times$, for any

$$\begin{pmatrix} 1 & u_{\mathbb{A}} \\ 0 & 1_{d-1} \end{pmatrix} \in N_{\Delta-\{\alpha\}}(\mathbb{A})$$

(resp.

$$\begin{pmatrix} 1 & 0 \\ \widetilde{u}_{\mathbb{A}} & 1_{d-1} \end{pmatrix} \in \widetilde{N}_{\Delta-\{\alpha\}}(\mathbb{A}))$$

we then have

$$\begin{pmatrix} 1 & u_{\mathbb{A}} \\ 0 & 1_{d-1} \end{pmatrix} = \begin{pmatrix} a_{\mathbb{A}} & 0 \\ 0 & 1_{d-1} \end{pmatrix} \begin{pmatrix} 1 & v_{\mathbb{A}} \\ 0 & 1_{d-1} \end{pmatrix} \begin{pmatrix} a_{\mathbb{A}} & 0 \\ 0 & 1_{d-1} \end{pmatrix}^{-1} \begin{pmatrix} 1 & v_{\mathbb{A}} \\ 0 & 1_{d-1} \end{pmatrix}^{-1}$$

(resp.

$$\begin{pmatrix} 1 & 0 \\ \widetilde{u}_{\mathbb{A}} & 1_{d-1} \end{pmatrix} = \begin{pmatrix} a_{\mathbb{A}} & 0 \\ 0 & 1_{d-1} \end{pmatrix} \begin{pmatrix} 1 & 0 \\ \widetilde{v}_{\mathbb{A}} & 1_{d-1} \end{pmatrix} \begin{pmatrix} a_{\mathbb{A}} & 0 \\ 0 & 1_{d-1} \end{pmatrix}^{-1} \begin{pmatrix} 1 & 0 \\ \widetilde{v}_{\mathbb{A}} & 1_{d-1} \end{pmatrix}^{-1})$$

where $v_{\mathbb{A}} = u_{\mathbb{A}}/(a_{\mathbb{A}} - 1)$ (resp. $\widetilde{v}_{\mathbb{A}} = a_{\mathbb{A}}\widetilde{u}_{\mathbb{A}}/(1 - a_{\mathbb{A}})$). Therefore any character of $GL_d(\mathbb{A})$ is trivial on $SL_d(\mathbb{A})$. □

Any $\xi \in \Xi_M$ may be viewed as a complex function on $M(F)\backslash M(\mathbb{A})$ and is then an automorphic form for M. More precisely, if we set $\chi = \xi | Z_M(\mathbb{A})$, we have $\chi \in \mathcal{X}_M$ and

$$\xi \in \mathcal{A}_{M,\chi}.$$

It follows that $(\mathbb{C}, \xi)$ is an automorphic irreducible representation of $M(\mathbb{A})$ with central character χ (it is isomorphic to the subrepresentation $(\mathbb{C}\xi, R_{M,\chi}|\mathbb{C}\xi)$ of $(\mathcal{A}_{M,\chi}, R_{M,\chi})$.

For each $\chi \in \mathcal{X}_M$ let $\Xi_{M,\chi}$ be the set of $\xi \in \Xi_M$ such that $\xi|Z_M(\mathbb{A}) = \chi$ and let

$$\mathcal{A}_{M,\chi,\mathrm{triv}} = \sum_{\xi \in \Xi_{M,\chi}} \mathbb{C}\xi \subset \mathcal{A}_{M,\chi}. \tag{9.4.2}$$

The sum is direct and $\mathcal{A}_{M,\chi,\mathrm{triv}}$ is stable under $R_{M,\chi}(M(\mathbb{A}))$. We denote by $R_{M,\chi,\mathrm{triv}}$ the restriction of $R_{M,\chi}$ to $\mathcal{A}_{M,\chi,\mathrm{triv}}$.

Lemma (9.4.3). — (i) $(\mathcal{A}_{M,\chi,\mathrm{triv}}, R_{M,\chi,\mathrm{triv}})$ *is an admissible representation.*

(ii) *For any* $\xi \in \Xi_{M,\chi}$ *we have*

$$\dim_{\mathbb{C}} \mathrm{Hom}_{\mathrm{Rep}_s(M(\mathbb{A}))}\big((\mathbb{C},\xi),(\mathcal{A}_{M,\chi},R_{M,\chi})\big) = 1.$$

(iii) *If* $M = T$, *i.e. if* $I = \Delta$, *we have*

$$\mathcal{A}_{M,\chi,\mathrm{triv}} = \mathcal{A}_{M,\chi} = \mathbb{C}\chi.$$

(iv) *If* $M \supsetneqq T$, *i.e. if* $I \subsetneqq \Delta$, *and if* χ *is unitary* $\mathcal{A}_{M,\chi,\mathrm{triv}}$ *is contained in* $\mathcal{A}^2_{M,\chi,\mathrm{res}}$ *and in fact is an orthogonal direct summand of* $\mathcal{A}^2_{M,\chi,\mathrm{res}}$. *Moreover, for any* $\xi \in \Xi_{M,\chi}$ *we have*

$$m^2(\xi) = m^2_{\mathrm{res}}(\xi) = 1.$$

Proof: For $j = 1, \ldots, s$ the quotient

$$F^\times\{(a_{\mathbb{A}})^d \mid a_{\mathbb{A}} \in \mathbb{A}^\times\}\backslash \mathbb{A}^\times/\mathcal{O}^\times$$

is finite. Therefore, for any compact open subgroup K' of $M(\mathbb{A})$ the quotient

$$(F^\times)^s \det_M(Z_M(\mathbb{A}))\backslash(\mathbb{A}^\times)^s/\det_M(K')$$

is finite and the set of $\xi \in \Xi_{M,\chi}$ which are trivial on K' is finite (see (9.4.2)). This proves part (i).

Parts (ii) and (iii) are obvious.

If $\chi \in \mathcal{X}_M$ is unitary any $\xi \in \Xi_{M,\chi}$ is also unitary (if D is the lowest common multiple of $(d_1, \ldots, d_s)$ and if we set $D_j = D/d_j$, $j = 1, \ldots, s$, we have

$$\xi^D(a_{\mathbb{A},1}, \ldots, a_{\mathbb{A},s}) = \chi((a_{\mathbb{A},1})^{D_1}, \ldots, (a_{\mathbb{A},s})^{D_s})$$

for every $(a_{\mathbb{A},1}, \ldots, a_{\mathbb{A},s}) \in (\mathbb{A}^\times)^s$, where we have identified $Z_M(\mathbb{A})$ with $(\mathbb{A}^\times)^s$ in the usual way). As

$$\mathrm{vol}(Z_M(\mathbb{A})M(F)\backslash M(\mathbb{A}), \frac{dm_{\mathbb{A}}}{dz_{\mathbb{A}}d\mu})$$

is finite it follows that $\Xi_{M,\chi} \subset \mathcal{A}^2_{M,\chi}$ for any unitary $\chi \in \mathcal{X}_M$. Therefore we have

$$\mathcal{A}_{M,\chi,\mathrm{triv}} \subset \mathcal{A}^2_{M,\chi,\mathrm{disc}}$$

for any unitary $\chi \in \mathcal{X}_M$.

Now if χ is unitary and if $\xi \in \Xi_{M,\chi}$ let ξ_{cusp} be the orthogonal projection of ξ into the orthogonal direct summand $\mathcal{A}_{M,\chi,\mathrm{cusp}}$ of $\mathcal{A}^2_{M,\chi,\mathrm{disc}}$ (see (9.3.5)). We have

$$R_{M,\chi,\mathrm{cusp}}(m_{\mathbb{A}})(\xi_{\mathrm{cusp}}) = \xi(m_{\mathbb{A}})\xi_{\mathrm{cusp}} \qquad (\forall m_{\mathbb{A}} \in M(\mathbb{A}))$$

(the orthogonal projection of $\mathcal{A}^2_{M,\chi,\mathrm{disc}}$ onto $\mathcal{A}_{M,\chi,\mathrm{cusp}}$ is $M(\mathbb{A})$-equivariant). Therefore, if there exists $m^0_{\mathbb{A}} \in M(\mathbb{A})$ such that $\xi_{\mathrm{cusp}}(m^0_{\mathbb{A}}) \neq 0$ we have $\xi_{\mathrm{cusp}}(m_{\mathbb{A}}) \neq 0$ for every $m_{\mathbb{A}} \in M(\mathbb{A})$. But this contradicts (9.2.6) unless $Z_M(\mathbb{A})M(F)\backslash M(\mathbb{A})$ is compact. If $M \supsetneq T$, $Z_M(\mathbb{A})M(F)\backslash M(\mathbb{A})$ cannot be compact and $\xi_{\mathrm{cusp}} \equiv 0$ for any $\xi \in \Xi_{M,\chi}$, so that $\mathcal{A}_{M,\chi,\mathrm{triv}} \subset \mathcal{A}^2_{M,\chi,\mathrm{res}}$ for any unitary $\chi \in \mathcal{X}_M$. Then the other assertions of part (iv) follows from parts (i) and (ii) which are already proved and from (D.6.6). □

(9.5) Trace of $f_{\mathbb{A}}$ into the discrete spectrum

For any admissible irreducible representation $(\mathcal{V}, \pi)$ of $M(\mathbb{A})$ we have a "unique" decomposition into a restricted tensor product

$$(\mathcal{V}, \pi) \cong \bigotimes_{x \in |X|}{}' (\mathcal{V}_x, \pi_x)$$

where, for each $x \in |X|$, $(\mathcal{V}_x, \pi_x)$ is an admissible irreducible representation of $M(F_x)$ and where $\dim_{\mathbb{C}}(\mathcal{V}_x^{M(\mathcal{O}_x)}) = 1$ for almost every $x \in |X|$ (for each $x \in |X|$ the $\mathbb{Q}$-algebra

$$C_c^\infty(M(F_x)//M(\mathcal{O}_x)) \cong \bigotimes_{j=1}^{s} C_c^\infty(GL_{d_j}(F_x)//GL_{d_j}(\mathcal{O}_x))$$

is commutative, see the proof of (D.7.1), (D.7.4) and (4.1.17)). For any finite set T of places of F,

$$(\mathcal{V}_T,\pi_T)=\bigotimes_{x\in T}(\mathcal{V}_x,\pi_x)$$

(resp.

$$(\mathcal{V}^T,\pi^T)=\bigotimes_{x\in|X|-T}(\mathcal{V}_x,\pi_x))$$

is an admissible irreducible representation of $M(F_T)$ (resp. $M(\mathbb{A}^T)$) and is called the **component** of $(\mathcal{V},\pi)$ at T (resp. outside T).

Let

$$f_{\mathbb{A}}=f_\infty f^{\infty,o}f_o\in C_c^\infty(F_\infty^\times\backslash GL_d(\mathbb{A})//\mathcal{B}_\infty^\circ K_{\mathcal{I}}^{\infty,o}K_o)$$

be the function introduced in (6.1) (see also (9.0)).

THEOREM (9.5.1). — *Let $(\mathcal{V},\pi)$ be a unitarizable admissible irreducible representation of $GL_d(\mathbb{A})$ which is trivial on $F_\infty^\times$, i.e. such that its central character $\omega_\pi:\mathbb{A}^\times\longrightarrow\mathbb{C}^\times$ is trivial on $F_\infty^\times\subset\mathbb{A}^\times$ (see* (D.1.12)*), and let us consider the following conditions on $(\mathcal{V},\pi)$:*

(i) *its component at ∞ is isomorphic to the Steinberg representation $(\mathrm{St}_\infty,\mathrm{st}_\infty)$ or to the trivial representation $(\mathbb{C},1_\infty)$ of $F_\infty^\times\backslash GL_d(F_\infty)$ (see* (8.1)*),*

(ii) *its component outside $\{\infty,o\}$ has a non-zero fixed vector under $K_{\mathcal{I}}^{\infty,o}$,*

(iii) *its component at o is spherical (see* (7.5)*).*

If they are simultaneously satisfied we have

$$\begin{aligned}\operatorname{tr}\pi(f_{\mathbb{A}})=\epsilon_\infty(\pi_\infty)\operatorname{tr}\big(\pi^{\infty,o}(f^{\infty,o})|(\mathcal{V}^{\infty,o})^{K_{\mathcal{I}}^{\infty,o}}\big)\\ \times p^{\deg(o)r(d-1)/2}(z_1(\pi_o)^r+\cdots+z_d(\pi_o)^r)\end{aligned}$$

where $\epsilon_\infty(\pi_\infty)=(-1)^{d-1}$ (resp. $\epsilon_\infty(\pi_\infty)=1$) if $(\mathcal{V}_\infty,\pi_\infty)$ is isomorphic to $(\mathrm{St}_\infty,\mathrm{st}_\infty)$ (resp. $(\mathbb{C},1_\infty)$), where $z_1(\pi_o),\ldots,z_d(\pi_o)$ are the Hecke eigenvalues of $(\mathcal{V}_o,\pi_o)$ (see (7.5)*) and where dz_∞ and $dg_{\mathbb{A}}=dg_\infty dg^{\infty,o}dg_o$ are normalized as in* (6.1).

If at least one of those conditions fails then $\operatorname{tr}\pi(f_{\mathbb{A}})$ vanishes.

Proof: We have

$$\operatorname{tr}\pi(f_{\mathbb{A}})=\operatorname{tr}\pi_\infty(f_\infty)\operatorname{tr}\pi^{\infty,o}(f^{\infty,o})\operatorname{tr}\pi_o(f_o)$$

(see (D.7.5.2)) and $(\mathcal{V}_\infty,\pi_\infty)$ is unitarizable (see (D.7.5.1)). Therefore (9.5.1) follows directly from (8.5.3), the definition of $\operatorname{tr}\pi^{\infty,o}$ (see (D.2)) and (7.5.6).

□

Let $\chi \in \mathcal{X}_{GL_d}$ (a smooth character of $F^\times\backslash\mathbb{A}^\times$) and let us assume that χ is trivial on $F_\infty^\times$. Then χ is automatically of finite order ($F_\infty^\times F^\times\backslash\mathbb{A}^\times$ is compact) and therefore unitary.

LEMMA (9.5.2). — *The set*

$$\{(\mathcal{V},\pi)\in\Pi^2_{GL_d,\chi}\mid \operatorname{tr}\pi(f_{\mathbb{A}})\neq 0\}$$

is finite.

Proof : Let $\mathcal{J}'_\infty$ (resp. $\mathcal{J}''_\infty$) be the annihilator in $\mathbb{C}\otimes C_c^\infty(GL_d(F_\infty))$ of the 1-dimensional subspace $(\mathrm{St}_\infty)^{\mathcal{B}_\infty^\circ}\subset \mathrm{St}_\infty$ of the Steinberg representation st_∞ (resp. of the vector $1\in\mathbb{C}$ of the trivial representation 1_∞) (see the first step of the proof of (8.1.2)). Then $\mathcal{J}'_\infty$ and $\mathcal{J}''_\infty$ are admissible left ideals and $\mathcal{C}^\infty_{GL_d}[\mathcal{J}'_\infty,K_\mathcal{I}^\infty]$ and $\mathcal{C}^\infty_{GL_d}[\mathcal{J}''_\infty,K_\mathcal{I}^\infty]$ are finite dimensional $\mathbb{C}$-vector subspaces of $\mathcal{C}^\infty_{GL_d}$ (see (9.3.9)).

But, applying (9.5.1), we obtain that, for any unitarizable admissible irreducible subrepresentation $(\mathcal{V},\pi)$ of $(\mathcal{C}^\infty_{GL_d},R_{GL_d})$ (with $\omega_\pi|F_\infty^\times\equiv 1$) such that

$$\operatorname{tr}\pi(f_{\mathbb{A}})\neq 0,$$

we have

$$(0)\neq \mathcal{V}^{\mathcal{B}_\infty^\circ K_\mathcal{I}^\infty}\subset \mathcal{C}^\infty_{GL_d}[\mathcal{J}'_\infty,K_\mathcal{I}^\infty]+\mathcal{C}^\infty_{GL_d}[\mathcal{J}''_\infty,K_\mathcal{I}^\infty]\subset\mathcal{C}^\infty_{GL_d}.$$

The lemma follows (use (9.3.8)(ii)). □

We can now intoduce the **formal trace**

$$\operatorname{ftr} R^2_{GL_d,\chi,\mathrm{disc}}(f_{\mathbb{A}})=\sum_{(\mathcal{V},\pi)\in\Pi^2_{GL_d,\chi}} m^2(\pi)\operatorname{tr}\pi(f_{\mathbb{A}}) \qquad (9.5.3)$$

(this sum is finite thanks to (9.5.3) and (9.3.8)(i)).

REMARK (9.5.4). — In fact Moeglin and Waldspurger have given a complete description of $(\mathcal{A}^2_{GL_d,\chi,\mathrm{disc}},R^2_{GL_d,\chi,\mathrm{disc}})$ in terms of the cuspidal automorphic forms for certain standard Levi subgroups M of GL_d (see [Mo–Wa 1] (Introduction, Théorème)). It follows from their result and from (9.2.10) that $(\mathcal{A}^2_{GL_d,\chi,\mathrm{disc}},R^2_{GL_d,\chi,\mathrm{disc}})$ is an admissible representation of $GL_d(\mathbb{A})$. In particular the operator $R^2_{GL_d,\chi,\mathrm{disc}}(f_{\mathbb{A}})$ has a well-defined trace (see (D.2)). Obviously this trace coincides with the above formal trace. It also follows from their result and from [Sha 1] (see (9.2.15)) that $m^2(\pi)=1$ for all $(\mathcal{V},\pi)\in\Pi^2_{GL_d,\chi}$. □

THEOREM (9.5.5). — *Let $\chi \in \mathcal{X}_{GL_d}$ be such that $\chi|F_\infty^\times \equiv 1$ and let $(\mathcal{V},\pi)$ be an (admissible) irreducible subrepresentation of $(\mathcal{A}^2_{GL_d,\chi}, R^2_{GL_d,\chi})$.*

(i) *If its component at ∞ is isomorphic to the Steinberg representation $(\mathrm{St}_\infty, \mathrm{st}_\infty)$ of $F_\infty^\times\backslash GL_d(F_\infty)$ then the subspace $\mathcal{V}$ of $\mathcal{A}^2_{GL_d,\chi}$ is automatically contained in $\mathcal{A}_{GL_d,\chi,\mathrm{cusp}}$. In particular $(\mathcal{V},\pi)$ is a cuspidal automorphic irreducible representation of $GL_d(\mathbb{A})$.*

(ii) *If its component at ∞ is isomorphic to the trivial representation $(\mathbb{C}, 1_\infty)$ of $F_\infty^\times\backslash GL_d(F_\infty)$ then the subspace $\mathcal{V}$ of $\mathcal{A}^2_{GL_d,\chi}$ is automatically equal to the 1-dimensional $\mathbb{C}$-vector subspace $\mathbb{C}\xi$ of $\mathcal{A}^2_{GL_d,\chi}$ for some $\xi \in \Xi_{GL_d,\chi}$ with $\xi|F_\infty^\times \equiv 1$. In particular $(\mathcal{V},\pi)$ is isomorphic to $(\mathbb{C},\xi)$ for some $\xi \in \Xi_{GL_d,\chi}$.*

COROLLARY (9.5.6). — *Let $\chi \in \mathcal{X}_{GL_d}$ be such that $\chi|F_\infty^\times \equiv 1$ and let $(\mathcal{V},\pi) \in \Pi^2_{GL_d,\chi}$.*

(i) *If the component at ∞ of $(\mathcal{V},\pi)$ is isomorphic to the Steinberg representation $(\mathrm{St}_\infty, \mathrm{st}_\infty)$ of $F_\infty^\times\backslash GL_d(F_\infty)$ then $(\mathcal{V},\pi)$ is automatically cuspidal and*

$$m^2(\pi) = m_{\mathrm{cusp}}(\pi).$$

(ii) *If the component at ∞ of $(\mathcal{V},\pi)$ is isomorphic to the trivial representation $(\mathbb{C}, 1_\infty)$ of $F_\infty^\times\backslash GL_d(F_\infty)$ then $(\mathcal{V},\pi)$ is automatically isomorphic to $(\mathbb{C},\xi)$ for some $\xi \in \Xi_{GL_d,\chi}$ and $m^2(\pi) = 1$.*

□

The proofs of the two parts of (9.5.5) are completely different and independent. The proof of the first part is based on the following local result. We again use the notations of chapters 7 and 8 for the statement and the proof of this result. In particular F is a non-archimedean local field.

LEMMA (9.5.7). — *For each subset $I \subset \Delta$ the smooth representation $r_\Delta^I(\mathrm{St}, \mathrm{st})$ of $M_I(F)$ (see (8.1) and (7.1)) is irreducible and its central character (see (D.1.12)) is equal to $\delta_{B(F)}^{1/2}|Z_I(F)$.*

Proof: We have proved in (8.1.2)(iii) that

$$r_\Delta^\emptyset(\mathrm{St}, \mathrm{st}) \cong (\mathbb{C}, \delta_{B(F)}^{1/2}).$$

On the one hand it follows that, if $r_\Delta^I(\mathrm{St}, \mathrm{st})$ admits a central character ω, we automatically have

$$(\delta_{B^I(F)}^{-1/2}|Z_I(F))\omega = \delta_{B(F)}^{1/2}|Z_I(F)$$

by definition of $r_I^\emptyset$, i.e.

$$\omega = \delta_{B(F)}^{1/2}|Z_I(F)$$

($\delta_{B^I(F)}$ is trivial on $Z_I(F) \subset T(F)$). On the other hand it follows that

$$r_I^{\emptyset}(r_{\Delta}^I(\mathrm{St},\mathrm{st})) \cong (\mathbb{C}, \delta_{B(F)}^{1/2})$$

(see (7.1.3)(iv)).

The representation $(\mathrm{St},\mathrm{st})$ is a quotient of $(I(\delta_{B(F)}^{-1/2}), i(\delta_{B(F)}^{-1/2}))$ (see (8.1)). Therefore $r_{\Delta}^I(\mathrm{St},\mathrm{st})$ is a quotient of $r_{\Delta}^I(I(\delta_{B(F)}^{-1/2}), i(\delta_{B(F)}^{-1/2}))$ (see (7.1.3)(i)). But the structure of the last smooth representation is known (see (7.3.4)). It follows that $r_{\Delta}^I(\mathrm{St},\mathrm{st})$ has finite length (see (7.3.3)(i)) and that, for any irreducible subquotient $(\mathcal{V},\pi)$ of $r_{\Delta}^I(\mathrm{St},\mathrm{st})$, there exists $w \in W$ such that $w(I) \subset R^+$ and such that $(\mathcal{V},\pi)$ is isomorphic to a subquotient of $i_{\emptyset}^I(\mathbb{C}, w^{-1}(\delta_{B(F)}^{-1/2}))$. As any irreducible subquotient of $i_{\emptyset}^I(\mathbb{C}, w^{-1}(\delta_{B(F)}^{-1/2}))$ is isomorphic to a subrepresentation of $i_{\emptyset}^I(\mathbb{C},\chi)$ for some χ in the W_I-orbit of $w^{-1}(\delta_{B(F)}^{-1/2})$ (see (7.3.5)) it follows by adjunction that $r_I^{\emptyset}(\mathcal{V},\pi) \neq (0)$ (see (7.1.3)(i)). Here we have used obvious generalizations of (7.3.3)(i) and (7.3.5) where G is replaced by M_I. These generalizations can be deduced from (7.3.3)(i) and (7.3.5) for $GL_{d_1},\dots,GL_{d_s}$ ($d_I = (d_1,\dots,d_s)$) and from (D.7.1).

Now the irreducibility of $r_{\Delta}^I(\mathrm{St},\mathrm{st})$ follows from the irreducibility of $r_I^{\emptyset}(r_{\Delta}^I(\mathrm{St},\mathrm{st}))$ and the lemma is proved. □

We will also need another basic result of the reduction theory.

LEMMA (9.5.8) (Harder). — *Let $I \subset \Delta$ and let c_2 be an integer. Then there exists an integer $c_1' \leq c_2$ having the following property: let $u_{\mathbb{A}}$, $u_{\mathbb{A}}' \in U^I(\mathbb{A})$, $t_{\mathbb{A}}$, $t_{\mathbb{A}}' \in T(\mathbb{A})_{]-\infty,c_2]}^I$, $m_{\mathcal{O}}$, $m_{\mathcal{O}}' \in M_I(\mathcal{O})$ and $\gamma \in M_I(F)$ be such that*

$$u_{\mathbb{A}}' t_{\mathbb{A}}' m_{\mathcal{O}}' = \gamma u_{\mathbb{A}} t_{\mathbb{A}} m_{\mathcal{O}}$$

and let $\alpha \in I$ be such that

$$\deg(\alpha(t_{\mathbb{A}})) < c_1',$$

then we have

$$\gamma \in P_{I-\{\alpha\}}^I(F) \subset M_I(F).$$

□

A proof of (9.5.8) is given in (E.1.1).

Proof of (9.5.5)(i) : Let $P' \subset P$ be standard parabolic subgroups of GL_d with standard Levi decompositions $P' = M'N'$, $P = MN$. Then for any $\varphi \in C_{GL_d}^{\infty}$ and any $g_{\mathbb{A}} \in GL_d(\mathbb{A})$ we have

$$\varphi_{P'}(g_{\mathbb{A}}) = \frac{1}{\mathrm{vol}\left(N''(F)\backslash N''(\mathbb{A}), \frac{dn_{\mathbb{A}}''}{dv''}\right)} \int_{N''(F)\backslash N''(\mathbb{A})} \varphi_P(n_{\mathbb{A}}'' g_{\mathbb{A}}) \frac{dn_{\mathbb{A}}''}{dv''}$$

where $N'' = M \cap N'$ is the unipotent radical of the standard parabolic subgroup $P'' = M \cap P'$ of M. In particular, if $\varphi \in \mathcal{C}_{GL_d}^{\infty}$ is not cuspidal, there exists a maximal proper standard parabolic subgroup $P = P_{\Delta - \{\alpha\}}$, $\alpha \in \Delta$, of GL_d such that φ_P is not identically zero.

Let us fix $\varphi \in \mathcal{V}$. We want to prove that φ is cuspidal. Let us assume that this is not the case and let us try to find a contradiction.

First step : We choose $\alpha \in \Delta$ such that $\varphi_{P_{\Delta-\{\alpha\}}}$ is not identically zero and we set $P = P_{\Delta-\{\alpha\}}$. We choose a compact open subgroup K' of $GL_d(\mathbb{A})$ such that $\varphi \in \mathcal{V}^{K'}$. We denote by $P = MN$ ($M = M_{\Delta-\{\alpha\}}$, $N = N_{\Delta-\{\alpha\}}$) the standard Levi decomposition of P. Then thanks to (9.2.7), (9.2.9) and (9.5.8) we can find integers $c_1'' \leq c_2$ having the following properties:

(1) $M(\mathbb{A}) = M(F)U^{\Delta-\{\alpha\}}(\mathbb{A})T(\mathbb{A})^{\Delta-\{\alpha\}}_{]-\infty,c_2]}M(\mathcal{O})$,

(2) if we set

$$\mathcal{S} = U(\mathbb{A})T(\mathbb{A})^{\Delta}_{]-\infty,c_2]}GL_d(\mathcal{O}),$$

then, for any $g_{\mathbb{A}} = u_{\mathbb{A}}t_{\mathbb{A}}k \in \mathcal{S}$ ($u_{\mathbb{A}} \in U(\mathbb{A})$, $t_{\mathbb{A}} \in T(\mathbb{A})^{\Delta}_{]-\infty,c_2]}$, $k \in GL_d(\mathcal{O})$) with $\deg(\alpha(t_{\mathbb{A}})) < c_1''$, we have

$$\varphi(g_{\mathbb{A}}) = \varphi_P(g_{\mathbb{A}}),$$

(3) for any $g_{\mathbb{A}} = u_{\mathbb{A}}t_{\mathbb{A}}k$, $g'_{\mathbb{A}} = u'_{\mathbb{A}}t'_{\mathbb{A}}k' \in \mathcal{S}$ with $\deg(\alpha(t_{\mathbb{A}})) < c_1''$ we have

$$g'_{\mathbb{A}}g_{\mathbb{A}}^{-1} \in GL_d(F)$$

if and only if

$$g'_{\mathbb{A}}g_{\mathbb{A}}^{-1} \in P(F).$$

Second step : Let us denote by

$$\delta : \mathcal{S} \longrightarrow \mathbb{Z}$$

the function given by

$$\delta(u_{\mathbb{A}}t_{\mathbb{A}}k) = \inf\{\deg(\beta(t_{\mathbb{A}})) - \deg(\alpha(t_{\mathbb{A}})) \mid \beta \in \Delta - \{\alpha\}\}$$

(we leave it to the reader to check that it is well-defined). Let us set

$$\widetilde{\mathcal{U}} = \{g_{\mathbb{A}} \in \mathcal{S} \mid \delta(g_{\mathbb{A}}) \geq 0\}$$

and

$$\mathcal{U} = \mathbb{A}^{\times}P(F)\backslash M(F)\widetilde{\mathcal{U}} \subset \mathbb{A}^{\times}P(F)\backslash GL_d(\mathbb{A}).$$

Then $\mathcal{U}$ is an open subset of $\mathbb{A}^{\times}P(F)\backslash GL_d(\mathbb{A})$.

The functions $|\varphi|$, $|\varphi_P|$ and $|\varphi - \varphi_P|$ may be viewed as functions on $\mathbb{A}^\times P(F)\backslash GL_d(\mathbb{A})$. Let us check that

$$\int_{\mathcal{U}} |\varphi_P(g_{\mathbb{A}})|^2 \frac{dg_{\mathbb{A}}}{dz_{\mathbb{A}} d\mu d\nu} < +\infty$$

(recall that $dg_{\mathbb{A}}$ and $dz_{\mathbb{A}}$ are the Haar measures on $GL_d(\mathbb{A})$ and $\mathbb{A}^\times$ respectively which are normalized by $\mathrm{vol}(K_{\mathcal{I}}, dg_{\mathbb{A}}) = 1$ and $\mathrm{vol}(\mathbb{A}^\times \cap K_{\mathcal{I}}, dz_{\mathbb{A}}) = 1$ and that $d\mu d\nu$ is the counting measure on $P(F) = M(F)N(F)$). Let us set

$$\widetilde{\mathcal{U}}' = \{u_{\mathbb{A}} t_{\mathbb{A}} k \in \widetilde{\mathcal{U}} \mid \deg(\alpha(t_{\mathbb{A}})) < c_1''\},$$

$$\mathcal{U}' = \mathbb{A}^\times P(F)\backslash M(F)\widetilde{\mathcal{U}}' \subset \mathcal{U}$$

and

$$\overline{\mathcal{U}}' = \mathbb{A}^\times GL_d(F)\backslash GL_d(F)\widetilde{\mathcal{U}}' \subset \mathbb{A}^\times GL_d(F)\backslash GL_d(\mathbb{A}).$$

By hypothesis φ is square-integrable on $\mathbb{A}^\times GL_d(F)\backslash GL_d(\mathbb{A})$ and therefore on the open subset $\overline{\mathcal{U}}'$ of $\mathbb{A}^\times GL_d(F)\backslash GL_d(\mathbb{A})$. But the canonical projection

$$\mathbb{A}^\times P(F)\backslash GL_d(\mathbb{A}) \twoheadrightarrow \mathbb{A}^\times GL_d(F)\backslash GL_d(\mathbb{A})$$

induces an isomorphism from $\mathcal{U}'$ onto $\overline{\mathcal{U}}'$ (see property (3) of the integers $c_1'' \leq c_2$) and the measures $dz_{\mathbb{A}} d\mu d\nu\backslash dg_{\mathbb{A}}$ and $dz_{\mathbb{A}} d\gamma\backslash dg_{\mathbb{A}}$ correspond by this isomorphism. Hence we obtain

$$\int_{\mathcal{U}'} |\varphi(g_{\mathbb{A}})|^2 \frac{dg_{\mathbb{A}}}{dz_{\mathbb{A}} d\mu d\nu} < +\infty,$$

i.e.

$$\int_{\mathcal{U}'} |\varphi_P(g_{\mathbb{A}})|^2 \frac{dg_{\mathbb{A}}}{dz_{\mathbb{A}} d\mu d\nu} < +\infty$$

(see property (2) of the integers $c_1'' \leq c_2$). Finally $\mathcal{U} - \mathcal{U}'$ is relatively compact (it is contained in the image of $U(\mathbb{A})T(\mathbb{A})^{\Delta}_{[c_1'', c_2]} GL_d(\mathcal{O})$ in $\mathbb{A}^\times P(F)\backslash GL_d(\mathbb{A})$ and this image is compact, see the proof of (9.2.6)). Our assertion follows.

Using that φ_P is invariant under left translation by $N(\mathbb{A})$ we may reformulate our assertion as

$$\int_{N(\mathbb{A})\backslash\mathcal{U}} |\varphi_P(g_{\mathbb{A}})|^2 \frac{dg_{\mathbb{A}}}{dz_{\mathbb{A}} d\mu dn_{\mathbb{A}}} < +\infty$$

(recall that $N(F)\backslash N(\mathbb{A})$ is compact).

Third step : If $\psi \in \mathbb{C} \otimes C_c^\infty(GL_d(\mathbb{A}))$ we have the integration formula (see (4.1.7))

$$\int_{GL_d(\mathbb{A})} \psi(g_{\mathbb{A}}) dg_{\mathbb{A}}$$
$$= \int_{GL_d(\mathcal{O})} \Big(\int_{M(\mathbb{A})} \Big(\int_{N(\mathbb{A})} \psi(n_{\mathbb{A}} m_{\mathbb{A}} k) dn_{\mathbb{A}} \Big) \delta_{P(\mathbb{A})}^{-1}(m_{\mathbb{A}}) dm_{\mathbb{A}} \Big) dk,$$

where, for this step only, the Haar measures $dg_{\mathbb{A}}$, dk, $dm_{\mathbb{A}}$ and $dn_{\mathbb{A}}$ are normalized in such a way that the volumes of the standard compact subgroups $GL_d(\mathcal{O})$, $GL_d(\mathcal{O})$, $M(\mathcal{O})$ and $N(\mathcal{O})$ are all equal to 1 (the modulus character $\delta_{P(\mathbb{A})}$ enters because we have permuted n and m). Thus, for any $\psi \in \mathbb{C} \otimes C_c^\infty(\mathbb{A}^\times M(F) N(\mathbb{A}) \backslash GL_d(\mathbb{A}))$, we have the integration formula

$$\int_{\mathbb{A}^\times M(F) N(\mathbb{A}) \backslash GL_d(\mathbb{A})} \psi(g_{\mathbb{A}}) \frac{dg_{\mathbb{A}}}{dz_{\mathbb{A}} d\mu dn_{\mathbb{A}}}$$
$$= \int_{GL_d(\mathcal{O})} \Big(\int_{\mathbb{A}^\times M(F) \backslash M(\mathbb{A})} \psi(m_{\mathbb{A}} k) \delta_{P(\mathbb{A})}^{-1}(m_{\mathbb{A}}) \frac{dm_{\mathbb{A}}}{dz_{\mathbb{A}} d\mu} \Big) dk.$$

By the Fubini theorem it follows that, for any continuous complex function ψ on $\mathbb{A}^\times M(F) N(\mathbb{A}) \backslash GL_d(\mathbb{A})$ such that

$$\int_{\mathbb{A}^\times M(F) N(\mathbb{A}) \backslash GL_d(\mathbb{A})} |\psi(g_{\mathbb{A}})| \frac{dg_{\mathbb{A}}}{dz_{\mathbb{A}} d\mu dn_{\mathbb{A}}} < +\infty,$$

we have

$$\int_{\mathbb{A}^\times M(F) \backslash M(\mathbb{A})} |\psi(m_{\mathbb{A}} g_{\mathbb{A}})| \delta_{P(\mathbb{A})}^{-1}(m_{\mathbb{A}}) \frac{dm_{\mathbb{A}}}{dz_{\mathbb{A}} d\mu} < +\infty$$

for almost every $g_{\mathbb{A}} = k \in GL_d(\mathcal{O})$ with respect to dk and therefore, for almost every $g_{\mathbb{A}} \in GL_d(\mathbb{A})$ with respect to $dg_{\mathbb{A}}$ (ψ is $N(\mathbb{A})$-invariant under left translation and we have $GL_d(\mathbb{A}) = M(\mathbb{A}) N(\mathbb{A}) GL_d(\mathcal{O})$ by the Iwasawa decomposition; see (4.1)).

If $\psi' \in \mathbb{C} \otimes C_c^\infty(\mathbb{A}^\times M(F) \backslash M(\mathbb{A}))$ we also have the integration formula

$$\int_{\mathbb{A}^\times M(F) \backslash M(\mathbb{A})} \psi'(m_{\mathbb{A}}) \frac{dm_{\mathbb{A}}}{dz_{\mathbb{A}} d\mu}$$
$$= \int_{Z_M(\mathbb{A}) M(F) \backslash M(\mathbb{A})} \Big(\int_{\mathbb{A}^\times Z_M(F) \backslash Z_M(\mathbb{A})} \psi'(z_{M,\mathbb{A}} m_{\mathbb{A}}) \frac{dz_{M,\mathbb{A}}}{dz_{\mathbb{A}} d\zeta_M} \Big) \frac{dm_{\mathbb{A}}}{dz_{M,\mathbb{A}} d\mu}$$

where $d\zeta_M$ is the counting measure on $Z_M(F)$. By the Fubini theorem again it follows that, for any continuous complex function ψ' on $\mathbb{A}^\times M(F) \backslash M(\mathbb{A})$ such that

$$\int_{\mathbb{A}^\times M(F) \backslash M(\mathbb{A})} |\psi'(m_{\mathbb{A}})| \frac{dm_{\mathbb{A}}}{dz_{\mathbb{A}} d\mu} < +\infty,$$

we have

$$\int_{\mathbb{A}^\times Z_M(F)\backslash Z_M(\mathbb{A})} |\psi'(z_{M,\mathbb{A}}m_{\mathbb{A}})| \frac{dz_{M,\mathbb{A}}}{dz_{\mathbb{A}}d\zeta_M} < +\infty$$

for almost every $m_{\mathbb{A}} \in M(\mathbb{A})$ with respect to $dm_{\mathbb{A}}$.

Therefore, for any continuous complex function ψ on $\mathbb{A}^\times M(F)N(\mathbb{A})\backslash GL_d(\mathbb{A})$ such that

$$\int_{\mathbb{A}^\times M(F)N(\mathbb{A})\backslash GL_d(\mathbb{A})} |\psi(g_{\mathbb{A}})| \frac{dg_{\mathbb{A}}}{dz_{\mathbb{A}}d\mu dn_{\mathbb{A}}} < +\infty,$$

we have

$$\int_{\mathbb{A}^\times Z_M(F)\backslash Z_M(\mathbb{A})} |\psi(z_{M,\mathbb{A}}g_{\mathbb{A}})|\delta_{P(\mathbb{A})}^{-1}(z_{M,\mathbb{A}}) \frac{dz_{M,\mathbb{A}}}{dz_{\mathbb{A}}d\zeta_M} < +\infty$$

for almost every $g_{\mathbb{A}} \in G(\mathbb{A})$ with respect to $dg_{\mathbb{A}}$.

Applying this result to

$$\psi = |\varphi_P|^2 1_{N(\mathbb{A})\backslash\mathcal{U}},$$

where $1_{N(\mathbb{A})\backslash\mathcal{U}}$ is the characteristic function of $N(\mathbb{A})\backslash\mathcal{U} \subset \mathbb{A}^\times M(F)N(\mathbb{A})\backslash GL_d(\mathbb{A})$, we obtain that, for almost every $g_{\mathbb{A}} \in \mathcal{S}$ with respect to $dg_{\mathbb{A}}$, we have

$$\int_{\mathbb{A}^\times Z_M(F)\backslash Z_M(\mathbb{A})_{]-\infty,\delta(g_{\mathbb{A}})]}} |\varphi_P(z_{M,\mathbb{A}}g_{\mathbb{A}})|^2\delta_{P(\mathbb{A})}^{-1}(z_{M,\mathbb{A}}) \frac{dz_{M,\mathbb{A}}}{dz_{\mathbb{A}}d\zeta_M} < +\infty$$

with

$$Z_M(\mathbb{A})_{]-\infty,c]} \stackrel{\text{dfn}}{=\!=} \{z_{M,\mathbb{A}} \in Z_M(\mathbb{A}) \mid \deg(\alpha(z_{M,\mathbb{A}})) \leq c\}$$

for each integer c. Indeed, if $g_{\mathbb{A}} = u_{\mathbb{A}}t_{\mathbb{A}}k \in \mathcal{S}$ and $z_{M,\mathbb{A}} \in Z_M(\mathbb{A})_{]-\infty,\delta(g_{\mathbb{A}})]}$, we have

$$z_{M,\mathbb{A}}g_{\mathbb{A}} = (z_{M,\mathbb{A}}u_{\mathbb{A}}z_{M,\mathbb{A}}^{-1})(z_{M,\mathbb{A}}t_{\mathbb{A}})k$$

and

$$\delta(z_{M,\mathbb{A}}g_{\mathbb{A}}) = \delta(g_{\mathbb{A}}) - \deg(\alpha(z_{M,\mathbb{A}})) \geq 0,$$

so that

$$z_{M,\mathbb{A}}g_{\mathbb{A}} \in \widetilde{\mathcal{U}}$$

and

$$\mathbb{A}^\times M(F)N(\mathbb{A})z_{M,\mathbb{A}}g_{\mathbb{A}} \in N(\mathbb{A})\backslash\mathcal{U}.$$

Fourth step : We have

$$T(\mathbb{A})^{\Delta-\{\alpha\}}_{]-\infty,c_2]} = Z_M(\mathbb{A})T(\mathbb{A})^{\Delta}_{]-\infty,c_2]},$$

so that

$$M(\mathbb{A}) = M(F)Z_M(\mathbb{A})U^{\Delta-\{\alpha\}}(\mathbb{A})T(\mathbb{A})^{\Delta}_{]-\infty,c_2]}M(\mathcal{O})$$

(see property (1) of the integers $c_1'' \leq c_2$). Therefore we have

$$P(\mathbb{A}) = M(F)Z_M(\mathbb{A})U(\mathbb{A})T(\mathbb{A})^{\Delta}_{]-\infty,c_2]}M(\mathcal{O})$$

and by the Iwasawa decomposition (see (4.1)) it follows that

$$GL_d(\mathbb{A}) = M(F)Z_M(\mathbb{A})\mathcal{S}.$$

But φ_P is invariant under left translation by $M(F)$, it is locally constant and it does not vanish identically. Therefore, by the third step there exists $g_{\mathbb{A}} \in \mathcal{S}$ such that φ_P does not vanish identically on $Z_M(\mathbb{A})g_{\mathbb{A}}$ and such that

$$\int_{\mathbb{A}^\times Z_M(F)\backslash Z_M(\mathbb{A})_{]-\infty,\delta(g_{\mathbb{A}})]}} |\varphi_P(z_{M,\mathbb{A}}g_{\mathbb{A}})|^2 \delta_{P(\mathbb{A})}^{-1}(z_{M,\mathbb{A}})\frac{dz_{M,\mathbb{A}}}{dz_{\mathbb{A}}d\zeta_M} < +\infty.$$

Let us fix such a $g_{\mathbb{A}}$ and let us replace φ by $\pi(g_{\mathbb{A}})(\varphi) \in \mathcal{V}$. Then we have

$$\varphi_P|Z_M(\mathbb{A}) \not\equiv 0$$

and

$$\int_{\mathbb{A}^\times Z_M(F)\backslash Z_M(\mathbb{A})_{]-\infty,c]}} |\varphi_P(z_{M,\mathbb{A}})|^2 \delta_{P(\mathbb{A})}^{-1}(z_{M,\mathbb{A}})\frac{dz_{M,\mathbb{A}}}{dz_{\mathbb{A}}d\zeta_M} < +\infty$$

for some and thus for all integers c (the function φ_P is continuous and

$$\mathbb{A}^\times Z_M(F)\backslash\big(Z_M(\mathbb{A})_{]-\infty,c']} - Z_M(\mathbb{A})_{]-\infty,c]}\big)$$

is compact for all integers $c \leq c'$).

Fifth step : Let us choose $z_M^\infty \in Z_M(\mathbb{A}^\infty)$ such that the function

$$z_{M,\infty} \longmapsto \varphi_P(z_{M,\infty}z_M^\infty)$$

is not identically zero. The multiplication by z_M^∞ induces an open embedding

$$(F_\infty^\times\backslash Z_M(F_\infty)) \times ((\mathcal{O}^\infty)^\times\backslash Z_M(\mathcal{O}^\infty)) \hookrightarrow \mathbb{A}^\times Z_M(F)\backslash Z_M(\mathbb{A}).$$

Restricting the function $|\varphi_P|^2\delta_{P(\mathbb{A})}^{-1}$ to the image of this embedding and applying the Fubini theorem once more we obtain that, for almost every $z_{M,\mathcal{O}^\infty} \in Z_M(\mathcal{O}^\infty)$ (with respect to some Haar measure) and for every integer c, we have

$$\int_{F_\infty^\times \backslash Z_M(F_\infty)_{]-\infty,c]}} |\varphi_P(z_{M,\infty}z_{M,\mathcal{O}^\infty}z_M^\infty)|^2\delta_{P(F_\infty)}^{-1}(z_{M,\infty})\frac{dz_{M,\infty}}{dz_\infty} < +\infty$$

with

$$Z_M(F_\infty)_{]-\infty,c]} \overset{\text{dfn}}{=\!=} \{z_{M,\infty} \in Z_M(F_\infty) \mid \deg(\infty)\infty(\alpha(z_{M,\infty})) \leq c\}$$

($dz_{M,\infty}$ and dz_∞ are the Haar measures on $Z_M(F_\infty)$ and $F_\infty^\times$ respectively which are normalized by $\mathrm{vol}(Z_M(\mathcal{O}_\infty), dz_{M,\infty}) = 1$ and $\mathrm{vol}(\mathcal{O}_\infty^\times, dz_\infty) = 1$). As before, choosing a suitable $z_{M,\mathcal{O}^\infty}$ and replacing z_M^∞ by $z_{M,\mathcal{O}^\infty}z_M^\infty$ we get that

$$\int_{F_\infty^\times \backslash Z_M(F_\infty)_{]-\infty,c]}} |\varphi_P(z_{M,\infty}z_M^\infty)|^2\delta_{P(F_\infty)}^{-1}(z_{M,\infty})\frac{dz_{M,\infty}}{dz_\infty} < +\infty$$

for every integer c, the function

$$Z_M(F_\infty) \to \mathbb{C},\ z_{M,\infty} \mapsto \varphi_P(z_{M,\infty}z_M^\infty)$$

not being identically zero.

Sixth step : The morphism

$$(\mathcal{V}, \pi|M(F_\infty)) \to (\mathcal{C}_M^\infty, R_M|M(F_\infty)),\ \psi \mapsto \psi_P|M(\mathbb{A}),$$

obviously factors through the quotient

$$(\mathcal{V}_\infty/\mathcal{V}_\infty(N(F_\infty)), \overline{\pi}_\infty) \otimes_{\mathbb{C}} \mathcal{V}^\infty$$

of

$$(\mathcal{V}, \pi|M(F_\infty)) \cong (\mathcal{V}_\infty, \pi_\infty) \otimes_{\mathbb{C}} \mathcal{V}^\infty,$$

where $\overline{\pi}_\infty$ is induced by $\pi_\infty|M(F_\infty)$. But the Jacquet module $(\mathcal{V}_\infty/\mathcal{V}_\infty(N(F_\infty)), \overline{\pi}_\infty)$ is isomorphic to $(r_\Delta^{\Delta-\{\alpha\}}(\mathcal{V}_\infty, \pi_\infty)) \otimes (\mathbb{C}, \delta_{P(F_\infty)}^{1/2})$ and by hypothesis $(\mathcal{V}_\infty, \pi_\infty)$ is isomorphic to $(\mathrm{St}_\infty, \mathrm{st}_\infty)$. Therefore, applying lemma (9.5.7) we get that

$$R_M(z_{M,\infty})(\psi_P|M(\mathbb{A})) = (\delta_{B(F_\infty)}^{1/2}\delta_{P(F_\infty)}^{1/2})(z_{M,\infty})(\psi_P|M(\mathbb{A}))$$

for every $z_{M,\infty} \in Z_M(F_\infty)$ and every $\psi \in \mathcal{V}$. In particular we have

$$\varphi_P(z_{M,\infty}z_M^\infty) = (\delta_{B(F_\infty)}^{1/2}\delta_{P(F_\infty)}^{1/2})(z_{M,\infty})\varphi_P(z_M^\infty)$$

for every $z_{M,\infty}$.

Finally, $\varphi_P(z_M^\infty)$ cannot be zero, we have

$$\delta_{B(F_\infty)}(z_{M,\infty}) = \delta_{P(F_\infty)}(z_{M,\infty}) = |\alpha(z_{M,\infty})|_\infty^{i(d-i)}$$

for every $z_{M,\infty} \in Z_M(F_\infty)$ if $\alpha = \epsilon_i - \epsilon_{i+1}$ for some $i \in \{1, \ldots, d-1\}$ and we have

$$\int_{F_\infty^\times \backslash Z_M(F_\infty)_{]-\infty,c]}} |\alpha(z_{M,\infty})|_\infty^{i(d-i)} \frac{dz_{M,\infty}}{dz_\infty} = \sum_{n=-\infty}^{[c/\deg(\infty)]} p^{-n\deg(\infty)i(d-i)} = +\infty.$$

Therefore we have obtained a contradiction. □

The proof of (9.5.5)(ii) is based on the strong approximation theorem (see [Moo] Lemma (13.1)).

LEMMA (9.5.9) (Strong approximation). — *Let x be a place of F. Then $SL_d(F)SL_d(F_x)$ is dense in $SL_d(\mathbb{A})$.*

□

A proof of (9.5.9) is given in (E.2.1)

Proof of (9.5.5)(ii) : Let $\varphi \in \mathcal{V} - \{0\}$. The fact that φ is an automorphy and the hypothesis $(\mathcal{V}_\infty, \pi_\infty) \cong (\mathbb{C}, 1_\infty)$ imply that

$$\varphi(\gamma h_\infty g_{\mathbb{A}}) = \varphi(g_{\mathbb{A}})$$

for every $\gamma \in GL_d(F)$, every $h_\infty \in GL_d(F_\infty)$ and every $g_{\mathbb{A}} \in GL_d(\mathbb{A})$ (we have $h_\infty g_{\mathbb{A}} = g_{\mathbb{A}}(g_\infty^{-1} h_\infty g_\infty)$). Then, using the continuity of the function φ and applying (9.5.9), we get that

$$\varphi(h_{\mathbb{A}} g_{\mathbb{A}}) = \varphi(g_{\mathbb{A}})$$

for every $h_{\mathbb{A}} \in SL_d(\mathbb{A})$ and every $g_{\mathbb{A}} \in GL_d(\mathbb{A})$. Therefore there exists a unique $\psi \in \mathcal{C}_{GL_1}^\infty$ such that

$$\varphi = \psi \circ \det_{GL_d}.$$

Let

$$(\mathcal{W}, \rho) \subset (\mathcal{C}_{GL_1}^\infty, R_{GL_1})$$

be the subrepresentation of $GL_1(\mathbb{A}) = \mathbb{A}^\times$ on the $\mathbb{C}$-linear span $\mathcal{W}$ of $R_{GL_1}(\mathbb{A}^\times)(\psi)$ in $\mathcal{C}_{GL_1}^\infty$. The morphism

$$(\mathcal{W}, \rho \circ \det_{GL_d}) \to (\mathcal{V}, \pi),\ \psi' \mapsto \psi' \circ \det_{GL_d},$$

is an isomorphism in $\mathrm{Rep}_{\mathrm{s}}(GL_d(\mathbb{A}))$. Therefore $(\mathcal{W}, \rho)$ is an admissible irreducible representation of $\mathbb{A}^\times$ and there exists a smooth character ξ of $\mathbb{A}^\times$ such that $(\mathcal{W}, \rho)$ is isomorphic to $(\mathbb{C}, \xi)$. Obviously we have

$$\psi(a_{\mathbb{A}}) = \xi(a_{\mathbb{A}})\psi(1) \qquad (\forall a_{\mathbb{A}} \in \mathbb{A}^\times)$$

and we have $\xi \in \Xi_{GL_{d,x}}$. This completes the proof of (9.5.5)(ii). □

(9.6) Main theorem

We will need the following local result. We again use the notations of chapters 7 and 8 for the statement and the proof of this result. In particular F is a non-archimedean local field.

LEMMA (9.6.1). — *For each unramified character ξ of $F^\times$, i.e. each complex character $\xi : F^\times \to \mathbb{C}^\times$ which is trivial on $\mathcal{O}^\times$, the 1-dimensional smooth representation $(\mathbb{C}, \xi \circ \det_{GL_d})$ of $GL_d(F)$ is spherical and its Hecke eigenvalues are*

$$\xi(\varpi)q^{(d-1)/2}, \xi(\varpi)q^{(d-3)/2}, \ldots, \xi(\varpi)q^{(1-d)/2}$$

where ϖ is any uniformizer of $\mathcal{O}$.

Proof : The first assertion is obvious. Let χ be the unramified character of $T(F)$ defined by

$$\chi = (\xi \circ \det_{GL_d})\delta_{B(F)}^{-1/2}.$$

Then the morphism

$$(I(\delta_{B(F)}^{-1/2}), (\xi \circ \det_{GL_d})i(\delta_{B(F)}^{-1/2})) \to (I(\chi), i(\chi)),\ f \mapsto (\xi \circ \det_{GL_d})f,$$

is an isomorphism in $\mathrm{Rep}_{\mathrm{s}}(GL_d(F))$ and therefore $(V(\chi), \pi(\chi))$ is isomorphic to

$$(V(\delta_{B(F)}^{-1/2}), (\xi \circ \det_{GL_d})\pi(\delta_{B(F)}^{-1/2}))$$

(see (7.5.2)). But it follows from (8.1.2)(ii) that

$$(V(\delta_{B(F)}^{-1/2}), \pi(\delta_{B(F)}^{-1/2}))$$

is isomorphic to $(\mathbb{C}, 1)$, so that $(V(\chi), \pi(\chi))$ is isomorphic to $(\mathbb{C}, \xi \circ \det_{GL_d})$ and the lemma is proved. □

MAIN THEOREM (9.6.2). — *Let $\chi \in \mathcal{X}_{GL_d}$ be unitary. Then the formal trace*

$$\mathrm{ftr}\, R^2_{GL_d,\chi,\mathrm{disc}}(f_{\mathbb{A}})$$

is equal to

$$(-1)^{d-1} \sum_{(\mathcal{V},\pi)} m_{\mathrm{cusp}}(\pi)\mathrm{tr}\big(\pi^{\infty,o}(f^{\infty,o})|(\mathcal{V}^{\infty,o})^{K_{\mathcal{I}}^{\infty,o}}\big)$$
$$\times p^{\deg(o)r(d-1)/2}(z_1(\pi_o)^r + \cdots + z_d(\pi_o)^r)$$
$$+\sum_{\xi}(\xi^{\infty,o} \circ \det_{GL_d})(f^{\infty,o})\xi_o(\varpi_o)^r(1 + p^{\deg(o)r} + \cdots + p^{\deg(o)r(d-1)})$$

where

(1) $(\mathcal{V}, \pi)$ runs through the set of the $(\mathcal{V}, \pi) \in \Pi_{GL_d,\chi,\mathrm{cusp}}$ such that the component $(\mathcal{V}_\infty, \pi_\infty)$ of $(\mathcal{V}, \pi)$ at ∞ is isomorphic to $(\mathrm{St}_\infty, \mathrm{st}_\infty)$, the

component $(\mathcal{V}^{\infty,o},\pi^{\infty,o})$ *of* $(\mathcal{V},\pi)$ *outside* $\{\infty,o\}$ *has non-zero fixed vectors under* $K_{\mathcal{I}}^{\infty,o}$ *and the component* $(\mathcal{V}_o,\pi_o)$ *of* $(\mathcal{V},\pi)$ *at* o *is spherical, and the sum over the* (V,π) *is finite,*

(2) $z_1(\pi_o),\ldots,z_d(\pi_o)$ *are the Hecke eigenvalues of* $(\mathcal{V}_o,\pi_o)$*,*

(3) ξ *runs through the set of the* $\xi\in\Xi_{GL_d,\chi}$ *such that* $\xi_\infty=1_\infty$*,* $\xi^{\infty,o}$ *is trivial on* $\det_{GL_d}(K_{\mathcal{I}}^{\infty,o})$ *and* ξ_o *is unramified, and the sum over the* ξ *is finite,*

(4) $(\xi^{\infty,o}\circ\det_{GL_d})(f^{\infty,o})$ *is the integral*

$$\int_{GL_d(\mathbb{A}^{\infty,o})} f^{\infty,o}(g^{\infty,o})\xi^{\infty,o}(\det_{GL_d}(g^{\infty,o}))dg^{\infty,o},$$

the Haar measure $dg^{\infty,o}$ *on* $GL_d(\mathbb{A}^{\infty,o})$ *being normalized by* $\mathrm{vol}(K_{\mathcal{I}}^{\infty,o}, dg^{\infty,o})=1$*, and* ϖ_o *is a uniformizer of* $\mathcal{O}_o$*.*

Proof: This is a direct consequence of (9.5.1), (9.5.6) and (9.6.1). □

REMARK (9.6.3). — The formal trace

$$\mathrm{ftr}\, R^2_{GL_d,\chi,\mathrm{disc}}(f_{\mathbb{A}})$$

is automatically zero unless $\chi_\infty=1_\infty$, $\chi^{\infty,o}$ is trivial on $\mathbb{A}^\times\cap K_{\mathcal{I}}^{\infty,o}$ and χ_o is unramified. In particular the number of χ such that this formal trace is non-zero is finite. □

(9.7) Comments and references

There are several standard expository papers of the theory of automorphic forms: [Ge–PS] and [H-C] in the number field case; [Har 2] in the function field case; [Bo–Ja] and [Mo–Wa 2] (Ch. I) where both cases are considered.

Theorem (9.5.5) is well known to the specialists. Related results may be found in [Mo–Wa 2] (I.4.11), in [Wal] and in [Cl] (4.10).

10

Non-invariant Arthur trace formula: the geometric side

(10.0) Introduction

We will use the same notations as in chapters 6 and 9: F is a function field, ∞ is a fixed place of F, $G = GL_d$, ...

The purpose of this chapter is to prove that, at least for r large enough, the Lefschetz number $\mathrm{Lef}_r(f^{\infty,o})$ is equal to the geometric side of the non-invariant Arthur trace formula for the function $f_{\mathbb{A}} = f_\infty f^{\infty,o} f_o$ (see (6.1) for the notations).

Here, thanks to the very-cuspidality of f_∞, this geometric side of the non-invariant Arthur trace formula is simply the integral over $F_\infty^\times G(F)\backslash G(\mathbb{A})$ of

$$\sum_{\gamma \in G(F)} f_{\mathbb{A}}(g^{-1}\gamma g).$$

In particular, it is independent of the truncation parameter.

We conclude this chapter with a proof of the Flicker–Kazhdan simple trace formula.

(10.1) The kernel $K(h, g)$

Let us fix a locally constant complex function with compact support f on $F_\infty^\times\backslash G(\mathbb{A})$ and a Haar measure dg on $G(\mathbb{A})$ with a splitting

$$dg = \prod_x dg_x$$

(for each place x of F, dg_x is a Haar measure on $G(F_x)$ and $\mathrm{vol}(K_x, dg_x) = 1$ for almost all x).

We denote by dz_∞ the Haar measure on $F_\infty^\times$ which is normalized by $\mathrm{vol}(\mathcal{O}_\infty^\times, dz_\infty) = 1$ and we denote by $d\gamma$ the counting measure on $G(F)$. Let

$$L^2_{G,1_\infty} = L^2(F_\infty^\times G(F)\backslash G(\mathbb{A}), \frac{dg}{dz_\infty d\gamma})$$

be the Hilbert space of measurable complex functions φ on $F_\infty^\times G(F)\backslash G(\mathbb{A})$ such that

$$\int_{F_\infty^\times G(F)\backslash G(\mathbb{A})} |\varphi(g)|^2 \frac{dg}{dz_\infty d\gamma} < +\infty.$$

For any $(h,g) \in (F_\infty^\times G(F)\backslash G(\mathbb{A})) \times (F_\infty^\times G(F)\backslash G(\mathbb{A}))$ there are only finitely many γ in $G(F)$ such that

$$f(h^{-1}\gamma g) \neq 0,$$

so that the sum

$$K(h,g) = \sum_{\gamma \in G(F)} f(h^{-1}\gamma g) \tag{10.1.1}$$

is finite and defines a complex function $K(h,g)$ on $(F_\infty^\times G(F)\backslash G(\mathbb{A})) \times (F_\infty^\times G(F)\backslash G(\mathbb{A}))$. If K' is a compact open subgroup of $G(\mathbb{A})$ such that f is K'-bi-invariant (there always exists such a K'), we have

$$K(h,gk') = K(hk',g) = K(h,g)$$

for all $g, h \in G(\mathbb{A})$ and every $k' \in K'$. In particular the function $K(h,g)$ is locally constant. For each fixed h (resp. g) in $G(\mathbb{A})$ the support of the function $K(h,g)$ of g (resp. h) in $F_\infty^\times G(F)\backslash G(\mathbb{A})$ is obviously contained in

$$G(F)\backslash G(F)h\mathrm{Supp}(f) \quad (\text{resp. } G(F)\backslash G(F)g\mathrm{Supp}(f)^{-1})$$

and therefore is compact.

LEMMA (10.1.2). — *For any $\varphi \in L^2_{G,1_\infty}$ and any $h \in G(\mathbb{A})$ the integrals*

$$\int_{F_\infty^\times\backslash G(\mathbb{A})} f(g)\varphi(hg)\frac{dg}{dz_\infty}$$

and

$$\int_{F_\infty^\times G(F)\backslash G(\mathbb{A})} K(h,g)\varphi(g)\frac{dg}{dz_\infty d\gamma}$$

are absolutely convergent and equal. If we denote by $R_{G,1_\infty}(f)(\varphi)(h)$ their common value the complex function $R_{G,1_\infty}(f)(\varphi)$ on $F_\infty^\times G(F)\backslash G(\mathbb{A})$ belongs to

$$L^2_{G,1_\infty} \cap C^\infty(F_\infty^\times G(F)\backslash G(\mathbb{A})).$$

The operator $R_{G,1_\infty}(f)$ on $L^2_{G,1_\infty}$ with kernel $K(h,g)$ is continuous and its norm is bounded above by the L^1-norm

$$\int_{F_\infty^\times\backslash G(\mathbb{A})} |f(g)|\frac{dg}{dz_\infty}.$$

Proof : We may assume that $f = 1_C$ for some compact open subset C of $F_\infty^\times\backslash G(\mathbb{A})$ and that φ is a non-negative real function. Then we have

$$\begin{aligned}\int_{F_\infty^\times\backslash G(\mathbb{A})} f(g)\varphi(hg)\frac{dg}{dz_\infty} &= \int_{F_\infty^\times\backslash G(\mathbb{A})} f(h^{-1}g)\varphi(g)\frac{dg}{dz_\infty}\\ &= \int_{F_\infty^\times G(F)\backslash G(\mathbb{A})}\Big(\sum_{\gamma\in G(F)} f(h^{-1}\gamma g)\varphi(\gamma g)\Big)\frac{dg}{dz_\infty d\gamma}\\ &= \int_{F_\infty^\times G(F)\backslash G(\mathbb{A})} K(h,g)\varphi(g)\frac{dg}{dz_\infty d\gamma}.\end{aligned}$$

By the Cauchy–Schwarz inequality the square of the last integral is bounded by

$$\Big(\int_{F_\infty^\times G(F)\backslash G(\mathbb{A})} K(h,g)^2\frac{dg}{dz_\infty d\gamma}\Big)\Big(\int_{F_\infty^\times G(F)\backslash G(\mathbb{A})} \varphi(g)^2\frac{dg}{dz_\infty d\gamma}\Big)$$

and therefore is finite.

Now it is clear that $R_{G,1_\infty}(f)(\varphi)$ is right K'-invariant as long as f is left K'-invariant for some compact open subgroup K' of $G(\mathbb{A})$, so that

$$R_{G,1_\infty}(f)(\varphi) \in C^\infty(F_\infty^\times G(F)\backslash G(\mathbb{A})).$$

Moreover by the Cauchy–Schwarz inequality we have

$$\begin{aligned}&\int_{F_\infty^\times G(F)\backslash G(\mathbb{A})} |R_{G,1_\infty}(f)(\varphi)(h)|^2\frac{dh}{dz_\infty d\gamma}\\ &\quad= \int_{F_\infty^\times G(F)\backslash G(\mathbb{A})}\Big(\int_C \varphi(hg)\frac{dg}{dz_\infty}\Big)^2\frac{dh}{dz_\infty d\gamma}\\ &\quad\leq \int_{F_\infty^\times G(F)\backslash G(\mathbb{A})} \mathrm{vol}(C,\frac{dg}{dz_\infty})\Big(\int_C \varphi(hg)^2\frac{dg}{dz_\infty}\Big)\frac{dh}{dz_\infty d\gamma}\\ &\quad\leq \mathrm{vol}(C,\frac{dg}{dz_\infty})^2\int_{F_\infty^\times G(F)\backslash G(\mathbb{A})}\varphi(h)^2\frac{dh}{dz_\infty d\gamma} < +\infty,\end{aligned}$$

so that $R_{G,1_\infty}(f)(\varphi) \in L^2_{G,1_\infty}$ and the norm of the operator $R_{G,1_\infty}(f)$ is bounded above by

$$\mathrm{vol}(C,\frac{dg}{dz_\infty}) = \int_{F_\infty^\times\backslash G(\mathbb{A})} |f(g)|\frac{dg}{dz_\infty}.$$

□

We consider the following equivalence relation on $G(F)$. Two elements γ_1, γ_2 in $G(F)$ are equivalent if and only if they have the same characteristic polynomial. We denote by $G(F)_\sharp$ the set of classes in $G(F)$ for this equivalence relation and for any $\mathcal{O} \in G(F)_\sharp$ we denote by $D_\mathcal{O}(T) \in F[T]$ the corresponding characteristic polynomial. Any $\mathcal{O} \in G(F)_\sharp$ is a finite union of conjugacy classes in $G(F)$. These conjugacy classes are in one-to-one correspondence with the sequences of unitary polynomials $(P_1(T), \ldots, P_d(T))$ in $F[T]$ such that

$$P_1(T)|P_2(T)|\cdots|P_d(T)$$

and

$$P_1(T)P_2(T)\cdots P_d(T) = D_\mathcal{O}(T)$$

(the elementary divisors of the conjugacy class). In particular, if $D_\mathcal{O}(T)$ is a product of distinct irreducible unitary polynomials in $F[T]$, $\mathcal{O}$ coincides with the conjugacy class with elementary divisors

$$\begin{cases} P_1(T) = \cdots = P_{d-1}(T) = 1, \\ P_d(T) = D_\mathcal{O}(T). \end{cases}$$

For each $\mathcal{O} \in G(F)_\sharp$,

$$K_\mathcal{O}(h,g) = \sum_{\gamma \in \mathcal{O}} f(h^{-1}\gamma g) \tag{10.1.3}$$

is also a locally constant function on $(F_\infty^\times G(F)\backslash G(\mathbb{A})) \times (F_\infty^\times G(F)\backslash G(\mathbb{A}))$ and we have

$$K(h,g) = \sum_{\mathcal{O} \in G(F)_\sharp} K_\mathcal{O}(h,g) \tag{10.1.4}$$

(the sums in (10.1.3) and (10.1.4) are finite).

(10.2) Integrability of $K(h,g)$ along the diagonal

Let us denote by $k(g)$ (resp. $k_\mathcal{O}(g)$) the restriction of $K(h,g)$ (resp. $K_\mathcal{O}(h,g)$ for any $\mathcal{O} \in G(F)_\sharp$) along the diagonal $h = g$.

Ideally the operator $R_G(f)$ would have a trace and this trace would be equal to the integral of $k(g)$ on $F_\infty^\times G(F)\backslash G(\mathbb{A})$. Unfortunately the situation is much more complicated and the integral diverges. To get his non-invariant trace formula (in the number field case) Arthur has replaced $k(g)$ by a suitable truncation $k^T(g)$ of it (see [Ar 1] (7.1)). For our purpose we may restrict ourselves to a class of functions f for which $k^T(g) = k(g)$.

Let us make the following assumptions on f:

(A) *for any* $g \in G(\mathbb{A})$ *and any* $k_\infty \in K_\infty$ *we have* $f(k_\infty^{-1} g k_\infty) = f(g)$,

(B) *for any subset* $I \subsetneq \Delta$, *any* $m_{I,\infty} \in M_I(F_\infty)$ *and any* $h^\infty \in G(\mathbb{A}^\infty)$ *we have*

$$\int_{N_I(F_\infty)} f(m_{I,\infty} n_{I,\infty} h^\infty) dn_{I,\infty} = 0.$$

Here $K_\infty = G(\mathcal{O}_\infty) \subset G(F_\infty) \subset G(\mathbb{A})$, $P_I \subset G$ is the standard parabolic subgroup corresponding to I, $P_I = M_I N_I$ is its standard Levi decomposition (see (5.1)) and $dn_{I,\infty}$ is an arbitrary Haar measure on $N_I(F_\infty)$.

REMARK (10.2.1). — Let $f_{\mathbb{A}} = f_\infty f^{\infty,o} f_o$ be the function introduced in chapter 6. Then, if we set

$$f(g) = \int_{K_\infty} f_{\mathbb{A}}(k_\infty^{-1} g k_\infty) dk_\infty \qquad (\forall g \in G(\mathbb{A}))$$

(dk_∞ is the Haar measure on K_∞ of total volume 1), f is a locally constant function with compact support on $F_\infty^\times \backslash G(\mathbb{A})$ which satisfies assumptions (A) and (B) (see (5.1.3)(ii)) and which have the same orbital integrals and traces as f. □

THEOREM (10.2.2) (Arthur). — *Under assumptions* (A) *and* (B) *on* f *we have*

$$\sum_{\mathcal{O} \in G(F)_\sharp} \int_{F_\infty^\times G(F) \backslash G(\mathbb{A})} |k_{\mathcal{O}}(g)| \frac{dg}{dz_\infty d\gamma} < +\infty.$$

The proof of this theorem will be given in (10.4). For the moment let us remark that (10.2.2) is a special case of the obvious function field analog of [Ar 1] (7.1). Indeed the next lemma implies that

$$k_{\mathcal{O}}^T(g) = k_{\mathcal{O}}(g) \qquad (\forall g \in G(\mathbb{A}))$$

for any $\mathcal{O} \in G(F)_\sharp$ and any truncation parameter T (notations of loc. cit.).

LEMMA (10.2.3). — *Under assumptions* (A) *and* (B) *on* f *we have*

$$\int_{N_I(\mathbb{A})} f(g^{-1} m_I n_I g) dn_I = 0$$

for any $I \subsetneq \Delta$, *any* $g \in G(\mathbb{A})$ *and any* $m_I \in M_I(\mathbb{A})$.

Proof : We have the Iwasawa decomposition $G(F_\infty) = P_I(F_\infty)K_\infty$. Therefore, thanks to assumption (A) we may assume that $g_\infty = n_\infty m_\infty$ with $n_\infty \in N_I(F_\infty)$ and $m_\infty \in M_I(F_\infty)$. But we have

$$(n_\infty m_\infty)^{-1} m_{I,\infty} n_{I,\infty} (n_\infty m_\infty) = m'_{I,\infty} n'_{I,\infty}$$

with

$$m'_{I,\infty} = m_\infty^{-1} m_{I,\infty} m_\infty \in M_I(F_\infty)$$

and

$$n'_{I,\infty} = m_\infty^{-1}\big((m_{I,\infty}^{-1} n_\infty m_{I,\infty})^{-1} n_{I,\infty} n_\infty\big) m_\infty \in N_I(F_\infty).$$

Then the lemma follows from assumption (B) (perform the change of variable $n_{I,\infty} \mapsto n'_{I,\infty}$). □

(10.3) Harder's reduction theory revisited

We set

$$\mathfrak{a}_\emptyset = \{H = (H_1, \ldots, H_d) \in \mathbb{Q}^d\}.$$

For each $i \in \Delta = \{1, \ldots, d-1\}$ we denote by α_i the $\mathbb{Q}$-linear form

$$H \mapsto \alpha_i(H) = H_i - H_{i+1}$$

on $\mathfrak{a}_\emptyset$. For each $I \subset \Delta$ we denote by

$$\mathfrak{a}_I \subset \mathfrak{a}_\emptyset$$

the $\mathbb{Q}$-vector subspace defined by the equations

$$\alpha_i(H) = 0 \qquad (i \in I).$$

If $d_I = (d_1, \ldots, d_s)$ is the partition of d corresponding to I, the projection

$$\mathfrak{a}_I \to \mathbb{Q}^s,\ H \mapsto (H_{d_1}, H_{d_1+d_2}, \ldots, H_{d_1+d_2+\cdots+d_s}),$$

is an isomorphism.

For each $i \in \Delta$ let

$$\varpi_i^\Delta = \varpi_i = \frac{d-i}{d}\sum_{k=1}^{i} k\alpha_k + \frac{i}{d}\sum_{k=i+1}^{d-1}(d-k)\alpha_k$$

be the corresponding fundamental weight, so that ϖ_i is the $\mathbb{Q}$-linear form on $\mathfrak{a}_\emptyset$ given by

$$\varpi_i(H) = H_1 + \cdots + H_i - \frac{i}{d}(H_1 + \cdots + H_d) \qquad (\forall H \in \mathfrak{a}_\emptyset).$$

Then for each $J \subset I \subset \Delta$ we have

$$\mathfrak{a}_I \subset \mathfrak{a}_J \subset \mathfrak{a}_\emptyset$$

and

$$\mathfrak{a}_J = \mathfrak{a}_I \oplus \mathfrak{a}_J^I$$

where we have set

$$\mathfrak{a}_J^I = \{H \in \mathfrak{a}_J \mid H_1 + \cdots + H_d = 0 \text{ and } \varpi_i(H) = 0, \forall i \in \Delta - I\}.$$

For any $H \in \mathfrak{a}_J$ we denote by H_I and H_J^I the canonical projections of H onto $\mathfrak{a}_I$ and $\mathfrak{a}_J^I$ respectively.

If $I \subset \Delta$ and if $d_I = (d_1, \ldots, d_s)$ is the corresponding partition of d we have

$$I = \coprod_{j=1}^{s} \{d_1 + \cdots + d_{j-1} + 1, \ldots, d_1 + \cdots + d_j - 1\}$$

and

$$\mathfrak{a}_\emptyset^I = \{H \in \mathfrak{a}_\emptyset \mid \sum_{k=1}^{d_j} H_{d_1+\cdots+d_{j-1}+k} = 0, \ \forall j = 1, \ldots, s\}.$$

For each $i = d_1 + \cdots + d_{j-1} + k \in I$, $1 \leq k \leq d_j - 1$, $1 \leq j \leq s$, we may also consider the $\mathbb{Q}$-linear form

$$\varpi_i^I = \frac{d_j - k}{d_j} \sum_{\ell=1}^{k} \ell \alpha_{d_1+\cdots+d_{j-1}+\ell} + \frac{k}{d_j} \sum_{\ell=k+1}^{d_j-1} (d_j - \ell) \alpha_{d_1+\cdots+d_{j-1}+\ell}$$

on $\mathfrak{a}_\emptyset$. Then, if we define the coroot $\alpha_i^\vee \in \mathfrak{a}_\emptyset$ by

$$(\alpha_i^\vee)_k = \begin{cases} 1 & \text{if } k = i, \\ -1 & \text{if } k = i+1, \\ 0 & \text{otherwise}, \end{cases}$$

for every $i \in \Delta$, $(\alpha_i^\vee)_{i \in I}$ is a basis of the $\mathbb{Q}$-vector subspace $\mathfrak{a}_\emptyset^I$ of $\mathfrak{a}_\emptyset$ and $(\varpi_i^I | \mathfrak{a}_\emptyset^I)_{i \in I}$ is its dual basis.

Following Arthur (see [Ar 1] § 2 and [La 2] § 3), for any $J \subset I \subset \Delta$ we denote by

(10.3.1) $$\tau_J^I \text{ (resp. } \hat{\tau}_J^I) : \mathfrak{a}_\emptyset \to \{0, 1\} \subset \mathbb{C}$$

the characteristic function of the subset

$$\mathfrak{a}_I + \{H \in \mathfrak{a}_J^I \mid \alpha_i(H) > 0, \ \forall i \in I - J\} + \mathfrak{a}_\emptyset^J$$

(resp.

$$\mathfrak{a}_I + \{H \in \mathfrak{a}_J^I \mid \varpi_i^I(H) > 0, \ \forall i \in I - J\} + \mathfrak{a}_\emptyset^J \,)$$

of $\mathfrak{a}_\emptyset$.

LEMMA (10.3.2) (Langlands). — *For any $K \subset I \subset \Delta$ we have*

$$\sum_{\substack{J \subset \Delta \\ K \subset J \subset I}} (-1)^{|J-K|} \tau_K^J(H) \widehat{\tau}_J^I(H) = \delta_K^I$$

for every $H \in \mathfrak{a}_\emptyset$. Here $\delta_K^I = 1$ if $K = I$ and $\delta_K^I = 0$ otherwise.

Proof : For any J such that $K \subset J \subset I$ the functions τ_K^J and $\widehat{\tau}_K^I$ are invariant under translation by $\mathfrak{a}_I$ and $\mathfrak{a}_\emptyset^K$ (we have

$$\mathfrak{a}_\emptyset = \underbrace{\mathfrak{a}_I \oplus \overbrace{\mathfrak{a}_J^I \oplus \mathfrak{a}_K^J}}_{\mathfrak{a}_K} \overbrace{\oplus \mathfrak{a}_\emptyset^K}^{\mathfrak{a}_\emptyset^I} \;).$$

Therefore we may assume that $H \in \mathfrak{a}_K^I$.

Let us fix $H \in \mathfrak{a}_K^I$ and let us set

$$L = K \cup \{i \in I - K \mid \varpi_i^I(H) \leq 0\}$$

and

$$M = K \cup \{i \in I - K \mid \alpha_i(H) > 0\}.$$

We have

$$\tau_K^J(H)\widehat{\tau}_J^I(H) \neq 0$$

for a set J with $K \subset J \subset I$ if and only if we have

$$L \subset J \subset M.$$

Moreover, if this is the case we have

$$\tau_K^J(H)\widehat{\tau}_J^I(H) = 1.$$

Therefore we may assume that $L \subset M$ (if $L \not\subset M$ we have $I - K \neq \emptyset$ and $\delta_K^I = 0$) and

$$\sum_{\substack{J \subset \Delta \\ K \subset J \subset I}} (-1)^{|J-K|} \tau_K^J(H)\widehat{\tau}_J^I(H) = \sum_{\substack{J \subset \Delta \\ L \subset J \subset M}} (-1)^{|J-K|}$$

is equal to

$$(-1)^{|L-K|}(1-1)^{|M-L|} = (-1)^{|L-K|}\delta_L^M.$$

If $L \neq M$ we have $K \neq I$ and the formula of the lemma is proved. If $K = I$ we have $L = M$ and the formula of the lemma is proved too. Consequently we may assume that $L = M$ and that $K \neq I$ and we search for a contradiction. Let $d_I = (d_1, \ldots, d_s)$ be the partition of d corresponding to I and let us fix $j \in \{1, \ldots, s\}$. We set

$$\begin{cases} I_j = \{1, \ldots, d_j - 1\}, \\ L_j = \{k \in I_j \mid d_1 + \cdots + d_{j-1} + k \in L\}, \\ K_j = \{k \in I_j \mid d_1 + \cdots + d_{j-1} + k \in K\}, \end{cases}$$

and

$$H_{j,k} = H_{d_1 + \cdots + d_{j-1} + k}$$

for any $k \in \{1, \ldots, d_j\}$. We have

$$\begin{cases} H_{j,1} + \cdots + H_{j,d_j} = 0, \\ H_{j,k} = H_{j,k+1} \quad (\forall k \in K_j), \\ H_{j,k} > H_{j,k+1} \quad (\forall k \in L_j - K_j), \end{cases}$$

and

$$\varpi^I_{d_1 + \cdots + d_{j-1} + k}(H) = H_{j,1} + \cdots + H_{j,k} > 0 \qquad (\forall k \in I_j - L_j).$$

It follows that

$$H_{j,1} + \cdots + H_{j,k} \geq 0 \qquad (\forall k \in I_j).$$

Indeed, if $k \in I_j - L_j$ there is nothing to prove; if

$$H_{j,1} + \cdots + H_{j,k} < 0$$

for some $k \in L_j$ and if $k \in L_j$ is minimal for this property we necessarily have $H_{j,k} < 0$; then, by an obvious induction on ℓ, we see that

$$H_{j,\ell} < 0$$

and

$$H_{j,1} + \cdots + H_{j,\ell} < 0$$

for every $\ell = k, \ldots, d_j$ (if $H_{j,1} + \cdots + H_{j,\ell} < 0$ for some $\ell \in I_j$ we have $\ell \in L_j$ and therefore $H_{j,\ell+1} \leq H_{j,\ell}$) and we get a contradiction (take $\ell = d_j$). We even have

$$H_{j,1} + \cdots + H_{j,k} > 0 \qquad (\forall k \in L_j - K_j).$$

Indeed, if $k \in L_j - K_j$ we have

$$H_{j,k} > H_{j,k+1}$$

and from the equality

$$H_{j,1}+\cdots+H_{j,k}=0$$

we get either

$$H_{j,1}+\cdots+H_{j,k-1}<0$$

if $H_{j,k}>0$, or

$$H_{j,1}+\cdots+H_{j,k+1}<0$$

if $H_{j,k}\leq 0$ and this contradicts the previous assertion. But, by definition of L_j, this means that

$$L_j=K_j$$

and, by definition of $M=L$, this implies that

$$H_{j,k}\leq H_{j,k+1}\qquad (\forall k\in I_j).$$

Now only $H=0$ satisfies at the same time

$$H_{j,1}+\cdots+H_{j,k}\geq 0\qquad (\forall k\in I_j),$$

$$H_{j,k}\leq H_{j,k+1}\qquad (\forall k\in I_j)$$

and

$$H_{j,1}+\cdots+H_{j,d_j}=0$$

for every $j=1,\ldots,s$. But for $H=0$ we have $L=I$ and $M=K$ and this contradicts the hypotheses $L=M$ and $K\neq I$. □

For any $K\subset I\subset\Delta$ let us introduce the function

$$\Phi_K^I=\sum_{\substack{J\subset\Delta\\ K\subset J\subset I}}(-1)^{|I-J|}\hat{\tau}_J^I:\mathfrak{a}_\emptyset\to\mathbb{C}. \tag{10.3.3}$$

LEMMA (10.3.4). — (i) *Φ_K^I is the characteristic function of the subset*

$$\mathfrak{a}_I\oplus\{H\in\mathfrak{a}_\emptyset^I\mid \varpi_i^I(H)\leq 0,\ \forall i\in I-K\}$$

of $\mathfrak{a}_\emptyset$.

(ii) *For any $K\subset L\subset\Delta$ we have*

$$\sum_{\substack{I\subset\Delta\\ K\subset I\subset L}}\Phi_K^I\tau_I^L=1.$$

Proof: Part (i) follows directly from the definition of $\hat{\tau}_J^I$ as a characteristic function.

Part (ii) follows from the equalities

$$\sum_{\substack{I\subset\Delta\\ J\subset I\subset L}} (-1)^{|L-I|}\hat{\tau}_J^I\tau_I^L = \delta_J^L \qquad (\forall J\subset L\subset\Delta)$$

which themselves follow from (10.3.2) (write (10.3.2) as an equality of matrices). □

If $g\in G(\mathbb{A})$ we can decompose g into

$$g = n_\emptyset m_\emptyset k$$

with $n_\emptyset \in N_\emptyset(\mathbb{A})$, $m_\emptyset \in M_\emptyset(\mathbb{A})$ and $k\in K$ (Iwasawa decomposition). Moreover $m_\emptyset$ is uniquely determined by g up to right multiplication by an element of $M_\emptyset(\mathbb{A})\cap K$. Let us write $m_\emptyset = (t_1,\ldots,t_d)$ in $M_\emptyset(\mathbb{A})\cong(\mathbb{A}^\times)^d$. Then we can set

$$H_\emptyset(g) = -\deg_{M_\emptyset}(m_\emptyset) = (-\deg(t_1),\ldots,-\deg(t_d))$$

(see (9.1)) and in this way we get a map

$$H_\emptyset : G(\mathbb{A})\to\mathfrak{a}_\emptyset \tag{10.3.5}$$

which is left $N_\emptyset(\mathbb{A})$-invariant and right K-invariant.

For any $J\subset I\subset\Delta$ let us denote by

$$H_I : G(\mathbb{A})\to\mathfrak{a}_I\subset\mathfrak{a}_\emptyset \tag{10.3.6}$$

and

$$H_J^I : G(\mathbb{A})\to\mathfrak{a}_\emptyset^I\subset\mathfrak{a}_\emptyset \tag{10.3.7}$$

the compositions of $H_\emptyset$ with the two canonical projections of $\mathfrak{a}_\emptyset = \mathfrak{a}_I\oplus\mathfrak{a}_J^I\oplus\mathfrak{a}_\emptyset^J$ onto the first two factors $\mathfrak{a}_I$ and $\mathfrak{a}_J^I$, so that

$$H_\emptyset = H_I\oplus H_J^I\oplus H_\emptyset^J.$$

If

$$g = n_I m_I k\in G(\mathbb{A})$$

with $n_I\in N_I(\mathbb{A})$, $m_I = (g_1,\ldots,g_s)\in M_I(\mathbb{A})\cong GL_{d_1}(\mathbb{A})\times\cdots\times GL_{d_s}(\mathbb{A})$ and $k\in K$, where $d_I = (d_1,\ldots,d_s)$ is the partition of d corresponding to I, we have

$$H_I(g)_{d_1+\cdots+d_{j-1}+1} = \cdots = H_I(g)_{d_1+\cdots+d_j} = -\frac{\deg(\det g_j)}{d_j}$$

for $j = 1, \ldots, s$. In other words, if we identify $\mathfrak{a}_I$ with $\mathbb{Q}^s$ as before we have

$$H_I(g) = -\deg_{M_I}(m_I)$$

(see (9.1)). It follows that H_I is left $(N_I(\mathbb{A})M_I(F))$-invariant and that $H_\emptyset^I$ is left $(N_I(\mathbb{A})Z_I(\mathbb{A}))$-invariant where Z_I is the center of M_I (H_I and $H_\emptyset^I$ being both left $N_\emptyset^I(\mathbb{A})$-invariant and right K-invariant).

For each $I \subset \Delta$ we denote by

$$\mathfrak{S}^I \subset G(\mathbb{A}) \tag{10.3.8}$$

the **Siegel set** of $g \in G(\mathbb{A})$ such that

$$\alpha_i(H_\emptyset(g)) \geq -2g_X \qquad (\forall i \in I),$$

where g_X is the geometric genus of the projective curve X.

LEMMA (10.3.9). — *Let $I \subset \Delta$.*

(i) *We have $G(\mathbb{A}) = M_I(F)\mathfrak{S}^I$.*

(ii) *There exists an integer $c_I \geq -2g_X$ having the following property: let $T \in \mathfrak{a}_\emptyset$ be such that*

$$\alpha_i(T) \geq c_I \qquad (\forall i \in I),$$

let $J \subset I$ and let $g \in G(\mathbb{A})$ and $\mu_I \in M_I(F)$; then, if g and $\mu_I g$ are both in $\mathfrak{S}^I$ and if

$$\alpha_i(H_\emptyset(g) - T) > 0 \qquad (\forall i \in I - J),$$

we have

$$\mu_I \in P_J^I(F) \subset M_I(F).$$

Proof: With the notations of (9.2) we have

$$\mathfrak{S}^I = U(\mathbb{A})T(\mathbb{A})^I_{]-\infty,2g_X]}K.$$

But we have seen in (9.2.7) that

$$M_I(\mathbb{A}) = M_I(F)U^I(\mathbb{A})T(\mathbb{A})^I_{]-\infty,2g_X]}M_I(\mathcal{O})$$

and, as $G(\mathbb{A}) = N_I(\mathbb{A})M_I(\mathbb{A})K$, part (i) of the lemma follows.

Thanks to (9.5.8) there exists an integer $c_1' \leq 2g_X$ (take $c_2 = 2g_X$) having the following property: let $u, u' \in U(\mathbb{A})$, $t, t' \in T(\mathbb{A})^I_{]-\infty,2g_X]}$, $k, k' \in K$ and $\mu_I \in M_I(F)$ be such that

$$u't'k' = \mu_I utk$$

and let $\alpha \in I$ be such that

$$\deg(\alpha(t)) < c_1';$$

then we have

$$\mu_I \in P_{I-\{\alpha\}}^I(F).$$

Let us take $c_I = c_1'$. As we have

$$P_J^I = \bigcap_{\alpha \in I-J} P_{I-\{\alpha\}}^I,$$

part (ii) of the lemma follows. □

For each $T \in \mathfrak{a}_\emptyset = \mathbb{Q}^d$ we set

$$d(T) \overset{\text{dfn}}{=\!=} \inf\{\alpha_i(T) \mid i = 1, \ldots, d-1\}. \tag{10.3.10}$$

For each $T \in \mathbb{Q}^d$ such that $d(T) \geq -2g_X$ and for each $I \subset \Delta$ we denote by

$$\mathfrak{S}^I(T) \subset \mathfrak{S}^I \subset G(\mathbb{A})$$

the set of $g \in \mathfrak{S}^I$ such that

$$\varpi_i^I(H_\emptyset(g) - T) \leq 0 \qquad (\forall i \in I)$$

and we denote by

$$F^I(\cdot, T) : G(\mathbb{A}) \to \{0, 1\} \subset \mathbb{C} \tag{10.3.11}$$

the characteristic function of the subset

$$M_I(F)\mathfrak{S}^I(T) \subset G(\mathbb{A}).$$

This function is left $(N_I(\mathbb{A})Z_I(F_\infty)M_I(F))$-invariant and right K-invariant.

PROPOSITION (10.3.12) (Arthur). — *Let $T \in \mathbb{Q}^d$ be such that $d(T) \geq -2g_X$ and let $I \subset \Delta$. Then, if $c_I \geq -2g_X$ is an integer having the property of* (10.3.9)(ii) *and if*

$$\alpha_i(T) \geq c_I \qquad (\forall i \in I),$$

we have

$$\sum_{J \subset I} \sum_{\pi_I \in P_J(F)\backslash P_I(F)} F^J(\pi_I g, T)\tau_J^I(H_\emptyset(\pi_I g) - T) = 1 \qquad (\forall g \in G(\mathbb{A})).$$

Let us remark that the functions $F^J(\cdot, T)$ and $\tau_J^I(H_\emptyset(\cdot) - T)$ are both left $P_J(F)$-invariant and both take non-negative values, so that the above double sum is well-defined (a priori it could be infinite).

Proof : Let us fix $g \in G(\mathbb{A})$.

Thanks to (10.3.9)(i) there exists $\mu_I \in M_I(F)$ such that $\mu_I g \in \mathfrak{S}^I$ and, thanks to (10.3.4)(ii), there exists $J \subset I$ such that

$$\Phi_\emptyset^J(H_\emptyset(\mu_I g) - T) = \tau_J^I(H_\emptyset(\mu_I g) - T) = 1.$$

For such a $\mu_I \in M_I(F)$ and such a $J \subset I$ we have

$$F^J(\mu_I g, T)\tau_J^I(H_\emptyset(\mu_I g) - T) = 1$$

$(\mathfrak{S}^I \subset \mathfrak{S}^J)$.

Let $\mu_{I,1}, \mu_{I,2} \in M_I(F)$ and $J_1, J_2 \subset I$ be such that, for $n = 1, 2$, we have

$$\mu_{I,n} g \in \mathfrak{S}^{J_n}$$

and

$$\Phi_\emptyset^{J_n}(H_\emptyset(\mu_{I,n} g) - T) = \tau_{J_n}^I(H_\emptyset(\mu_{I,n} g) - T) = 1.$$

Then for $n = 1, 2$ we have

$$\mu_{I,n} g \in \mathfrak{S}^I.$$

Indeed, for any $J \subset \Delta$, any $k \in \{1, \dots, t-1\}$ and any $H \in \mathfrak{a}_\emptyset$ we have

$$\alpha_{e_1+\cdots+e_k}(H_\emptyset^J) = -\varpi_{e_1+\cdots+e_k-1}^J(H) - \varpi_{e_1+\cdots+e_k+1}^J(H),$$

where $(e_1, \dots, e_t)$ is the partition of d corresponding to J (with the convention $\varpi_j^J \equiv 0$, $\forall j \notin J$) ; therefore, for $n = 1, 2$ we have

$$\alpha_i(H_\emptyset(\mu_{I,n} g) - T) \geq \alpha_i(H_{J_n}(\mu_{I,n} g) - T_{J_n}) > 0$$

and thus

$$\alpha_i(H_\emptyset(\mu_{I,n} g)) > \alpha_i(T) \geq -2g_X$$

for every $i \in I - J_n$. Now, thanks to (10.3.9)(ii) and to our choice of T, we have

$$\mu_{I,2}\mu_{I,1}^{-1} \in P_{J_1}^I(F)$$

and

$$\mu_{I,1}\mu_{I,2}^{-1} \in P_{J_2}^I(F),$$

so that

$$\mu_{I,2}\mu_{I,1}^{-1} = (\mu_{I,1}\mu_{I,2}^{-1})^{-1} \in P_J^I(F)$$

where we have set $J = J_1 \cap J_2$. Therefore we have

$$H_J(\mu_{I,1} g) = H_J(\mu_{I,2} g)$$

and, if we denote by H_J this element of $\mathfrak{a}_J \subset \mathfrak{a}_\emptyset$, for $n = 1, 2$ we have

$$\Phi_J^{J_n}(H_J - T) = \Phi_J^{J_n}(H_\emptyset(\mu_{I,n} g) - T) = 1$$

and

$$\tau_{J_n}^I(H_J - T) = \tau_{J_n}^I(H_\emptyset(\mu_{I,n}g) - T) = 1$$

$(H_\emptyset(\mu_{I,n}g) - H_J \in \mathfrak{a}_\emptyset^J \subset \mathfrak{a}_\emptyset^{J_n})$. But we have

$$\sum_{\substack{K\subset\Delta \\ J\subset K\subset I}} \Phi_J^K \tau_K^I = 1$$

(see (10.3.4)(ii)), so that $J_1 = J_2$, and the proposition follows. □

LEMMA (10.3.13). — *For any $T \in \mathfrak{a}_\emptyset$ such that $d(T) \geq -2g_X$ and any $I \subset \Delta$ the projection of $\mathfrak{S}^I(T)$ into*

$$N_I(\mathbb{A})Z_I(F_\infty)M_I(F)\backslash G(\mathbb{A})$$

is a compact set. In other words, if we regard $F^I(\cdot, T)$ as a function on

$$N_I(\mathbb{A})Z_I(F_\infty)M_I(F)\backslash G(\mathbb{A}),$$

it has compact support.

Proof: We have

$$\mathfrak{S}^I(T) = U(\mathbb{A})(T(\mathbb{A}) \cap \mathfrak{S}^I(T))K$$

and the quotient

$$N_I(\mathbb{A})U^I(F)\backslash U(\mathbb{A}) \cong U^I(F)\backslash U^I(\mathbb{A})$$

is compact (see (9.2.1)). Therefore it is sufficient to check that

$$Z_I(F_\infty)T(F)\backslash(T(\mathbb{A}) \cap \mathfrak{S}^I(T))/T(\mathcal{O})$$

is finite. But there exists an integer $c_1 \leq 2g_X$ such that

$$T(\mathbb{A}) \cap \mathfrak{S}^I(T) \subset T(\mathbb{A})^I_{[c_1, 2g_X]}$$

(take c_1 such that

$$c_1 \leq -\frac{d_j}{k(d_j - k)}\varpi^I_{d_1+\cdots+d_{j-1}+k}(T) - (d_j - 2)g_X$$

for every $k = 1, \ldots, d_j - 1$ and every $j = 1, \ldots, s$, where $(d_1, \ldots, d_s)$ is the partition of d corresponding to I) and the sets $F_\infty^\times F^\times\backslash\mathbb{A}^\times/\mathcal{O}^\times$ and $F^\times\backslash(\mathbb{A}^\times)_{[c_1,2g_X]}/\mathcal{O}^\times$ are finite (see the proof of (9.2.6)). The lemma follows. □

(10.4) Proof of the integrability of $k(g)$

Let $f : F_\infty^\times\backslash G(\mathbb{A}) \to \mathbb{C}$ be a locally constant function with compact support satisfying hypotheses (A) and (B) of (10.2). Thanks to (10.3.12) and to the left $G(F)$-invariance of $k_{\mathcal{O}}$, for each $T \in \mathfrak{a}_\emptyset$ such that $d(T) \geq c_\Delta$ and each $\mathcal{O} \in G(F)_\sharp$ we have

$$k_{\mathcal{O}}(g) = \sum_{I\subset\Delta} \sum_{\delta\in P_I(F)\backslash G(F)} F^I(\delta g, T)\tau_I^\Delta(H_\emptyset(\delta g) - T)k_{\mathcal{O}}(\delta g).$$

Therefore, to prove (10.2.2) it is sufficient to prove the following proposition.

PROPOSITION (10.4.1). — *For each $I \subset \Delta$ there exists an integer $C_I^\Delta \geq -2g_X$ (depending on the support of f) such that*

$$\sum_{\mathcal{O}\in G(F)_\sharp} \int_{F_\infty^\times P_I(F)\backslash G(\mathbb{A})} F^I(g, T)\tau_I^\Delta(H_\emptyset(g) - T)|k_{\mathcal{O}}(g)|\frac{dg}{dz_\infty d\pi_I} < +\infty$$

for every $T \in \mathfrak{a}_\emptyset$ with $d(T) \geq C_I^\Delta$ ($d\pi_I$ is the counting measure on $P_I(F)$).

Let $J \subset I \subset \Delta$ and let $(d_1, \ldots, d_s)$ and $(e_1, \ldots, e_t)$ be the partitions of d corresponding to I and J respectively.

LEMMA (10.4.2). — *Let $\Omega \subset G(\mathbb{A})$ be a compact subset and let a be an integer. There exists an integer $C_J^I(\Omega, a) \geq a$ having the following property: let $\pi_I \in P_I(F)$ and let us assume that there exist $g \in G(\mathbb{A})$ and $T \in \mathfrak{a}_\emptyset$ with*

$$\alpha_i(T) \geq C_J^I(\Omega, a) \qquad (\forall i = 1, \ldots, d-1)$$

such that

$$g^{-1}\pi_I g \in \Omega$$

and

$$\begin{cases} \alpha_i(H_\emptyset(g)) \geq a & (\forall i \in J), \\ \varpi_i^J(H_\emptyset(g) - T_\emptyset) \leq 0 & (\forall i \in J), \\ \alpha_i(H_J(g) - T_J) > 0 & (\forall i \in I - J); \end{cases}$$

then we have

$$\pi_I \in P_J(F) \subset P_I(F).$$

COROLLARY (10.4.3). — *There exists an integer $C_J^I \geq -2g_X$ having the following property: let $T \in \mathfrak{a}_\emptyset$ with $d(T) \geq C_J^I$, let $\pi_I \in P_I(F)$ and let us assume that there exists $g \in G(\mathbb{A})$ such that*

$$F^J(g, T)\tau_J^I(H_\emptyset(g) - T)f(g^{-1}\pi_I g) \neq 0;$$

then we have

$$\pi_I \in P_J(F) \subset P_I(F).$$

Proof of the corollary (*assuming the lemma*) : Let us set

$$C_J^I = C_J^I(\mathrm{Supp}(f), -2g_X).$$

Then, replacing π_I and g by $\mu_J^{-1}\pi_I\mu_J$ and $\mu_J^{-1}g$ respectively for some $\mu_J \in M_J(F)$ if it is necessary, we see that the hypotheses of the corollary imply the hypotheses of the lemma, hence its conclusion. □

Proof of the lemma : Let us fix π_I, g and T with $d(T) \geq a$ satisfying the hypotheses of the lemma.

Replacing Ω by the compact subset

$$\bigcup_{k\in K} k\Omega k^{-1}$$

of $G(\mathbb{A})$ we may assume that

$$g = ut$$

where $u \in U(\mathbb{A})$ and $t = (t_1, \ldots, t_d) \in T(\mathbb{A}) \cong (\mathbb{A}^\times)^d$, so that

$$H_\emptyset(g) = (-\deg(t_1), \ldots, -\deg(t_d)).$$

Now, for every $H \in \mathfrak{a}_\emptyset$ and every $i \in \Delta - J$ (i.e. $i = e_1 + \cdots + e_k$ for some $k \in \{1, \ldots, t-1\}$), we have

$$\alpha_i(H) = \alpha_i(H_J) - \varpi_{i-1}^J(H_\emptyset) - \varpi_{i+1}^J(H_\emptyset)$$

(with the convention $\varpi_j^J \equiv 0$, $\forall j \in \Delta - I$). Therefore our hypotheses imply that

$$\alpha_i(H_\emptyset(g) - T) > 0 \qquad (\forall i \in I - J),$$

so that

$$\alpha_i(H_\emptyset(g)) > \alpha_i(T) \geq a \qquad (\forall i \in I - J).$$

In particular we have

$$\deg(t_\ell) - \deg(t_m) > \alpha_{e_1+\cdots+e_k}(T) + (\ell - m - 1)a$$

for all ℓ, $m \in \{1, \ldots, d\}$ and every $k \in \{1, \ldots, t-1\}$ such that there exists $j \in \{1, \ldots, s\}$ with

$$d_1 + \cdots + d_{j-1} + 1 \leq m \leq e_1 + \cdots + e_k < \ell \leq d_1 + \cdots + d_j.$$

But for all ℓ, $m \in \{1, \ldots, d\}$ there exists an integer $C_{\ell m}(\Omega)$ such that

$$\deg(h_{\ell m}) \geq C_{\ell m}(\Omega) \qquad (\forall h \in \Omega)$$

($h_{\ell m}$ is the (ℓ, m)-th entry of the matrix h). Therefore we obtain the inequalities

$$\deg((u^{-1}\pi_I u)_{\ell m}) > C_{\ell m}(\Omega) + \alpha_{e_1+\cdots+e_k}(T) + (\ell - m - 1)a$$

for all ℓ, $m \in \{1, \ldots, d\}$ and every $k \in \{1, \ldots, t-1\}$ such that there exists $j \in \{1, \ldots, s\}$ with

$$d_1 + \cdots + d_{j-1} + 1 \leq m \leq e_1 + \cdots + e_k < \ell \leq d_1 + \cdots + d_j$$

(we have $(t^{-1}u^{-1}\pi_I ut)_{\ell m} = t_\ell^{-1}(u^{-1}\pi_I u)_{\ell m} t_m$ for any $\ell, m \in \{1, \ldots, d\}$).

Let us take for $C_J^I(\Omega, a)$ any integer $\geq a$ satisfying the inequality

$$C_{\ell m}(\Omega) + C_J^I(\Omega, a) + (\ell - m - 1)a \geq 0$$

for all ℓ, $m \in \{1, \ldots, d\}$ such that there exist $j \in \{1, \ldots, s\}$ and $k \in \{1, \ldots, t-1\}$ with

$$d_1 + \cdots + d_{j-1} + 1 \leq m \leq e_1 + \cdots + e_k < \ell \leq d_1 + \cdots d_j.$$

Then our hypotheses imply that

$$\deg((u^{-1}\pi_I u)_{\ell m}) > 0$$

for every ℓ and m as before. But by an easy induction on (ℓ, m) the last inequalities imply that

$$\pi_I \in P_J(F) \subset P_I(F).$$

Indeed, if $(\pi_I)_{\ell' m'} = 0$ for every $\ell' \in \{\ell, \ldots, d_1 + \cdots + d_j\}$ and every $m' \in \{d_1 + \cdots + d_{j-1} + 1, \ldots, m\}$ such that $\ell' \neq \ell$ or $m' \neq m$ and if

$$\deg((u^{-1}\pi_I u)_{\ell m}) > 0,$$

we have

$$(\pi_I)_{\ell m} = (u^{-1}\pi_I u)_{\ell m} = 0$$

as this element belongs to F and has positive degree. □

Let us fix $I \subset \Delta$. What makes (10.4.3) very useful is that we have

$$\mathcal{O} \cap P_I(F) = (\mathcal{O} \cap M_I(F))N_I(F)$$

for any $\mathcal{O} \in G(F)_\sharp$, so that

$$\begin{aligned}\int_{F_\infty^\times P_I(F)\backslash G(\mathbb{A})} & F^I(g,T)\tau_I^\Delta(H_\emptyset(g) - T)|k_{\mathcal{O}}(g)| \frac{dg}{dz_\infty d\pi_I} \\ & \leq \sum_{\mu_I \in \mathcal{O}\cap M_I(F)} \int_{F_\infty^\times P_I(F)\backslash G(\mathbb{A})} F^I(g,T)\tau_I^\Delta(H_\emptyset(g) - T) \\ & \qquad \Big| \sum_{\nu_I \in N_I(F)} f(g^{-1}\mu_I \nu_I g)\Big| \frac{dg}{dz_\infty d\pi_I}\end{aligned}$$

for every $\mathcal{O} \in G(F)_\sharp$ and every $T \in \mathfrak{a}_\emptyset$ with $d(T) \geq C_I^\Delta$.

Now let us fix a compact open subset ω' of $U(\mathbb{A})$ such that

$$U(F)\omega' = U(\mathbb{A})$$

(recall that $U(F)\backslash U(\mathbb{A})$ is compact; see (9.2.1)) and a system of representatives $\mathcal{T}^\infty \subset T(\mathbb{A}^\infty)$ of the finite quotient

$$T(F)\backslash T(\mathbb{A}^\infty)/T(\mathcal{O}^\infty)$$

(see the proof of (9.2.6)). Moreover let us fix $T_0 \in \mathfrak{a}_\emptyset$ such that, for each $t^\infty \in \mathcal{T}^\infty$, we have

$$\begin{cases} \varpi_i^I(T_0) \leq \varpi_i^I(H_\emptyset(t^\infty)) & (\forall i \in I), \\ \alpha_j((T_0)_I) \geq \alpha_j(H_I(t^\infty)) & (\forall j \in \Delta - I), \end{cases}$$

let us set

$$\epsilon_0 = \sup\{|\alpha_i(t^\infty)| \mid i = 1, \dots, d-1 \text{ and } t^\infty \in \mathcal{T}^\infty\}$$

and, for each $T \in \mathfrak{a}_\emptyset$, let us fix a system of representatives

$$\mathcal{T}_{I,\infty}^T \subset T(F_\infty)$$

of the set of classes $F_\infty^\times t_\infty T(\mathcal{O}_\infty) \in F_\infty^\times\backslash T(F_\infty)/T(\mathcal{O}_\infty)$ which satisfy

$$\begin{cases} \alpha_i(H_\emptyset(t_\infty)) \geq -2g_X - \epsilon_0 & (\forall i \in I), \\ \varpi_i^I(H_\emptyset(t_\infty)) \leq \varpi_i^I(T - T_0) & (\forall i \in I), \\ \alpha_j(H_\emptyset(t_\infty)) > \alpha_j(T - T_0) & (\forall j \in \Delta - I). \end{cases}$$

Then, for each $T \in \mathfrak{a}_\emptyset$ with $d(T) \geq -2g_X$, the image of the map

$$\begin{aligned} \omega' \times \mathcal{T}_{I,\infty}^T \times \mathcal{T}^\infty \times K &\to F_\infty^\times P_I(F)\backslash G(\mathbb{A}), \\ (u', t_\infty, t^\infty, k) &\mapsto F_\infty^\times P_I(F) u' t_\infty t^\infty k, \end{aligned}$$

contains

$$\{F_\infty^\times P_I(F)g \mid F^I(g,T)\tau_I^\Delta(H_\emptyset(g) - T) \neq 0\}.$$

Indeed, for any $j \in \Delta - I$ and any $H \in \mathfrak{a}_\emptyset$ we have

$$\alpha_j(H) = \alpha_j(H_I) - \varpi_{j-1}^I(H) - \varpi_{j+1}^I(H)$$

(with the convention $\varpi_i^I \equiv 0$, $\forall i \notin I$), so that, for each $T \in \mathfrak{a}_\emptyset$ with $d(T) \geq -2g_X$, each $t^\infty \in \mathcal{T}^\infty$ and each $t_\infty \in T(F_\infty)$ which satisfy

$$\begin{cases} \alpha_i(H_\emptyset(t_\infty t^\infty)) \geq -2g_X & (\forall i \in I), \\ \varpi_i^I(H_\emptyset(t_\infty t^\infty) - T) \leq 0 & (\forall i \in I), \\ \alpha_j(H_I(t_\infty t^\infty) - T_I) > 0 & (\forall j \in \Delta - I), \end{cases}$$

we have

$$t_\infty \in \mathcal{T}_{I,\infty}^T.$$

It follows that, for any locally constant non-negative function ψ on $F_\infty^\times P_I(F)\backslash G(\mathbb{A})$, we can bound the integral

$$\int_{F_\infty^\times P_I(F)\backslash G(\mathbb{A})} F^I(g,T)\tau_I^\Delta(H_\emptyset(g)-T)\psi(g)\frac{dg}{dz_\infty d\pi_I}$$

above by

$$\int_K\Big(\sum_{t_\infty\in\mathcal{T}_{I,\infty}^T}\delta_\emptyset^{-1}(t_\infty)\Big(\sum_{t^\infty\in\mathcal{T}^\infty}\delta_\emptyset^{-1}(t^\infty)\int_{\omega'}\psi(u't_\infty t^\infty k)du'\Big)\Big)dk$$

for any $T\in\mathfrak{a}_\emptyset$ with $d(T)\geq -2g_X$ (du' and dk are the Haar measures on $U(\mathbb{A})$ and K which are normalized by $\mathrm{vol}(U(\mathcal{O}),du')=1$ and $\mathrm{vol}(K,dk)=\mathrm{vol}(K,dg)$ respectively; see (4.1.7)).

Finally, let us set

$$\epsilon_1=\sup\{\epsilon_0,\alpha_j(T_0)\mid j\in\Delta-I\} \tag{10.4.4}$$

and let $\omega\subset U(\mathbb{A})$ be the set of $u\in U(\mathbb{A})$ which may be written as

$$u=(t^\infty)^{-1}t_\infty^{-1}u't_\infty t^\infty$$

with $u'\in\omega'$, $t^\infty\in\mathcal{T}^\infty$ and $t_\infty\in T(F_\infty)$ such that

$$d(H_\emptyset(t_\infty))\geq -2g_X-\epsilon_1.$$

It is easy to see that ω is a compact subset of $U(\mathbb{A})$. Then, for any function ψ as before and any $t_\infty\in\mathcal{T}_{I,\infty}^T$, we can bound the integrals

$$\int_{\omega'}\psi(u't_\infty t^\infty k)du'\qquad(t^\infty\in\mathcal{T}^\infty,k\in K)$$

above by

$$\mathrm{vol}(\omega',du')\mathrm{Sup}\{\psi(t_\infty g)\mid g\in\mathcal{T}^\infty\omega K\}$$

and therefore the integral

$$\int_K\Big(\sum_{t^\infty\in\mathcal{T}^\infty}\delta_\emptyset^{-1}(t^\infty)\int_{\omega'}\psi(u't_\infty t^\infty k)du'\Big)dk$$

by

$$\mathrm{vol}(K,dg)\,\mathrm{vol}(\omega',du')\Big(\sum_{t^\infty\in\mathcal{T}^\infty}\delta_\emptyset^{-1}(t^\infty)\Big)\sup\{\psi(t_\infty g)\mid g\in\mathcal{T}^\infty\omega K\}.$$

In particular we have proved the following proposition.

PROPOSITION (10.4.5). — *For each* $\mathfrak{o} \in G(F)_\sharp$ *and each* $T \in \mathfrak{a}_\emptyset$ *with* $d(T) \geq C_I^\Delta$ *the integral*

$$\int_{F_\infty^\times P_I(F)\backslash G(\mathbb{A})} F^I(g,T)\tau_I^\Delta(H_\emptyset(g)-T)|k_\mathfrak{o}(g)|\frac{dg}{dz_\infty d\pi_I}$$

is bounded above by the product of

$$\mathrm{vol}(K,dg)\,\mathrm{vol}(\omega',du')\Big(\sum_{t^\infty\in T^\infty}\delta_\emptyset^{-1}(t^\infty)\Big)$$

and

$$\sup\Big\{\sum_{\mu_I\in\mathfrak{o}\cap M_I(F)}\ \sum_{t_\infty\in T_{I,\infty}^T}\delta_\emptyset^{-1}(t_\infty)\Big|\sum_{\nu_I\in N_I(F)} f(g^{-1}t_\infty^{-1}\mu_I\nu_I t_\infty g)\Big|$$
$$\mid g\in T^\infty\omega K\Big\}.$$

□

In order to establish an upper-bound for the sum over $\nu_I \in N_I(F)$ which occurs in proposition (10.4.5) we will use the Poisson summation formula. Let $\mathfrak{n}_I$ be the Lie algebra of N_I. We identify $N_I(\mathbb{A})$ with $\mathfrak{n}_I(\mathbb{A})$ (and therefore $N_I(F)$ with $\mathfrak{n}_I(F)$) by

$$N_I(\mathbb{A}) \xrightarrow{\sim} \mathfrak{n}_I(\mathbb{A}),\ n_I \mapsto n_I - 1.$$

LEMMA (10.4.6). — *Under the above identification the Haar measure* dn_I *on* $N_I(\mathbb{A})$ *which is normalized by* $\mathrm{vol}(N_I(\mathcal{O}),dn_I)=1$ *corresponds to the Haar measure* dX_I *on* $\mathfrak{n}_I(\mathbb{A})$ *which is normalized by* $\mathrm{vol}(\mathfrak{n}_I(\mathcal{O}),dX_I)=1$.

Proof : It is sufficient to prove that the measure $d(n_I-1)$ on $N_I(\mathbb{A})$ is invariant under left translation, i.e. that

$$d(X_I + X_I^0 + X_I^0X_I) = dX_I$$

for each $X_I^0 \in \mathfrak{n}_I(\mathbb{A})$. But we have

$$d(X_I + X_I^0 + X_I^0X_I) = d(X_I + X_I^0X_I) = d((1+X_I^0)X_I)$$

and $d((1+X_I^0)X_I)$ is obviously a Haar measure on $\mathfrak{n}_I(\mathbb{A})$ which gives the volume 1 to $\mathfrak{n}_I(\mathcal{O})$, so that

$$d((1+X_I^0)X_I) = dX_I.$$

□

For each locally constant function Φ with compact support on $\mathfrak{n}_I(\mathbb{A})$ let

(10.4.7) $$\widehat{\Phi}(Y_I) = \int_{\mathfrak{n}_I(\mathbb{A})} \Phi(X_I)\Psi(\mathrm{tr}({}^tX_IY_I))dX_I$$

be its **Fourier transform** with respect to a fixed non-trivial smooth character

$$\Psi : F\backslash\mathbb{A} \to \mathbb{C}^\times.$$

It is well known that $\widehat{\Phi}$ is also a locally constant function with compact support on $\mathfrak{n}_I(\mathbb{A})$ and that

$$\widehat{\widehat{\Phi}}(X_I) = V_I^2\Phi(-X_I)$$

for every $X_I \in \mathfrak{n}_I(\mathbb{A})$, where we have set

$$V_I = \mathrm{vol}(\mathfrak{n}_I(F)\backslash\mathfrak{n}_I(\mathbb{A}), dX_I) = q^{(g_X-1)\sum_{1\leq j<k\leq s} d_jd_k}$$

(see [We 2] (Ch. VII, § 2); the self-dual Haar measure gives the volume 1 to $\mathfrak{n}_I(F)\backslash\mathfrak{n}_I(\mathbb{A})$).

LEMMA (10.4.8) (Poisson summation formula). — *For each locally constant function Φ with compact support on $\mathfrak{n}_I(\mathbb{A})$ we have*

$$\sum_{\xi_I\in\mathfrak{n}_I(F)} \widehat{\Phi}(\xi_I) = V_I \sum_{\xi_I\in\mathfrak{n}_I(F)} \Phi(\xi_I),$$

both sums being finite.

Proof: See [We 2] (Ch. VII, § 2). □

KEY LEMMA (10.4.9). — *Let φ be a locally constant function with compact support on $N_I(\mathbb{A})$ such that*

$$\int_{N_J(\mathbb{A})} \varphi(n_In'_J)dn'_J = 0$$

for every J with $I \subset J \subsetneqq \Delta$ and every $n_I \in N_I(\mathbb{A})$ (dn'_J is the Haar measure on $N_J(\mathbb{A})$ which is normalized by $\mathrm{vol}(N_J(\mathcal{O}), dn'_J) = 1$). Let us set

$$\Phi(X_I) = \varphi(1 + X_I) \qquad (\forall X_I \in \mathfrak{n}_I(\mathbb{A})),$$

so that Φ is a locally constant function with compact support on $\mathfrak{n}_I(\mathbb{A})$. Then we have

$$\sum_{\nu_I\in N_I(F)} \varphi(\nu_I) = V_I^{-1} \sum_{\eta_I\in\mathfrak{n}_I^\circ(F)} \widehat{\Phi}(\eta_I)$$

where $\mathfrak{n}_I^\circ \subset \mathfrak{n}_I$ is the Zariski open subset

$$\mathfrak{n}_I - \bigcup_{I\subset J\subsetneqq\Delta} \mathfrak{n}_I^J$$

($\mathfrak{n}_I^J$ is the Lie algebra of $N_I^J = M_J \cap N_I$).

Proof : Thanks to the Poisson summation formula it is sufficient to prove that the support of $\widehat{\Phi}$ is contained in $\mathfrak{n}_I^\circ(\mathbb{A})$.

By assumption on φ we have

$$\int_{\mathfrak{n}_J(\mathbb{A})} \Phi(X_I + X'_J + X_I X'_J) dX'_J = 0$$

for every $X_I \in \mathfrak{n}_I(\mathbb{A})$ and every J with $I \subset J \subsetneq \Delta$, where dX'_J is the Haar measure on $\mathfrak{n}_J(\mathbb{A})$ which is normalized by $\mathrm{vol}(\mathfrak{n}_J(\mathcal{O}), dX'_J) = 1$ (use (10.4.6)). But it is easy to see that

$$X_I X'_J \in \mathfrak{n}_J(\mathbb{A})$$

for every $X_I \in \mathfrak{n}_I(\mathbb{A})$ and every $X'_J \in \mathfrak{n}_J(\mathbb{A})$ and that

$$d((1 + X_I)X'_J) = dX'_J$$

for every $X_I \in \mathfrak{n}_I(\mathbb{A})$. Therefore we have

$$\int_{\mathfrak{n}_J(\mathbb{A})} \Phi(X_I + X'_J) dX'_J = 0$$

for every $X_I \in \mathfrak{n}_I(\mathbb{A})$ and every J with $I \subset J \subsetneq \Delta$.

Moreover we have

$$\mathrm{tr}({}^t X_J Y_I^J) = 0$$

for every $X_J \in \mathfrak{n}_J(\mathbb{A})$, every $Y_I^J \in \mathfrak{n}_I^J(\mathbb{A})$ and every J with $I \subset J \subset \Delta$. Therefore we have

$$\widehat{\Phi}(Y_I^J) = \int_{\mathfrak{n}_I^J(\mathbb{A})} \Big(\int_{\mathfrak{n}_J(\mathbb{A})} \Phi(X_I^J + X_J) dX_J \Big) \Psi(\mathrm{tr}({}^t X_I^J Y_I^J)) dX_I^J = 0$$

for every J with $I \subset J \subsetneq \Delta$ and every $Y_I^J \in \mathfrak{n}_I^J(\mathbb{A}) \subset \mathfrak{n}_I(\mathbb{A})$. □

For each $g \in G(\mathbb{A})$ and each $m_I \in M_I(\mathbb{A})$ let us set

$$\Phi_{g,m_I}(X_I) = f(g^{-1} m_I (1 + X_I) g) \quad (\forall X_I \in \mathfrak{n}_I(\mathbb{A})). \tag{10.4.10}$$

Then it follows from lemmas (10.2.3) and (10.4.9) that we have

$$\begin{aligned} &\Big| \sum_{\nu_I \in N_I(F)} f(g^{-1} t_\infty^{-1} \mu_I \nu_I t_\infty g) \Big| \\ &\qquad \leq V_I^{-1} \delta_I(t_\infty) \sum_{\eta_I \in \mathfrak{n}_I^\circ(F)} |\widehat{\Phi}_{g, t_\infty^{-1} \mu_I t_\infty}(t_\infty \eta_I t_\infty^{-1})| \end{aligned} \tag{10.4.11}$$

for every $g \in G(\mathbb{A})$, every $t_\infty \in T(F_\infty)$ and every $\mu_I \in M_I(F)$: recall that, for any locally constant complex function Φ with compact support on $\mathfrak{n}_I(\mathbb{A})$ and for any $m_I \in M_I(\mathbb{A})$, we have

$$(X_I \mapsto \Phi(m_I^{-1} X_I m_I))^\wedge = (Y_I \mapsto \delta_I(m_I) \widehat{\Phi}({}^t m_I Y_I {}^t m_I^{-1})). \tag{10.4.12}$$

Proof of proposition (10.4.1) *and consequently of theorem* (10.2.2) : Firstly we apply (10.4.5). Then we see that it is sufficient to prove that

$$\sum_{\mu_I \in M_I(F)} \sum_{t_\infty \in \mathcal{T}_{I,\infty}^T} \delta_\emptyset^{-1}(t_\infty) \Big| \sum_{\nu_I \in N_I(F)} f(g^{-1}t_\infty^{-1}\mu_I \nu_I t_\infty g) \Big| < +\infty$$

for any fixed g in the compact set $\mathcal{T}^\infty \omega K$.

Secondly, for all but finitely many $\mu_I \in M_I(F)$, the corresponding summand is actually zero. Indeed, if Ω is the compact support of f in $F_\infty^\times \backslash G(\mathbb{A})$ the projection of

$$(F_\infty^\times \backslash P_I(\mathbb{A})) \cap \Big(\bigcup_{t_\infty \in \mathcal{T}_{I,\infty}^T} t_\infty g \Omega g^{-1} t_\infty^{-1} \Big)$$

into $F_\infty^\times \backslash M_I(\mathbb{A})$ is compact too (the quotient

$$Z_I(F_\infty)\backslash Z_I(F_\infty)\mathcal{T}_{I,\infty}^T$$

is finite) and $M_I(F)$ is discrete in $F_\infty^\times \backslash M_I(\mathbb{A})$. Therefore it is sufficient to prove that

$$\sum_{t_\infty \in \mathcal{T}_{I,\infty}^T} \delta_\emptyset^{-1}(t_\infty) \Big| \sum_{\nu_I \in N_I(F)} f(g^{-1}t_\infty^{-1}\mu_I \nu_I t_\infty g) \Big| < +\infty$$

for any fixed $g \in \mathcal{T}^\infty \omega K$ and $\mu_I \in M_I(F)$.

Thirdly, let $\widehat{\omega}$ be a compact open subset of $\mathfrak{n}_I(\mathbb{A})$ such that

$$\mathrm{Supp}(\widehat{\Phi}_{g, t_\infty^{-1}\mu_I t_\infty}) \subset \widehat{\omega}$$

for all $t_\infty \in \mathcal{T}_{I,\infty}^T$ (recall that $Z_I(F_\infty)\backslash Z_I(F_\infty)\mathcal{T}_{I,\infty}^T$ is finite) and let $\Sigma \subset \mathcal{T}_{I,\infty}^T$ be the set of $t_\infty \in \mathcal{T}_{I,\infty}^T$ such that

$$(t_\infty \mathfrak{n}_I^\circ(F) t_\infty^{-1}) \cap \widehat{\omega} \neq \emptyset.$$

If we prove that Σ is finite, then

$$\bigcup_{t_\infty \in \Sigma} t_\infty^{-1} \widehat{\omega} t_\infty$$

will also be a compact subset of $\mathfrak{n}_I(\mathbb{A})$ and its intersection with $\mathfrak{n}_I^\circ(F)$ will be finite ($\mathfrak{n}_I(F)$ is discrete in $\mathfrak{n}_I(\mathbb{A})$).Thus by (10.4.11) the sum

$$\sum_{t_\infty \in \mathcal{T}_{I,\infty}^T} \frac{\delta_I(t_\infty)}{\delta_\emptyset(t_\infty)} \sum_{\eta_I \in \mathfrak{n}_I^\circ(F)} |\widehat{\Phi}_{g, t_\infty^{-1}\mu_I t_\infty}(t_\infty \eta_I t_\infty^{-1})|$$

will be finite and the proof of (10.4.1) will be completed.

Let us choose an integer D such that for each $X_I \in \widehat{\omega}$ the degrees of all the entries of X_I are larger than or equal to D. Then, for any $t_\infty = (t_{\infty,1}, \dots, t_{\infty,d}) \in (F_\infty^\times)^d \cong T(F_\infty)$ such that

$$(t_\infty \mathfrak{n}_I^\circ(F) t_\infty^{-1}) \cap \widehat{\omega} \neq \emptyset$$

and any $\ell \in \{1, \dots, s-1\}$ $(d_I = (d_1, \dots, d_s))$, there exist $i, j \in \{1, \dots, d\}$ with

$$i \leq d_1 + \dots + d_\ell < j$$

and

$$\frac{\deg(\infty)}{f}(\infty(t_{\infty,i}) - \infty(t_{\infty,j})) \geq D.$$

Indeed, if $\eta_I \in \mathfrak{n}_I^\circ(F) \cap (t_\infty^{-1} \widehat{\omega} t_\infty)$ and if $\ell \in \{1, \dots, s-1\}$, there exist $i, j \in \{1, \dots, d\}$ with

$$i \leq d_1 + \dots + d_\ell < j$$

such that the entry η_{ij} of η_I is non-zero and we then have

$$\frac{\deg(\infty)}{f}(\infty(t_{\infty,i}) - \infty(t_{\infty,j})) = \deg(t_{\infty,i}\eta_{ij}t_{\infty,j}^{-1}) \geq D$$

as $t_{\infty,i}\eta_{ij}t_{\infty,j}^{-1}$ is an entry of $t_\infty \eta_I t_\infty^{-1} \in \widehat{\omega}$. Therefore Σ is contained in the set of $t_\infty \in T_{I,\infty}^T$ such that, for each $\ell \in \{1, \dots, s-1\}$, there exist $i, j \in \{1, \dots, d\}$ with

$$i \leq d_1 + \dots + d_\ell < j$$

and

$$(\alpha_i + \alpha_{i+1} + \dots + \alpha_{j-1})(H_\emptyset(t_\infty)) \leq -D.$$

It is now clear that Σ is finite. □

REMARK (10.4.13). — In the case $I = \Delta$ our proof of (10.4.1) simplifies drastically. In fact, in this case, it is sufficient to remark that $F^\Delta(\cdot, T)$ has a compact support in $F_\infty^\times G(F) \backslash G(\mathbb{A})$. □

(10.5) The distributions J_{geom} and $J_{\mathcal{O}}$

For any locally constant complex function with compact support f on $F_\infty^\times \backslash G(\mathbb{A})$ satisfying hypotheses (A) and (B) of (10.2) we set

$$J_{\mathcal{O}}(f) = \int_{F_\infty^\times G(F) \backslash G(\mathbb{A})} k_{\mathcal{O}}(g) \frac{dg}{dz_\infty d\gamma} \tag{10.5.1}$$

for each $\mathcal{O} \in G(F)_\sharp$ and

$$J_{\mathrm{geom}}(f) = \sum_{\mathcal{O} \in G(F)_\sharp} J_{\mathcal{O}}(f). \tag{10.5.2}$$

It follows from (10.2.2) that the integrals $J_{\mathcal{O}}(f)$, $\mathcal{O} \in G(F)_\sharp$, are all absolutely convergent and that the series $J_{\mathrm{geom}}(f)$ is absolutely convergent. In fact, for a given f, $k_{\mathcal{O}}$ is identically zero for all but finitely many $\mathcal{O} \in G(F)_\sharp$, so that the sum $J_{\mathrm{geom}}(f)$ is finite. Indeed, for any given compact subset Ω of $F_\infty^\times\backslash G(\mathbb{A})$ there are only finitely many $\mathcal{O} \in G(F)_\sharp$ such that

$$\widetilde{\Omega} \cap \{g\gamma g^{-1} \mid g \in G(\mathbb{A}), \gamma \in \mathcal{O}\} \neq \emptyset$$

where $\widetilde{\Omega} \subset G(\mathbb{A})$ is the inverse image of Ω under the canonical projection of $G(\mathbb{A})$ onto $F_\infty^\times\backslash G(\mathbb{A})$.

Among the $\mathcal{O}$'s the most important ones are those for which $D_{\mathcal{O}}(T)$ is irreducible. For such an $\mathcal{O}$ the integral $J_{\mathcal{O}}(f)$ is absolutely convergent without hypotheses (A) and (B) on f and it may be computed in terms of orbital integrals. More precisely and more generally we have

LEMMA (10.5.3). — *Let f be any locally constant function with compact support on $F_\infty^\times\backslash G(\mathbb{A})$ and let δ be an elliptic element in $G(F)$ (i.e. such that its minimal polynomial is irreducible over F). Then we have*

(i)
$$\int_{F_\infty^\times G(F)\backslash G(\mathbb{A})} \Big(\sum_{\gamma \in O_G(\delta)(F)} |f(g^{-1}\gamma g)|\Big) \frac{dg}{dz_\infty d\gamma} < +\infty$$

where $O_G(\delta)(F) = \{\gamma^{-1}\delta\gamma \mid \gamma \in G(F)\}$ is the orbit of δ in $G(F)$,

(ii)
$$\mathrm{vol}(F_\infty^\times G_\delta(F)\backslash G_\delta(\mathbb{A}), \frac{dg_\delta}{dz_\infty d\gamma_\delta}) < +\infty$$

where dg_δ is an arbitrary Haar measure on the centralizer $G_\delta(\mathbb{A})$ of δ in $G(\mathbb{A})$ and $d\gamma_\delta$ is the counting measure on the centralizer $G_\delta(F)$ of δ in $G(F)$,

(iii)
$$\int_{G_\delta(\mathbb{A})\backslash G(\mathbb{A})} |f(g^{-1}\delta g)| \frac{dg}{dg_\delta} < +\infty,$$

(iv)
$$\int_{F_\infty^\times G(F)\backslash G(\mathbb{A})} \Big(\sum_{\gamma \in O_G(\delta)(F)} f(g^{-1}\gamma g)\Big) \frac{dg}{dz_\infty d\gamma}$$
$$= \mathrm{vol}(F_\infty^\times G_\delta(F)\backslash G_\delta(\mathbb{A}), \frac{dg_\delta}{dz_\infty d\gamma_\delta}) \int_{G_\delta(\mathbb{A})\backslash G(\mathbb{A})} f(g^{-1}\delta g) \frac{dg}{dg_\delta}.$$

Proof : We may assume that f takes only non-negative real values. Then both sides of equality (iv) are well-defined in $\mathbb{R}_{\geq 0} \cup \{+\infty\}$ and are equal to

$$\int_{F_\infty^\times G_\delta(F)\backslash G(\mathbb{A})} f(g^{-1}\delta g)\frac{dg}{dz_\infty d\gamma_\delta}.$$

Therefore, to finish the proof of the lemma it is sufficient to prove assertions (ii) and (iii). For (ii) we remark that

$$F_\infty^\times G_\delta(F)\backslash G_\delta(\mathbb{A}) \cong F_\infty^\times GL_{d'}(F')\backslash GL_{d'}(\mathbb{A}')$$

where $F' = F[\delta]$, $d' = d/[F' : F]$ and $\mathbb{A}'$ is the ring of adeles of F'. For (iii) we remark that the orbit

$$O_G(\delta)(\mathbb{A}) = \{g^{-1}\delta g \mid g \in G(\mathbb{A})\}$$

is closed in $G(\mathbb{A})$ for the adelic topology (for each place x of F, F_x is a separable extension of F, so that $\mathcal{O}_G(\delta)(F_x)$ is closed in $G(F_x)$; see (4.3.2)(i)). □

Without any further assumptions on f the $J_{\mathcal{O}}(f)$ for which $D_{\mathcal{O}}(T)$ is reducible are very difficult to compute (compare with [Ar 1], [Ar 2], [Ar 3], [Ar 4] and [Ar 5] in the number field case).

(10.6) Kazhdan's trick

Let o be a place of F distinct from ∞ and let f^o be a locally constant function with compact support on $F_\infty^\times\backslash G(\mathbb{A}^o)$. We make the following assumptions:

(A^o) *for any* $g^o \in G(\mathbb{A}^o)$ *and any* $k_\infty \in K_\infty = G(\mathcal{O}_\infty) \subset G(F_\infty) \subset G(\mathbb{A}^o)$ *we have* $f^o(k_\infty^{-1}g^o k_\infty) = f^o(g^o)$,

(B^o) *for any* $I \subsetneq \Delta$, *any* $m_{I,\infty} \in M_I(F_\infty)$ *and any* $h^{\infty,o} \in G(\mathbb{A}^{\infty,o})$ *we have*

$$\int_{N_I(F_\infty)} f^o(m_{I,\infty}n_{I,\infty}h^{\infty,o})dn_{I,\infty} = 0,$$

$dn_{I,\infty}$ *being an arbitrary Haar measure on* $N_I(F_\infty)$.

REMARK (10.6.1). — If φ_o is any locally constant function with compact support on $G(F_o)$, $\varphi = f^o\varphi_o$ satisfies hypotheses (A) and (B) of (10.2). □

We will say that $\mathcal{O} \in G(F)_\sharp$ is **elliptic over** F if its characteristic polynomial $D_{\mathcal{O}}(T)$ is a power of an irreducible polynomial in $F[T]$, i.e. if

$$D_{\mathcal{O}}(T) = P(T)^{d'}$$

with $P(T) \in F[T]$ irreducible, the leading coefficient of $P(T)$ being equal to 1 and d' being a positive integer which divides d. Let x be a place of F. We will say that $\mathcal{O} \in G(F)_\sharp$ is **elliptic over** F_x if its characteristic polynomial $D_{\mathcal{O}}(T)$ is a power of an irreducible polynomial in $F_x[T]$. Obviously, if $\mathcal{O} \in G(F)_\sharp$ is elliptic over F_x it is elliptic over F.

THEOREM (10.6.2). — *There exists a function*

$$R : \{\text{compact open subsets in } F_\infty^\times \backslash G(\mathbb{A}^o)\} \to \mathbb{Z}_{>0}$$

having the following property. Let f^o be a locally constant function with compact support on $T_\infty^\times \backslash G(\mathbb{A}^o)$ which satisfies hypotheses (A^o) *and* (B^o), *let r be a positive integer, let*

$$f_o = f_o^{(r)} \in \mathcal{C}_c^\infty(G(F_o)/\!\!/G(\mathcal{O}_o))$$

be the Hecke function with Satake transform

$$f_o^\vee(z) = p^{\deg(o)r(d-1)/2}(z_1^r + \cdots + z_d^r)$$

and let $\mathcal{o} \in G(F)_\sharp$. Let us assume that

$$r \geq R(\mathrm{Supp}(f^o))$$

and that

$$J_{\mathcal{o}}(f^o f_o^{(r)}) \neq 0.$$

Then $\mathcal{o}$ is elliptic over F_o and therefore over F.

Let Z be a totally disconnected locally compact topological space and let $\mathcal{C}_c^\infty(Z, \mathbb{C})$ be the $\mathbb{C}$-vector space of locally constant complex valued functions with compact support on Z.

A **distribution** D on Z is simply a $\mathbb{C}$-linear form

$$D : \mathcal{C}_c^\infty(Z, \mathbb{C}) \to \mathbb{C}.$$

The **support** of such a distribution D is the smallest closed subset $\mathrm{Supp}(D)$ in Z such that

$$D(\varphi) = 0$$

for each $\varphi \in \mathcal{C}_c^\infty(Z, \mathbb{C})$ which vanishes identically on $\mathrm{Supp}(D)$. If H is a totally disconnected locally compact topological group, a distribution D on H is said to be **invariant** if

$$D(\varphi^h) = D(\varphi)$$

for each $\varphi \in \mathcal{C}_c^\infty(Z, \mathbb{C})$ and each $h \in H$, where we have set

$$\varphi^h(h') = \varphi(h^{-1}h'h) \qquad (\forall h' \in H).$$

The proof of theorem (10.6.2) relies on the following key observation.

KEY OBSERVATION (10.6.3). — *Let f^o be a locally constant function with compact support on $T_\infty^\times\backslash G(\mathbb{A}^o)$ which satisfies hypotheses* (A^o) *and* (B^o), *let $\mathfrak{o} \in G(F)_\sharp$ and let $\mathfrak{o}_{F_o} \subset G(F_o)$ be the set of $\gamma_o \in G(F_o)$ the characteristic polynomial of which is $D_\mathfrak{o}(T)$ (a closed subset of $G(F_o)$). Then the distribution*

$$\varphi_o \mapsto J_\mathfrak{o}(f^o\varphi_o)$$

on $G(F_o)$ is invariant and its support is contained in $\mathfrak{o}_{F_o}$. □

For each $\gamma_o \in G(F_o)$ and each Haar measure dg_{γ_o} on the centralizer $G_{\gamma_o}(F_o)$ of γ_o in $G(F_o)$ (see (4.8.6)), we have an invariant distribution

$$\varphi_o \mapsto O_{\gamma_o}(\varphi_o, dg_{\gamma_o}) = \sum_{G_{\gamma_o}(F_o)\backslash G(F_o)} \varphi_o(g_o^{-1}\gamma_o g_o)\frac{dg_o}{dg_{\gamma_o}}$$

on $G(F_o)$ (see (4.8.9)) the support of which is the closure in $G(F_o)$ (for the ϖ_o-adic topology) of the orbit of γ_o,

$$O_G(\gamma_o)(F_o) = \{g_o^{-1}\gamma_o g_o \mid g_o \in G(F_o)\}.$$

Moreover this closure is the disjoint union of finitely many orbits (see (4.3.2)(ii)).

Conversely we have the following lemma.

LEMMA (10.6.4). — *Let $\gamma_o^{(1)},\ldots,\gamma_o^{(N)}$ be finitely many elements in $G(F_o)$ and for each $n = 1,\ldots,N$ let $dg_{\gamma_o^{(n)}}$ be a Haar measure on $G_{\gamma_o^{(n)}}(F_o)$. If D is an invariant distribution on $G(F_o)$ such that*

$$\mathrm{Supp}(D) \subset \bigcup_{n=1}^{N} O_G(\gamma_o^{(n)})(F_o),$$

then for suitable constants $c_1,\ldots,c_N \in \mathbb{C}$ we have

$$D(\varphi_o) = \sum_{n=1}^{N} c_n O_{\gamma_o^{(n)}}(\varphi_o, dg_{\gamma_o^{(n)}}) \qquad (\forall \varphi_o \in C_c^\infty(G(F_o),\mathbb{C})).$$

Moreover, if we assume that

$$O_G(\gamma_o^{(n')})(F_o) \cap O_G(\gamma_o^{(n'')})(F_o) = \emptyset$$

for any $n' \neq n''$ in $\{1,\ldots,N\}$ the constants $c_1,\ldots,c_N$ are uniquely determined by D. □

A proof of this lemma is given in appendix F.

For any $o \in G(F)_\sharp$ there are only finitely many sequences $(P_1(T), \dots, P_d(T))$ of unitary polynomials in $F_o[T]$ such that

$$P_1(T)|P_2(T)|\cdots|P_d(T)$$

and

$$P_1(T)P_2(T)\cdots P_d(T) = D_o(T),$$

so that o_{F_o} is a disjoint union of finitely many orbits. Therefore our key observation (10.6.3) and the above lemma imply

PROPOSITION (10.6.5). — *Let f^o be a locally constant function with compact support on $T_\infty^\times\backslash G(\mathbb{A}^o)$ which satisfies hypotheses* (A^o) *and* (B^o), *let $o \in G(F)_\sharp$, let $o_{F_o,\natural}$ be a system of representatives of the $G(F_o)$-conjugacy classes in o_{F_o} (a finite set) and, for each $\gamma_o \in o_{F_o,\natural}$, let dg_{γ_o} be a Haar measure on the centralizer $G_{\gamma_o}(F_o)$ of γ_o in F_o. Then there exists a unique family of complex numbers $(a(o, f^o, \gamma_o, dg_{\gamma_o}))_{\gamma_o \in o_{F_o,\natural}}$ such that*

$$J_o(f^o\varphi_o) = \sum_{\gamma_o \in o_{F_o,\natural}} a(o, f^o, \gamma_o, dg_{\gamma_o}) O_{\gamma_o}(\varphi_o, dg_{\gamma_o}) \quad (\forall \varphi_o \in C_c^\infty(G(F_o), \mathbb{C})).$$

□

Proof of theorem (10.6.2), *construction of the function R* : Let $\Omega^o \subset F_\infty^\times\backslash G(\mathbb{A}^o)$ be a compact open subset and let $\widetilde{\Omega}^o \subset G(\mathbb{A}^o)$ be its inverse image under the canonical projection $G(\mathbb{A}^o) \twoheadrightarrow F_\infty^\times\backslash G(\mathbb{A}^o)$. Let us consider the characteristic polynomial map

$$\begin{aligned} G(\mathbb{A}^o) &\to T^d + \mathbb{A}^o T^{d-1} + \cdots + \mathbb{A}^o T + (\mathbb{A}^o)^\times \subset \mathbb{A}^o[T], \\ \gamma^o &\mapsto D_{\gamma^o}(T) = T^d - \operatorname{tr}\gamma^o T^{d-1} + \cdots + (-1)^d \det\gamma^o, \end{aligned}$$

and let $\widetilde{\Sigma}^o$ be the image of $\widetilde{\Omega}^o$ under this map. We have

$$D_{z_\infty\gamma^o}(T) = z_\infty^d D_{\gamma^o}(z_\infty^{-1}T)$$

and therefore, if we let $F_\infty^\times$ act on $T^d + \mathbb{A}^o T^{d-1} + \cdots + \mathbb{A}^o T + (\mathbb{A}^o)^\times$ by

$$(z_\infty, D(T)) \mapsto z_\infty \cdot D(T) = z_\infty^d D(z_\infty^{-1}T),$$

we have

$$F_\infty^\times \cdot \widetilde{\Sigma}^o \subset \widetilde{\Sigma}^o$$

and the quotient

$$\Sigma^o = F_\infty^\times\backslash\widetilde{\Sigma}^o$$

is compact.

Next let us introduce a height function

$$\mathrm{ht}^o : T^d + \mathbb{A}^o T^{d-1} + \cdots + \mathbb{A}^o T + (\mathbb{A}^o)^\times \to [1, +\infty[.$$

We fix an algebraic closure $\overline{F}$ of F and for each place x of F we choose an algebraic closure $\overline{F}_x$ of F_x which contains $\overline{F}$. For each place x of F the normalized absolute value $|\ |_x$ of F_x (recall that

$$|a|_x = p^{-\deg(x)x(a)} \qquad (\forall a \in F_x))$$

admits a unique extension to $\overline{F}_x$ (see [Se 1] (Ch. II, § 2)) and we still let $|\ |_x$ denote this extension. For each place x of F and each

$$D_x(T) \in T^d + F_x T^{d-1} + \cdots + F_x T + F_x^\times$$

we set

$$\mathrm{ht}_x(D_x(T)) = \sup\{\Big|\frac{\alpha_i}{\alpha_j}\Big|_x \mid 1 \leq i, j \leq d\}$$

where $\alpha_1, \ldots, \alpha_d$ are the roots of $D_x(T)$ in $\overline{F}_x$. Then for each

$$D^o(T) = (D_x(T))_{x \neq o} \in T^d + \mathbb{A}^o T^{d-1} + \cdots + \mathbb{A}^o T + (\mathbb{A}^o)^\times$$

we set

$$\mathrm{ht}^o(D^o(T)) = \prod_{x \neq o} \mathrm{ht}_x(D_x(T))$$

(for almost all $x \neq o$ we have

$$D_x(T) \in T^d + \mathcal{O}_x T^{d-1} + \cdots + \mathcal{O}_x T + \mathcal{O}_x^\times$$

and

$$\mathrm{ht}_x(D_x(T)) = 1\,).$$

This height has the following two properties.

Firstly ht^o is bounded above on $\widetilde{\Sigma}^o$. Indeed, for each place $x \neq o$, let us introduce the polynomial

$$\prod_{1 \leq i,j \leq d} (T - \frac{\alpha_{i,x}}{\alpha_{j,x}}) = T^{d^2} + b_{1,x} T^{d^2-1} + \cdots + b_{d^2-1,x} T + b_{d^2,x}$$

where $\alpha_{1,x}, \ldots, \alpha_{d,x}$ are the roots of $D_x(T) = T^d + a_{1,x} T^{d-1} + \cdots + a_{d,x}$ in $\overline{F}_x$; for each $n = 1, \ldots, d^2$ we have

$$b_{n,x} = \frac{P_n(a_{1,x}, \ldots, a_{d,x})}{a_{d,x}^n}$$

for some universal polynomial

$$P_n(A_1, \ldots, A_d) \in \mathbb{Z}[A_1, \ldots, A_d]$$

which is homogeneous of degree dn if we have assigned the degree i to the variable A_i, $i = 1, \ldots, d$; but, for any $i, j \in \{1, \ldots, d\}$ such that $|\alpha_{i,x}/\alpha_{j,x}|_x \geq 1$, we have

$$\left|\frac{\alpha_{i,x}}{\alpha_{j,x}}\right|_x = \left|-\sum_{n=1}^{d^2} b_{n,x}\left(\frac{\alpha_{i,x}}{\alpha_{j,x}}\right)^{1-n}\right|_x \leq \sup\{|b_{n,x}| \mid 1 \leq n \leq d^2\};$$

therefore we have

$$\mathrm{ht}_x(D_x(T)) \leq \sup\{\frac{|P_n(a_{1,x}, \ldots, a_{d,x})|}{|a_{d,x}|^n} \mid 1 \leq n \leq d^2\}$$

and ht^o is bounded above on $\widetilde{\Sigma}^o$ by the compactness of Σ^o.

Secondly, for any

$$D(T) \in T^d + FT^{d-1} + \cdots + FT + F^\times$$

we have

$$\mathrm{ht}_o(D(T)) \leq \mathrm{ht}^o(D(T)).$$

Indeed let $\alpha_1, \ldots, \alpha_d$ be the roots of $D(T)$ in $\overline{F}$; by the product formula we have

$$\left|\frac{\alpha_i}{\alpha_j}\right|_o = \prod_{x \neq o}\left|\frac{\alpha_j}{\alpha_i}\right|_x$$

for any $i, j \in \{1, \ldots, d\}$ and the property follows.

Now we take for $R(\Omega^o)$ any integer such that

$$\mathrm{ht}^o(D^o(T)) < p^{\deg(o)R(\Omega^o)/d} \qquad (\forall D^o(T) \in \widetilde{\Sigma}^o).$$

With this choice of $R(\Omega^o)$ we have

$$\mathrm{ht}_o(D_{\mathcal{O}}(T)) < p^{\deg(o)R(\Omega^o)/d}$$

for any $\mathcal{O} \in G(F)_\sharp$ such that there exist at least one $\gamma \in \mathcal{O}$ and one $g^o \in G(\mathbb{A}^o)$ with

$$(g^o)^{-1}\gamma g^o \in \widetilde{\Omega}^o.$$

□

End of the proof of theorem (10.6.2) : Let $\mathcal{O} \in G(F)_\sharp$ be such that $J_{\mathcal{O}}(f^o f_o^{(r)}) \neq 0$ for some $r \geq R(\mathrm{Supp}(f^o))$. Then on the one hand there exist at least one $\gamma \in \mathcal{O}$ and one $g^o \in G(\mathbb{A}^o)$ such that

$$f^o((g^o)^{-1}\gamma g^o) \neq 0$$

and it follows that

$$\mathrm{ht}_o(D_{\mathcal{O}}(T)) < p^{\deg(o)R(\mathrm{Supp}(f^o))/d}.$$

On the other hand there exists at least one $\gamma_o \in \mathcal{O}_{F_o,\natural}$ such that

$$O_{\gamma_o}(f_o^{(r)}, dg_{\gamma_o}) \neq 0$$

(see (10.6.5)) and it follows that

$$D_{\mathcal{O}}(T) = D_o'(T)D_o''(T)$$

where $D_o'(T)$ and $D_o''(T)$ are two polynomials with coefficients in $\mathcal{O}_o$, the leading coefficients of which are equal to 1, such that $D_o'(T)$ is a power of an irreducible polynomial in $F_o[T]$, where

$$o(D_o'(0)) = r$$

and

$$D_o''(0) \in \mathcal{O}_o^\times$$

(see (4.8.13)).

Now either $D_o''(T) = 1$ and $\mathcal{O}$ is elliptic over F_o or the degree of $D_o''(T)$ is positive and we have

$$\mathrm{ht}_o(D_{\mathcal{O}}(T)) = p^{\deg(o)r/\deg(D_o'(T))} \geq p^{\deg(o)r/d}$$

(if α' and α'' are roots in $\overline{F}_o$ of $D_o'(T)$ and $D_o''(T)$ respectively we have

$$|\alpha'|_o = p^{\deg(o)r/\deg(D_o'(T))}$$

and

$$|\alpha''|_o = 1\,).$$

But the inequalities

$$p^{\deg(o)r/d} \leq \mathrm{ht}_o(D_{\mathcal{O}}(T)) < p^{\deg(o)R(\mathrm{Supp}(f^o))/d}$$

contradict the inequality $r \geq R(\mathrm{Supp}(f^o))$. □

(10.7) The distributions J_δ

Let $\mathcal{O}$ be an elliptic class in $G(F)_\sharp$ and let $\mathcal{O}_\natural$ be a system of representatives of the $G(F)$-conjugacy classes in $\mathcal{O}$. If

$$D_{\mathcal{O}}(T) = Q_{\mathcal{O}}(T)^{d'_{\mathcal{O}}}$$

where $D_{\mathcal{O}}(T)$ is an irreducible polynomial in $F[T]$ with leading coefficient 1 and where $d'_{\mathcal{O}}$ is a positive integer which divides d, $\mathcal{O}_\natural$ is in one-to-one correspondence with the sequences of integers

$$0 \leq d'_1 \leq d'_2 \leq \cdots \leq d'_d \leq d'_{\mathcal{O}}$$

such that $d'_1 + d'_2 + \cdots + d'_d = d'_{\mathcal{O}}$ (the elementary divisors of any element of $\mathcal{O}$ are

$$Q_{\mathcal{O}}(T)^{d'_1} | Q_{\mathcal{O}}(T)^{d'_2} | \cdots | Q_{\mathcal{O}}(T)^{d'_d}$$

for such a sequence of integers). We denote by $\delta_{\mathcal{O}}$ the unique elliptic element of $\mathcal{O}_\natural$: it corresponds to the sequence

$$d'_i = \begin{cases} 0 & \text{if } 1 \leq i \leq d - d'_{\mathcal{O}}, \\ 1 & \text{if } d - d'_{\mathcal{O}} + 1 \leq i \leq d. \end{cases}$$

For this elliptic element $\delta_{\mathcal{O}}$ we set

$$\text{(10.7.1)} \quad J_{\delta_{\mathcal{O}}}(f) = \int_{F_\infty^\times G(F)\backslash G(\mathbb{A})} \Big(\sum_{\gamma \in O_G(\delta_{\mathcal{O}})(F)} f(g^{-1}\gamma g) \Big) \frac{dg}{dz_\infty d\gamma}$$
$$= \mathrm{vol}\big(F_\infty^\times G_{\delta_{\mathcal{O}}}(F)\backslash G_{\delta_{\mathcal{O}}}(\mathbb{A}), \frac{dg_{\delta_{\mathcal{O}}}}{dz_\infty d\gamma_{\delta_{\mathcal{O}}}}\big)$$
$$\int_{G_{\delta_{\mathcal{O}}}(\mathbb{A})/G(\mathbb{A})} f(g^{-1}\delta_{\mathcal{O}} g) \frac{dg}{dg_{\delta_{\mathcal{O}}}}$$

for any locally constant complex function f with compact support on $F_\infty^\times \backslash G(\mathbb{A})$ (see (10.5.3) for the notations and for the absolute convergence of these integrals and their equality).

If $d'_{\mathcal{O}} = 1$, so that $D_{\mathcal{O}}(T)$ is irreducible, $J_{\mathcal{O}}(f)$ is defined for any f as above and

$$\text{(10.7.2)} \qquad J_{\mathcal{O}}(f) = J_{\delta_{\mathcal{O}}}(f)$$

(we have $\mathcal{O} = O_G(\delta_{\mathcal{O}})(F)$). But, if $d'_{\mathcal{O}} > 1$, this is not true any more (even if f satisfies hypotheses (A) and (B) of (10.2), so that $J_{\mathcal{O}}(f)$ is defined). In the number field case, if $\mathcal{O}$ is the unipotent class ($D_{\mathcal{O}}(T) = (T-1)^d$) Arthur has introduced new distributions J_δ, $\delta \in \mathcal{O}$, $\delta \sim \delta_{\mathcal{O}}$, to measure the difference between $J_{\mathcal{O}}(f)$ and $J_{\delta_{\mathcal{O}}}(f)$ (see [Ar 2] Thm. 4.2). Our next goal is to adapt Arthur's construction of the J_δ to a general elliptic $\mathcal{O}$ in the function field case.

Let

$$\mathcal{U}_{\mathcal{O}} \subset G$$

be the closed F-subscheme defined by the condition: "the characteristic polynomial is equal to $D_{\mathcal{O}}(T)$". If F^{sep} is a separable closure of F there are only finitely many $G(F^{\mathrm{sep}})$-conjugacy classes in $\mathcal{U}_{\mathcal{O}}(F^{\mathrm{sep}})$ and they are permuted by the Galois group $\mathrm{Gal}(F^{\mathrm{sep}}/F)$ of F^{sep} over F. The union of the $G(F^{\mathrm{sep}})$-conjugacy classes in a given $\mathrm{Gal}(F^{\mathrm{sep}}/F)$-orbit is equal to $U(F^{\mathrm{sep}})$ for some irreducible locally closed subscheme U of $\mathcal{U}_{\mathcal{O}}$ which is smooth over F and G-invariant by conjugation (see [Bo 2] (II, 6.7) and (AG, 14.4)). If $U(F) \neq \emptyset$ and if $\delta \in U(F)$, U is nothing else than the G-conjugacy class $O_G(\delta)$ of δ in G. In general there exists a finite Galois extension F' of F such that $U(F') \neq \emptyset$ and, if $\{\delta'_1, \ldots, \delta'_n\}$ is a $\mathrm{Gal}(F'/F)$-orbit in $U(F')$, $F' \otimes_F U$ is the disjoint union of the $(F' \otimes_F G)$-conjugacy classes of $\delta'_1, \ldots, \delta'_n$ in $F' \otimes_F G$.

Let us denote by

$$(\mathcal{U}_{\mathcal{O}})$$

the finite collection of the above subschemes U of $\mathcal{U}_{\mathcal{O}}$. For any separable field extension E of F, $\mathcal{U}_{\mathcal{O}}(E)$ is the disjoint union of the $U(E)$, $U \in (\mathcal{U}_{\mathcal{O}})$.

On $(\mathcal{U}_{\mathcal{O}})$ we have the order relation

$$(U' \preceq U) \Longleftrightarrow (U' \subset \overline{U})$$

where $\overline{U}$ is the Zariski closure of U in G (an irreducible closed subscheme of G which contains U as a dense Zariski open subset). If $U' \preceq U$ in $(\mathcal{U}_{\mathcal{O}})$, we have

$$\dim(U') \leq \dim(U)$$

with equality if and only if $U' = U$ ($\dim(\cdot)$ is the dimension as scheme over F).

For each $U \in (\mathcal{U}_{\mathcal{O}})$ let us set

$$k_U(g) = \sum_{\gamma \in U(F)} f(g^{-1}\gamma g) \qquad (\forall g \in G(\mathbb{A})). \tag{10.7.3}$$

Then $k_U(\cdot)$ is a well-defined locally constant function on $F_\infty^\times G(F)\backslash G(\mathbb{A})$. For each $T \in \mathfrak{a}_\emptyset$ with $d(T) \geq -2g_X$ the integral

$$\int_{F_\infty^\times G(F)\backslash G(\mathbb{A})} F^\Delta(g,T) k_U(g) \frac{dg}{dz_\infty d\gamma}$$

is absolutely convergent and, in fact, can be reduced to a finite sum (the function $F^\Delta(\cdot,T)$ has compact support in $F_\infty^\times G(F)\backslash G(\mathbb{A})$).

For each $U \in (\mathcal{U}_{\mathcal{O}})$ let us fix a finite system of generators

$$\{q_{\overline{U},1}, \ldots, q_{\overline{U},n_{\overline{U}}}\} \tag{10.7.4}$$

of the ideal in

$$\Gamma(G, \mathcal{O}_G) = F[(g_{ij})_{1\leq i,j\leq d}, \frac{1}{\det g}]$$

defining the closed subscheme $\overline{U}$ of G. Then for each place $x \neq \infty$ of F and each non-negative integer N let

$$\chi_{\overline{U},x,N} : G(F_x) \to \{0,1\} \subset \mathbb{C} \tag{10.7.5}$$

be the characteristic function of the subset

$$\{g_x \in G(F_x) \mid x(q_{\overline{U},\nu}(g_x)) \geq N, \ \forall \nu = 1, \ldots, n_{\overline{U}}\}$$

of $G(F_x)$ and let us simply denote by $f\chi_{\overline{U},x,N}$ the locally constant function with compact support

$$F_\infty^\times g \mapsto f(g)\chi_{\overline{U},x,N}(g_x)$$

on $F_\infty^\times\backslash G(\mathbb{A})$. Let us observe that $f\chi_{\overline{U},x,N}$ is equal to f in a neighborhood of $\overline{U}(\mathbb{A})$ for the adelic topology.

THEOREM (10.7.6) (Arthur). — *On the $\mathbb{C}$-vector space of all locally constant complex functions f with compact support on $F_\infty^\times\backslash G(\mathbb{A})$ which satisfy hypotheses* (A) *and* (B) *of* (10.2) *there is a unique family $J_U(f)$, $U \in (\mathcal{U}_{\mathcal{O}})$, of $\mathbb{C}$-linear forms such that*

$$J_{\mathcal{O}}(f) = \sum_{U\in(\mathcal{U}_{\mathcal{O}})} J_U(f)$$

which have the following properties:

(i) *for each $U \in (\mathcal{U}_{\mathcal{O}})$, each f and each $\epsilon \in \,]0,1[$ the integral*

$$\int_{F_\infty^\times G(F)\backslash G(\mathbb{A})} F^\Delta(g,T)k_U(g)\frac{dg}{dz_\infty d\gamma}$$

admits a limit

$$J_U(f)$$

when $d(T) = \inf\{\alpha_i(T) \mid i = 1, \ldots, d-1\}$ goes to $+\infty$ and is subject to the relation

$$d(T) \geq \epsilon\|T\| \stackrel{\text{dfn}}{=\!=} \epsilon\sup\{|\alpha_i(T)| \mid i = 1, \ldots, d-1\},$$

(ii) *for each* $U \in (\mathcal{U}_{\mathcal{O}})$, *each place* $x \neq \infty$ *of* F *and each* f *the limit when* N *goes to* $+\infty$ *of*

$$J_{\mathcal{O}}(f\chi_{\overline{U},x,N})$$

exists and is equal to

$$J_{\overline{U}}(f) \stackrel{\text{dfn}}{=\!=} \sum_{\substack{U' \in (\mathcal{U}_{\mathcal{O}}) \\ U' \preceq U}} J_{U'}(f).$$

The proof of this theorem will be given in (10.8) and (10.9).

REMARKS (10.7.7.1). — If $U(F) = \emptyset$ we have $J_U \equiv 0$.

(10.7.7.2). — If $U = O_G(\delta_{\mathcal{O}})$ is the unique elliptic G-conjugacy class in $\mathcal{U}_{\mathcal{O}}$, for any f the integral

$$\int_{F_\infty^\times G(F)\backslash G(\mathbb{A})} k_U(g) \frac{dg}{dz_\infty d\gamma}$$

is absolutely convergent and is equal to $J_{\delta_{\mathcal{O}}}(f)$ (see (10.7.1) and (10.5.3)). □

The above remarks allow us to define the $\mathbb{C}$-linear forms J_δ, $\delta \in \mathcal{O}_\natural$, by

(10.7.8) $$J_\delta = J_{O_G(\delta)}$$

and to replace the collection of the J_U, $U \in (\mathcal{U}_{\mathcal{O}})$, by the collection of the J_δ, $\delta \in \mathcal{O}_\natural$.

For us the main consequence of theorem (10.7.6) is

COROLLARY (10.7.9). — *Let* $o \neq \infty$ *be a place of* F, *let* f^o *be a locally constant complex function with compact support on* $F_\infty^\times\backslash G(\mathbb{A}^o)$ *satisfying hypotheses* (A^o) *and* (B^o) *of* (10.6) *and, for each* $\delta \in \mathcal{O}_\natural$, *let* $dg_{\delta,o}$ *be a Haar measure on the centralizer* $G_\delta(F_o)$ *of* δ *in* $G(F_o)$. *Then there is a unique family of complex numbers* $\big(a(o,f^o,\delta,dg_{\delta,o})\big)_{\delta\in\mathcal{O}_\natural}$ *such that, for any* $\varphi_o \in C_c^\infty(G(F_o),\mathbb{C})$, *we have*

$$J_{\mathcal{O}}(f^o\varphi_o) = \sum_{\delta\in\mathcal{O}_\natural} a(o,f^o,\delta,dg_{\delta,o}) O_\delta(\varphi_o,dg_{\delta,o})$$

and even

$$J_\delta(f^o\varphi_o) = a(o,f^o,\delta,dg_{\delta,o}) O_\delta(\varphi_o,dg_{\delta,o}) \qquad (\forall \delta \in \mathcal{O}_\natural).$$

In particular, for any Haar measure $dg_{\delta_{\mathcal{O}}}^o$ on $G_{\delta_{\mathcal{O}}}(\mathbb{A}^o)$ we have

$$a(o,f^o,\delta_{\mathcal{O}},dg_{\delta_{\mathcal{O}},o}) = \mathrm{vol}\big(F_\infty^\times G_{\delta_{\mathcal{O}}}(F)\backslash G_{\delta_{\mathcal{O}}}(\mathbb{A}), \frac{dg_{\delta_{\mathcal{O}}}^o dg_{\delta_{\mathcal{O}},o}}{dz_\infty d\gamma_{\delta_{\mathcal{O}}}}\big) O_{\delta_{\mathcal{O}}}(f^o,dg_{\delta_{\mathcal{O}}}^o)$$

where we have set

$$O_{\delta_{\mathcal{O}}}(f^o, dg^o_{\delta_{\mathcal{O}}}) = \int_{G_{\delta_{\mathcal{O}}}(\mathbb{A}^o)} f^o((g^o)^{-1}\delta_{\mathcal{O}} g^o)\frac{dg^o}{dg^o_{\delta_{\mathcal{O}}}}$$

(the $G(\mathbb{A}^o)$-conjugacy class of $\delta_{\mathcal{O}}$ in $G(\mathbb{A}^o)$ is closed for the adelic topology, so that the above orbital integral is absolutely convergent).

Proof of (10.7.9) (*assuming* (10.7.6)) : Let $\mathcal{O}_{F_o}$, $\mathcal{O}_{F_o,\natural}$ and dg_{γ_o}, $\gamma_o \in \mathcal{O}_{F_o,\natural}$, be as in (10.6.5). We may and we do assume that $\mathcal{O}_\natural \subset \mathcal{O}_{F_o,\natural}$ and that $dg_{\gamma_o} = dg_{\delta,o}$ if $\gamma_o = \delta \in \mathcal{O}_\natural$. Then there are uniquely determined complex numbers $a(o, f^o, \gamma_o, dg_{\gamma_o})$, $\gamma_o \in \mathcal{O}_{F_o,\natural}$, such that

$$J_{\mathcal{O}}(f^o\varphi_o) = \sum_{\gamma_o \in \mathcal{O}_{F_o,\natural}} a(o, f^o, \gamma_o, dg_{\gamma_o}) O_{\gamma_o}(f_o, dg_{\gamma_o})$$

for every $\varphi_o \in C_c^\infty(G(F_o), \mathbb{C})$ (see loc. cit.).

Now the problem is essentially to prove that

$$a(o, f^o, \gamma_o, dg_{\gamma_o}) = 0 \qquad (\forall \gamma_o \in \mathcal{O}_{F_o,\natural} - \mathcal{O}_\natural).$$

Let $U \in (\mathcal{U}_{\mathcal{O}})$, $\gamma_o \in \mathcal{O}_{F_o,\natural}$ and $\varphi_o \in C_c^\infty(G(F_o), \mathbb{C})$. Then by the Lebesgue dominated convergence theorem we have

$$\lim_{N\to+\infty} O_{\gamma_o}(\varphi_o \chi_{\overline{U},o,N}, dg_{\gamma_o}) = \begin{cases} O_{\gamma_o}(\varphi_o, dg_{\gamma_o}) & \text{if } \gamma_o \in \overline{U}(F_o), \\ 0 & \text{otherwise.} \end{cases}$$

Therefore, for any $U \in (\mathcal{U}_{\mathcal{O}})$ and any $\varphi_o \in C_c^\infty(G(F_o), \mathbb{C})$, theorem (10.7.6) tells us that

$$J_{\overline{U}}(f^o\varphi_o) = \sum_{\gamma_o \in \mathcal{O}_{F_o,\natural} \cap \overline{U}(F_o)} a(o, f^o, \gamma_o, dg_{\gamma_o}) O_{\gamma_o}(\varphi_o, dg_{\gamma_o})$$

and by induction on $\dim(U)$ we obtain

$$J_U(f^o\varphi_o) = \sum_{\gamma_o \in \mathcal{O}_{F_o,\natural} \cap U(F_o)} a(o, f^o, \gamma_o, dg_{\gamma_o}) O_{\gamma_o}(\varphi_o, dg_{\gamma_o}).$$

But J_U is identically zero unless $U = O_G(\delta)$ for some $\delta \in \mathcal{O}_\natural$ and the distributions $O_{\gamma_o}(\cdot, dg_{\gamma_o})$, $\gamma_o \in \mathcal{O}_{F_o,\natural}$, are linearly independent (see (10.6.4)). Thus we get that

$$a(o, f^o, \gamma_o, dg_{\gamma_o}) = 0 \qquad (\forall \gamma_o \in \mathcal{O}_{F_o,\natural} - \mathcal{O}_\natural)$$

and this completes the proof of the corollary. □

Now we can easily prove the main result of this chapter

THEOREM (10.7.10). — *Let $o \neq \infty$ be a place of F, let f^o be a locally constant complex function with compact support on $F_\infty^\times \backslash G(\mathbb{A}^o)$ satisfying hypotheses* (A^o) *and* (B^o) *of* (10.6), *let r be some positive integer and let*

$$f_o = f_o^{(r)} \in C_c^\infty(G(F_o)/\!\!/G(\mathcal{O}_o))$$

be the Hecke function with Satake transform

$$f_o^\vee(z) = p^{\deg(o)r(d-1)/2}(z_1^r + \cdots + z_d^r).$$

We set

$$f = f^o f_o^{(r)}.$$

Then, if r is large enough with respect to the support of f^o, more precisely if $r \geq R(\mathrm{Supp}(f^o))$ with the notations of (10.6.2), *$J_{\mathrm{geom}}(f)$ is equal to*

$$\sum_{\delta \in G(F)_{\natural,\mathrm{ell}}} \mathrm{vol}\Big(F_\infty^\times G_\delta(F)\backslash G_\delta(\mathbb{A}), \frac{dg_\delta}{dz_\infty d\gamma_\delta}\Big) \int_{G_\delta(\mathbb{A})\backslash G(\mathbb{A})} f(g^{-1}\delta g)\frac{dg}{dg_\delta}.$$

Here $G(F)_{\natural,\mathrm{ell}}$ is a system of representatives of the $G(F)$-conjugacy classes of elliptic elements in $G(F)$ and, for each $\delta \in G(F)$, dg_δ is an arbitrary Haar measure on $G_\delta(\mathbb{A})$.

Proof: Let us assume that $r \geq R(\mathrm{Supp}(f^o))$. Then we have

$$J_{\mathcal{O}}(f) = 0$$

if $\mathcal{O}$ is not elliptic over F_o. Moreover, if $\mathcal{O}$ is elliptic over F_o it follows from (10.7.9) and (4.8.13) that

$$J_{\mathcal{O}}(f) = J_{\delta_{\mathcal{O}}}(f).$$

Finally, if $\delta \in G(F)_{\natural,\mathrm{ell}}$ is not elliptic over F_o the same argument as in the last part of the proof of (10.6.2) shows that

$$\int_{G_\delta(\mathbb{A})\backslash G(\mathbb{A})} f(g^{-1}\delta g)\frac{dg}{dg_\delta} = 0.$$

So the proof is completed. □

(10.8) Reduction of (10.7.6)

Let us fix $\omega \subset U(\mathbb{A})$ and $\mathcal{T}^\infty \subset T(\mathbb{A}^\infty)$ as in (10.4.5). Let Ω be a compact open subset of $F_\infty^\times\backslash G(\mathbb{A})$ and let $K' \subset K \subset G(\mathbb{A})$ be a compact open subgroup such that

$$K'\Omega K' = \Omega.$$

For each $g \in \mathcal{T}^\infty \omega K \subset G(\mathbb{A})$ and each $I \subsetneqq \Delta$ we denote by

$$\Omega_g^{M_I} \subset F_\infty^\times\backslash M_I(\mathbb{A})$$

(resp.

$$\Omega_g^{N_I} \subset N_I(\mathbb{A}))$$

the projection of

$$(g\Omega g^{-1}) \cap (F_\infty^\times\backslash M_I(\mathbb{A})N_I(\mathbb{A}))$$

into $F_\infty^\times\backslash M_I(\mathbb{A})$ (resp. $N_I(\mathbb{A})$) and we set

$$\omega_g^{N_I} = \{X_I \in \mathfrak{n}_I(\mathbb{A}) \mid 1 + X_I \in \Omega_g^{N_I}\}$$

and

$$(\Omega, K')_g^{N_I} = \{X_I \in \mathfrak{n}_I(\mathbb{A}) \mid 1 \pm n_I^{-1}X_I \in gK'g^{-1},\ \forall n_I \in \Omega_g^{N_I}\}.$$

We let the reader check that $\Omega_g^{M_I}$ (resp. $\Omega_g^{N_I}$, $\omega_g^{N_I}$) is a compact open subset of $F_\infty^\times\backslash M_I(\mathbb{A})$ (resp. $N_I(\mathbb{A})$, $\mathfrak{n}_I(\mathbb{A})$) and that $(\Omega, K')_g^{N_I}$ is a compact neighborhood of 0 in $\mathfrak{n}_I(\mathbb{A})$. Then we set

$$\Omega^{M_I} = \bigcup_{g \in \mathcal{T}^\infty \omega K} \Omega_g^{M_I} \subset F_\infty^\times\backslash M_I(\mathbb{A}),$$

$$\omega^{N_I} = \bigcup_{g \in \mathcal{T}^\infty \omega K} \omega_g^{N_I} \subset N_I(\mathbb{A})$$

and

$$(\Omega, K')^{N_I} = \bigcap_{g \in \mathcal{T}^\infty \omega K} (\Omega, K')_g^{N_I} \subset \mathfrak{n}_I(\mathbb{A}).$$

Again, Ω^{M_I} (resp. ω^{N_I}) is a compact open subset of $F_\infty^\times\backslash M_I(\mathbb{A})$ (resp. $\mathfrak{n}_I(\mathbb{A})$) and $(\Omega, K')^{N_I}$ is a compact neighborhood of 0 in $\mathfrak{n}_I(\mathbb{A})$ ($\mathcal{T}^\infty \omega K/K'$ is finite). There exists at least one $a \in \mathbb{A}^\times$ such that

$$a\mathfrak{n}_I(\mathcal{O}) \subset (\Omega, K')^{N_I}$$

and the degrees of such elements a are bounded below. Let us set

$$\deg(\Omega, K')^{N_I} = \inf\{\deg(a) \mid a \in \mathbb{A}^\times,\ a\mathfrak{n}_I(\mathcal{O}) \subset (\Omega, K')^{N_I}\}.$$

The following proposition improves (10.4.1).

PROPOSITION (10.8.1). — *Let f be a locally constant complex function with compact support on $F_\infty^\times\backslash G(\mathbb{A})$ satisfying hypotheses* (A) *and* (B) *of* (10.2), *let Ω be its support and let $K' \subset K \subset G(\mathbb{A})$ be a compact open subgroup such that f is K'-bi-invariant. Let C_I^Δ, $I \subsetneqq \Delta$, be as in* (10.4.1) *and let ϵ_1 be as in* (10.4.4).

Then, for any $I \subsetneqq \Delta$ and any $T \in \mathfrak{a}_\emptyset$ such that $d(T) \geq C_I^\Delta$ and such that there exists at least one $j \in \Delta - I$ with

$$\alpha_j(T) \geq \deg(\Omega, K')^{N_I} + (d-1)(2g_X + \epsilon_1) - 2,$$

we have

$$\int_{F_\infty^\times G(F)\backslash G(\mathbb{A})} F^I(g,T)\tau_I^\Delta(H_\emptyset(g) - T)|k_{\mathfrak{o}}(g)|\frac{dg}{dz_\infty d\pi_I} = 0 \qquad (\forall \mathfrak{o} \in G(F)_\sharp).$$

In particular, for each $T \in \mathfrak{a}_\emptyset$ such that $d(T) \geq \sup\{C_I^\Delta \mid I \subsetneqq \Delta\}$ and that

$$d(T) \geq \deg(\Omega, K')^{N_I} + (d-1)(2g_X + \epsilon_1) - 2 \qquad (\forall I \subsetneqq \Delta),$$

we have

$$J_{\mathfrak{o}}(f) = \int_{F_\infty^\times G(F)\backslash G(\mathbb{A})} F^\Delta(g,T)k_{\mathfrak{o}}(g)\frac{dg}{dz_\infty d\gamma}.$$

Let V be a closed subscheme of G over F and let $(q_1, \ldots, q_n)$ be a system of generators of the ideal of

$$H^0(G, \mathcal{O}_G) = F\big[(g_{ij})_{1\leq i,j\leq d}, \frac{1}{\det g}\big]$$

defining V in G. We assume that V is G-invariant under conjugation, i.e. that there exist polynomials

$$p_{\nu,\nu'} \in H^0(G \times_F G, \mathcal{O}_{G\times_F G}) \qquad (\nu, \nu' = 1, \ldots, n)$$

such that

$$q_\nu(ghg^{-1}) = \sum_{\nu'=1}^n p_{\nu,\nu'}(g,h)q_{\nu'}(h) \qquad (\forall \nu = 1, \ldots, n,\ \forall g, h \in G).$$

Let $x \neq \infty$ be a place of F. For each non-negative integer N, let us denote by $\chi_{V,x,N}$ the characteristic function of the subset

$$\{g_x \in G(F_x) \mid x(q_\nu(g_x)) \geq N,\ \forall \nu = 1, \ldots, n\}$$

of $G(F_x)$.

With these notations and assumptions we have

PROPOSITION (10.8.2). — *For each non-negative real function* $\varphi \in C_c^\infty(F_\infty^\times\backslash G(\mathbb{A}),\mathbb{C})$ *there exist positive real numbers A and B such that, for each non-negative integer N and each $T \in \mathfrak{a}_\emptyset$ with $d(T) \geq -2g_X$, we have*

$$\int_{F_\infty^\times G(F)\backslash G(\mathbb{A})} F^\Delta(g,T)\Big(\sum_{\gamma\in G(F)-V(F)} \varphi\chi_{V,x,N}(g^{-1}\gamma g)\Big)\frac{dg}{dz_\infty d\gamma}$$
$$\leq Ap^{-BN}(1+\|T\|)^{d-1}.$$

Before proving propositions (10.8.1) and (10.8.2) let us deduce (10.7.6).

Proof of theorem (10.7.6) (*assuming* (10.8.1) *and* (10.8.2)) : Let us fix $U \in (\mathcal{U}_\mathcal{O})$ and a place $x \neq \infty$ of F. Let $T \in \mathfrak{a}_\emptyset$ with $d(T) \geq \sup\{C_I^\Delta \mid I \subsetneqq \Delta\}$ and let N be a non-negative integer. For each $\varphi \in C_c^\infty(F_\infty^\times\backslash G(\mathbb{A}),\mathbb{C})$ we set

$$J_{\overline{U}}^T(\varphi) = \int_{F_\infty^\times G(F)\backslash G(\mathbb{A})} F^\Delta(g,T)\Big(\sum_{\gamma\in\overline{U}(F)} \varphi(g^{-1}\gamma g)\Big)\frac{dg}{dz_\infty d\gamma}.$$

The key to the proof of theorem (10.7.6) is to obtain a good upper bound for

$$|J_\mathcal{O}(f\chi_{\overline{U},x,N}) - J_{\overline{U}}^T(f)|, \tag{10.8.3}$$

at least if T is large enough.

Since

$$J_{\overline{U}}^T(f) = J_{\overline{U}}^T(f\chi_{\overline{U},x,N}),$$

an upper bound of (10.8.3) is given by the sum of the two expressions

$$(*)\quad \Big|J_\mathcal{O}(f\chi_{\overline{U},x,N}) - \int_{F_\infty^\times G(F)\backslash G(\mathbb{A})} F^\Delta(g,T)\Big(\sum_{\gamma\in\mathcal{O}} f\chi_{\overline{U},x,N}(g^{-1}\gamma g)\Big)\frac{dg}{dz_\infty d\gamma}\Big|$$

and

$$(**)\quad \int_{F_\infty^\times G(F)\backslash G(\mathbb{A})} F^\Delta(g,T)\Big(\sum_{\gamma\in\mathcal{O}-\overline{U}(F)} |f\chi_{\overline{U},x,N}(g^{-1}\gamma g)|\Big)\frac{dg}{dz_\infty d\gamma}.$$

If T is large enough we will successively prove the vanishing of $(*)$ using proposition (10.8.1) and bound $(**)$ above using proposition (10.8.2).

Let Ω (resp. $\Omega_N \subset \Omega$) be the compact support of f (resp. $f\chi_{\overline{U},x,N}$) in $F_\infty^\times\backslash G(\mathbb{A})$. Let K' be a compact open subgroup of $K \subset G(\mathbb{A})$ such that f is K'-bi-invariant. There exists a positive integer N_0, which does not depend on N, such that

$$\chi_{\overline{U},x,N}(k_{x,1}g_xk_{x,2}) = \chi_{\overline{U},x,N}(g_x)$$

for every $g = g^x g_x \in \Omega_N \subset \Omega$ and every $k_{x,1}$, $k_{x,2}$ in the principal congruence subgroup
$$1 + \varpi_x^{N+N_0} gl_d(\mathcal{O}_x) \subset GL_d(\mathcal{O}_x) = K_x$$
(as usual ϖ_x is a uniformizer of $\mathcal{O}_x$). Indeed let Ω_x be a compact subset of $G(F_x)$ and let
$$q_x \in \mathcal{O}_x[(g_{ij})_{1\leq i,j\leq d}];$$
then there exists a positive integer N_0' such that
$$\varpi^{N_0'} g_x \in gl_d(\mathcal{O}_x) \qquad (\forall g_x \in \Omega_x)$$
and if we set
$$N_0 = \deg(q)N_0'$$
we have
$$q_x(k_{x,1}g_x k_{x,2}) - q_x(g_x) \in \varpi_x^N \mathcal{O}_x \quad (\forall g_x \in \Omega_x,\ k_{x,1}, k_{x,2} \in 1 + \varpi_x^{N+N_0} gl_d(\mathcal{O}_x)).$$
Now let us set
$$K_N' = K' \cap K^x(1 + \varpi_x^{N+N_0} gl_d(\mathcal{O}_x)),$$
so that $f\chi_{\overline{U},x,N}$ is K_N'-bi-invariant. Then for each $I \subsetneqq \Delta$ and each $a \in \mathbb{A}^\times$ such that
$$a\mathfrak{n}_I(\mathcal{O}) \subset (\Omega, K')^{N_I}$$
there exists a positive integer N_1' such that
$$\varpi_x^{N_1'} a_x g_x^{-1} n_{I,x}^{-1} \mathfrak{n}_I(\mathcal{O}_x) g_x \in \varpi_x^{N_0} gl_d(\mathcal{O}_x) \qquad (\forall n_I \in \Omega_g^{N_I},\ g \in T^\infty \omega K).$$
It follows that there exists a positive integer N_1, which does not depend on N, such that
$$\deg(\Omega, K_N')^{N_I} \leq \deg(\Omega, K')^{N_I} + \frac{\deg(x)}{f}(N + N_1)$$
for every $I \subsetneqq \Delta$. But we have
$$(\Omega, K_N')^{N_I} \subset (\Omega_N, K_N')^{N_I}$$
and
$$\deg(\Omega_N, K_N')^{N_I} \leq \deg(\Omega, K_N')^{N_I}$$
for every $I \subsetneqq \Delta$. Therefore, if we assume moreover that
$$d(T) \geq \deg(\Omega, K')^{N_I} + \frac{\deg(x)}{f}(N + N_1) + (d-1)(2g_X + \epsilon_1) - 2 \qquad (\forall I \subsetneqq \Delta)$$
we get the vanishing of the expression $(*)$ by applying (10.8.1).

Next we remark that

$$\sum_{\gamma \in \mathcal{O}-\overline{U}(F)} |f\chi_{\overline{U},x,N}(g^{-1}\gamma g)| \leq \sum_{\gamma \in G(F)-\overline{U}(F)} |f\chi_{\overline{U},x,N}(g^{-1}\gamma g)| \quad (\forall g \in G(\mathbb{A})).$$

Thus by applying (10.8.2) we get the upper bound

$$\int_{F_\infty^\times G(F)\backslash G(\mathbb{A})} F^{\Delta}(g,T)\Big(\sum_{\gamma \in \mathcal{O}-\overline{U}(F)} |f\chi_{\overline{U},x,N}(g^{-1}\gamma g)|\Big)\frac{dg}{dz_\infty d\gamma}$$
$$\leq Ap^{-BN}(1+\|T\|^{d-1})$$

for the expression (**) (recall that $C_I^{\Delta} \geq -2g_X$ for each $I \subsetneqq \Delta$). Here A and B are positive real numbers which do not depend on N and T.

We have proved that there exist positive real numbers A, B and $C \geq \sup\{C_I^{\Delta} \mid I \subsetneqq \Delta\}$ such that

$$(10.8.4) \qquad |J_{\mathcal{O}}(f\chi_{\overline{U},x,N}) - J_{\overline{U}}^{T}(f)| \leq Ap^{-BN}(1+\|T\|^{d-1})$$

as long as $T \in \mathfrak{a}_{\emptyset}$ satisfies the inequality

$$d(T) \geq \frac{\deg(x)}{f}N + C.$$

Now we can easily conclude the proof of theorem (10.7.6) using the key estimate (10.8.4). Firstly, if we assume that

$$d(T) \geq \frac{\deg(x)}{f}(N+1) + C,$$

(10.8.4) gives us the upper bound

$$|J_{\mathcal{O}}(f\chi_{\overline{U},x,N}) - J_{\mathcal{O}}(f\chi_{\overline{U},x,N+1})| \leq A(p^{-BN}+p^{-B(N+1)})(1+\|T\|^{d-1}).$$

Specializing T in such way that

$$d(T) = \|T\| = \frac{\deg(x)}{f}(N+1) + C$$

we obtain

$$|J_{\mathcal{O}}(f\chi_{\overline{U},x,N}) - J_{\mathcal{O}}(f\chi_{\overline{U},x,N+1})|$$
$$\leq A(p^{-BN}+p^{-B(N+1)})(1+(\frac{\deg(x)}{f}(N+1)+C)^{d-1}).$$

It follows that $(J_{\mathcal{O}}(f\chi_{\overline{U},x,N}))_{N\geq 0}$ is a Cauchy sequence. Let $J_{\overline{U}}(f)$ denote its limit.

Secondly let us fix $\epsilon \in]0,1[$. In (10.8.4) let us restrict ourself to truncation parameters $T \in \mathfrak{a}_{\emptyset}$ such that

$$d(T) \geq C$$

and

$$d(T) \geq \epsilon\|T\|$$

and let us take

$$N = N(T) = \Big[\frac{f}{\deg(x)}(d(T) - C)\Big].$$

Then we have

$$|J_{\mathcal{O}}(f\chi_{\overline{U},x,N(T)}) - J_{\overline{U}}^{T}(f)| \leq Ap^{-Bf(d(T)-C)/\deg(x)}\Big(1 + \Big(\frac{d(T)}{\epsilon}\Big)^{d-1}\Big).$$

Moreover, if $d(T)$ goes to $+\infty$, $N(T)$ goes to $+\infty$ too, so that $J_{\mathcal{O}}(f\chi_{\overline{U},x,N(T)})$ converges to $J_{\overline{U}}(f)$. Therefore, if $d(T) \geq \epsilon\|T\|$ and if $d(T)$ goes to $+\infty$, $J_{\overline{U}}^{T}(f)$ converges to $J_{\overline{U}}(f)$.

Thirdly, by induction on $\dim(U)$ we define $J_{U}^{T}(f)$ and $J_{U}(f)$, $\forall U \in (\mathcal{U}_{\mathcal{O}})$, $\forall T \in \mathfrak{a}_{\emptyset}$ with $d(T) \geq \sup\{C_{I}^{\Delta} \mid I \subsetneq \Delta\}$, in such way that

$$J_{\overline{U}}^{T}(f) = \sum_{\substack{U' \in (\mathcal{U}_{\mathcal{O}}) \\ U' \preceq U}} J_{U'}^{T}(f)$$

and

$$J_{\overline{U}}(f) = \sum_{\substack{U' \in (\mathcal{U}_{\mathcal{O}}) \\ U' \preceq U}} J_{U'}(f).$$

It is clear that

$$J_{U}^{T}(f) = \int_{F_{\infty}^{\times}G(F)\backslash G(\mathbb{A})} F^{\Delta}(g,T)k_{U}(g)\frac{dg}{dz_{\infty}d\gamma}$$

and that, for each $\epsilon \in]0,1[$,

$$\lim_{\substack{d(T)\to+\infty \\ d(T)\geq\epsilon\|T\|}} J_{U}^{T}(f)$$

exists and is equal to $J_{U}(f)$. This concludes the proof of theorem (10.7.6).

□

Recall that we have, in the course of the proof of proposition (10.4.5), defined the positive real number ϵ_0 by

$$\epsilon_0 = \sup\{|\alpha_i(t^\infty)| \mid i = 1, \ldots, d-1 \text{ and } t^\infty \in \mathcal{T}^\infty\}$$

where $\mathcal{T}^\infty \subset T(\mathbb{A}^\infty)$ is a system of representatives for $T(F)\backslash T(\mathbb{A}^\infty)/T(\mathcal{O}^\infty)$.

The main new ingredient that we will need to prove propositions (10.8.1) and (10.8.2) is

LEMMA (10.8.5). — (i) *For each $I \subset \Delta$ and each compact open subset Σ_I in $F_\infty^\times\backslash M_I(\mathbb{A})$ there exists a positive real number $C_I(\Sigma_I)$ such that*

$$|M_I(F) \cap (t_\infty \widetilde{\Sigma}_I t_\infty^{-1})| \leq C_I(\Sigma_I)\delta_\emptyset^I(t_\infty)$$

for every $t_\infty \in T(F_\infty)$ with

$$\alpha_i(H_\emptyset(t_\infty)) \geq -2g_X - \epsilon_0 \qquad (\forall i \in I).$$

Here $\delta_\emptyset^I = \delta_\emptyset/\delta_I$ is the modulus character of $P_\emptyset^I(\mathbb{A})$ and $\widetilde{\Sigma}_I \subset M_I(\mathbb{A})$ is the inverse image of Σ_I under the canonical projection of $M_I(\mathbb{A})$ onto $F_\infty^\times\backslash M_I(\mathbb{A})$.

(ii) *Let Σ be a compact open subset in $F_\infty^\times\backslash G(\mathbb{A})$, let $V \subset G$ be a G-invariant (under conjugation) closed subscheme over F, let $\{q_1, \ldots, q_n\}$ be a system of generators of the ideal of $H^0(G, \mathcal{O}_G)$ defining V in G and let $x \neq \infty$ be a place of F. For each non-negative integer N let us set*

$$\Sigma_{V,x,N} = \{g = g^x g_x \in G(\mathbb{A}) \mid F_\infty^\times g \in \Sigma \text{ and } x(q_\nu(g_x)) \geq N,\ \forall \nu = 1, \ldots, n\}.$$

Then there exist positive real numbers $A' = A'(\Sigma, V, x)$ and $B' = B'(\Sigma, V, x)$ such that

$$|(G(F) - V(F)) \cap (t_\infty \Sigma_{V,x,N} t_\infty^{-1})| \leq A' p^{-B'N} \delta_\emptyset(t_\infty)$$

for every $N \geq 0$ and every $T \in \mathfrak{a}_\emptyset$ with $d(H_\emptyset(t)) \geq -2g_X - \epsilon_0$.

The proof of this lemma will be given in (10.9).

Proof of proposition (10.8.1) (*assuming* (10.8.5)) : Let $I \subsetneqq \Delta$ and let $T \in \mathfrak{a}_\emptyset$ with $d(T) \geq C_I^\Delta$. Combining (10.4.5) and (10.4.11) as in the proof of (10.4.1) we obtain that the integral

$$\int_{F_\infty^\times P_I(F)\backslash G(\mathbb{A})} F^I(g, T)\tau_I^\Delta(H_\emptyset(g) - T)|k_{\mathcal{O}}(g)| \frac{dg}{dz_\infty d\gamma}$$

is bounded above by the product of

$$\mathrm{vol}(K, dg)\,\mathrm{vol}(\omega', du')\Big(\sum_{t^\infty \in \mathcal{T}^\infty} \delta_\emptyset^{-1}(t^\infty)\Big) V_I^{-1}$$

and

$$\sum_{t_\infty \in T^T_{I,\infty}} (\delta^I_\emptyset)^{-1}(t_\infty) \sum_{\mu_I \in \mathfrak{o} \cap M_I(F)} \sum_{\eta_I \in \mathfrak{n}^\circ_I(F)} \sup\{|\widehat{\Phi}_{g,t_\infty^{-1}\mu_I t_\infty}(t_\infty \eta_I t_\infty^{-1})| \\ \mid g \in T^\infty \omega K\}$$

with the notations of loc. cit.

But the function

$$\Phi_{g,m_I}(X_I) = f(g^{-1}m_I(1+X_I)g)$$

satisfies the properties

(a) $$|\Phi_{g,m_I}(X_I)| \leq \|f\| \overset{\text{dfn}}{=\!=} \sup\{|f(h)| \mid h \in G(\mathbb{A})\},$$

(b) $$(\Phi_{g,m_I}(X_I) \neq 0) \Longrightarrow (F_\infty^\times m_I \in \Omega^{M_I} \text{ and } X_I \in \omega^{N_I})$$

and

(c) $$\Phi_{g,m_I}(X_I + X_I') = \Phi_{g,m_I}(X_I) \qquad (\forall X_I' \in (\Omega, K')^{N_I})$$

for every $g \in T^\infty \omega K \subset G(\mathbb{A})$, every $m_I \in M_I(\mathbb{A})$ and every $X_I \in \mathfrak{n}_I(\mathbb{A})$. Indeed properties (a) and (b) are obvious; if $X_I \in \omega_g^{N_I}$ we have

$$g^{-1}m_I(1+X_I+X_I')g \in g^{-1}m_I(1+X_I)gK'$$

and

$$\Phi_{g,m_I}(X_I + X_I') = \Phi_{g,m_I}(X_I)$$

for every $X_I' \in (\Omega, K')_g^{N_I}$; similarly, if $X_I + X_I' \in \omega_g^{N_I}$ for some $X_I' \in (\Omega, K')_g^{N_I}$ we have

$$\Phi_{g,m_I}(X_I) = \Phi_{g,m_I}(X_I + X_I')$$

as $-X_I' \in (\Omega, K')_g^{N_I}$; finally, if $X_I \notin \omega_g^{N_I}$ and $X_I + X_I' \notin \omega_g^{N_I}$ we have

$$\Phi_{g,m_I}(X_I + X_I') = 0 = \Phi_{g,m_I}(X_I).$$

Therefore the function

$$\widehat{\Phi}_{g,m_I}(Y_I) = \int_{\mathfrak{n}_I(\mathbb{A})} \Phi_{g,m_I}(X_I)\Psi(\mathrm{tr}({}^tX_I Y_I))dX_I$$

satisfies the properties

(d) $$|\widehat{\Phi}_{g,m_I}(Y_I)| \leq \|f\| \operatorname{vol}(\omega^{N_I}, dX_I)$$

and

$$\text{(e)} \qquad (\widehat{\Phi}_{g,m_I}(Y_I) \neq 0) \Longrightarrow (m_I \in \Omega^{M_I} \text{ and } Y_I \in c_\Psi^{-1} a^{-1} \mathfrak{n}_I(\mathcal{O}))$$

for every $g \in \mathcal{T}^\infty \omega K$, every $m_I \in M_I(\mathbb{A})$, every $Y_I \in \mathfrak{n}_I(\mathbb{A})$ and every $a \in \mathbb{A}^\times$ such that

$$a\mathfrak{n}_I(\mathcal{O}) \subset (\Omega, K')^{N_I}.$$

Here $c_\Psi \in \mathbb{A}^\times$ is any idele such that

$$c_\Psi^{-1}\mathcal{O} = \{v \in \mathbb{A} \mid \Psi(uv) = 1, \ \forall u \in \mathcal{O}\}$$

and consequently such that

$$c_\Psi^{-1} a^{-1} \mathfrak{n}_I(\mathcal{O}) = \{Y_I \in \mathfrak{n}_I(\mathbb{A}) \mid \Psi(\mathrm{tr}({}^t X_I Y_I)) = 1, \ \forall X_I \in a\mathfrak{n}_I(\mathcal{O})\}$$

(see [We 2] (Ch. VII, § 2) for a proof of (e)).

Now thanks to (10.8.5)(i) we have

$$|o \cap M_I(F) \cap (t_\infty \widetilde{\Omega}^{M_I} t_\infty^{-1})| \leq C_I(\Omega^{M_I}) \delta_\emptyset^I(t_\infty)$$

for every $t_\infty \in \mathcal{T}_{I,\infty}^T$ and every $T \in \mathfrak{a}_\emptyset$ with $d(T) \geq -2g_X - \epsilon_0$. It follows that, for each $a \in \mathbb{A}^\times$ with $a\mathfrak{n}_I(\mathcal{O}) \subset (\Omega, K')^{N_I}$ and each $T \in \mathfrak{a}_\emptyset$ with $d(T) \geq C_I^\Delta$, we can estimate from above the integral

$$\int_{F_\infty^\times G(F) \backslash G(\mathbb{A})} F^I(g,T) \tau_I^\Delta(H_\emptyset(g) - T) |k_o(g)| \frac{dg}{dz_\infty d\gamma}$$

by the product of

$$\mathrm{vol}(K, dg)\, \mathrm{vol}(\omega', du') \Big(\sum_{t^\infty \in \mathcal{T}^\infty} \delta_\emptyset^{-1}(t_\infty) \Big) V_I^{-1} \|f\| \, \mathrm{vol}(\omega^{N_I}, dX_I) C_I(\Omega^{M_I})$$

and

$$\sum_{t_\infty \in \mathcal{T}_{I,\infty}^T} |\mathfrak{n}_I^\circ(F) \cap (c_\Psi^{-1} a^{-1} t_\infty^{-1} \mathfrak{n}_I(\mathcal{O}) t_\infty)|.$$

Let $\mathcal{S}_I$ be the collection of subsets S of

$$\coprod_{1 \leq k < \ell \leq s} [d_1 + \cdots + d_{k-1} + 1, \ldots, d_1 + \cdots + d_k] \times [d_1 + \cdots + d_{\ell-1} + 1, \ldots, d_1 + \cdots + d_\ell]$$

such that, for each $k = 1, \ldots, s-1$, there exists at least one $(i,j) \in S$ with

$$1 \leq i \leq d_1 + \cdots + d_k < j \leq d.$$

For each $S \in \mathcal{S}_I$ let

$$\mathfrak{n}_I^S \subset \mathfrak{n}_I^\circ \subset \mathfrak{n}_I$$

be the Zariski locally closed subset defined by the conditions

$$(X_I)_{ij} \neq 0 \qquad (\forall (i,j) \in S)$$

and

$$(X_I)_{ij} = 0 \qquad (\forall (i,j) \notin S).$$

Then $\mathfrak{n}_I^\circ$ is set theoretically the disjoint union of the $\mathfrak{n}_I^S$, $S \in \mathcal{S}_I$, and we have

$$|\mathfrak{n}_i^\circ(F) \cap (c_\Psi^{-1} a^{-1} t_\infty^{-1} \mathfrak{n}_I(\mathcal{O}) t_\infty)| = \sum_{S \in \mathcal{S}_I} \prod_{(i,j) \in S} |F^\times \cap (c_\Psi^{-1} a^{-1} t_{\infty,i}^{-1} t_{\infty,j} \mathcal{O})|$$

for any $a \in \mathbb{A}^\times$ and any $t_\infty \in T(F_\infty)$. But we have

$$F \cap u\mathcal{O} = \{0\}$$

for any $u \in \mathbb{A}^\times$ such that $\deg(u) > 0$, so that

$$|\mathfrak{n}_I^\circ(F) \cap (c_\Psi^{-1} a^{-1} t_\infty^{-1} \mathfrak{n}_I(\mathcal{O}) t_\infty)| = 0$$

for any $a \in \mathbb{A}^\times$ and any $t_\infty \in T(F_\infty)$ such that, for each $S \in \mathcal{S}_I$, there exists at least one $(i,j) \in S$ with

$$(\epsilon_i - \epsilon_j)(H_\emptyset(t_\infty)) > \deg(a) + \deg(c_\Psi).$$

In particular, if $a \in \mathbb{A}^\times$, $t_\infty \in T(F_\infty)$ and if we assume that $d(H_\emptyset(t_\infty)) \geq -2g_X - \epsilon_1$ and that

$$\alpha_j(H_\emptyset(t_\infty)) > \deg(a) + \deg(c_\Psi) + (d-2)(2g_X + \epsilon_1)$$

for at least one $j \in \Delta - I$ we have

$$|\mathfrak{n}_I^\circ(F) \cap (c_\Psi^{-1} a^{-1} t_\infty^{-1} \mathfrak{n}_I(\mathcal{O}) t_\infty)| = 0.$$

Therefore, if $a \in \mathbb{A}^\times$, $T \in \mathfrak{a}_\emptyset$ and if we assume that $d(T) \geq -2g_X$ and that

$$\alpha_j(T) \geq \deg(a) + \deg(c_\Psi) + (d-2)(2g_X + \epsilon_1)$$

for at least one $j \in \Delta - I$ we have

$$|\mathfrak{n}_I^\circ(F) \cap (c_\Psi^{-1} a^{-1} t_\infty^{-1} \mathfrak{n}_I(\mathcal{O}) t_\infty)| = 0 \qquad (\forall t_\infty \in T_{I,\infty}^T).$$

To finish the proof of proposition (10.8.1) it is sufficient to choose $a \in \mathbb{A}^\times$ in such a way that

$$a\mathfrak{n}_I(\mathcal{O}) \subset (\Omega, K')^{N_I}$$

and

$$\deg(a) = \deg(\Omega, K')^{N_I}$$

and to remark that

$$\deg(c_\Psi) = 2g_X - 2$$

(see [We 2] (Ch. VII, § 2)). □

Proof of proposition (10.8.2) (*assuming* (10.8.5)) : We have the following estimate:

$$\int_{F_\infty^\times G(F)\backslash G(\mathbb{A})} F^\Delta(g,T)\Big(\sum_{\gamma\in G(F)-V(F)} \varphi\chi_{V,x,N}(g^{-1}\gamma g)\Big)\frac{dg}{dz_\infty d\gamma}$$
$$\leq \mathrm{vol}(K,dg)\,\mathrm{vol}(\omega',du')\Big(\sum_{t^\infty\in\mathcal{T}^\infty}\delta_\emptyset^{-1}(t^\infty)\Big)\sum_{t_\infty\in\mathcal{T}_{\Delta,\infty}^T}\delta_\emptyset^{-1}(t_\infty)$$
$$\times\sum_{\gamma\in G(F)-V(F)}\sup\{\varphi\chi_{V,x,N}(g^{-1}t_\infty^{-1}\gamma t_\infty g)\mid g\in\mathcal{T}^\infty\omega K\}$$

(see the argument preceding proposition (10.4.5)). Let us set

$$\Sigma=\bigcup_{g\in\mathcal{T}^\infty\omega K} g\mathrm{Supp}(\varphi)g^{-1}$$

(a compact open subset of $F_\infty^\times\backslash G(\mathbb{A})$) and

$$\|\varphi\|=\sup\{\varphi(g)\mid g\in G(\mathbb{A})\}.$$

By hypothesis there are polynomials $p_{\nu,\nu'}\in H^0(G\times_F G,\mathcal{O}_{G\times_F G})$, $\nu,\nu'=1,\ldots,n$, such that

$$q_\nu(h)=\sum_{\nu'=1}^n p_{\nu,\nu'}(g,g^{-1}hg)q_{\nu'}(g^{-1}hg)\qquad(\forall\nu=1,\ldots,n,\ \forall g,h\in G).$$

Let us set

$$N_0=\inf\{x(p_{\nu,\nu'}(g_x,h'_x))\mid g=g^xg_x\in\mathcal{T}^\infty\omega K,$$
$$h'=h'^xh'_x\in\mathrm{Supp}(\varphi),\ \nu,\nu'=1,\ldots,n\}.$$

Then we have

$$\varphi\chi_{V,x,N}(g^{-1}hg)\leq\|\varphi\|1_{\Sigma_{V,x,N+N_0}}(h)$$

for every $g\in\mathcal{T}^\infty\omega K$, every $h\in G(\mathbb{A})$ and every non-negative integer N such that $N+N_0\geq 0$.

Now we can apply (10.8.5) and we get that

$$\sum_{t_\infty\in\mathcal{T}_{\Delta,\infty}^T}\delta_\emptyset^{-1}(t_\infty)\sum_{\gamma\in G(F)-V(F)}\sup\{\varphi\chi_{V,x,N}(g^{-1}t_\infty^{-1}\gamma t_\infty g)\mid g\in\mathcal{T}^\infty\omega K\}$$
$$\leq\|\varphi\|A'p^{-B'(N+N_0)}|\mathcal{T}_{\Delta,\infty}^T|$$

for every $T\in\mathfrak{a}_\emptyset$ with $d(T)\geq -2g_X$.

Finally, for each $i = 1, \ldots, d-1$ we have

$$\alpha_i = \frac{d}{i(d-i)}\varpi_i - \frac{1}{i}\sum_{k=1}^{i-1} k\alpha_k - \frac{1}{d-i}\sum_{k=i+1}^{d-1}(d-k)\alpha_k,$$

so that

$$-2g_X - \epsilon_0 \leq \alpha_i(H_\emptyset(t_\infty)) \leq \frac{d}{i(d-i)}\varpi_i(T-T_0) + \frac{d-2}{2}(2g_X + \epsilon_0)$$

for every $t_\infty \in \mathcal{T}_{\Delta,\infty}^T$. It follows that

$$|\mathcal{T}_{\Delta,\infty}^T| \leq \frac{1}{\deg(\infty)^{d-1}}\prod_{i=1}^{d-1}\Big(\frac{d}{i(d-i)}\varpi_i(T-T_0) + \frac{d}{2}(2g_X+\epsilon_0) + \deg(\infty)\Big)$$

and there exists a positive real number A'' such that

$$|\mathcal{T}_{\Delta,\infty}^T| \leq A''(1+\|T\|)^{d-1}$$

for every $T \in \mathfrak{a}_\emptyset$ with $d(T) \geq -2g_X$. We set

$$A = \mathrm{vol}(K, dg)\,\mathrm{vol}(\omega', du')\Big(\sum_{t^\infty \in \mathcal{T}^\infty} \delta_\emptyset^{-1}(t^\infty)\Big)\|\varphi\|A'p^{-B'N_0}A''$$

and

$$B = B'$$

and we have finished the proof of proposition (10.8.2). □

(10.9) Proof of lemma (10.8.5)

Firstly let us remark that it is sufficient to prove the first part of (10.8.5) for $I = \Delta$ but for an arbitrary $G = GL_d$ (identify M_I with $GL_{d_1} \times \cdots \times GL_{d_s}$ where $d_I = (d_1, \ldots, d_s)$ is the partition of d corresponding to I and cover $\widetilde{\Sigma}_I$ by $\widetilde{\Sigma}_1 \times \cdots \times \widetilde{\Sigma}_s$ where Σ_j is a compact open subset of $F_\infty^\times \backslash GL_{d_j}(\mathbb{A})$ and $\widetilde{\Sigma}_j$ is its inverse image in $GL_{d_j}(\mathbb{A})$ for $j = 1, \ldots, s$).

Secondly let us remark that the first part of (10.8.5) for $I = \Delta$ is a particular case of the second one (take $V = \emptyset$, $n = 1$, $q_1 = 1$, $x \neq \infty$ arbitrary and $N = 0$).

Therefore it is sufficient to prove the second part of (10.8.5). Nevertheless let us give a direct proof of (10.8.5)(i) for $I = \Delta$.

Proof of (10.8.5)(i) *for* $I = \Delta$: Let $\widetilde{\Sigma} = \widetilde{\Sigma}_\Delta$ be the inverse image of $\Sigma = \Sigma_\Delta$ under the canonical projection of $G(\mathbb{A})$ onto $F_\infty^\times \backslash G(\mathbb{A})$. Then

$$\widetilde{\Sigma}^1 = \{g \in \widetilde{\Sigma} \mid \deg(\det g) = 0\}$$

is a compact open subset of $G(\mathbb{A})$ and we have

$$G(F) \cap (t_\infty \widetilde{\Sigma} t_\infty^{-1}) = G(F) \cap (t_\infty \widetilde{\Sigma}^1 t_\infty^{-1})$$

for any $t_\infty \in T(F_\infty)$. Let us set $\mathfrak{g} = gl_d$, so that

$$G = \{g \in \mathfrak{g} \mid \det g \text{ is invertible}\} \subset \mathfrak{g}.$$

There exist elements $a \in \mathbb{A}^\times$ such that

$$\widetilde{\Sigma}^1 \subset a\mathfrak{g}(\mathcal{O}).$$

Let us fix one. Then we have

$$G(F) \cap (t_\infty \widetilde{\Sigma}^1 t_\infty^{-1}) \subset \mathfrak{g}(F) \cap (a t_\infty \mathfrak{g}(\mathcal{O}) t_\infty^{-1})$$

for any $t_\infty \in T(F_\infty)$. Therefore, to conclude the proof it is sufficient to find a positive real number $C(a)$ such that

$$|\mathfrak{g}(F) \cap (a t_\infty \mathfrak{g}(\mathcal{O}) t_\infty^{-1})| \leq C(a) \delta_\emptyset(t_\infty)$$

for every $t_\infty \in T(F_\infty)$ with $d(H_\emptyset(t_\infty)) \geq -2g_X - \epsilon_0$.

But for any $b \in \mathbb{A}^\times$ we have

$$F \cap b\mathcal{O} = \{0\}$$

if $\deg(b) > 0$ and

$$\dim_{\mathbb{F}_p}(F \cap b\mathcal{O}) \leq f(\delta_X - \deg(b))$$

if $\deg(b) \leq 0$ where we have set

$$\delta_X = \inf\{\frac{\deg(x)}{f} \mid x \text{ a place of } F\}.$$

Indeed, if $c \in \mathbb{A}^\times$ and if x is a place of F we have an exact sequence of $\mathbb{F}_p$-vector spaces

$$0 \to F \cap c\mathcal{O} \to F \cap \varpi_x^{-1} c\mathcal{O} \to \kappa(x)$$

where ϖ_x is a uniformizer of $\mathcal{O}_x$, $\kappa(x) = \mathcal{O}_x/\varpi_x\mathcal{O}_x$ is the residue field of $\mathcal{O}_x$ and the image of $\varpi_x^{-1}cu \in F \cap \varpi_x^{-1}c\mathcal{O}$ under the last map is $u_x + \varpi_x\mathcal{O}_x \in \kappa(x)$. Therefore we have

$$|\mathfrak{g}(F) \cap (at_\infty \mathfrak{g}(\mathcal{O})t_\infty^{-1})| = \prod_{1\leq i,j\leq d} |F \cap (at_{\infty,i}t_{\infty,j}^{-1}\mathcal{O})| \leq p^{S(t_\infty)}$$

for every $t_\infty \in T(F_\infty)$, where we have set

$$S(t_\infty) = f \sum_{\substack{1\leq i,j\leq d \\ (\epsilon_i-\epsilon_j)(H_\emptyset(t_\infty))\geq \deg(a)}} (\delta_X + (\epsilon_i - \epsilon_j)(H_\emptyset(t_\infty)) - \deg(a)).$$

The expression $S(t_\infty)$ is equal to

$$S'(t_\infty) + \frac{d(d-1)}{2} f(\delta_X - \deg(a)) + S''(t_\infty) + \log_p \delta_\emptyset(t_\infty)$$

where we have set

$$S'(t_\infty) = f \sum_{\substack{1\leq j<i\leq d \\ (\epsilon_i-\epsilon_j)(H_\emptyset(t_\infty))>-\deg(a)}} (-\delta_X + (\epsilon_i - \epsilon_j)(H_\emptyset(t_\infty)) + \deg(a))$$

and

$$S''(t_\infty) = f \sum_{\substack{1\leq j<i\leq d \\ (\epsilon_i-\epsilon_j)(H_\emptyset(t_\infty))\geq \deg(a)}} (\delta_X + (\epsilon_i - \epsilon_j)(H_\emptyset(t_\infty)) - \deg(a)).$$

Now if $1 \leq j < i \leq d$ we have

$$(\epsilon_i - \epsilon_j)(H_\emptyset(t_\infty)) \leq (d-1)(2g_X + \epsilon_0)$$

for every $t_\infty \in T(F_\infty)$ with $d(H_\emptyset(t_\infty)) \geq -2g_X - \epsilon_0$, so that

$$S'(t_\infty) \leq S'(a) \qquad (\text{resp. } S''(t_\infty) \leq S''(a))$$

for every $t_\infty \in T(F_\infty)$ with $d(H_\emptyset(t_\infty)) \geq -2g_X - \epsilon_0$, where we have set

$$S'(a) = \frac{d(d-1)}{2} f(-\delta_X + (d-1)(2g_X + \epsilon_0) + \deg(a))$$

(resp.

$$S''(a) = \frac{d(d-1)}{2} f(\delta_X + (d-1)(2g_X + \epsilon_0) - \deg(a)))$$

if $\deg(a) > -(d-1)(2g_X + \epsilon_0)$ (resp. if $\deg(a) \leq (d-1)(2g_X + \epsilon_0)$) and $S'(a) = 0$ (resp. $S''(a) = 0$) otherwise. It follows that

$$|\mathfrak{g}(F) \cap (at_\infty \mathfrak{g}(\mathcal{O})t_\infty^{-1})| \leq C(a)\delta_\emptyset(t_\infty) \tag{10.9.1}$$

for every $t_\infty \in T(F_\infty)$ with $d(H_\emptyset(t_\infty)) \geq -2g_X - \epsilon_0$, where we have set

$$C(a) = p^{S'(a) + \frac{d(d-1)}{2} f(\delta_X - \deg(a)) + S''(a)}$$

and the proof of (10.8.5)(i) is completed. □

REMARK (10.9.2). — Let $I \subset \Delta$ and let $\mathfrak{m}_I$ be the Lie algebra of M_I. Let us apply (10.9.1) to $G = GL_{d_1}, \ldots, GL_{d_s}$ where $d_I = (d_1, \ldots, d_s)$ is the partition of d corresponding to I. Then we get a positive real number $C_I(a)$ $(= C_{d_1}(a) \cdots C_{d_s}(a))$ such that

$$|\mathfrak{m}_I(F) \cap (at_\infty \mathfrak{m}_I(\mathcal{O})t_\infty^{-1})| \leq C_I(a)\delta_\emptyset^I(t_\infty)$$

for every $t_\infty \in T(F_\infty)$ with $\alpha_i(H_\emptyset(t_\infty)) \geq -2g_X - \epsilon_0$, $\forall i \in I$. □

We will deduce (10.8.5)(ii) from

LEMMA (10.9.3). — *Let $a \in \mathbb{A}^\times$, let $q \in H^0(\mathfrak{g}, \mathcal{O}_\mathfrak{g})$ and let $x \neq \infty$ be a place of F. We assume that there exists a rational character of T over F,*

$$\chi : T \to \mathbb{G}_{m,F},$$

such that

$$q(tgt^{-1}) = \chi(t)q(g) \qquad (\forall t \in T,\ g \in \mathfrak{g}).$$

Then there exist positive real numbers $A'(a,q,x)$ and $B'(a,q,x)$ such that the number of elements in the set

$$\{g \in \mathfrak{g}(F) \cap (at_\infty \mathfrak{g}(\mathcal{O})t_\infty^{-1}) \mid q(g) \neq 0 \text{ and } x(q(g)) \geq N\}$$

is bounded above by

$$A'(a,q,x)p^{-B'(a,q,x)N}\delta_\emptyset(t_\infty)$$

for every non-negative integer N and every $t_\infty \in T(F_\infty)$ such that $d(H_\emptyset(t_\infty)) \geq -2g_X - \epsilon_0$.

Proof of part (ii) *of lemma* (10.8.5) (*assuming* (10.9.3)) : Firstly let us remark that we may replace the system of generators $\{q_1, \ldots, q_n\}$ of the ideal $\mathcal{I}_V \subset H^0(G, \mathcal{O}_G)$ defining V by any other one $\{q'_1, \ldots, q'_{n'}\}$. Indeed there are polynomials $r_{\nu',\nu} \in H^0(G, \mathcal{O}_G)$, $\nu = 1, \ldots, n$, $\nu' = 1, \ldots, n'$, such that

$$q'_{\nu'} = \sum_{\nu=1}^{n} r_{\nu',\nu} q_\nu \qquad (\forall \nu' = 1, \ldots, n')$$

and, if we set

$$N_0 = \inf\{x(r_{\nu',\nu}(g_x)) \mid F_\infty^\times g = F_\infty^\times g^x g_x \in \Sigma,\ \nu = 1, \ldots, n,\ \nu' = 1, \ldots, n'\}$$

and

$$\Sigma'_{V,x,N+N_0}$$
$$= \{g = g^x g_x \in G(\mathbb{A}) \mid F_\infty^\times g \in \Sigma \text{ and } x(q'_{\nu'}(g_x)) \geq N + N_0,\ \forall \nu' = 1, \ldots, n'\},$$

we have

$$\Sigma_{V,x,N} \subset \Sigma'_{V,x,N+N_0}$$

for every non-negative integer N. In particular we may assume that $\{q_1, \ldots, q_n\}$ is a system of generators of the ideal

$$\mathcal{J}_V = H^0(\mathfrak{g}, \mathcal{O}_{\mathfrak{g}}) \cap \mathcal{I}_V$$

of $H^0(\mathfrak{g}, \mathcal{O}_{\mathfrak{g}})$.

Secondly, for each rational character χ of T over F let

$$H^0(\mathfrak{g}, \mathcal{O}_{\mathfrak{g}})_\chi = \{q \in H^0(\mathfrak{g}, \mathcal{O}_{\mathfrak{g}}) \mid q(tgt^{-1}) = \chi(t)q(g), \ \forall t \in T, \ \forall g \in \mathfrak{g}\}$$

be the corresponding eigenspace. We have

$$H^0(\mathfrak{g}, \mathcal{O}_{\mathfrak{g}}) = \bigoplus_{\chi \in X^*(T)} H^0(\mathfrak{g}, \mathcal{O}_{\mathfrak{g}})_\chi$$

where $X^*(T)$ is the group of rational characters of T over F. As

$$q(tgt^{-1}) \in \mathcal{J}_V \qquad (\forall q \in \mathcal{J}_V, \ \forall t \in T),$$

we also have

$$\mathcal{J}_V = \bigoplus_{\chi \in X^*(T)} (H^0(\mathfrak{g}, \mathcal{O}_{\mathfrak{g}})_\chi \cap \mathcal{J}_V).$$

Therefore we may assume that, for each $\nu = 1, \ldots, n$, there exists $\chi_\nu \in X^*(T)$ such that

$$q_\nu(tgt^{-1}) = \chi_\nu(t)q_\nu(g) \qquad (\forall t \in T, \ \forall g \in G).$$

Thirdly we can find $a \in \mathbb{A}^\times$ such that

$$\{g \in G(\mathbb{A}) \mid F_\infty^\times g \in \Sigma \text{ and } \deg(\det g) = 0\} \subset a\mathfrak{g}(\mathcal{O})$$

and, if $\mathcal{V}$ is the closed subscheme of $\mathfrak{g}$ defined by the ideal $\mathcal{J}_V$ of $H^0(\mathfrak{g}, \mathcal{O}_{\mathfrak{g}})$, the set

$$(G(F) - V(F)) \cap (t_\infty \Sigma_{V,x,N} t_\infty^{-1})$$

is contained in the set

$$\{g \in (\mathfrak{g}(F) - \mathcal{V}(F)) \cap (at_\infty \mathfrak{g}(\mathcal{O}) t_\infty^{-1}) \mid x(q_\nu(g)) \geq N, \ \forall \nu = 1, \ldots, n\}$$

for every non-negative integer N and every $t_\infty \in T(F_\infty)$.

But this last set is contained in the union over $\nu = 1, \ldots, n$ of the sets

$$\{g \in \mathfrak{g}(F) \cap (at_\infty \mathfrak{g}(\mathcal{O}) t_\infty^{-1}) \mid q_\nu(g) \neq 0 \text{ and } x(q_\nu(g)) \geq N\}$$

for every non-negative integer N and every $t_\infty \in T(F_\infty)$. Therefore, to conclude the proof it is sufficient to apply (10.9.2) to $q_1, \ldots, q_n$ and to take

$$A'(\Sigma, V, x) = \sup\{A'(a, q_\nu, x) \mid \nu = 1, \ldots, n\}$$

and

$$B'(\Sigma, V, x) = \inf\{B'(a, q_\nu, x) \mid \nu = 1, \ldots, n\}.$$

□

In order to prove lemma (10.9.3) it is more convenient to generalize it a little bit.

LEMMA (10.9.4). — *Let $I \subset \Delta$ and let $\mathfrak{m}_I$ be the Lie algebra of M_I. Let $a \in \mathbb{A}^\times$, let $q \in H^0(\mathfrak{m}_I, \mathcal{O}_{\mathfrak{m}_I})$ and let $x \neq \infty$ be a place of F. We assume that there exists $\chi \in X^*(T)$ such that*

$$q(tm_It^{-1}) = \chi(t)q(m_I) \qquad (\forall t \in T,\ m_I \in \mathfrak{m}_I).$$

Then there exist positive real numbers $A'_I(a,q,x)$ and $B'_I(a,q,x)$ such that the number of elements in the set

$$\{m_I \in \mathfrak{m}_I(F) \cap (at_\infty \mathfrak{m}_I(\mathcal{O})t_\infty^{-1}) \mid q(m_I) \neq 0 \text{ and } x(q(m_I)) \geq N\}$$

is bounded above by

$$A'_I(a,q,x)p^{-B'_I(a,q,x)N}\delta_\emptyset^I(t_\infty)$$

for every non-negative integer N and every $t_\infty \in T(F_\infty)$ with $\alpha_i(H_\emptyset(t_\infty)) \geq -2g_X - \epsilon_0$, $\forall i \in I$ (recall that $\delta_\emptyset^I = \delta_\emptyset/\delta_I$ is the modulus character of $P_\emptyset^I(\mathbb{A})$).

Obviously lemma (10.9.3) is the particular case $I = \Delta$ of (10.9.4).

The proof of lemma (10.9.4), which is due to Arthur (see [Ar 2] § 6), is quite long and tricky. It is done by induction on $|I|$. Let us start with the particular case $I = \emptyset$.

Proof of (10.9.4) *for* $I = \emptyset$: In this case

$$|\mathfrak{m}_\emptyset(F) \cap a\mathfrak{m}_\emptyset(\mathcal{O})| = |F \cap a\mathcal{O}|^d$$

is finite and there exists a non-negative integer N_0 such that

$$x(q(m_\emptyset)) \leq N_0$$

for all $m_\emptyset \in \mathfrak{m}_\emptyset(F) \cap a\mathfrak{m}_\emptyset(\mathcal{O})$ such that $q(m_\emptyset) \neq 0$. Then we can simply take

$$A'_\emptyset(a,q,x) = |F \cap a\mathcal{O}|^d p^{N_0}$$

and

$$B'_\emptyset(a,q,x) = 1.$$

□

From now on we assume that $I \neq \emptyset$. The case $q \equiv 0$ being obvious, we may assume that $q \not\equiv 0$. Then we have

$$\chi(Z_I) = \{1\}$$

and we can write

$$\chi(t) = \prod_{i \in I} \Big(\frac{t_i}{t_{i+1}}\Big)^{\chi_i} \qquad (\forall t \in T)$$

for some uniquely determined family of integers $(\chi_i)_{i \in I}$. Let us set

$$I_{>0} = \{i \in I \mid \chi_i > 0\}$$

and

$$I_{\leq 0} = \{i \in I \mid \chi_i \leq 0\}.$$

Proof of (10.9.4) *if $I_{>0}$ is empty*: Let N be a non-negative integer and let $t_\infty \in T(F_\infty)$ with $\alpha_i(H_\emptyset(t_\infty)) \geq -2g_X - \epsilon_0$, $\forall i \in I$, be such that the set

$$\{m_I \in \mathfrak{m}_I(F) \cap (at_\infty \mathfrak{m}_I(\mathcal{O}) t_\infty^{-1}) \mid q(m_I) \neq 0 \text{ and } x(q(m_I)) \geq N\}$$

is not empty. Then without any assumption on $I_{>0}$ there exists a real number C_1 such that

$$\sum_{i \in I_{>0}} \chi_i \alpha_i(H_\emptyset(t_\infty)) \geq C_1 + \frac{\deg(x)}{f} N. \tag{10.9.5}$$

Indeed let $m_I \in \mathfrak{m}_I(\mathcal{O})$ be such that $at_\infty m_I t_\infty^{-1}$ belongs to the above set; then we have

$$\deg(q(at_\infty m_I t_\infty^{-1})) = 0,$$

i.e.

$$\begin{aligned}\sum_{i \in I_{>0}} \chi_i \alpha_i(H_\emptyset(t_\infty)) = &\sum_{i \in I_{\leq 0}} (-\chi_i)\alpha_i(H_\emptyset(t_\infty)) \\ &+ \deg^x(q(a^x m_I^x)) \\ &+ \frac{\deg(x)}{f} x(q(a_x m_{I,x})),\end{aligned}$$

and it is sufficient to take

$$C_1 = \Big(\sum_{i \in I_{\leq 0}} \chi_i (2g_X + \epsilon_0)\Big) + N^x$$

where N^x is any integer such that

$$\deg^x(q(a^x m_I^x)) \geq N^x \qquad (\forall m_I^x \in \mathfrak{m}_I(\mathcal{O}^x)).$$

Now, if $I_{>0}$ is empty, either

$$N > -fC_1/\deg(x)$$

and the set

$$\{m_I \in \mathfrak{m}_I(F) \cap (at_\infty \mathfrak{m}_I(\mathcal{O})t_\infty^{-1}) \mid q(m_I) \neq 0 \text{ and } x(q(m_I)) \geq N\}$$

is empty, or

$$N \leq -fC_1/\deg(x)$$

and an upper bound of the number of elements in that set is given by

$$|\mathfrak{m}_I(F) \cap (at_\infty \mathfrak{m}_I(\mathcal{O})t_\infty^{-1})|$$

and then by

$$C_I(a)\delta_\emptyset^I(t_\infty)$$

for some positive real number $C_I(a)$ (see (10.9.2)). Therefore, if $I_{>0} = \emptyset$ we can simply take

$$A'_I(a,q,x) = C_I(a)p^{-fC_1/\deg(x)}$$

and

$$B'_I(a,q,x) = 1$$

□

From now on we assume that $I_{>0} \neq \emptyset$.

Proof of (10.9.4): *reductions* : We fix C_1 as in inequality (10.9.5). It is sufficient to prove (10.9.4) for any $t_\infty \in T(F_\infty)$ with $\alpha_i(H_\emptyset(t_\infty)) \geq -2g_X - \epsilon_0$, $\forall i \in I$, and any non-negative integer N satisfying (10.9.5).

Moreover, if N'_0 is any real number it is sufficient to prove (10.9.4) for all non-negative integers N such that $N > N'_0$ (if it is necessary replace $A'_I(a,q,x)$ by

$$\sup\{A'_I(a,q,x), C_I(a)p^{B'_I(a,q,x)N'_0}\}$$

with the notations of (10.9.2)). □

Now, if $t_\infty \in T(F_\infty)$ and $N \in \mathbb{Z}_{\geq 0}$ satisfy inequality (10.9.5), there exists at least one $i_0 \in I_{>0}$ such that

$$\alpha_{i_0}(H_\emptyset(t_\infty)) \geq \frac{1}{\left(\sum_{i \in I_{>0}} \chi_i\right)}(C_1 + \frac{\deg(x)}{f}N). \tag{10.9.6}$$

Let us fix such an $i_0 \in I_{>0}$, let us set

$$J = I - \{i_0\},$$

let us denote by $\mathfrak{p}_J^I$ the Lie algebra of the maximal parabolic subgroup P_J^I of M_I and let us set

$$N_0 = \frac{f}{\deg(x)}\Big[\Big(\sum_{i\in I_{>0}} \chi_i\Big)((d-2)(2g_X+\epsilon_0) - \deg(a)) - C_1\Big].$$

Then, for any $t_\infty \in T(F_\infty)$ with $\alpha_i(H_\emptyset(t_\infty)) \geq -2g_X - \epsilon_0$, $\forall i \in J$, and any non-negative integer $N > N_0$ which satisfy inequality (10.9.6), we have

$$F \cap (at_{\infty,j}t_{\infty,k}^{-1}\mathcal{O}) = \{0\}$$

for any root $\epsilon_j - \epsilon_k$ of (M_I, T) such that $k \leq i_0 < j$ and consequently we have

$$\mathfrak{m}_I(F) \cap (at_\infty \mathfrak{m}_I(\mathcal{O})t_\infty^{-1}) = \mathfrak{p}_J^I(F) \cap (at_\infty \mathfrak{p}_J^I(\mathcal{O})t_\infty^{-1}).$$

Let us consider the canonical splitting $\mathfrak{p}_J^I = \mathfrak{m}_J \oplus \mathfrak{n}_J^I$. For each $m_J \in \mathfrak{m}_J(F)$ let

$$q_{m_J} \in H^0(\mathfrak{n}_J^I, \mathcal{O}_{\mathfrak{n}_J^I})$$

be defined by

$$q_{m_J}(n_J^I) = q(m_J + n_J^I) \qquad (\forall n_J^I \in \mathfrak{n}_J^I).$$

Let us set

$$\mathfrak{m}_J(F)^\circ = \{m_J \in \mathfrak{m}_J(F) \mid q_{m_J} \not\equiv 0\}$$

and for each $m_J \in \mathfrak{m}_J(F)^\circ$ and each non-negative integer N let us set

$$\omega_{m_J,N} = \{n_J^I \in a\mathfrak{n}_J^I(\mathcal{O}) \mid x(q_{m_J}(n_J^I)) \geq N\}$$

and

$$\omega_{m_J,N}^\circ = \{n_J^I \in \omega_{m_J,N} \mid q_{m_J}(n_J^I) \neq 0\} \subset \omega_{m_J,N}.$$

Then for every $t_\infty \in T(F_\infty)$ and every $N \in \mathbb{Z}_{\geq 0}$ the number of elements in

$$\{p_J^I \in \mathfrak{p}_J^I(F) \cap (at_\infty \mathfrak{p}_J^I(\mathcal{O})t_\infty^{-1}) \mid q(p_J^I) \neq 0 \text{ and } x(q(p_J^I)) \geq N\}$$

is equal to

$$\sum_{m_J \in \mathfrak{m}_J(F)^\circ} 1_{a\mathfrak{m}_J(\mathcal{O})}(t_\infty^{-1} m_J t_\infty) \sum_{n_J^I \in \mathfrak{n}_J^I(F)} 1_{\omega_{m_J,N}^\circ}(t_\infty^{-1} n_J^I t_\infty)$$

and is thus bounded above by

$$(10.9.7)_N \qquad \sum_{m_J \in \mathfrak{m}_J(F)^\circ} 1_{a\mathfrak{m}_J(\mathcal{O})}(t_\infty^{-1} m_J t_\infty) \sum_{n_J^I \in \mathfrak{n}_J^I(F)} 1_{\omega_{m_J,N}}(t_\infty^{-1} n_J^I t_\infty).$$

Summarizing the above discussion we get that, for every $t_\infty \in T(F_\infty)$ with $\alpha_i(H_\emptyset(t_\infty)) \geq -2g_X - \epsilon_0$, $\forall i \in J$, and every non-negative integer $N > N_0$ which satisfy inequality (10.9.6), the number of elements in the set

$$\{m_I \in \mathfrak{m}_I(F) \cap (at_\infty \mathfrak{m}_I(\mathcal{O})t_\infty^{-1}) \mid q(m_I) \neq 0 \text{ and } x(q(m_I)) \geq N\}$$

is bounded above by $(10.9.7)_N$.

Proof of (10.9.4): *Arthur's trick* : The sets $\omega_{m_J,N}$, $m_J \in \mathfrak{m}_J(F)^\circ$, and the expression $(10.9.7)_N$ make sense for an arbitrary non-negative real number N and we have

$$\omega_{m_J,N} \subset \omega_{m_J,N'}$$

and

$$(10.9.7)_N \leq (10.9.7)_{N'}$$

for any real numbers N, N' such that $0 \leq N' \leq N$. In particular, if ϵ is an arbitrarily fixed real number with $0 < \epsilon \leq 1$, we have

$$(10.9.7)_N \leq (10.9.7)_{\epsilon N}$$

for every non-negative integer N. □

In order to estimate $(10.9.7)_{\epsilon N}$ we will use the Poisson summation formula. For each $m_J \in \mathfrak{m}_J(F)^\circ$, each $\epsilon \in]0,1]$ and each $N \in \mathbb{Z}_{\geq 0}$ let

$$\widehat{1}_{\omega_{m_J,\epsilon N}}(Y_J^I) = \int_{\mathfrak{n}_J^I(\mathbb{A})} 1_{\omega_{m_J,\epsilon N}}(X_J^I)\Psi(\mathrm{tr}({}^tX_J^I Y_J^I))dX_J^I \qquad (Y_J^I \in \mathfrak{n}_J^I(\mathbb{A}))$$

be the Fourier transform of the characteristic function of $\omega_{m_J,\epsilon N}$ (see (10.4.7)). Then for every $t_\infty \in T(F_\infty)$ we have

$$\sum_{n_J^I \in \mathfrak{n}_J^I(F)} 1_{\omega_{m_J,\epsilon N}}(t_\infty^{-1} n_J^I t_\infty) = (V_J^I)^{-1}\delta_J^I(t_\infty) \sum_{n_J^I \in \mathfrak{n}_J^I(F)} \widehat{1}_{\omega_{m_J,\epsilon N}}(t_\infty n_J^I t_\infty^{-1})$$

where we have set

$$V_J^I = \mathrm{vol}(\mathfrak{n}_J^I(F)\backslash \mathfrak{n}_J^I(\mathbb{A}), dX_J^I)$$

and

$$\delta_J^I = \delta_J/\delta_I$$

(see (10.4.8) and (10.4.12)).

Proof of (10.9.4): *an upper bound of* $(10.9.7)_{\epsilon N}$: Obviously, for each $m_J \in \mathfrak{m}_J(F)^\circ$, each $\epsilon \in]0,1]$ and each $N \in \mathbb{Z}_{\geq 0}$ we have

$$|\widehat{1}_{\omega_{m_J,\epsilon N}}(Y_J^I)| \leq \mathrm{vol}(\omega_{m_J,\epsilon N}, dX_J^I) \qquad (\forall Y_J^I \in \mathfrak{n}_J^I(\mathbb{A})).$$

Moreover, if $b_{m_J,\epsilon N} \in \mathbb{A}^\times$ is any idele such that

$$\omega_{m_J,\epsilon N} + b_{m_J,\epsilon N}\mathfrak{n}_J^I(\mathcal{O}) = \omega_{m_J,\epsilon N},$$

we have

$$\mathrm{Supp}(\widehat{1}_{\omega_{m_J,\epsilon N}}) \subset c_\Psi^{-1} b_{m_J,\epsilon N}^{-1} \mathfrak{n}_J^I(\mathcal{O})$$

where $c_\Psi \in \mathbb{A}^\times$ is any idele such that

$$c_\Psi^{-1}\mathcal{O} = \{v \in \mathbb{A} \mid \Psi(uv) = 1,\ \forall u \in \mathcal{O}\}$$

(see [We 2] (Ch. VII, § 2)). But we can find a non-negative integer N_1 having the following property: for every $p_{J,x}^I \in \mathfrak{p}_J^I(\mathcal{O}_x)$ the polynomial

$$q(a_x(p_{J,x}^I + \varpi_x^{N_1} n_J^I)) - q(a_x p_{J,x}^I)$$

in the entries of $n_J^I \in \mathfrak{n}_J^I$ has all its coefficients in $\mathcal{O}_x$ (and its constant term is zero). Therefore, for each $m_J \in \mathfrak{m}_J(F)^\circ$ such that $m_{J,x} \in a_x\mathfrak{m}_J(\mathcal{O}_x)$, each $\epsilon \in]0,1]$ and each $N \in \mathbb{Z}_{\geq 0}$ we can take

$$b_{m_J,\epsilon N} = b_{\epsilon N} \overset{\text{dfn}}{=\!=} a\varpi_x^{N_1+[\epsilon N]+1}.$$

It follows that, for each $t_\infty \in T(F_\infty)$, each $\epsilon \in]0,1]$ and each $N \in \mathbb{Z}_{\geq 0}$, an upper bound of $(10.9.7)_{\epsilon N}$ is given by

$$(V_J^I)^{-1}\delta_J^I(t_\infty) \sum_{m_J \in \mathfrak{m}_J(F)^\circ} 1_{a\mathfrak{m}_J(\mathcal{O})}(t_\infty^{-1} m_J t_\infty)\, \mathrm{vol}(\omega_{m_J,\epsilon N}, dX_J^I)$$
$$\times |\mathfrak{n}_J^I(F) \cap (c_\Psi^{-1} b_{\epsilon N}^{-1} t_\infty^{-1} \mathfrak{n}_J^I(\mathcal{O}) t_\infty)|$$

and, consequently, by the sum of

$$(10.9.8)_{\epsilon N} \quad (V_J^I)^{-1}\delta_J^I(t_\infty)\, \mathrm{vol}(a\mathfrak{n}_J^I(\mathcal{O}), dX_J^I) \sum_{m_J \in \mathfrak{m}_J(F)^\circ} 1_{a\mathfrak{m}_J(\mathcal{O})}(t_\infty^{-1} m_J t_\infty)$$
$$\times (|\mathfrak{n}_J^I(F) \cap (c_\Psi^{-1} b_{\epsilon N}^{-1} t_\infty^{-1} \mathfrak{n}_J^I(\mathcal{O}) t_\infty)| - 1)$$

and

$$(10.9.9)_{\epsilon N} \quad (V_J^I)^{-1}\delta_J^I(t_\infty) \sum_{m_J \in \mathfrak{m}_J(F)^\circ} 1_{a\mathfrak{m}_J(\mathcal{O})}(t_\infty^{-1} m_J t_\infty)\, \mathrm{vol}(\omega_{m_J,\epsilon N}, dX_J^I)$$

(note that $|\mathfrak{n}_J^I(F) \cap (c_\Psi^{-1} b_{\epsilon N}^{-1} t_\infty^{-1} \mathfrak{n}_J^I(\mathcal{O}) t_\infty)| \geq 1$ and recall that $\omega_{m_J,\epsilon N} \subset a\mathfrak{n}_J^I(\mathcal{O})$). □

End of the proof of (10.9.4). *First step: vanishing of* $(10.9.8)_{\epsilon N}$ *for an adequate choice of* ϵ : Let us take

$$\epsilon = \Big(1 + \sum_{i \in I_{>0}} \chi_i\Big)^{-1}.$$

Then we can find a real number $N_0' \geq N_0$ such that

$$\frac{C_1 + \deg(x)N/f}{\sum_{i\in I_{>0}} \chi_i} - (d-2)(2g_X + \epsilon_0) > \deg(c_\Psi) + \deg(b_{\epsilon N})$$

for all non-negative integers $N > N_0'$ (we have

$$\deg(b_{\epsilon N}) = \deg(a) + \frac{\deg(x)}{f}(N_1 + [\epsilon N] + 1)).$$

Having fixed such an N_0' we obtain the inequality

$$(\epsilon_i - \epsilon_j)(H_\emptyset(t_\infty)) \geq \alpha_{i_0}(H_\emptyset(t_\infty)) - (d-2)(2g_X + \epsilon_0) > \deg(c_\Psi) + \deg(b_{\epsilon N})$$

for each root $\epsilon_i - \epsilon_j$ of (G,T) which occurs in N_J^I, each $t_\infty \in T(F_\infty)$ with $\alpha_i(H_\emptyset(t_\infty)) \geq -2g_X - \epsilon_0$, $\forall i \in J$, and each non-negative integer $N > N_0'$ satisfying inequality (10.9.6). Indeed any such root $\epsilon_i - \epsilon_j$ is the sum of α_{i_0} and at most $d-2$ simple roots in $J = I - \{i_0\}$.

It follows that

$$F \cap (c_\Psi^{-1} b_{\epsilon N}^{-1} t_{\infty,i}^{-1} t_{\infty,j} \mathcal{O}) = \{0\}$$

for all $\epsilon_i - \epsilon_j$, t_∞ and N as above. Therefore we have proved that

$$\mathfrak{n}_J^I(F) \cap (c_\Psi^{-1} b_{\epsilon N}^{-1} t_\infty^{-1} \mathfrak{n}_J^I(\mathcal{O}) t_\infty) = \{0\}$$

and that $(10.9.8)_{\epsilon N}$ vanishes for any $t_\infty \in T(F_\infty)$ with $\alpha_i(H_\emptyset(t_\infty)) \geq -2g_X - \epsilon_0$, $\forall i \in J$, and any non-negative integer $N > N_0'$ satisfying inequality (10.9.6).

Second step: an upper bound for $(10.9.9)_{\epsilon N}$ *using the induction hypothesis*: Let us choose some total order $(\beta_1, \ldots, \beta_\ell)$ on the set of roots $\beta = \epsilon_i - \epsilon_j$ of (G,T) which occur in N_J^I. It induces a total order $(n_1, \ldots, n_\ell)$ on the set of (non-trivial) entries of $n_J^I \in \mathfrak{n}_J^I$ such that, for any $t \in T$, the (non-trivial) entries of $tn_J^I t^{-1}$ are

$$(t^{\beta_1} n_1, \ldots, t^{\beta_\ell} n_\ell)$$

where we have set

$$t^{\epsilon_i - \epsilon_j} = t_i / t_j.$$

Then for any $m_J \in \mathfrak{m}_J$ and any $n_J^I \in \mathfrak{n}_J^I$ we can write

$$q(m_J + n_J^I) = q_{m_J}(n_J^I) = \sum_{\mu \in (\mathbb{Z}_{\geq 0})^\ell} q_\mu(m_J) n^\mu,$$

where we have set

$$n^\mu = n_1^{\mu_1} \ldots n_\ell^{\mu_\ell}.$$

For each $\mu \in (\mathbb{Z}_{\geq 0})^{\ell}$ we clearly have

$$q_{\mu}(tm_J t^{-1}) = \chi_{\mu}(t) q_{\mu}(m_J) \qquad (\forall t \in T,\ \forall m_J \in \mathfrak{m}_J)$$

where we have set

$$\chi_{\mu}(t) = \chi(t) t^{-(\mu_1\beta_1+\cdots+\mu_\ell\beta_\ell)}.$$

Let $\overline{F}_x$ be an algebraic closure of F_x. The valuation x of F_x can be extended in a unique way to a valuation of $\overline{F}_x$ that we still denote by x (see [Se 1] (Ch. II, § 2)). For each

$$r = \sum_{\mu\in(\mathbb{Z}_{\geq 0})^{\ell}} r_{\mu} n^{\mu} \in \overline{F}_x[n_1, \ldots, n_\ell]$$

let us set

$$v_x(r) = \inf\{x(r_\mu) \mid \mu \in (\mathbb{Z}_{\geq 0})^{\ell}\} \in \mathbb{Q} \cup \{+\infty\}.$$

For each non-negative integer δ and each real number M let

$$\mathcal{R}_x^{\delta}(M) \subset \overline{F}_x[n_1, \ldots, n_\ell]$$

be the set of polynomials $r \in \overline{F}_x[n_1, \ldots, n_\ell]$ of total degree at most δ such that

$$v_x(r) < M.$$

Let us fix Haar measures $dn_{x,1}, \ldots, dn_{x,\ell}$ on F_x and let

$$dX_{J,x}^{I} = dn_x = dn_{x,1} \cdots dn_{x,\ell}$$

be the product Haar measure on $\mathfrak{n}_J^I(F_x) \cong F_x^{\ell}$.

Claim. — *For any $a_x \in F_x^{\times}$ and any non-negative integer δ there exists a positive real number $C(a_x, \delta)$ having the following property: for all real numbers $N' \geq M \geq 0$ and every $r \in \mathcal{R}_x^{\delta}(M)$ we have*

$$\mathrm{vol}(\{n_x \in (a_x \mathcal{O}_x)^{\ell} \mid x(r(n_x)) \geq N'\}, dn_x) \leq C(a_x, \delta) p^{\deg(x)(M-N')/2^{\delta+\ell-1}}.$$

We will prove the claim at the end of the section. For the moment let us assume it and let us continue the second step of the proof of (10.9.4).

For each $t_\infty \in T(F_\infty)$, each $\epsilon \in]0,1]$ and each $N \in \mathbb{Z}_{\geq 0}$ we can estimate from above the expression

$$\sum_{m_J\in\mathfrak{m}_J(F)^{\circ}} 1_{a\mathfrak{m}_J(\mathcal{O})}(t_\infty^{-1} m_J t_\infty) \,\mathrm{vol}(\omega_{m_J,\epsilon N}, dX_J^I)$$

by the sum of the expressions

$$(*) \qquad \sum_{\substack{m_J \in \mathfrak{m}_J(F) \\ v_x(q_{m_J}) < [\epsilon N/2]}} 1_{\mathfrak{a}\mathfrak{m}_J(\mathcal{O})}(t_\infty^{-1} m_J t_\infty) \operatorname{vol}(\omega_{m_J,\epsilon N}, dX_J^I)$$

and

$$(**) \qquad \operatorname{vol}(\mathfrak{a}\mathfrak{n}_J^I(\mathcal{O}), dX_J^I) \sum_{\mu \in (\mathbb{Z}_{\geq 0})^\ell} \sum_{\substack{m_J \in \mathfrak{m}_J(F) \\ q_\mu(m_J) \neq 0 \\ x(q_\mu(m_J)) \geq [\epsilon N/2]}} 1_{\mathfrak{a}\mathfrak{m}_J(\mathcal{O})}(t_\infty^{-1} m_J t_\infty)$$

(recall that $\omega_{m_J,\epsilon N} \subset \mathfrak{a}\mathfrak{n}_J^I(\mathcal{O})$).

For each $m_J \in \mathfrak{m}_J(F)$ the total degree of q_{m_J} is bounded above by the total degree $\mathrm{d}^\circ(q)$ of $q \in H^0(\mathfrak{m}_I, \mathcal{O}_{\mathfrak{m}_I})$. Therefore, on the one hand, using part (i) of (10.8.5) which is already proved (or the induction hypothesis), and the claim, we obtain that, expression $(*)$ is bounded above by

$$C_J(\mathfrak{a}\mathfrak{m}_J(\mathcal{O}))\delta_\emptyset^J(t_\infty) \operatorname{vol}(a^x \mathfrak{n}_J^I(\mathcal{O}^x), dX_J^{I,x}) C(a_x, \mathrm{d}^\circ(q)) \\ p^{\deg(x)([\epsilon N/2] - \epsilon N)/2^{\mathrm{d}^\circ(q)+\ell-1}}.$$

Here $dX_J^{I,x}$ is defined by the splitting $dX_J^I = dX_J^{I,x} dX_{J,x}^I$. On the other hand, using the induction hypothesis we obtain that the expression $(**)$ is bounded above by

$$\operatorname{vol}(\mathfrak{a}\mathfrak{n}_J^I(\mathcal{O}), dX_J^I) \sum_{\substack{\mu \in (\mathbb{Z}_{\geq 0})^\ell \\ \mu_1 + \cdots + \mu_\ell \leq \mathrm{d}^\circ(q)}} A_J'(a, q_\mu, x) p^{-B_J'(a, q_\mu, x)[\epsilon N/2]} \delta_\emptyset^J(t_\infty).$$

Consequently, expression $(10.9.8)_{\epsilon N}$ is bounded above by

$$(V_J^I)^{-1} C_J(\mathfrak{a}\mathfrak{m}_J(\mathcal{O})) \operatorname{vol}(a^x \mathfrak{n}_J^I(\mathcal{O}^x), dX_J^{I,x}) C(a_x, \mathrm{d}^\circ(q)) \\ \times p^{\deg(x)([\epsilon N/2] - \epsilon N)/2^{\mathrm{d}^\circ(q)+\ell-1}} \delta_\emptyset^I(t_\infty) \\ + (V_J^I)^{-1} \operatorname{vol}(\mathfrak{a}\mathfrak{n}_J^I(\mathcal{O}), dX_J^I) \\ \sum_{\substack{\mu \in (\mathbb{Z}_{\geq 0})^\ell \\ \mu_1 + \cdots + \mu_\ell \leq \mathrm{d}^\circ(q)}} A_J'(a, q_\mu, x) p^{-B_J'(a, q_\mu, x)[\epsilon N/2]} \delta_\emptyset^I(t_\infty).$$

In particular, we can find positive real numbers $A''_{i_0,I}(a, q, x)$ and $B''_{i_0,I}(a, q, x)$ such that expression $(10.9.8)_{\epsilon N}$ is bounded above by

$$A''_{i_0,I}(a, q, x) p^{-B''_{i_0,I}(a,q,x)\epsilon N} \delta_\emptyset^I(t_\infty)$$

for every $t_\infty \in T(F_\infty)$ with $\alpha_i(H_\emptyset(t_\infty)) \geq -2g_X - \epsilon_0$, $\forall i \in J$, every $\epsilon \in]0, 1]$ and every $N \in \mathbb{Z}_{\geq 0}$.

Third step: conclusion : Let us choose ϵ and N_0' as in the first step, so that $(10.9.8)_{\epsilon N}$ vanishes for any $t_\infty \in T(F_\infty)$ with $\alpha_i(H_\emptyset(t_\infty)) \geq -2g_X - \epsilon_0$, $\forall i \in J$, and any non-negative integer $N > N_0'$ satisfying inequality (10.9.6).

Then $(10.9.7)_N$ is bounded above by

$$A''_{i_0,I}(a,q,x)p^{-B''_{i_0,I}(a,q,x)\epsilon N}\delta_\emptyset^I(t_\infty)$$

for every $t_\infty \in T(F_\infty)$ with $\alpha_i(H_\emptyset(t_\infty)) \geq -2g_X - \epsilon_0$, $\forall i \in J$, and every non-negative integer $N > N_0'$ satisfying inequality (10.9.6).

Therefore, if we set

$$A'_I(a,q,x) = \sup\{A''_{i_0,I}(a,q,x) \mid i_0 \in I_{>0}\}$$

and

$$B'_I(a,q,x) = \inf\{B''_{i_0,I}(a,q,x)\epsilon \mid i \in I_{>0}\},$$

the number of elements in the set

$$\{m_I \in \mathfrak{m}_I(F) \cap (at_\infty \mathfrak{m}_I(\mathcal{O})t_\infty^{-1}) \mid q(m_I) \neq 0 \text{ and } x(q(m_I)) \geq N\}$$

is bounded above by

$$A'_I(a,q,x)p^{-B'_I(a,q,x)N}\delta_\emptyset^I(t_\infty)$$

for every $t_\infty \in T(F_\infty)$ with $\alpha_i(H_\emptyset(t_\infty)) \geq -2g_X - \epsilon_0$, $\forall i \in I$, and every non-negative integer $N > N_0'$ which satisfy inequality (10.9.5). This concludes the proof of (10.9.4) and (10.9.3). □

Proof of the claim (see [Ar 5] (7.1)) : Let us suppose that the claim is proved for $a_x = 1$. Then the claim holds for an arbitrary a_x. Indeed, if $a_x \in \mathcal{O}_x$, we can simply take $C(a_x, \delta) = C(1,\delta)$ (remark that $(a_x\mathcal{O}_x)^\ell \subset \mathcal{O}_x^\ell$) and, if $x(a_x) < 0$, we have

$$v_x(r(a_x n_x)) \leq v_x(r) \qquad (\forall r \in \overline{F}_x[n_1,\ldots,n_\ell])$$

and $d(a_x n_x) = |a_x|^\ell dn_x$, so that we can take $C(a_x,\delta) = |a_x|^\ell C(1,\delta)$. Therefore we may assume that $a_x = 1$. It will be convenient to normalize the Haar measure $dn_x = dn_{x,1}\cdots dn_{x,\ell}$ on F_x^ℓ in such way that

$$\mathrm{vol}(\mathcal{O}_x, dn_{x,\lambda}) = 1 \qquad (\forall \lambda = 1,\ldots,\ell).$$

Let us begin the proof of the claim with the case $\ell = 1$. For $\delta = 0$ the claim is obvious (take $C(1,0) = 1$). Let us assume that $\delta \geq 1$ and let us assume inductively that the claim has been proved when δ is replaced by $\delta - 1$. Let $r \in \overline{F}_x[n]$ be of degree δ and let $\nu \in \overline{F}_x$ be a root of r. We can write

$$r = (n - \nu)r'$$

with $r' \in \overline{F}_x[n]$ of degree $\delta - 1$. It is easy to see that

$$v_x(r') \leq v_x(r) - \inf\{0, x(\nu)\}.$$

If we have $x(\nu) < 0$ we have

$$v_x(r') \leq v_x(r) - x(\nu)$$

and

$$x(r'(n_x)) = x(r(n_x)) - x(\nu) \qquad (\forall n_x \in \mathcal{O}_x),$$

so that the volume of

$$\{n_x \in \mathcal{O}_x \mid x(r(n_x)) \geq N'\}$$

is bounded above by

$$C(1, \delta - 1)p^{\deg(x)((M - x(\nu)) - (N' - x(\nu)))/2^{\delta - 1}} = C(1, \delta - 1)p^{\deg(x)(M - N')/2^{\delta - 1}}$$

as long as $v_x(r) < M$. Otherwise we have

$$v_x(r') \leq v_x(r)$$

and, for any $n_x \in \mathcal{O}_x$, either

$$x(r'(n_x)) \geq x(r(n_x)) - \frac{N' - M}{2}$$

or

$$x(n_x - \nu) > \frac{N' - M}{2},$$

so that the volume of

$$\{n_x \in \mathcal{O}_x \mid x(r(n_x)) \geq N'\}$$

is bounded above by the sum of

$$C(1, \delta - 1)p^{\deg(x)(M - (N' - \frac{N' - M}{2}))/2^{\delta - 1}} = C(1, \delta - 1)p^{\deg(x)(M - N')/2^{\delta}}$$

and

$$\mathrm{vol}\Big(\Big\{n_x \in \mathcal{O}_x \mid x(n_x - \nu) > \frac{N' - M}{2}\Big\}, dn_x\Big) \leq p^{\deg(x)(M - N')/2}$$

as long as $v_x(r) < M$. Therefore, to conclude the proof in the case $\ell = 1$ it is sufficient to take

$$C(1,\delta) = C(1,\delta-1) + 1.$$

Let us now assume that $\ell \geq 2$ and let us assume inductively that the claim has been proved when ℓ is replaced by $\ell - 1$. We will write

$$C(a_x,\delta) = C_\ell(a_x,\delta)$$

to indicate the dependence on ℓ. If $r \in \mathcal{R}_x^\delta(M)$ we can write

$$r(n_1,\ldots,n_\ell) = \sum_{\lambda=0}^{\delta} r_\lambda(n_1,\ldots,n_{\ell-1})n_\ell^\lambda$$

and there exists at least one $\lambda \in \{0,1,\ldots,\delta\}$ such that $v_x(r_\lambda) < M$. Let us fix such a λ. Then the set

$$\{n_x \in \mathcal{O}_x^\ell \mid x(r(n_x)) \geq N'\}$$

is contained in the union of the sets

$$\left\{\widehat{n}_x \in \mathcal{O}_x^{\ell-1} \mid x(r_\lambda(\widehat{n}_x)) \geq \frac{N'+M}{2}\right\} \times \mathcal{O}_x$$

and

$$\{n \in \mathcal{O}_x^\ell \mid x(r_\lambda(\widehat{n}_x)) < \frac{N'+M}{2} \text{ and } x(r(n_x)) \geq N'\}$$

where we have set

$$\widehat{n}_x = (n_{x,1},\ldots,n_{x,\ell-1}).$$

Therefore its volume is bounded above by the sum of

$$\begin{aligned} &C_{\ell-1}(1,\delta)p^{\deg(x)\left(M-\frac{N'+M}{2}\right)/2^{\delta+\ell-2}}\operatorname{vol}(\mathcal{O}_x, dn_{x,\ell}) \\ &= C_{\ell-1}(1,\delta)p^{\deg(x)(M-N')/2^{\delta+\ell-1}} \end{aligned}$$

(by the induction hypothesis) and

$$\int_{\mathcal{O}_x^{\ell-1}} C_1(1,\delta)p^{\deg(x)\left(\frac{N'+M}{2}-N'\right)/2^\delta}\,d\widehat{n}_x = C_1(1,\delta)p^{\deg(x)(M-N')/2^{\delta+1}}$$

(by the case $\ell = 1$ which is already proved). Therefore, to conclude the proof it is sufficient to take

$$C_\ell(1,\delta) = C_{\ell-1}(1,\delta) + C_1(1,\delta).$$

□

(10.10) Flicker–Kazhdan simple trace formula

As an application of our main theorem (10.7.10) let us give a proof of the Flicker–Kazhdan simple trace formula (see [Fl–Ka]).

THEOREM (10.10.1). — *Let $o \neq \infty$ be a place of F, let f^o be a locally constant complex function with compact support on $F_\infty^\times \backslash G(\mathbb{A}^o)$, let r be some positive integer and let*

$$f_o = f_o^{(r)} \in \mathcal{C}_c^\infty(G(F_o)/\!\!/G(\mathcal{O}_o))$$

be the Hecke function with Satake transform

$$f_o^\vee(z) = p^{\deg(o)r(d-1)/2}(z_1^r + \cdots + z_d^r).$$

We assume that f^o is supercuspidal at ∞, i.e. that, for any subset $I \subsetneqq \Delta$ and any $h_1^o, h_2^o \in G(\mathbb{A}^o)$, we have

$$\int_{N_I(F_\infty)} f^o(h_1^o n_{I,\infty} h_2^o) dn_{I,\infty} = 0$$

where $dn_{I,\infty}$ is an arbitrary Haar measure on $N_I(F_\infty)$. We set

$$f = f^o f_o^{(r)}.$$

Then if r is large enough with respect to the support of f^o we have

$$\sum_{\substack{\chi \in \mathcal{X}_G \\ \chi_\infty = 1_\infty}} \sum_{(\mathcal{V},\pi) \in \Pi_{G,\chi,\mathrm{cusp}}} m_{\mathrm{cusp}}(\pi) \operatorname{tr} \pi(f)$$
$$= \sum_{\delta \in G(F)_{\natural,\mathrm{ell}}} \mathrm{vol}\Big(F_\infty^\times G_\delta(F) \backslash G_\delta(\mathbb{A}), \frac{dg_\delta}{dz_\infty d\gamma_\delta}\Big) \int_{G_\delta(\mathbb{A}) \backslash G(\mathbb{A})} f(g^{-1}\delta g) \frac{dg}{dg_\delta}.$$

Here $\mathcal{X}_G$ is the abelian group of smooth complex characters of $F^\times \backslash \mathbb{A}^\times$, $\Pi_{G,\chi,\mathrm{cusp}}$ is a system of representatives of the isomorphism classes of cuspidal automorphic irreducible representations of $G(\mathbb{A})$ which admit χ as central character, for each $(\mathcal{V},\pi) \in \Pi_{G,\chi,\mathrm{cusp}}$ we have denoted by $m_{\mathrm{cusp}}(\pi)$ its multiplicity in $(\mathcal{A}_{G,\chi,\mathrm{cusp}}, R_{G,\chi,\mathrm{cusp}})$ and the trace $\operatorname{tr} \pi(f)$ is computed with respect to the Haar measure $dz_\infty \backslash dg$ on $F_\infty^\times \backslash G(\mathbb{A})$ (see (9.1) and (9.2)), $G(F)_{\natural,\mathrm{ell}}$ is a system of representatives of the $G(F)$-conjugacy classes of elliptic elements in $G(F)$ and, for each $\delta \in G(F)_{\natural,\mathrm{ell}}$, dg_δ is an arbitrary Haar measure on $G_\delta(\mathbb{A})$ and $d\gamma_\delta$ is the counting measure on $G_\delta(F)$.

Proof: Let dk_∞ be the Haar measure on $K_\infty = GL_d(\mathcal{O}_\infty)$ of total mass 1. If we set

$$\widetilde{f^o}(g^o) = \int_{K_\infty} f^o(k_\infty^{-1} g^o k_\infty) dk_\infty \qquad (\forall g^o \in G(\mathbb{A})),$$

then $\widetilde{f^o}$ is again a locally constant complex function with compact support on $F_\infty^\times \backslash G(\mathbb{A}^o)$ which is supercuspidal at ∞. Moreover it satisfies hypotheses (A^o) and (B^o) of (10.6) and both sides of the trace formula that we want to prove remain unchanged if we replace f^o by $\widetilde{f^o}$. Therefore we may assume that f^o satisfies hypotheses (A^o) and (B^o) of (10.6). Then, thanks to (10.7.10), it is sufficient to prove that

$$\sum_{\substack{\chi \in \mathcal{X}_G \\ \chi_\infty = 1_\infty}} \sum_{(V,\pi) \in \Pi_{G,\chi,\mathrm{cusp}}} m_{\mathrm{cusp}}(\pi) \mathrm{tr}\, \pi(f) = J_{\mathrm{geom}}(f).$$

Next let us fix a compact open subgroup K' of $G(\mathbb{A})$ such that f is K'-bi-invariant. Let us set

$$\mathcal{X}_G^{\infty,K'} = \{\chi \in \mathcal{X}_G \mid \chi_\infty = 1_\infty \text{ and } \chi(Z_G(\mathbb{A}) \cap K') = (1)\}.$$

As the quotient group

$$Z_G(F_\infty) Z_G(F) \backslash Z_G(\mathbb{A}) / (Z_G(\mathbb{A}) \cap K')$$

is finite (see (9.1)), $\mathcal{X}_G^{\infty,K'}$ is finite and any $\chi \in \mathcal{X}_G^{\infty,K'}$ is unitary. By (9.3.3), for each unitary χ in $\mathcal{X}_G$, $\mathcal{A}_{G,\chi,\mathrm{cusp}}$ is a $G(\mathbb{A})$-subrepresentation of

$$L^2_{G,1_\infty} = L^2(F_\infty^\times G(F) \backslash G(\mathbb{A}), \frac{dg}{dz_\infty d\gamma})$$

and by (9.2.10) (or (9.2.14) combined with (9.1.3)) $\mathcal{A}_{G,\chi,\mathrm{cusp}}$ is admissible. Therefore

$$\bigoplus_{\chi \in \mathcal{X}_G^{\infty,K'}} (\mathcal{A}_{G,\chi,\mathrm{cusp}})^{K'}$$

is a finite dimensional subspace of the $\mathbb{C}$-vector space $L^2_{G,1_\infty}$. Let us check that

$$R_{G,1_\infty}(f)(L^2_{G,1_\infty}) \subset \bigoplus_{\chi \in \mathcal{X}_G^{\infty,K'}} (\mathcal{A}_{G,\chi,\mathrm{cusp}})^{K'} \subset L^2_{G,1_\infty},$$

so that $R_{G,1_\infty}(f)$ is of finite rank. Let $\varphi \in L^2_{G,1_\infty}$ and let us set

$$\varphi'(g) = \int_{K'} \varphi(gk')dk' \qquad (\forall g \in G(\mathbb{A}))$$

where dk' is the Haar measure on K' of total mass 1. Then it is easy to see that

$$\varphi' \in (\mathcal{C}_G^\infty)^{K'} \cap L^2_{G,1_\infty}$$

and that

$$R_{G,1_\infty}(f)(\varphi) = R_{G,1_\infty}(f)(\varphi') \in \Big(\bigoplus_{\chi\in\mathcal{X}_G^{\infty,K'}} (\mathcal{C}_{G,\chi}^\infty)^{K'}\Big) \cap L^2_{G,1_\infty}$$

($\mathcal{C}_G^\infty$ and $\mathcal{C}_{G,\chi}^\infty$ have been defined in (9.1)). Therefore, to prove that

$$R_{G,1_\infty}(f)(\varphi) \in \bigoplus_{\chi\in\mathcal{X}_G^{\infty,K'}} (\mathcal{A}_{G,\mathrm{cusp},\chi})^{K'}$$

it is sufficient to prove that $R_{G,1_\infty}(f)(\varphi')$ is cuspidal (see (9.2.10)). Let $I \subsetneqq \Delta$, let dn_I be a Haar measure on $N_I(\mathbb{A})$, let $d\nu_I$ be the counting measure on $N_I(F)$ and let $h \in G(\mathbb{A})$. Then we have

$$\begin{aligned}
&\int_{N_I(F)\backslash N_I(\mathbb{A})} R_{G,1_\infty}(f)(\varphi')(n_I h)\frac{dn_I}{d\nu_I}\\
&= \int_{N_I(F)\backslash N_I(\mathbb{A})}\Big(\int_{F_\infty^\times\backslash G(\mathbb{A})} f(g)\varphi'(n_I hg)\frac{dg}{dz_\infty}\Big)\frac{dn_I}{d\nu_I}\\
&= \int_{N_I(F)\backslash N_I(\mathbb{A})}\Big(\int_{F_\infty^\times\backslash G(\mathbb{A})} f(h^{-1}n_I^{-1}g)\varphi'(g)\frac{dg}{dz_\infty}\Big)\frac{dn_I}{d\nu_I}\\
&= \int_{N_I(F)\backslash N_I(\mathbb{A})}\Big(\int_{F_\infty^\times N_I(F)\backslash G(\mathbb{A})} \varphi'(g)\Big(\sum_{\nu_I'\in N_I(F)} f(h^{-1}n_I^{-1}\nu_I' g)\Big)\frac{dg}{dz_\infty d\nu_I}\Big)\frac{dn_I}{d\nu_I}.
\end{aligned}$$

But f has compact support in $F_\infty^\times\backslash G(\mathbb{A})$ and $N_I(F)\backslash N_I(\mathbb{A})$ is compact (see (9.2.1)), so that the above integrals are all absolutely convergent and we can permute the order of the integrations. Therefore we get

$$\begin{aligned}
&\int_{N_I(F)\backslash N_I(\mathbb{A})} R_{G,1_\infty}(f)(\varphi')(n_I h)\frac{dn_I}{d\nu_I}\\
&= \int_{F_\infty^\times N_I(F)\backslash G(\mathbb{A})} \varphi'(g)\Big(\int_{N_I(F)\backslash N_I(\mathbb{A})}\Big(\sum_{\nu_I'\in N_I(F)} f(h^{-1}n_I^{-1}\nu_I' g)\Big)\frac{dn_I}{d\nu_I}\Big)\frac{dg}{dz_\infty d\nu_I}\\
&= \int_{F_\infty^\times N_I(F)\backslash G(\mathbb{A})} \varphi'(g)\Big(\int_{N_I(\mathbb{A})} f(h^{-1}n_I' g)dn_I'\Big)\frac{dg}{dz_\infty d\nu_I}
\end{aligned}$$

($n_I' = n_I^{-1}\nu_I'$). Let us fix a splitting

$$dn_I' = dn_I'^{\infty} dn_{I,\infty}'$$

of the Haar measure dn_I' on $N_I(\mathbb{A})$. By hypothesis on f^o the integral

$$\begin{aligned}
&\int_{N_I(\mathbb{A})} f(h^{-1}n_I'g)dn_I' \\
&= \int_{N_I(\mathbb{A}^\infty)} \Big(\int_{N_I(F_\infty)} f(h^{-1}n_I'^{\infty}n_{I,\infty}'g)dn_{I,\infty}'\Big)dn_I'^{\infty} \\
&= \int_{N_I(\mathbb{A}^\infty)} f_o(h_o^{-1}n_{I,o}'g_o)\Big(\int_{N_I(F_\infty)} f^o((h^o)^{-1}n_I'^{\infty,o}n_{I,\infty}'g^o)dn_{I,\infty}'\Big)dn_I'^{\infty}
\end{aligned}$$

vanishes for every $g, h \in G(\mathbb{A})$. The cuspidality of $R_{G,1_\infty}(f)(\varphi')$ follows.

Now the operator of finite rank $R_{G,1_\infty}(f)$ has a trace and its trace is equal to

$$\sum_{\chi \in \mathcal{X}_G^{\infty,K'}} \sum_{\substack{(\mathcal{V},\pi)\in\Pi_{G,\chi,\mathrm{cusp}} \\ \mathcal{V}^{K'}\neq\{0\}}} m_{\mathrm{cusp}}(\pi)\mathrm{tr}(\pi(f)|\mathcal{V}^{K'}),$$

i.e. to

$$\sum_{\substack{\chi\in\mathcal{X}_G \\ \chi_\infty = 1_\infty}} \sum_{(\mathcal{V},\pi)\in\Pi_{G,\chi,\mathrm{cusp}}} m_{\mathrm{cusp}}(\pi)\mathrm{tr}\,\pi(f)$$

(we have $\mathrm{tr}\,\pi(f) = \mathrm{tr}(\pi(f)|\mathcal{V}^{K'})$ and $\chi(Z_G(\mathbb{A}) \cap K') = (1)$ if $\mathcal{V}^{K'} \neq \{0\}$). Therefore the problem is reduced to proving that

$$\mathrm{tr}\, R_{G,1_\infty}(f) = J_{\mathrm{geom}}(f).$$

Let us fix an orthonormal basis $\{\varphi_1, \ldots, \varphi_n\}$ of $R_{G,1_\infty}(f)(L^2_{G,1_\infty}) \subset L^2_{G,1_\infty}$ for the inner product

$$\langle\psi, \varphi\rangle = \int_{F_\infty^\times G(F)\backslash G(\mathbb{A})} \overline{\psi(g)}\varphi(g)\frac{dg}{dz_\infty d\gamma} \qquad (\phi, \psi \in L^2_{G,1_\infty}).$$

Then there exist a matrix $A = (a_{ij})_{1\leq i,j\leq n} \in gl_n(\mathbb{C})$ and vectors $\psi_1, \ldots, \psi_n$ in the orthogonal subspace of $R_{G,1_\infty}(f)(L^2_{G,1_\infty})$ in $L^2_{G,1_\infty}$ such that

$$R_{G,1_\infty}(f)(\varphi) = \sum_{i,j=1}^n a_{ij}\langle\varphi_j, \varphi\rangle\varphi_i + \sum_{i=1}^n \langle\psi_i, \varphi\rangle\varphi_i$$

for every $\varphi \in L^2_{G,1_\infty}$. Since

$$\varphi_1, \ldots, \varphi_n \in R_{G,1_\infty}(f)(L^2_{G,1_\infty}) \subset (\mathcal{C}_G^\infty)^{K'},$$

$$R_{G,1_\infty}(f)(\varphi) = R_{G,1_\infty}(f)(\varphi')$$

and

$$\langle \psi, \varphi' \rangle = \langle \psi', \varphi \rangle$$

for every $\varphi, \psi \in L^2_{G,1_\infty}$, where $\varphi', \psi' \in (\mathcal{C}_G^\infty)^{K'}$ are defined by

$$\varphi'(g) = \int_{K'} \varphi(gk')dk' \qquad (\forall g \in G(\mathbb{A}))$$

and

$$\psi'(g) = \int_{K'} \psi(gk')dk' \qquad (\forall g \in G(\mathbb{A})),$$

we have

$$\langle \varphi_j, \varphi \rangle = \langle \varphi_j, \varphi' \rangle \qquad (\forall j = 1, \ldots, n)$$

and

$$\langle \psi_i, \varphi \rangle = \langle \psi_i, \varphi' \rangle = \langle \psi_i', \varphi \rangle \qquad (\forall i = 1, \ldots, n)$$

for every $\varphi \in L^2_{G,1_\infty}$. In other words we have proved that $R_{G,1_\infty}(f)$ is the integral operator with kernel

$$\widetilde{K}(h,g) = \sum_{i,j=1}^{n} a_{ij}\varphi_i(h)\overline{\varphi_j(g)} + \sum_{i=1}^{n} \varphi_i(h)\overline{\psi_i'(g)}$$

and that

$$\widetilde{K}(hk'', gk') = \widetilde{K}(h,g)$$

for every $g, h \in G(\mathbb{A})$ and every $k', k'' \in K'$.

Then the trace of $R_{G,1_\infty}(f)$, which is nothing else than the trace of the matrix A, is obviously equal to

$$\int_{F_\infty^\times G(F)\backslash G(\mathbb{A})} \widetilde{K}(g,g)\frac{dg}{dz_\infty d\gamma}$$

(we have $\langle \varphi_i, \psi_i' \rangle = \langle \varphi_i, \psi_i \rangle = 0$ for all $i = 1, \ldots, n$). But we know already that $R_{G,1_\infty}(f)$ is an integral operator with kernel $K(h,g)$ (see (10.1.2)) and that

$$J_{\text{geom}}(f) = \int_{F_\infty^\times G(F)\backslash G(\mathbb{A})} K(g,g)\frac{dg}{dz_\infty d\gamma}$$

(see (10.5.2)). Therefore, to conclude the proof it is sufficient to compare the kernels $\widetilde{K}(h,g)$ and $K(h,g)$.

For any $g, h \in G(\mathbb{A})$ we have

$$R_{G,1_\infty}(f)(1_{F_\infty^\times G(F)gK'})(h) = \mathrm{vol}\big(F_\infty^\times G(F)\backslash F_\infty^\times G(F)gK', \frac{dg}{dz_\infty d\gamma}\big)\widetilde{K}(h,g)$$

and we have the same equality where $\widetilde{K}(h,g)$ is replaced by $K(h,g)$. Consequently, for every $g, h \in G(\mathbb{A})$ we have

$$\widetilde{K}(h,g) = K(h,g)$$

and the theorem is proved. □

(10.11) Comments and references

All the results of this chapter have been proved in greater generality by Arthur in the number field case (see [Ar 1], [Ar 2], [Ar 3], [Ar 4] and [Ar 5]) and Lafforgue in the function field case (see [Laf 1]). We have closely followed Arthur's paper. The modifications which are needed in the function field case are routine work.

Due to our strong hypotheses on f our results and our proofs look simpler than Arthur's ones. In particular, thanks to a trick due to Kazhdan we do not need to introduce the weighted orbital integrals. Nevertheless Arthur's most difficult lemmas are there.

Our hypotheses on f (f_∞ is very cuspidal and $f_o^\vee(z)$ is equal to $z_1^r + \cdots + z_d^r$ up to a constant for some $r \gg 0$) imply a lot of cancellations in the geometric side of the non-invariant Arthur trace formula. All the contributions coming from the proper parabolic subgroups of G disappear as in the cocompact case. Under a stronger assumption on f_∞, namely f_∞ supercuspidal, this phenomenon was discovered by Kazhdan (see [Fl–Ka]) and has been interpreted by him as the counterpart in harmonic analysis of Deligne's conjecture (see chapter 12). Our contribution has just been in realizing that the supercuspidality of f_∞ is not needed to get these cancellations: only very cuspidality matters.

11

Non-invariant Arthur trace formula: the spectral side

(11.0) Introduction

We will use the same notations as in chapters 9 and 10: F is a function field, ∞ is a fixed place of F, $G = Gl_d$, ...

The purpose of this chapter is to give Arthur's spectral expression for $J_{\mathrm{geom}}(f)$ and to prove Arthur's expansion of this spectral expression following the cuspidal data

$$J_{\mathrm{geom}}(f) = J^T_{\mathrm{spec}}(f) = \sum_{c \in C} J^T_c(f)$$

where T is a sufficiently regular truncation parameter.

Moreover, for a regular cuspidal datum c we will give Arthur's computation of $J^T_c(f)$ in terms of logarithmic derivatives of intertwining operators.

(11.1) Arthur's truncation operators

We set

$$\mathfrak{a}^+_\emptyset = \{T \in \mathfrak{a}_\emptyset \mid \alpha_i(T) \geq 0,\ \forall i = 1, \ldots, d-1\}.$$

LEMMA (11.1.1). — *For each $I \subset \Delta$, each $T \in \mathfrak{a}^+_\emptyset$ and each $g \in G(\mathbb{A})$ the set*

$$\{\delta \in P_I(F)\backslash G(F) \mid \widehat{\tau}^\Delta_I(H_\emptyset(\delta g) - T) \neq 0\}$$

is finite.

Proof: Firstly we will check that there is $c(g) \in \mathbb{Q}$ such that

$$\varpi_i(H_\emptyset(\delta g)) \leq c(g) \qquad (\forall \delta \in G(F),\ \forall i = 1, \dots, d-1).$$

According to the Bruhat decomposition any element $\delta \in G(F)$ may be written as $\pi_\emptyset \dot{w} \nu_\emptyset$ for some $\pi_\emptyset \in P_\emptyset(F)$, some $w \in W$ and some $\nu_\emptyset \in N_\emptyset(F)$. It follows that

$$H_\emptyset(\delta g) = H_\emptyset(\dot{w}\nu_\emptyset g).$$

According to the Iwasawa decomposition g may be written as $n_\emptyset m_\emptyset k$ for some $n_\emptyset \in N_\emptyset(\mathbb{A})$, some $m_\emptyset \in M_\emptyset(\mathbb{A})$ and some $k \in K$. Then we have

$$\dot{w}\nu_\emptyset g = (\dot{w} m_\emptyset \dot{w}^{-1})(\dot{w} m_\emptyset^{-1} \nu_\emptyset n_\emptyset m_\emptyset \dot{w}^{-1})(\dot{w}k)$$

and

$$H_\emptyset(\dot{w}\nu_\emptyset g) = w(H_\emptyset(g)) + H_\emptyset(\dot{w} n_\emptyset' \dot{w}^{-1})$$

where we have set

$$n_\emptyset' = m_\emptyset^{-1} \nu_\emptyset n_\emptyset m_\emptyset.$$

Now for each $i = 1, \dots, d-1$, each $w \in W$ and each $n_\emptyset' \in N_\emptyset(\mathbb{A})$ we have

$$\varpi_i(H_\emptyset(\dot{w} n_\emptyset' \dot{w}^{-1})) \leq 0.$$

Indeed we have

$$\dot{w} N_\emptyset \dot{w}^{-1} = (N_\emptyset \cap (\dot{w} N_\emptyset \dot{w}^{-1}))(\overline{N}_\emptyset \cap (\dot{w} N_\emptyset \dot{w}^{-1}))$$

(see [Ca] (2.5.12)) and it is sufficient to prove that

$$\varpi_i(H_\emptyset(\overline{n}_\emptyset)) \leq 0$$

for all $i = 1, \dots, d-1$ and all $\overline{n}_\emptyset \in \overline{N}_\emptyset(\mathbb{A})$; but, if we have

$$(\overline{N}_\emptyset(\mathbb{A}) m_\emptyset N_\emptyset(\mathbb{A})) \cap K \neq \emptyset$$

for some $m_\emptyset = (t_1, \dots, t_d) \in (\mathbb{A}^\times)^d \cong M_\emptyset(\mathbb{A})$, we have

$$t_1 \cdots t_i \in \mathcal{O} \cap \mathbb{A}^\times \qquad (\forall i = 1, \dots, d-1)$$

and

$$t_1 \cdots t_d \in \mathcal{O}^\times$$

(consider the minors of sizes $\{1, \dots, i\} \times \{1, \dots, i\}$ of any element of the above intersection), so that

$$\varpi_i(H_\emptyset(m_\emptyset)) \leq 0 \qquad (\forall i = 1, \dots, d-1).$$

Therefore we can take

$$c(g) = \sup\{\varpi_i(w(H_\emptyset(g))) \mid w \in W,\ i = 1, \dots, d-1\}.$$

Secondly we fix a system of representatives $\Gamma(g) \subset G(F)$ of $P_I(F)\backslash G(F)$ in such a way that, for any $\delta \in \Gamma(g)$, δg belongs to the Siegel set $\mathfrak{S}^I$ (see (10.3.9)(i)). Then for each $\delta \in \Gamma(g)$ we have

$$\alpha_i(H_\emptyset(\delta g)) \geq -2g_X \qquad (\forall i \in I).$$

Finally we are interested in those $\delta \in \Gamma(g)$ such that $\varpi_i(H_\emptyset(\delta g)) > \varpi_i(T)$ for all $i \in \Delta - I$. But the set of points $F_\infty^\times T(F)t \in F_\infty^\times T(F)\backslash T(\mathbb{A})$ which satisfy

$$\begin{cases} \varpi_i(H_\emptyset(t)) \leq c(g) & (\forall i \in \Delta), \\ \alpha_i(H_\emptyset(t)) \geq -2g_X & (\forall i \in I), \\ \varpi_i(H_\emptyset(t)) \geq \varpi_i(T) & (\forall i \in \Delta - I) \end{cases}$$

is a compact set. Indeed any such point $F_\infty^\times T(F)t$ also satisfies

$$\alpha_i(H_\emptyset(t)) \geq 2(\varpi_i(T) - c(g)) \qquad (\forall i \in \Delta - I)$$

(if we set $\varpi_0 \equiv \varpi_d \equiv 0$, we have $\alpha_i = 2\varpi_i - \varpi_{i-1} - \varpi_{i+1}$ for each $i \in \Delta$) and the sets $F_\infty^\times F^\times\backslash\mathbb{A}^\times/\mathcal{O}^\times$ and $F^\times\backslash(\mathbb{A}^\times)^1/\mathcal{O}^\times$ are finite. Since $N_\emptyset(F)\backslash N_\emptyset(\mathbb{A})$ is compact (see (9.2.1)), the set of points $F_\infty^\times P_\emptyset(F)h \in F_\infty^\times P_\emptyset(F)\backslash G(\mathbb{A})$ which satisfy the above conditions where we replace $H_\emptyset(t)$ by $H_\emptyset(g)$ is compact too. As $G(F)$ is discrete in $F_\infty^\times\backslash G(\mathbb{A})$ the lemma follows. □

Let us recall that

$$\mathcal{C}_G^\infty = \mathcal{C}^\infty(G(F)\backslash G(\mathbb{A}), \mathbb{C})$$

is the $\mathbb{C}$-vector space of complex functions φ on $G(\mathbb{A})$ which are left $G(F)$-invariant and right K'-invariant for some compact open subgroup K' of $K \subset G(\mathbb{A})$ depending on φ (see (9.1)). Let us also recall that, for each $\varphi \in \mathcal{C}_G^\infty$ and each $I \subset \Delta$, the constant term of φ along P_I is the function φ_I on $G(\mathbb{A})$ defined by

$$\varphi_I(g) = \int_{N_I(F)\backslash N_I(\mathbb{A})} \varphi(n_I g) \frac{dn_I}{d\nu_I} \qquad (\forall g \in G(\mathbb{A})).$$

Here $d\nu_I$ is the counting measure on $N_I(F)$ and dn_I is the Haar measure on $N_I(\mathbb{A})$ normalized by $\mathrm{vol}(N_I(F)\backslash N_I(\mathbb{A}), d\nu_I\backslash dn_I) = 1$. The function φ_I is left $(N_I(\mathbb{A})M_I(F))$-invariant and it is right K'-invariant if φ is right K'-invariant.

For each $T \in \mathfrak{a}_\emptyset^+$ and each $\varphi \in \mathcal{C}_G^\infty$ we set

(11.1.2)

$$(\Lambda^T\varphi)(g) = \sum_{I \subset \Delta} (-1)^{|\Delta - I|} \sum_{\delta \in P_I(F)\backslash G(F)} \widehat{\tau}_I^\Delta(H_\emptyset(\delta g) - T)\varphi_I(\delta g) \quad (\forall g \in G(\mathbb{A})).$$

Obviously $\Lambda^T\varphi \in \mathcal{C}_G^\infty$ and $\Lambda^T : \mathcal{C}_G^\infty \to \mathcal{C}_G^\infty$ is a $\mathbb{C}$-linear operator, the so called **Arthur truncation operator** (see [Ar 6]).

Let

$$\mathcal{C}_{G,1_\infty}^\infty = \mathcal{C}^\infty(F_\infty^\times G(F)\backslash G(\mathbb{A})) \subset \mathcal{C}_G^\infty$$

be the space of $\varphi \in \mathcal{C}_G^\infty$ which are left $F_\infty^\times$-invariant. Obviously we have

$$\Lambda^T(\mathcal{C}_{G,1_\infty}^\infty) \subset \mathcal{C}_{G,1_\infty}^\infty.$$

Proposition (11.1.3) (Arthur). — (i) $\Lambda^T \circ \Lambda^T = \Lambda^T$.

(ii) *Let $K' \subset K \subset G(\mathbb{A})$ be a compact open subgroup. Then there exists a subset $\Omega(K')$ of $G(F)\backslash G(\mathbb{A})$, which is compact modulo $F_\infty^\times$, such that*

$$\mathrm{Supp}(\Lambda^T\varphi) \subset \Omega(K')$$

for any right K'-invariant $\varphi \in \mathcal{C}_G^\infty$.

(iii) *Let $\varphi_1, \varphi_2 \in \mathcal{C}_{G,1_\infty}^\infty$. If φ_1 or φ_2 has compact support in $F_\infty^\times G(F)\backslash G(\mathbb{A})$ we have*

$$\langle \Lambda^T\varphi_1, \varphi_2\rangle = \langle \varphi_1, \Lambda^T\varphi_2\rangle$$

where

$$\langle \psi_1, \psi_2\rangle = \int_{F_\infty^\times G(F)\backslash G(\mathbb{A})} \overline{\psi_1(g)}\psi_2(g)\frac{dg}{dz_\infty d\gamma}$$

is the natural inner product on $L^2(F_\infty^\times G(F)\backslash G(\mathbb{A}), dz_\infty d\gamma\backslash dg)$ (see (10.1)).

We will need the following lemma.

Lemma (11.1.4). — *Let $\varphi \in \mathcal{C}_G^\infty$ and let $J \subset \Delta$. Then the support of $(\Lambda^T\varphi)_J$ is contained in*

$$\{g \in G(\mathbb{A}) \mid \varpi_i(H_\emptyset(g) - T) \le 0,\ \forall i \in \Delta - J\}.$$

Proof : Using the Bruhat decomposition

$$\begin{cases} G(F) = \coprod_w P_I(F)\dot{w}N_\emptyset(F), \\ \dot{w}^{-1}P_I(F)\dot{w} \cap N_\emptyset(F) = \dot{w}^{-1}N_\emptyset(F)\dot{w} \cap N_\emptyset(F), \end{cases}$$

where w runs through

$$D_{I,\emptyset} = \{w \in W \mid w^{-1}(I) \subset R^+\}$$

(see (5.4.1) and (5.4.3) or [Ca] (2.8.1), (2.8.6) and (2.8.9)), we see that $(\Lambda^T\varphi)_J(g)$ is equal to the sum over $I \subset \Delta$ and $w \in D_{I,\emptyset}$ of the integral over $n_I \in N_I(F)\backslash N_I(\mathbb{A})$ of the product of the sign

$$(-1)^{|\Delta - I|}$$

with the expression

$$\int_{N_J(F)\backslash N_J(\mathbb{A})} \sum_{\nu_\emptyset} \varphi(n_I \dot{w} \nu_\emptyset n_J g) \widehat{\tau}_I^\Delta (H_\emptyset(\dot{w}\nu_\emptyset n_J g) - T) \frac{dn_J}{d\nu_J}$$

where $\nu_\emptyset$ runs through

$$(\dot{w}^{-1} N_\emptyset(F) \dot{w} \cap N_\emptyset(F)) \backslash N_\emptyset(F).$$

Since we have

$$N_J(F)\backslash N_J(\mathbb{A}) = N_\emptyset(F)\backslash N_\emptyset^J(F) N_J(\mathbb{A})$$

the last expression is equal to

$$\int_{(\dot{w}^{-1}N_\emptyset(F)\dot{w}\cap N_\emptyset(F))\backslash N_\emptyset^J(F)N_J(\mathbb{A})} \varphi(n_I \dot{w} \nu_\emptyset^J n_J g) \widehat{\tau}_I^\Delta (H_\emptyset(\dot{w}\nu_\emptyset^J n_J g) - T) \frac{d\nu_\emptyset^J dn_J}{d\nu_\emptyset^w}$$

where $d\nu_\emptyset^J$ and $d\nu_\emptyset^w$ are the counting measures on $N_\emptyset^J(F)$ and $\dot{w}^{-1}N_\emptyset(F)\dot{w} \cap N_\emptyset(F)$ respectively.

Let us rearrange this integral. For each $w \in W$ and each $J \subset \Delta$ we have

$$\dot{w}^{-1} N_\emptyset \dot{w} \cap N_\emptyset = (\dot{w}^{-1} N_\emptyset \dot{w} \cap N_\emptyset^J)(\dot{w}^{-1} N_\emptyset \dot{w} \cap N_J).$$

Therefore we may replace $\nu_\emptyset^J n_J$ by $n_J^w \nu_\emptyset^J n_J$ and then firstly integrate over

$$(\dot{w}^{-1} N_\emptyset(F) \dot{w} \cap N_J(F)) \backslash (\dot{w}^{-1} N_\emptyset(\mathbb{A}) \dot{w} \cap N_J(\mathbb{A}))$$

with respect to the measure $d\nu_J^w \backslash dn_J^w$ and secondly integrate over

$$(\dot{w}^{-1} N_\emptyset(\mathbb{A}) \dot{w} \cap N_\emptyset^J(F) N_J(\mathbb{A})) \backslash N_\emptyset^J(F) N_J(\mathbb{A})$$

with respect to the Haar measure $d\nu_\emptyset^{J,w} dn_J^w \backslash d\nu_\emptyset^J dn_J$. Here $d\nu_J^w$ and $d\nu_\emptyset^{J,w}$ are the counting measures on $\dot{w}^{-1}N_\emptyset(F)\dot{w} \cap N_J(F)$ and $\dot{w}^{-1}N_\emptyset(F)\dot{w} \cap N_\emptyset^J(F)$ respectively and dn_J^w is the Haar measure on $\dot{w}^{-1}N_\emptyset(\mathbb{A})\dot{w} \cap N_J(\mathbb{A})$ which is normalized by

$$\mathrm{vol}\Big((\dot{w}^{-1} N_\emptyset(F) \dot{w} \cap N_J(F)) \backslash (\dot{w}^{-1} N_\emptyset(\mathbb{A}) \dot{w} \cap N_J(\mathbb{A})), \frac{dn_J^w}{d\nu_J^w}\Big) = 1.$$

But we have

$$\dot{w} n_J^w \nu_\emptyset^J n_J g = {}^w n_J \dot{w} \nu_\emptyset^J n_J g$$

with ${}^w n_J = \dot{w} n_J^w \dot{w}^{-1} \in N_\emptyset(\mathbb{A}) \cap \dot{w} N_J(\mathbb{A}) \dot{w}^{-1}$. Moreover, as $w \in D_{I,\emptyset}$ we have

$$N_\emptyset \cap \dot{w} N_J \dot{w}^{-1} = (N_I \cap \dot{w} N_J \dot{w}^{-1})(M_I \cap \dot{w} N_J \dot{w}^{-1})$$

and $M_I \cap \dot{w}N_J\dot{w}^{-1}$ is the unipotent radical $N_{K_I^w}^I$ of the standard parabolic subgroup $P_{K_I^w}^I$ of M_I for some $K_I^w \subset I$. Indeed we have

$$M_I \cap \dot{w}P_\emptyset\dot{w}^{-1} = P_\emptyset^I$$

with

$$M_I \cap \dot{w}N_\emptyset\dot{w}^{-1} = N_\emptyset^I$$

(see [Ca] (2.8.9)) and it follows that $M_I \cap \dot{w}P_J\dot{w}^{-1} \supset P_\emptyset^I$ is a standard parabolic subgroup of M_I with unipotent radical

$$M_I \cap \dot{w}N_J\dot{w}^{-1} \subset N_\emptyset^I \subset N_\emptyset$$

(it is easy to see that $M_I \cap \dot{w}M_J\dot{w}^{-1}$ is reductive). Therefore we have

$$N_I(N_\emptyset \cap \dot{w}N_J\dot{w}^{-1}) = N_{K_I^w}$$

and if we combine the integral over ${}^w n_J$, i.e. over n_J^w, with the integral over n_I we obtain that $(\Lambda^T\varphi)_J(g)$ is equal to the sum over $I \subset \Delta$ and $w \in D_{I,\emptyset}$ of the product of the sign

$$(-1)^{|\Delta - I|}$$

with the integral over the compact space

$$(N_{K_I^w}(F)\backslash N_{K_I^w}(\mathbb{A})) \times ((\dot{w}^{-1}N_\emptyset(\mathbb{A})\dot{w} \cap N_\emptyset^J(F)N_J(\mathbb{A}))\backslash N_\emptyset^J(F)N_J(\mathbb{A}))$$

of

$$\varphi(n_{K_I^w}\dot{w}\nu_\emptyset^J n_J g)\widehat{\tau}_I^\Delta(H_\emptyset(\dot{w}\nu_\emptyset^J n_J g) - T)$$

with respect to the Haar measure

$$\frac{dn_{K_I^w}}{d\nu_{K_I^w}} \times \frac{d\nu_\emptyset^J dn_J}{d\nu_\emptyset^{J,w} dn_J^w}$$

($d\nu_{K_I^w}$ is the counting measure on $N_{K_I^w}(F)$ and $dn_{K_I^w}$ is the Haar measure on $N_{K_I^w}(\mathbb{A})$ which is normalized by

$$\mathrm{vol}\Big(N_{K_I^w}(F)\backslash N_{K_I^w}(\mathbb{A}), \frac{dn_{K_I^w}}{d\nu_{K_I^w}}\Big) = 1.$$

Now let us fix $w \in W$ and let us consider the set of $I \subset \Delta$ such that $w \in D_{I,\emptyset}$ which give rise to a fixed K_I^w. We denote by $R_J \subset R$ the set of roots which vanish identically on $\mathfrak{a}_J \subset \mathfrak{a}_\emptyset$, i.e. the set of roots of (M_J, T), and we set

$$R_J^+ = R^+ \cap R_J,$$

$$\Delta_J^w = \{\alpha \in \Delta \mid w^{-1}(\alpha) \in R_J^+\}$$

and

$$\Delta^{w,J} = \{\alpha \in \Delta \mid w^{-1}(\alpha) \in R^+ - R_J^+\}.$$

Then, for any $I \subset \Delta$ such that $w \in D_{I,\emptyset}$, we have

$$K_I^w = I \cap \Delta_J^w$$

(we have

$$M_I \cap \dot{w} M_J \dot{w}^{-1} = M_{K_I^w}$$

as both sides are Levi subgroups containing T of the standard parabolic subgroup $P_{K_I^w}^I$ of M_I). Therefore, for any subset K^w of Δ_J^w, those $I \subset \Delta$ such that $w \in D_{I,\emptyset}$ and $K_I^w = K^w$ are exactly the subsets of Δ of the form

$$I = K^w \amalg I'$$

where I' is an arbitrary subset of $\Delta^{w,J}$. It follows that in our last expression for $(\Lambda^T \varphi)_J(g)$ we may replace the sums over $I \in \Delta$ and $w \in D_{I,\emptyset}$ by sums over $w \in W$, $K^w \subset \Delta_J^w$ and $I' \subset \Delta^{w,J}$. Then, if we set

$$\Phi_{w,K^w}(H) = \sum_{I' \subset \Delta^{w,J}} (-1)^{|\Delta^{w,J} - I'|} \hat{\tau}_{K^w \amalg I'}^{\Delta}(H) \qquad (\forall H \in \mathfrak{a}_\emptyset),$$

we obtain that $(\Lambda^T \varphi)_J(g)$ is the sum over $w \in W$ and $K^w \subset \Delta_J^w$ of the product of the sign

$$(-1)^{|\Delta - (\Delta_J^w \amalg \Delta^{w,J})|}$$

with the integral over $n_{K_I^w}$ and $\nu_\emptyset^J n_J$ of

$$\varphi(n_{K_I^w} \dot{w} \nu_\emptyset^J n_J g) \Phi_{w,K^w}(H_\emptyset(\dot{w} \nu_\emptyset^J n_J g) - T).$$

Clearly, to conclude the proof of (11.1.4) it is sufficient to check that, if

$$\Phi_{w,K^w}(H_\emptyset(\dot{w} n_\emptyset g) - T) \neq 0$$

for some $w \in W$, some $K^w \subset \Delta_J^w$ and some $n_\emptyset \in N_\emptyset(\mathbb{A})$, we have

$$\varpi_i(H_\emptyset(g) - T) \leq 0 \qquad (\forall i \in \Delta - J).$$

But, on the one hand, the function Φ_{w,K^w} is the characteristic function of the subset

$$\mathfrak{a}_\Delta \oplus \{H \in \mathfrak{a}_\emptyset^\Delta \mid \varpi_i(H) > 0, \ \forall i \in \Delta - (K^w \amalg \Delta^{w,J}),$$
$$\text{and } \varpi_i(H) \leq 0, \ \forall i \in \Delta^{w,J}\}$$

of $\mathfrak{a}_\emptyset$ (compare to (10.3.4)(i)) and, if we suppose that $\Phi_{w,K^w}(H) \neq 0$ for some $w \in W$, some $K^w \subset \Delta_J^w$ and some $H \in \mathfrak{a}_\emptyset$, we have

$$\varpi_i(w^{-1}(H)) \leq 0 \qquad (\forall i \in \Delta - I)$$

(decompose H^Δ into

$$H^\Delta = \sum_{j\in\Delta} \varpi_j(H^\Delta)\alpha_j^\vee$$

and remark that, for any $i \in \Delta - J$, we have

$$\varpi_i(w^{-1}(\alpha_j^\vee)) = 0 \qquad (\forall j \in \Delta_J^w),$$

$$\varpi_i(w^{-1}(\alpha_j^\vee)) \geq 0 \qquad (\forall j \in \Delta^{w,J})$$

and

$$\varpi_i(w^{-1}(\alpha_j^\vee)) \leq 0 \qquad (\forall j \in \Delta - (\Delta_J^w \amalg \Delta^{w,J}))).$$

On the other hand, as in the proof of (11.1.1) we can find $n'_\emptyset \in N_\emptyset(\mathbb{A})$ such that

$$H_\emptyset(\dot{w}n_\emptyset g) = w(H_\emptyset(g)) + H_\emptyset(\dot{w}n'_\emptyset\dot{w}^{-1})$$

and for this element $n'_\emptyset$ we have

$$H_\emptyset(g) - T = w^{-1}(H_\emptyset(\dot{w}n_\emptyset g) - T) - (T - w^{-1}(T)) - w^{-1}(H_\emptyset(\dot{w}n'_\emptyset\dot{w}^{-1})).$$

Finally, since $T \in \mathfrak{a}_\emptyset^+$ we have

$$\varpi_i(T - w^{-1}(T)) \geq 0 \qquad (\forall i \in \Delta).$$

Moreover we have

$$\varpi_i(w^{-1}(H_\emptyset(\dot{w}n'_\emptyset\dot{w}^{-1}))) \geq 0 \qquad (\forall i \in \Delta).$$

Indeed we have

$$\dot{w}N_\emptyset\dot{w}^{-1} = (N_\emptyset \cap \dot{w}N_\emptyset\dot{w}^{-1})(\overline{N}_\emptyset \cap \dot{w}N_\emptyset\dot{w}^{-1})$$

(see [Ca] (2.5.12)) and, if

$$N_\emptyset(\mathbb{A})m_\emptyset\overline{N}_\emptyset(\mathbb{A}) \cap K \neq \emptyset$$

for some $m_\emptyset = (t_1, \ldots, t_d) \in (\mathbb{A}^\times)^d \cong M_\emptyset(\mathbb{A})$, we have

$$t_1 \cdots t_d \in \mathcal{O}^\times$$

and

$$t_{d-i+1} \cdots t_d \in \mathcal{O} \cap \mathbb{A}^\times \qquad (\forall i = 1, \ldots, d-1)$$

(consider the minors of sizes $\{d-i+1, \ldots, d\} \times \{d-i+1, \ldots, d\}$ of any element in that intersection). Therefore, if $\Phi_{w,K^w}(H_\emptyset(\dot{w}n_\emptyset g) - T) \neq 0$ for some $w \in W$, some $K^w \subset \Delta_J^w$ and some $n_\emptyset \in N_\emptyset(\mathbb{A})$ we have

$$\varpi_i(H_\emptyset(g) - T) \leq 0 \qquad (\forall i \in \Delta - J)$$

and the proof of the lemma is completed. □

Proof of (11.1.3)(i) : If $J \subsetneq \Delta$ lemma (11.1.4) implies that

$$\widehat{\tau}_J^{\Delta}(H_\emptyset(h) - T)(\Lambda^T\varphi)_J(h) = 0 \qquad (\forall h \in G(\mathbb{A})).$$

The assertion follows. □

We will also need the following lemma (see [Mo–Wa 2] (I.2.8)).

LEMMA (11.1.5). — *Let K' be a compact open subgroup of $K \subset G(\mathbb{A})$. Then there exists a rational number $c(K') \geq -2g_X$ having the following property. If $T \in \mathfrak{a}_\emptyset$ is such that $\alpha_i(T) \geq c(K')$ for all $i = 1, \ldots, d-1$ and if $\varphi \in \mathcal{C}_G^\infty$ is right K'-invariant, for each $I \in \Delta$ and for each $g \in \mathfrak{G}$ such that*

$$\begin{cases} \alpha_i(H_I(g) - T_I) > 0 & (\forall i \in \Delta - I), \\ \varpi_i^I(H_\emptyset(g) - T) \leq 0 & (\forall i \in I), \end{cases}$$

we have $\varphi_I(g) = \varphi(g)$.

Proof : Replacing K' by a subgroup of finite index we may assume that K' is a normal subgroup of K. For each $I \subset \Delta$ let us fix a compact open subset $\omega_I \subset N_I(\mathbb{A})$ such that the projection

$$\omega_I \to N_I(F)\backslash N_I(\mathbb{A})$$

is bijective (see (9.2.1)). If

$$g = \nu_\emptyset n_\emptyset m_\emptyset k \in N_\emptyset(F)\omega_\emptyset M_\emptyset(\mathbb{A})^{\Delta}_{]-\infty, 2g_X]}K = \mathfrak{G}$$

we have

$$\begin{aligned} \varphi_I(g) &= \int_{N_I(F)\backslash N_I(\mathbb{A})} \varphi_I(n_I n_\emptyset m_\emptyset k)\frac{dn_I}{d\nu_I} = \int_{\omega_I} \varphi_I(n_I n_\emptyset m_\emptyset k)dn_I \\ &= \frac{1}{\mathrm{vol}(m_\emptyset^{-1}n_\emptyset^{-1}\omega_I n_\emptyset m_\emptyset, dn_I)} \int_{m_\emptyset^{-1}n_\emptyset^{-1}\omega_I n_\emptyset m_\emptyset} \varphi_I(n_\emptyset m_\emptyset n_I k)dn_I \end{aligned}$$

(recall that dn_I is normalized by $\mathrm{vol}(\omega_I, dn_I) = 1$). In particular, if

$$m_\emptyset^{-1}\omega_I' m_\emptyset \subset N_I(\mathbb{A}) \cap K' \subset K'$$

where we have set

$$\omega_I' = \{n_\emptyset^{-1} n_I n_\emptyset \mid n_\emptyset \in \omega_\emptyset,\ n_I \in \omega_I\},$$

we have $\varphi_I(g) = \varphi(g)$.

But ω_I' is a compact open subset of $N_I(\mathbb{A})$. Therefore there exists a rational number $c(K') \geq -2g_X$ such that

$$m_\emptyset^{-1}\omega_I' m_\emptyset \subset N_I(\mathbb{A}) \cap K'$$

for every $m_\emptyset \in M_\emptyset(\mathbb{A})^{\Delta}_{]-\infty,2g_X]}$ with

$$\alpha_i(H_\emptyset(m_\emptyset)) > c(K') \qquad (\forall i \in \Delta - I)$$

(if $\alpha_i + \cdots + \alpha_{j-1}$, $1 \leq i < j \leq d$, is a root of $M_\emptyset$ in N_I there exists at least one $k \in \Delta - I$ such that $i \leq k \leq j-1$). Since

$$H_\emptyset(g) = H_\emptyset(m_\emptyset)$$

and

$$\alpha_i(H) = \alpha_i(H_I) - \varpi_{i-1}^I(H) - \varpi_{i+1}^I(H)$$

for every $i \in \Delta - I$ and every $H \in \mathfrak{a}_\emptyset$, where we have set $\varpi_j^I \equiv 0$ if $j \notin I$, for any $T \in \mathfrak{a}_\emptyset$ with $\alpha_i(T) \geq c(K')$, $i = 1, \ldots, d-1$, the conditions

$$\begin{cases} \alpha_i(H_I(g) - T_I) > 0 & (\forall i \in \Delta - I), \\ \varpi_i^I(H_\emptyset(g) - T) \leq 0 & (\forall i \in I) \end{cases}$$

imply the conditions

$$\alpha_i(H_\emptyset(m_\emptyset)) > \alpha_i(T) \geq c(K') \qquad (\forall i \in \Delta - I).$$

The lemma follows. □

Proof of (11.1.3)(ii) : Let us fix $T_0 \in \mathfrak{a}_\emptyset$ with $\alpha_i(T_0) \geq c(K')$, $i = 1, \ldots, d-1$, where $c(K')$ is the constant of lemma (11.1.5). Then if we apply this lemma to $\Lambda^T\varphi$ we get that, for each $J \subset \Delta$, we have

$$\Lambda^T\varphi(g) = (\Lambda^T\varphi)_J(g)$$

as long as g belongs to the subset $\mathfrak{S}(J, T_0)$ of $\mathfrak{S}$ defined by the conditions

$$\begin{cases} \alpha_i(H_J(g) - (T_0)_J) > 0 & (\forall i \in \Delta - J), \\ \varpi_i^J(H_\emptyset(g) - T_0) \leq 0 & (\forall i \in J). \end{cases}$$

But by lemma (11.1.4) the support of $(\Lambda^T\varphi)_J$ is contained in

$$\{g \in G(\mathbb{A}) \mid \varpi_i(H_\emptyset(g) - T) \leq 0,\ \forall i \in \Delta - J\}$$

and, by (10.3.4)(ii), $\mathfrak{S}$ is the disjoint union of the $\mathfrak{S}(J, T_0)$, $J \subset \Delta$. Thus the support of $(\Lambda^T \varphi)|\mathfrak{S}$ is contained in the subset

$$\coprod_{J \subset \Delta} \{g \in \mathfrak{S}(J, T_0) \mid \varpi_i(H_\emptyset(g) - T) \leq 0, \ \forall i \in \Delta - J\}$$

of $\mathfrak{S}$ which is independent of φ. Then to conclude the proof of (11.1.3)(ii) it is sufficient to check that the projection of this subset into $F_\infty^\times P_\emptyset(F)\backslash\mathfrak{S}$ is compact. Since $N_\emptyset(F)\backslash N_\emptyset(\mathbb{A})$ is compact (see (9.2.1)) we only need to check that, for each $J \subset \Delta$, the set of classes

$$F_\infty^\times M_\emptyset(F) m_\emptyset M_\emptyset(\mathcal{O}) \qquad (m_\emptyset \in M_\emptyset(\mathbb{A}))$$

such that

$$\begin{cases} \alpha_i(H_\emptyset(m_\emptyset)) \geq -2g_X & (\forall i \in \Delta), \\ \alpha_i(H_J(m_\emptyset) - (T_0)_J) > 0 & (\forall i \in \Delta - J), \\ \varpi_i^J(H_\emptyset(m_\emptyset) - T_0) \leq 0 & (\forall i \in J), \\ \varpi_i(H_\emptyset(m_\emptyset) - T) \leq 0 & (\forall i \in \Delta - J) \end{cases}$$

is finite. But, if $H \in \mathfrak{a}_\emptyset$ satisfies the conditions

$$\begin{cases} \alpha_i(H) \geq -2g_X & (\forall i \in \Delta), \\ \varpi_i^J(H - T_0) \leq 0 & (\forall i \in J), \\ \varpi_i(H - T) \leq 0 & (\forall i \in \Delta - J), \end{cases}$$

we have

$$-2g_X \leq \alpha_i(H) \leq \frac{d_j}{k(d_j - k)} \varpi_i^J(T_0) + (d_j - 2)g_X$$

for each $i = d_1 + \cdots + d_{j-1} + k \in J$, $1 \leq k \leq d_j - 1$ (as usual $d_J = (d_1, \ldots, d_s)$ is the partition of d corresponding to J), and

$$-2g_X \leq \alpha_i(H) \leq \frac{d}{i(d-i)} \varpi_i(T) + (d-2)g_X$$

for each $i \in \Delta - J$. As $F_\infty^\times F^\times\backslash\mathbb{A}^\times/\mathcal{O}^\times$ and $F^\times\backslash(\mathbb{A}^\times)^1/\mathcal{O}^\times$ are finite the proof of (11.1.3)(ii) is completed. □

Proof of (11.1.3)(iii) : The inner product $\langle\Lambda^T\varphi_1, \varphi_2\rangle$ is defined by an absolutely convergent integral for any $\varphi_1, \varphi_2 \in C_{G,1_\infty}^\infty$ ($\Lambda^T\varphi_1$ has a compact support in $F_\infty^\times G(F)\backslash G(\mathbb{A})$ by (11.1.3)(ii) which is already proved). If φ_1 or φ_2 has a compact support in $F_\infty^\times G(F)\backslash G(\mathbb{A})$, for each $I \subset \Delta$ the integral over the compact space $N_I(F)\backslash N_I(\mathbb{A})$ with respect to the Haar measure $d\nu_I\backslash dn_I$ of

$$\int_{F_\infty^\times G(F)\backslash G(\mathbb{A})}|\varphi_2(g)|\Big(\sum_{\delta\in P_I(F)\backslash G(F)}|\varphi_1(n_I\delta g)|\widehat{\tau}_I^\Delta(H_\emptyset(\delta g)-T)\Big)\frac{dg}{dz_\infty d\gamma}$$

$$=\int_{F_\infty^\times P_I(F)\backslash G(\mathbb{A})}|\varphi_1(n_Ig)\varphi_2(g)|\widehat{\tau}_I^\Delta(H_\emptyset(g)-T)\frac{dg}{dz_\infty d\pi_I}$$

$$=\int_{F_\infty^\times P_I(F)\backslash G(\mathbb{A})}|\varphi_1(g)\varphi_2(n_I^{-1}g)|\widehat{\tau}_I^\Delta(H_\emptyset(g)-T)\frac{dg}{dz_\infty d\pi_I}$$

$$=\int_{F_\infty^\times G(F)\backslash G(\mathbb{A})}|\varphi_1(g)|\Big(\sum_{\delta\in P_I(F)\backslash G(F)}|\varphi_2(n_I^{-1}\delta g)|\widehat{\tau}_I^\Delta(H_\emptyset(\delta g)-T)\Big)\frac{dg}{dz_\infty d\gamma}$$

is finite ($d\pi_I$ is the counting measure on $P_I(F)$; use (11.1.1)). Therefore, if φ_1 or φ_2 has a compact support in $F_\infty^\times G(F)\backslash G(\mathbb{A})$, the inner product $\langle\Lambda^T\varphi_1,\varphi_2\rangle$ is equal to

$$\sum_{I\subset\Delta}(-1)^{|\Delta-I|}\int_{N_I(F)\backslash N_I(\mathbb{A})}\Big(\int_{F_\infty^\times P_I(F)\backslash G(\mathbb{A})}\overline{\varphi_1(n_Ig)}\varphi_2(g)\widehat{\tau}_I^\Delta(H_\emptyset(g)-T)\frac{dg}{dz_\infty d\pi_I}\Big)\frac{dn_I}{d\nu_I}$$

and thus to

$$\sum_{I\subset\Delta}(-1)^{|\Delta-I|}\int_{N_I(F)\backslash N_I(\mathbb{A})}\Big(\int_{F_\infty^\times P_I(F)\backslash G(\mathbb{A})}\overline{\varphi_1(g)}\varphi_2(n_I^{-1}g)\widehat{\tau}_I^\Delta(H_\emptyset(g)-T)\frac{dg}{dz_\infty d\pi_I}\Big)\frac{dn_I}{d\nu_I}$$

which is nothing else than the inner product $\langle\varphi_1,\Lambda^T\varphi_2\rangle$ and (11.1.3)(iii) is proved. □

In the next section we will need generalizations of the operators Λ^T for the standard parabolic subgroups of G. Let us fix $I\subset\Delta$. An obvious extension of the proof of lemma (11.1.1) shows that, for any $J\subset I$, any $T\in\mathfrak{a}_\emptyset^+$ and any $g\in G(\mathbb{A})$, the set

$$\{\pi_I\in P_J(F)\backslash P_I(F)\mid\widehat{\tau}_J^I(H_\emptyset(\pi_Ig)-T)\neq 0\}$$

is finite. Then, for each $T\in\mathfrak{a}_\emptyset^+$ and each complex function φ on $N_I(\mathbb{A})M_I(F)\backslash G(\mathbb{A})$ which is right K'-invariant for some compact open subgroup K' of $K\subset G(\mathbb{A})$, we define the function $\Lambda^{T,I}\varphi$ on $N_I(\mathbb{A})M_I(F)\backslash G(\mathbb{A})$ by

$$(\Lambda^{T,I}\varphi)(g)=\sum_{J\subset I}(-1)^{|I-J|}\sum_{\pi_I\in P_J(F)\backslash P_I(F)}\widehat{\tau}_J^I(H_\emptyset(\pi_Ig)-T)\varphi_J(\pi_Ig)\qquad(\forall g\in G(\mathbb{A})).\tag{11.1.6}$$

LEMMA (11.1.7). — *Let $\varphi \in \mathcal{C}_G^\infty$. Then for each $T \in \mathfrak{a}_\emptyset^+$ and each $g \in G(\mathbb{A})$ we have*

$$\varphi(g) = \sum_{I \subset \Delta} \sum_{\delta \in P_I(F)\backslash G(F)} \tau_I^\Delta(H_\emptyset(\delta g) - T)(\Lambda^{T,I}\varphi_I)(\delta g).$$

Proof: We have $(\varphi_I)_J = \varphi_J$, $\forall J \subset I \subset \Delta$, and

$$\sum_{J \subset I \subset \Delta} (-1)^{|\Delta - I|}\tau_I^\Delta \hat{\tau}_J^I = \delta_J^\Delta \qquad (\forall J \subset \Delta)$$

(write (10.3.2) as an equality of matrices). □

(11.2) $J_{\mathrm{geom}}(f)$ as the integral over the diagonal of the truncated kernel

Let us fix a locally constant complex function with compact support f on $F_\infty^\times\backslash G(\mathbb{A})$ and a Haar measure dg on $G(\mathbb{A})$ with a splitting

$$dg = \prod_x dg_x$$

as in (10.1). Then for any fixed $h \in G(\mathbb{A})$ the function of $g \in G(\mathbb{A})$,

$$K(h, g) = \sum_{\gamma \in G(F)} f(h^{-1}\gamma g),$$

belongs to $\mathcal{C}_{G,1_\infty}^\infty$ and we may apply to it the operators Λ^T, $(\cdot)_I$ and $\Lambda_2^{T,I}(\cdot)_I$ with $T \in \mathfrak{a}_\emptyset^+$ and $I \subset \Delta$. We will denote by

$$\Lambda_2^T K(h, g),\ K(h, g)_I \text{ and } \Lambda_2^{T,I} K(h, g)_I$$

the resulting functions.

LEMMA (11.2.1). — *For each $T \in \mathfrak{a}_\emptyset^+$ and each $(h, g) \in G(\mathbb{A})^2$ we have*

$$K(h, g) = \sum_{I \subset \Delta} \sum_{\delta \in P_I(F)\backslash G(F)} \tau_I^\Delta(H_\emptyset(\delta g) - T)\Lambda_2^{T,I} K(\delta h, \delta g)_I.$$

Proof: Apply lemma (11.1.7) to the function $K(h, g)$ of g and remark that $\Lambda_2^{T,I} K(h, g)_I$ is left $G(F)$-invariant as a function of h. □

Let us fix $I \subset \Delta$. Motivated by the last lemma we will study the constant term

$$K(h,g)_I = \int_{N_I(F)\backslash N_I(\mathbb{A})} K(h, n_I g)\frac{dn_I}{d\nu_I} \qquad (h, g \in G(\mathbb{A})).$$

Let

$$\mathrm{L}^2_{G,1_\infty,I} = \mathrm{L}^2(F_\infty^\times N_I(\mathbb{A})M_I(F)\backslash G(\mathbb{A}), \frac{dg}{dz_\infty dn_I d\mu_I})$$

be the Hilbert space of measurable complex functions φ on $F_\infty^\times N_I(\mathbb{A})M_I(F)\backslash G(\mathbb{A})$ such that

$$\int_{F_\infty^\times N_I(\mathbb{A})M_I(F)\backslash G(\mathbb{A})} |\varphi(g)|^2 \frac{dg}{dz_\infty dn_I d\mu_I} < +\infty$$

($d\mu_I$ is the counting measure on $M_I(F)$). For any $\varphi \in \mathrm{L}^2_{G,1_\infty,I}$ and almost every $h \in G(\mathbb{A})$ the integral

$$R_{G,1_\infty,I}(f)(\varphi)(h) = \int_{F_\infty^\times\backslash G(\mathbb{A})} f(g)\varphi(hg)\frac{dg}{dz_\infty}$$

is absolutely convergent and we have

$$R_{G,1_\infty,I}(f)(\varphi) \in \mathrm{L}^2_{G,1_\infty,I}$$

(compare to (10.1)). It is easy to see that, for each $(h,g) \in G(\mathbb{A})^2$, the sum of integrals,

$$K_I(h,g) = \sum_{\mu_I \in M_I(F)} \int_{N_I(\mathbb{A})} f(h^{-1}\mu_I n_I g) dn_I,$$

is absolutely convergent and defines a locally constant function on

$$(F_\infty^\times N_I(\mathbb{A})M_I(F)\backslash G(\mathbb{A}))^2.$$

LEMMA (11.2.2). — *The operator $R_{G,1_\infty,I}(f)$ on $L^2_{G,1_\infty,I}$ is an operator with kernel $K_I(h,g)$.*

Proof: See the proof of (10.1.2). □

LEMMA (11.2.3). — *For each $(h,g) \in G(\mathbb{A})^2$ we have*

$$K(h,g)_I = \sum_{\gamma \in P_I(F)\backslash G(F)} K_I(\gamma h, g),$$

the sum being absolutely convergent.

Proof: Since $N_I(F)\backslash N_I(\mathbb{A})$ is compact (see (9.2.1)) and since $h\mathrm{Supp}(f)g^{-1}$ is compact modulo $F_\infty^\times$ there are only finitely many $\gamma N_I(F)$ in $G(F)/N_I(F)$ such that

$$\gamma N_I(\mathbb{A}) \cap h\mathrm{Supp}(f)g^{-1} \neq \emptyset.$$

Moreover, for any fixed $h, g \in G(\mathbb{A})$ and $\gamma \in G(F)$ the integral

$$\int_{N_I(\mathbb{A})} f(h^{-1}\gamma n_I g)dn_I$$

is absolutely convergent. Therefore we can split the sum over $\gamma \in G(F)$ in the definition of $K(h,g)_I$ into a sum over $\nu_I \in N_I(F)$, a sum over $\mu_I \in M_I(F)$ and a sum over $\gamma^{-1} \in P_I(F)\backslash G(F)$. Then we can permute the sum over (γ^{-1}, μ_I) and the integral over $n_I \in N_I(F)\backslash N_I(\mathbb{A})$ and we can combine this last integral with the sum over ν_I. □

For any $T \in \mathfrak{a}_\emptyset^+$ we set

$$J_{\mathrm{spec}}^T(f) = \int_{F_\infty^\times G(F)\backslash G(\mathbb{A})} \Lambda_2^T K(g,g)\frac{dg}{dz_\infty d\gamma}. \tag{11.2.4}$$

LEMMA (11.2.5). — *If f satisfies hypotheses* (A) *and* (B) *of* (10.2), *we have*

$$\int_{N_I(\mathbb{A})} f(g^{-1}\delta^{-1}\gamma^{-1}n_I\delta g)dn_I = 0$$

for any $I \subsetneqq \Delta$, any $g \in G(\mathbb{A})$, any $\delta \in G(F)$ and any $\gamma \in G(F) - G_I^\circ(F)$, where G_I° is the Zariski open subset

$$G - \bigcup_{\substack{K \\ I\subset K\subsetneqq\Delta}} P_K$$

of G.

Proof : If $I \subset K \subsetneqq \Delta$ and if $\gamma = \nu_K\mu_K \in P_K(F) = N_K(F)M_K(F)$ we have

$$n_I = n_I^K n_K,$$
$$\gamma^{-1}n_I = \mu_K^{-1}n_I^K n_K'$$

with $n_K' = ((n_I^K)^{-1}\nu_K^{-1}n_I^K)n_K$, and

$$\int_{N_I(\mathbb{A})} f(g^{-1}\delta^{-1}\gamma^{-1}n_I\delta g)dn_I$$
$$= \int_{N_I^K(\mathbb{A})}\Big(\int_{N_K(\mathbb{A})} f(g^{-1}\delta^{-1}\mu_K^{-1}n_I^K n_K'\delta g)dn_K'\Big)dn_I^K.$$

Therefore the lemma follows from (10.2.3). □

Now, applying the truncation operator $\Lambda_2^{T,I}$ to both sides of the equality in (11.2.3) and combining the result with (11.2.1), we obtain the formula (11.2.6)

$$k(g) = \sum_{I\subset\Delta} \sum_{\delta\in P_I(F)\backslash G(F)} \tau_I^{\Delta}(H_\emptyset(\delta g) - T) \sum_{\gamma\in P_I(F)\backslash G_I^\circ(F)} \Lambda_2^{T,I} K_I(\gamma\delta g, \delta g)$$

for every $g \in G(\mathbb{A})$.

THEOREM (11.2.7). — *Let us assume that f satisfies hypotheses* (A) *and* (B) *of* (10.2). *Then*

(i) *for each $I \subset \Delta$ we have*

$$\int_{F_\infty^\times P_I(F)\backslash G(\mathbb{A})} \tau_I^{\Delta}(H_\emptyset(g) - T)\Big(\sum_{\gamma\in P_I(F)\backslash G_I^\circ(F)} |\Lambda_2^{T,I} K_I(\gamma g, g)|\Big) \frac{dg}{dz_\infty d\pi_I} < +\infty,$$

(ii) *for each $I \subsetneqq \Delta$ there exists a non-negative rational number C_I such that, for any $T \in \mathfrak{a}_\emptyset^+$ with $\alpha_i(T) \geq C_I$, $\forall i = 1, \ldots, d-1$, the (absolutely convergent) integral*

$$\int_{F_\infty^\times P_I(F)\backslash G(\mathbb{A})} \tau_I^{\Delta}(H_\emptyset(g) - T)\Big(\sum_{\gamma\in P_I(F)\backslash G_I^\circ(F)} \Lambda_2^{T,I} K_I(\gamma g, g)\Big) \frac{dg}{dz_\infty d\pi_I}$$

vanishes.

COROLLARY (11.2.8). — *Let us assume that f satisfies hypotheses* (A) *and* (B) *of* (10.2). *If $T \in \mathfrak{a}_\emptyset^+$ and if*

$$\alpha_i(T) \geq \sup\{C_I \mid I \subsetneqq \Delta\} \qquad (\forall i = 1, \ldots, d-1)$$

we have

$$J_{\mathrm{geom}}(f) = J_{\mathrm{spec}}^T(f).$$

□

REMARK (11.2.9). — Part (i) of (11.2.7) gives another proof of the absolute convergence of the integral

$$\int_{F_\infty^\times G(F)\backslash G(\mathbb{A})} k(g) \frac{dg}{dz_\infty d\gamma}.$$

□

Proof of (11.2.7)(i) : For any non-negative real function ψ on $F_\infty^\times P_I(F)\backslash G(\mathbb{A})$ which is right K'-invariant for some compact open subgroup K' of $K \subset G(\mathbb{A})$ we have the integration formula

$$\int_{F_\infty^\times P_I(F)\backslash G(\mathbb{A})} \psi(g)\frac{dg}{dz_\infty d\pi_I}$$
$$= \int_{N_I(F)\backslash N_I(\mathbb{A})}\Big(\int_{F_\infty^\times M_I(F)\backslash M_I(\mathbb{A})} \delta_I^{-1}(m_I)\Big(\int_K \psi(n_I m_I k)dk\Big)\frac{dm_I}{dz_\infty d\mu_I}\Big)\frac{dn_I}{d\nu_I}$$

where the Haar measures dn_I, dm_I and dk on $N_I(\mathbb{A})$, $M_I(\mathbb{A})$ and K are now normalized by $\mathrm{vol}(N_I(\mathbb{A})\cap K, dn_I) = 1$, $\mathrm{vol}(M_I(\mathbb{A})\cap K, dm_I) = 1$ and $\mathrm{vol}(K, dk) = \mathrm{vol}(K, dg)$ and where $d\nu_I$ and $d\mu_I$ are the counting measures on $N_I(F)$ and $M_I(F)$ respectively (recall that $\delta_I = \delta_{P_I(\mathbb{A})}$ is the modulus character of $P_I(\mathbb{A})$). Therefore, since $N_I(F)\backslash N_I(\mathbb{A})$ and K are compact, in order to prove that

$$\int_{F_\infty^\times P_I(F)\backslash G(\mathbb{A})} \psi(g)\frac{dg}{dz_\infty d\pi_I} < +\infty$$

it is sufficient to prove that, for any $n_I \in N_I(\mathbb{A})$ and any $k \in K$, the integral

$$\int_{Z_I(F_\infty)M_I(F)\backslash M_I(\mathbb{A})}\Big(\int_{F_\infty^\times\backslash Z_I(F_\infty)} \delta_I^{-1}(z_{I,\infty}m_I)\psi(n_I z_{I,\infty} m_I k) \frac{dz_{I,\infty}}{dz_\infty}\Big)\frac{dm_I}{dz_{I,\infty}d\mu_I}$$

is finite.

Now let us consider the function

$$\psi(g) = \tau_I^\Delta(H_\emptyset(g) - T)\sum_{\gamma\in P_I(F)\backslash G_I^\circ(F)} |\Lambda_2^{T,I}K_I(\gamma g, g)| \qquad (g \in G(\mathbb{A})).$$

According to the Bruhat decomposition we have

$$G_I^\circ(F) = \coprod_w P_I(F)\dot{w}N_\emptyset(F)$$

where w runs through

$$D_{I,\emptyset}^\circ \overset{\mathrm{dfn}}{=\!=} D_{I,\emptyset}\cap\Big(W - \bigcup_{\substack{K\\ I\subset K\subsetneq\Delta}} W_K\Big).$$

For each such w let us set

$$\psi_w(g) = \tau_I^\Delta(H_\emptyset(g) - T)\sum_{\nu_\emptyset} |\Lambda_2^{T,I}K_I(\dot{w}\nu_\emptyset g, g)| \qquad (g \in G(\mathbb{A}))$$

where $\nu_\emptyset$ runs through $(\dot{w}^{-1}P_I(F)\dot{w}\cap N_\emptyset(F))\backslash N_\emptyset(F)$, so that

$$\psi = \sum_w \psi_w.$$

If K' is a compact open subgroup of $K\subset G(\mathbb{A})$ such that f is K'-bi-invariant we have

$$K_I(hl', gk') = K_I(h,g)$$

for all $g,h\in G(\mathbb{A})$ and all $l',k'\in K'$ and each ψ_w is right K'-invariant. To finish the proof of (11.2.7)(i) let us show that, for each w as above, each $n_I\in N_I(\mathbb{A})$ and each $k\in K$, the function

$$m_I\mapsto \int_{F_\infty^\times\backslash Z_I(F_\infty)} \delta_I^{-1}(z_{I,\infty}m_I)\psi_w(n_Iz_{I,\infty}m_Ik)\frac{dz_{I,\infty}}{dz_\infty}$$

takes only finite values and has a compact support in $Z_I(F_\infty)M_I(F)\backslash M_I(\mathbb{A})$.

We begin with the following remarks. If $i\in\Delta$ and $w\in W$ the map

$$\mathfrak{a}_\emptyset\to\mathbb{Q},\ H\mapsto \varpi_i(H-w(H))$$

is a linear combination of simple roots with non-negative integral coefficients

$$\varpi_i(H-w(H))=\sum_{j\in\Delta} a_{ij}(w)\alpha_j(H)\qquad(\forall H\in\mathfrak{a}_\emptyset).$$

Indeed, if $\alpha\in\Delta$ and if $w=w's_\alpha$ for some $w'\in W$ of length $\ell(w)-1$ we have

$$\varpi_i(H-w(H))=\varpi_i(H-w'(H))+\varpi_i(w'(H-s_\alpha(H)))$$

and we have $H-s_\alpha(H)=\alpha(H)\alpha^\vee$, $w'(\alpha^\vee)\in R^{\vee+}$ and $\varpi_i(\beta^\vee)\in\mathbb{Z}_{\geq 0}$, $\forall\beta^\vee\in R^{\vee+}$ (s_α is the simple reflexion corresponding to α, $\alpha^\vee$ is the simple coroot corresponding to α and $R^{\vee+}$ is the set of positive coroots; recall that $\{\alpha^\vee\mid\alpha\in\Delta\}$ is the dual basis of $\{\varpi_i\mid i=1,\dots,d-1\}$). If we assume furthermore that $w\notin W_{\Delta-\{i\}}$ we have

$$a_{ii}(w)=\varpi_i(\varpi_i^\vee-w(\varpi_i^\vee))>0$$

where

$$\varpi_i^\vee=\frac{d-i}{d}\sum_{k=1}^{i}k\alpha_k^\vee+\frac{i}{d}\sum_{k=i+1}^{d-1}(d-k)\alpha_k^\vee$$

is the coweight corresponding to ϖ_i ($\{\varpi_i^\vee\mid i=1,\dots,d-1\}$ is the dual basis of Δ).

Secondly let us denote by a (resp. b) the minimum (resp. the maximum) of

$$-\varpi_i(H_\emptyset(kg))\qquad(\text{resp. } -\varpi_i(H_\emptyset(kg^{-1}))$$

for $k \in K$, $g \in \mathrm{Supp}(f)$ and $i = 1, \dots, d-1$ (recall that $F_\infty^\times \backslash \mathrm{Supp}(f)$ is compact). If $g, h \in G(\mathbb{A})$, $w \in W$ and $\nu_\emptyset \in N_\emptyset(F)$ are such that

$$\Lambda_2^{T,I} K_I(\dot{w}\nu_\emptyset h, g) \neq 0$$

there exist at least one $J \subset \Delta$, one $\pi_I \in P_I(F)$, one $\mu_I' \in M_I(F)$, one $n_I' \in N_I'(\mathbb{A})$ and one $n_J \in N_J(\mathbb{A})$ such that

$$h^{-1}\nu_\emptyset^{-1}\dot{w}^{-1}\mu_I' n_I' n_J \pi_I g \in \mathrm{Supp}(f)$$

$(K_I(\dot{w}\nu_\emptyset h, \pi_I g)_J \neq 0)$. But, on the one hand, there exists $k \in K$ such that

$$\begin{aligned}\varpi_i(H_I(\dot{w}\nu_\emptyset h) - H_I(g)) &= \varpi_i(H_I(\dot{w}\nu_\emptyset h) - H_I(\mu_I' n_I' n_J \pi_I g)) \\ &= -\varpi_i(H_I(kh^{-1}\nu_\emptyset^{-1}\dot{w}^{-1}\mu_I' n_I' n_J \pi_I g)) \\ &= -\varpi_i(H_\emptyset(kh^{-1}\nu_\emptyset^{-1}\dot{w}^{-1}\mu_I' n_I' n_J \pi_I g))\end{aligned}$$

for all $i \in \Delta - I$ and it follows that

$$\varpi_i(H_I(\dot{w}\nu_\emptyset h) - H_I(g)) \geq a \qquad (\forall i \in \Delta - I).$$

On the other hand, there exist $t, \tilde{t} \in T(\mathbb{A})$ and $u, \tilde{u} \in U(\mathbb{A})$ such that

$$\begin{cases} \mu_I' n_I' n_J \pi_I g \in tuK, \\ \nu_\emptyset h \in \tilde{t}\tilde{u}K, \end{cases}$$

and

$$\begin{cases} H_\emptyset(t) = H_\emptyset(\mu_I' \mu_I g), \\ H_\emptyset(\tilde{t}) = H_\emptyset(h) \end{cases}$$

$(\pi_I = \mu_I \nu_I,\ \mu_I \in M_I(F),\ \nu_I \in N_I(F))$. Moreover we have

$$H_\emptyset(u^{-1}t^{-1}\dot{w}\tilde{t}\tilde{u}) = w(H_\emptyset(\tilde{t})) - H_\emptyset(t) + H_\emptyset(\dot{w}\tilde{u}\dot{w}^{-1})$$

and

$$\varpi_i(H_\emptyset(\dot{w}\tilde{u}\dot{w}^{-1})) \leq 0 \qquad (\forall i = 1, \dots, d-1)$$

(see the proof of (11.1.1)). Thus it follows that

$$\varpi_i\big(H_\emptyset(t) - w(H_\emptyset(\tilde{t}))\big) \leq b \qquad (\forall i = 1, \dots, d-1)$$

and consequently that

$$\varpi_i\big(H_I(g) - w(H_I(h))\big) \leq b + \varpi_i\big(w(H_\emptyset^I(h))\big) \qquad (\forall i \in \Delta - I).$$

Thirdly let us fix $n_I \in N_I(\mathbb{A})$, $m_I \in M_I(\mathbb{A})$, $k \in K$ and $w \in D_{I,\emptyset}^{\circ}$. From what we have just shown it follows that any $z_{I,\infty} \in Z_I(F_\infty)$ and any $\nu_\emptyset \in N_\emptyset(F)$ such that

$$\Lambda_2^{T,I} K_I(\dot{w}\nu_\emptyset n_I z_{I,\infty} m_I k, n_I z_{I,\infty} m_I k) \neq 0$$

satisfy the conditions

$$\varpi_i\big(H_I(\dot{w}\nu_\emptyset n_I z_{I,\infty} m_I k) - H_I(z_{I,\infty} m_I)\big) \geq a$$

and

$$\sum_{j \in \Delta - I} a_{ij}(w)\alpha_j(H_I(z_{I,\infty} m_I)) \leq b + \varpi_i\big(w(H_\emptyset^I(m_I))\big)$$

for all $i \in \Delta - I$ (we have $\alpha_j(H_I(z_{I,\infty} m_I)) = 0$ for every $j \in I$). Therefore any $z_{I,\infty} \in Z_I(F_\infty)$ and any $\nu_\emptyset \in N_\emptyset(F)$ such that

$$\tau_I^\Delta(H_\emptyset(n_I z_{I,\infty} m_I k) - T)\Lambda_2^{T,I} K_I(\dot{w}\nu_\emptyset n_I z_{I,\infty} m_I k, n_I z_{I,\infty} m_I k) \neq 0$$

satisfy the conditions

$$\varpi_i\big(H_I(\dot{w}\nu_\emptyset n_I z_{I,\infty} m_I k) - H_I(z_{I,\infty} m_I)\big) \geq a$$

and

$$\alpha_i(T_I) \leq \alpha_i(H_I(z_{I,\infty} m_I)) \leq \frac{b + \varpi_i\big(w(H_\emptyset^I(m_I))\big)}{a_{ii}(w)} - \sum_{\substack{j \in \Delta - I \\ j \neq i}} \frac{a_{ij}(w)}{a_{ii}(w)}\alpha_j(T_I)$$

for all $i \in \Delta - I$. But the second of these conditions implies that $F_\infty^\times z_{I,\infty}$ belongs to a compact subset of $F_\infty^\times \backslash Z_I(F_\infty)$ which is independent of $\nu_\emptyset$ and then the first of these conditions implies that $\dot{w}\nu_\emptyset$ belongs to a fixed finite subset of $P_I(F)\backslash G(F)$ (see (11.1.1)), i.e. that $\nu_\emptyset$ belongs to a finite subset of $(\dot{w}^{-1}P_I(F)\dot{w} \cap N_\emptyset(F))\backslash N_\emptyset(F)$ which is independent of $z_{I,\infty}$. It follows that

$$\int_{F_\infty^\times \backslash Z_I(F_\infty)} \delta_I^{-1}(z_{I,\infty} m_I)\psi_w(n_I z_{I,\infty} m_I k)\frac{dz_{I,\infty}}{dz_\infty} < +\infty.$$

Finally, for any $h \in G(\mathbb{A})$ and any $k \in K$ let us set

$$\varphi_{h,k}(m_I) = K_I(h, m_I k) \qquad (\forall m_I \in M_I(\mathbb{A})).$$

Then $\varphi_{h,k}$ is right $(M_I(\mathbb{A}) \cap kK'k^{-1})$-invariant and, by an obvious extension of (11.1.3)(ii) (replace G by M_I and use the same arguments), there exists a subset

$$\Omega_I = \Omega_I(K', k) \subset M_I(F)\backslash M_I(\mathbb{A})$$

which is independent of h and which is compact modulo $Z_I(F_\infty)$ such that the support of the function

$$\Lambda_2^{T,I}K_I(h,m_Ik)=\sum_{J\subset I}(-1)^{|I-J|}$$
$$\sum_{\mu_I\in P_J^I(F)\backslash M_I(F)}\widehat{\tau}_J^I(H_\emptyset(\mu_Im_I)-T)(\varphi_{h,k})_J(\mu_Im_I)$$

of m_I is contained in Ω_I. Since $\Lambda_2^{T,I}K_I(h,g)$ is left $N_I(\mathbb{A})$-invariant as a function of g it follows that, for any fixed w, n_I and k, the support of the function

$$m_I\mapsto\psi_w(n_Im_Ik)$$

is contained in $\Omega_I(K',k)$. We have thus shown that the function

$$m_I\mapsto\int_{F_\infty^\times\backslash Z_I(F_\infty)}\delta_I^{-1}(z_{I,\infty}m_I)\psi_w(n_Iz_{I,\infty}m_Ik)\frac{dz_{I,\infty}}{dz_\infty}$$

is compactly supported in $Z_I(F_\infty)M_I(F)\backslash M_I(\mathbb{A})$ and the proof of (11.2.7)(i) is completed. □

Proof of (11.2.7)(ii) : Let us recall that, for each $I\subset\Delta$, we have the Bruhat decomposition

$$G(F)=\coprod_{w\in D_{I,I}}P_I(F)\dot{w}P_I(F)$$

where we have set

$$D_{I,I}=\{w\in W\mid w^{-1}(I)\subset R^+\text{ and }w(I)\subset R^+\}$$

(see (5.4.1) and (5.4.3) or [Ca] (2.8.1), (2.8.6) and (2.8.9)). According to this decomposition, for each $I\subsetneq\Delta$ we have the decomposition

$$G^\circ(F)=\coprod_w P_I(F)\dot{w}P_I(F)$$

where w now runs through

$$D_{I,I}^\circ\overset{\text{dfn}}{=\!=}D_{I,I}\cap\Big(W-\bigcup_{\substack{K\\ I\subset K\subsetneq\Delta}}W_K\Big)$$

and (11.2.7)(ii) is a consequence of the following stronger statement.

For each $I \subsetneq \Delta$ and each $w \in D_{I,I}^{\circ}$ the integral

$$(*) \qquad \int_{F_\infty^\times P_I(F)\backslash G(\mathbb{A})} \tau_I^\Delta(H_\emptyset(g)-T)\Big(\sum_{\pi_I}\Lambda_2^{T,I}K_I(\dot{w}\pi_I g, g)\Big)\frac{dg}{dz_\infty d\pi_I},$$

where π_I runs through $(\dot{w}^{-1}P_I(F)\dot{w}\cap P_I(F))\backslash P_I(F)$, vanishes if $\alpha_i(T)$ is large enough for every $i = 0,\ldots,d-1$.

Since

$$\tau_I^\Delta(H_\emptyset(g)-T)\Lambda_2^{T,I}K_I(\dot{w}\pi_I g, g) = \tau_I^\Delta(H_\emptyset(\pi_I g)-T)\Lambda_2^{T,I}K_I(\dot{w}\pi_I g, \pi_I g)$$

for all $g \in G(\mathbb{A})$ and all $\pi_I \in P_I(F)$ the integral $(*)$ is equal to

$$(**) \qquad \int_{F_\infty^\times(\dot{w}^{-1}P_I(F)\dot{w}\cap P_I(F))\backslash G(\mathbb{A})} \tau_I^\Delta(H_\emptyset(g)-T)\Lambda_2^{T,I}K_I(\dot{w}g,g)\frac{dg}{dz_\infty d\pi_I^w}$$

where $d\pi_I^w$ is the counting measure on $\dot{w}^{-1}P_I(F)\dot{w}\cap P_I(F)$.

As $w^{-1}\in D_{I,I}$ we have

$$\dot{w}^{-1}P_I\dot{w}\cap P_I = (\dot{w}^{-1}P_I\dot{w}\cap N_I)(\dot{w}^{-1}P_I\dot{w}\cap M_I)$$

and

$$\dot{w}^{-1}P_I\dot{w}\cap M_I = (\dot{w}^{-1}N_I\dot{w}\cap M_I)(\dot{w}^{-1}M_I\dot{w}\cap M_I) = N_K^I M_K = P_K^I$$

where we have put

$$K = I\cap w^{-1}(I)\subset I$$

(see [Ca] (2.8.6), (2.8.7) and (2.8.9)). Then the projection

$$N_K(\mathbb{A})M_K(F)\to N_K(\mathbb{A}),\ n_K\mu_K\mapsto n_K,$$

induces an isomorphism

$$(\dot{w}^{-1}P_I(F)\dot{w}\cap P_I(F))\backslash N_K(\mathbb{A})M_K(F) \xrightarrow{\sim} (\dot{w}^{-1}P_I(F)\dot{w}\cap N_I(F))N_K^I(F)\backslash N_I(\mathbb{A})N_K^I(\mathbb{A}).$$

Therefore, if ψ is a complex function on $F_\infty^\times(\dot{w}^{-1}P_I(F)\dot{w}\cap P_I(F))\backslash G(\mathbb{A})$ which is absolutely integrable and right K'-invariant for some compact open subgroup K' of $K\subset G(\mathbb{A})$ we can decompose the integral

$$\int_{F_\infty^\times(\dot{w}^{-1}P_I(F)\dot{w}\cap P_I(F))\backslash G(\mathbb{A})}\psi(g)\frac{dg}{dz_\infty d\pi_I^w}$$

into

$$\int_{F_\infty^\times N_K(\mathbb{A})M_K(F)\backslash G(\mathbb{A})}\left(\int_{N_K^I(F)\backslash N_K^I(\mathbb{A})}\left(\int_{(\dot{w}^{-1}P_I(F)\dot{w}\cap N_I(F))\backslash N_I(\mathbb{A})}\psi(n_In_K^Ig)\right.\right.$$
$$\left.\left.\times\frac{dn_I}{d\nu_I^w}\right)\frac{dn_K^I}{d\nu_K^I}\right)\frac{dg}{dz_\infty dn_K d\mu_K}.$$

Here $d\nu_K^I$, $d\nu_I^w$ and $d\mu_K$ are the counting measures on $N_K^I(F)$, $\dot{w}^{-1}P_I(F)\dot{w}\cap N_I(F)$ and $M_K(F)$ respectively and the Haar measures dn_K^I on $N_K^I(\mathbb{A})$, dn_I on $N_I(\mathbb{A})$ and dn_K on $N_K(\mathbb{A})$ are normalized by

$$\mathrm{vol}\big(N_K^I(F)\backslash N_K^I(\mathbb{A}),\frac{dn_K^I}{d\nu_K^I}\big)=\mathrm{vol}\big(N_I(F)\backslash N_I(\mathbb{A}),\frac{dn_I}{d\nu_I}\big)=1,$$

where $d\nu_I$ is the counting measure on $N_I(F)$, and by

$$dn_K=dn_Idn_K^I.$$

As $w\in D_{I,I}$ we have

$$\dot{w}^{-1}P_I\dot{w}\cap N_I=(\dot{w}^{-1}N_I\dot{w}\cap N_I)(\dot{w}^{-1}N_L^I\dot{w})$$

where we have put

$$L=I\cap w(I)=w(K)\subset I$$

(see [Ca] (2.8.6) and (2.8.9)). Therefore, for any ψ as above we can decompose the integral

$$\int_{(\dot{w}^{-1}P_I(F)\dot{w}\cap N_I(F))\backslash N_I(\mathbb{A})}\psi(n_In_K^Ig)\frac{dn_I}{d\nu_I^w}$$

into

$$\int_{(\dot{w}^{-1}P_I(\mathbb{A})\dot{w}\cap N_I(\mathbb{A}))\backslash N_I(\mathbb{A})}\left(\int_{(\dot{w}^{-1}P_I(F)\dot{w}\cap N_I(F))\backslash(\dot{w}^{-1}P_I(\mathbb{A})\dot{w}\cap N_I(\mathbb{A}))}\psi(n_I^wn_In_K^Ig)\right.$$
$$\left.\times\frac{dn_I^w}{d\nu_I^w}\right)\frac{dn_I}{dn_I^w}$$

and the integral

$$\int_{(\dot{w}^{-1}P_I(F)\dot{w}\cap N_I(F))\backslash(\dot{w}^{-1}P_I(\mathbb{A})\dot{w}\cap N_I(\mathbb{A}))}\psi(n_I^wn_In_K^Ig)\frac{dn_I^w}{d\nu_I^w}$$

into

$$\int_{N_L^I(F)\backslash N_L^I(\mathbb{A})}\left(\int_{(\dot{w}^{-1}N_I(F)\dot{w}\cap N_I(F))\backslash(\dot{w}^{-1}N_I(\mathbb{A})\dot{w}\cap N_I(\mathbb{A}))}\psi(\tilde{n}_I^w(\dot{w}^{-1}n_L^I\dot{w})n_In_K^Ig)\right.$$
$$\left.\times\frac{d\tilde{n}_I^w}{d\tilde{\nu}_I^w}\right)\frac{dn_L^I}{d\nu_L^I}$$

for every $n_K^I \in N_K^I(\mathbb{A})$ and every $g \in G(\mathbb{A})$. Here $d\nu_I^w$, $d\nu_L^I$ and $d\tilde{\nu}_I^w$ are the counting measures and the Haar measures dn_I^w , dn_L^I and $d\tilde{n}_I^w$ are normalized by

$$\mathrm{vol}\big((\dot{w}^{-1}P_I(F)\dot{w}\cap N_I(F))\backslash(\dot{w}^{-1}P_I(\mathbb{A})\dot{w}\cap N_I(\mathbb{A})), \frac{dn_I^w}{d\nu_I^w}\big) = 1,$$

$$\mathrm{vol}\big(N_L^I(F)\backslash N_L^I(\mathbb{A}), \frac{dn_L^I}{d\nu_L^I}\big) = 1$$

and

$$dn_I^w = d\tilde{n}_I^w d(\dot{w}^{-1}\nu_L^I\dot{w}).$$

But, for each $\tilde{n}_I^w \in \dot{w}^{-1}N_I(\mathbb{A})\dot{w}\cap N_I(\mathbb{A})$, each $n_L^I \in N_L^I(\mathbb{A})$, each $n_I \in N_I(\mathbb{A})$, each $n_K^I \in N_K^I(\mathbb{A})$ and each $g \in G(\mathbb{A})$, we have

$$\begin{aligned}&\Lambda_2^{T,I}K_I(\dot{w}\tilde{n}_I^w(\dot{w}^{-1}n_L^I\dot{w})n_In_K^Ig, \tilde{n}_I^w(\dot{w}^{-1}n_L^I\dot{w})n_In_K^Ig)\\ &= \Lambda_2^{T,I}K_I(n_L^I\dot{w}n_I'g, n_K^Ig)\end{aligned}$$

where we have put

$$n_I' = (n_K^I)^{-1}n_In_K^I$$

(we have $\dot{w}\tilde{n}_I^w\dot{w}^{-1} \in N_I(\mathbb{A})$, $n_L^I(\dot{w}n_K^I\dot{w}^{-1})(n_L^I)^{-1} \in N_I(\mathbb{A})$ and $\tilde{n}_I^w(\dot{w}^{-1}n_L^I\dot{w})n_I \in N_I(\mathbb{A})$).

We have thus shown that the integral $(*) = (**)$ is equal to the integral over $g \in F_\infty^\times N_K(\mathbb{A})M_K(F)\backslash G(\mathbb{A})$ of the product of $\tau_I^\Delta(H_\emptyset(g) - T)$ with

$$\int_{(\dot{w}^{-1}P_I(\mathbb{A})\dot{w}\cap N_I(\mathbb{A}))\backslash N_I(\mathbb{A})}\Big(\int_{N_K^I(F)\backslash N_K^I(\mathbb{A})}\Big(\int_{N_L^I(F)\backslash N_L^I(\mathbb{A})}\Lambda_2^{T,I}K_I(n_L^I\dot{w}n_I'g, n_K^Ig)\times\frac{dn_L^I}{d\nu_L^I}\Big)\frac{dn_K^I}{d\nu_K^I}\Big)\frac{dn_I'}{dn_I^w}$$

(we have

$$\mathrm{vol}\big((\dot{w}^{-1}N_I(F)\dot{w}\cap N_I(F))\backslash(\dot{w}^{-1}N_I(\mathbb{A})\dot{w}\cap N_I(\mathbb{A})), \frac{d\tilde{n}_I^w}{d\tilde{\nu}_I^w}\big) = 1,$$

$$(n_K^I)^{-1}(\dot{w}^{-1}P_I(\mathbb{A})\dot{w}\cap N_I(\mathbb{A}))n_K^I \subset \dot{w}^{-1}P_I(\mathbb{A})\dot{w}\cap N_I(\mathbb{A})$$

and

$$H_\emptyset(\tilde{n}_I^w(\dot{w}^{-1}n_L^I\dot{w})n_In_K^Ig) = H_\emptyset(g)$$

and we have $dn_I' = dn_I$ and $d((n_K^I)^{-1}n_I^wn_K^I) = dn_I^w$). In particular the non-vanishing of the integral $(*)$ implies that there exist $g \in G(\mathbb{A})$, $n_L^I \in N_L^I(\mathbb{A})$ and $n_I' \in N_I(\mathbb{A})$ satisfying the following conditions:

(a) $$\tau_I^\Delta(H_\emptyset(g) - T) \neq 0$$

and

$$\text{(b)} \qquad \int_{N_K^I(F)\backslash N_K^I(\mathbb{A})} \Lambda_2^{T,I} K_I(n_L^I \dot{w} n_I' g, n_K^I g) \frac{dn_K^I}{d\nu_K^I} \neq 0.$$

But, on the one hand, by an obvious extension of lemma (11.1.4) (replace G by M_I) condition (b) implies that

$$\varpi_i^I(H_\emptyset(g) - T) \leq 0 \qquad (\forall i \in I - K)$$

(decompose g according to the Iwasawa decomposition $G(\mathbb{A}) = N_I(\mathbb{A})M_I(\mathbb{A})K$ and note that $\Lambda^{T,I}$ may be interpreted as a truncation operator on M_I). On the other hand condition (b) implies that

$$\Lambda_2^{T,I} K_I(n_L^I \dot{w} n_I' g, n_K^I g) \neq 0$$

for some $n_K^I \in N_K^I(\mathbb{A})$ and we have seen in the proof of (11.2.7)(i) that this implies that

$$\varpi_j(H_I(g) - w(H_I(g))) \leq b + \varpi_j(w(H_\emptyset^I(g))) \qquad (\forall j \in \Delta - I),$$

where

$$b = \sup\{-\varpi_i(H_\emptyset(kg^{-1})) \mid k \in K,\ g \in \mathrm{Supp}(f) \text{ and } i = 1, \ldots, d-1\}.$$

Therefore, to prove the vanishing of $(*)$ for T large enough it is sufficient to prove that the three conditions

$$\begin{cases} \alpha_j(H_I(g) - T_I) > 0 & (\forall j \in \Delta - I), \\ \varpi_i^I(H_\emptyset^I(g) - T^I) \leq 0 & (\forall i \in I - K), \\ \varpi_j(H_I(g) - w(H_I(g))) \leq b + \varpi_j(w(H_\emptyset^I(g))) & (\forall j \in \Delta - I) \end{cases}$$

are not compatible if $K = I \cap w^{-1}(I)$ and

$$\alpha_i(T) \geq C_I \qquad (\forall i = 1, \ldots, d-1)$$

for some sufficiently large $C_I \in \mathbb{Q}$.

Let us assume that these three conditions are satisfied and let us search for a contradiction. As in the proof of (11.2.7)(i) we have

$$\varpi_j(H - w(H)) = \sum_{k \in \Delta - I} a_{jk}(w)\alpha_k(H) \qquad (\forall H \in \mathfrak{a}_\emptyset)$$

where $a_{jk}(w) \in \mathbb{Z}_{\geq 0}$ for all $j, k \in \Delta$. Then, combining the first and the third conditions, we get the inequalities

$$\varpi_j(T_I - w(T_I)) = \sum_{k \in \Delta - I} a_{jk}(w)\alpha_k(T_I) \leq b + \varpi_j(w(H_\emptyset^I(g))) \qquad (\forall j \in \Delta - I)$$

(we have $\alpha_k(H_I) = 0$ for each $k \in I$ and each $H_I \in \mathfrak{a}_I$). Since $(\alpha_i^\vee)_{i \in I}$ is a basis of the $\mathbb{Q}$-vector space $\mathfrak{a}_\emptyset^I$ with dual basis $(\varpi_i^I)_{i \in I}$ we can write

$$H_\emptyset^I = \sum_{i \in I} \varpi_i^I(H_\emptyset^I)\alpha_i^\vee \qquad (\forall H_\emptyset^I \in \mathfrak{a}_\emptyset^I).$$

Since $w(K) \subset I$ we have

$$\varpi_j(w(\alpha_i^\vee)) = 0 \qquad (\forall j \in \Delta - I,\ \forall i \in K).$$

Since $w(I) \subset R^+$ we have

$$\varpi_j(w(\alpha_i^\vee)) \geq 0 \qquad (\forall j \in \Delta - I,\ \forall i \in I).$$

Therefore we have

$$\begin{aligned}\varpi_j(w(H_\emptyset^I(g))) &= \sum_{i \in I - K} \varpi_i^I(H_\emptyset^I(g))\varpi_j(w(\alpha_i^\vee)) \\ &\leq \sum_{i \in I - K} \varpi_i^I(T^I)\varpi_j(w(\alpha_i^\vee)) = \varpi_j(w(T^I))\end{aligned}$$

for every $j \in \Delta - I$ (the inequality follows from the second condition). Thus we have

$$\varpi_j(T_I - w(T_I)) \leq b + \varpi_j(w(T^I)) \qquad (\forall j \in \Delta - I)$$

and then

$$\varpi_j(T - w(T)) \leq b \qquad (\forall j \in \Delta - I)$$

as $\varpi_j(T^I) = 0$ for any $j \in \Delta - I$. But we are assuming that $I \subsetneqq \Delta$ and $w \notin W_K$ for every K such that $I \subset K \subsetneqq \Delta$. In particular there exists $j \in \Delta - I$ such that w does not belong to $W_{\Delta - \{j\}}$. For such a j, we have

$$\varpi_j(T - w(T)) = \sum_{k \in \Delta} a_{jk}(w)\alpha_k(T)$$

with $a_{jk}(w) \in \mathbb{Z}_{\geq 0}$ for all $k \in \Delta$ and

$$a_{jj}(w) = \varpi_j(\varpi_j^\vee - w(\varpi_j^\vee)) > 0.$$

Let us take C_I larger than $b/a_{jj}(w)$. Then, for any $T \in \mathfrak{a}_\emptyset^+$ such that

$$\alpha_i(T) \geq C_I \qquad (\forall i = 1, \ldots, d-1),$$

we have

$$\varpi_j(T - w(T)) > b$$

and we have obtained a contradiction. □

(11.3) Expansion of the kernel following the cuspidal data
As in section (11.2) we fix a locally constant function f with compact support on $F_\infty^\times\backslash G(\mathbb{A})$ and a Haar measure dg on $G(\mathbb{A})$ with a splitting

$$dg = \prod_x dg_x.$$

Langlands has shown how to decompose the Hilbert space

$$L^2_{G,1_\infty} = L^2(F_\infty^\times G(F)\backslash G(\mathbb{A}), \frac{dg}{dz_\infty d\gamma})$$

and the kernel $K(h,g)$ of the operator $R_{G,1_\infty}(f)$ following the cuspidal data. Let us recall that a **cuspidal pair** is a pair (I,π) where I is a subset of Δ and π is an (isomorphism class of) irreducible cuspidal automorphic representation(s) of $F_\infty^\times\backslash M_I(\mathbb{A})$. Two such cuspidal pairs (I',π') and (I,π) are said to be **equivalent** if there exist $w\in W$ and a complex character λ of the abelian group $\mathbb{A}^\times M_I(\mathbb{A})^1\backslash M_I(\mathbb{A})$ such that

$$w(I) = I'$$

and

$$w(\pi(\lambda)) = \pi'$$

(here by definition $\pi(\lambda)$ is equal to $\lambda\otimes\pi$). A **cuspidal datum** is an equivalence class of cuspidal pairs. We will denote by $\mathcal{C}$ the set of all cuspidal data. Then for each $c\in\mathcal{C}$ Langlands has defined a closed $(F_\infty^\times\backslash G(\mathbb{A}))$-invariant subspace

$$L^2_{G,1_\infty,c}\subset L^2_{G,1_\infty}$$

in such a way that $L^2_{G,1_\infty,c'}$ is orthogonal to $L^2_{G,1_\infty,c}$ for any two distinct cuspidal data c' and c and that

$$L^2_{G,1_\infty} = \widehat{\bigoplus_{c\in\mathcal{C}}} L^2_{G,1_\infty,c}. \tag{11.3.1}$$

A complete proof of this decomposition and references are given in appendix G.

For each $c\in\mathcal{C}$ let us denote by

$$P_c : L^2_{G,1_\infty} \twoheadrightarrow L^2_{G,1_\infty,c}\subset L^2_{G,1_\infty}$$

the orthogonal projection onto $L^2_{G,1_\infty,c}$ (in particular we have $P_c^2 = P_c$ and P_c is selfadjoint) and let us set

$$R_{G,1_\infty,c}(f) = P_c\circ R_{G,1_\infty}(f) = R_{G,1_\infty}(f)\circ P_c.$$

PROPOSITION (11.3.2). — (i) *For each* $c \in C$ *the operator* $R_{G,1_\infty,c}(f)$ *on* $L^2_{G,1_\infty}$ *is given by a kernel* $K_c(h,g)$ *which is a locally constant complex function on* $(F_\infty^\times G(F)\backslash G(\mathbb{A}))^2$. *This kernel has the following properties:*

(a) *let* $K' \subset K \subset G(\mathbb{A})$ *be a compact open subgroup such that* f *is right* K'*-invariant, then we have*

$$K_c(hk',g) = K_c(h,gk') = K_c(h,g)$$

for all $g,h \in G(\mathbb{A})$ *and every* $k' \in K'$,

(b) *for any fixed* h *(resp.* g*) in* $G(\mathbb{A})$ *the function* $K_c(h,g)$ *of* g *(resp.* h*) is square-integrable on* $F_\infty^\times G(F)\backslash G(\mathbb{A})$, *i.e. belongs to* $L^2_{G,1_\infty}$.

(ii) *The kernels* $K_c(h,g)$ *are identically zero for all but finitely many* $c \in C$ *and we have*

$$K(h,g) = \sum_{c\in C} K_c(h,g) \qquad (\forall g,h \in G(\mathbb{A})).$$

Proof : Let us fix $c \in C$. We have seen in (10.1.2) (and its proof) that, for any $\varphi \in L^2_{G,1_\infty}$, the function $R_{G,1_\infty}(f)(\varphi)$ belongs to $L^2_{G,1_\infty} \cap C^\infty(F_\infty^\times G(F)\backslash G(\mathbb{A}))$ and even is right K'-invariant, that the operator $R_{G,1_\infty}(f)$ is bounded and that its norm is less than or equal to

$$\int_{F_\infty^\times\backslash G(\mathbb{A})} |f(g)| \frac{dg}{dz_\infty}.$$

Obviously the operator $R_{G,1_\infty,c}(f)$ has the same properties.

Now, if $\psi \in L^2_{G,1_\infty} \cap C^\infty(F_\infty^\times G(F)\backslash G(\mathbb{A}))$ is right K'-invariant, we have

$$\int_{F_\infty^\times G(F)\backslash G(\mathbb{A})} |\psi(g)|^2 \frac{dg}{dz_\infty d\gamma} = \sum_{h\in F_\infty^\times G(F)\backslash G(\mathbb{A})/K'} v(h)|\psi(h)|^2$$

where we have set

$$v(h) = \mathrm{vol}\big(F_\infty^\times G(F)\backslash F_\infty^\times G(F)hK', \frac{dg}{dz_\infty d\gamma}\big)$$

for each $h \in G(\mathbb{A})$, so that

$$|\psi(h)| \leq \frac{1}{\sqrt{v(h)}}\Big(\int_{F_\infty^\times G(F)\backslash G(\mathbb{A})} |\psi(g)|^2 \frac{dg}{dz_\infty d\gamma}\Big)^{1/2} \qquad (\forall h \in G(\mathbb{A})).$$

Therefore, for each $h \in G(\mathbb{A})$ the linear form

$$L^2_{G,1_\infty} \longrightarrow \mathbb{C},\ \varphi \longmapsto R_{G,1_\infty,c}(f)(\varphi)(h),$$

is continuous and there is a unique $\psi_h \in L^2_{G,1_\infty}$ such that

$$R_{G,1_\infty,c}(f)(\varphi)(h) = \int_{F_\infty^\times G(F)\backslash G(\mathbb{A})} \overline{\psi_h(g)}\varphi(g)\frac{dg}{dz_\infty d\gamma}.$$

Obviously we have

$$\psi_{z_\infty \gamma h k'}(g) = \psi_h(g) \qquad (\forall g \in G(\mathbb{A}))$$

for every $z_\infty \in F_\infty^\times$, every $\gamma \in G(F)$ and every $k' \in K'$.

For each $(h, g) \in (F_\infty^\times G(F)\backslash G(\mathbb{A}))^2$ let us set

$$K_c(h, g) = \int_{K'} \overline{\psi_h(gk')}dk'$$

where dk' is the normalized Haar measure of the compact group K'. It is obvious that $K_c(h, g)$ is a locally constant complex function on $(F_\infty^\times G(F)\backslash G(\mathbb{A}))^2$ which satisfies property (a) of the proposition and that, for any fixed h, $K_c(h, g)$ is square-integrable as a function of g (apply the Cauchy–Schwarz inequality). Moreover, as

$$R_{G,1_\infty}(f)(\varphi) = R_{G,1_\infty}(f)\Big(g \mapsto \int_{K'} \varphi(gk')dk'\Big)$$

for any $\varphi \in L^2_{G,1_\infty}$, $K_c(h, g)$ is a kernel for the operator $R_{G,1_\infty,c}(f)$ and, for any fixed g,

$$K_c(h, g) = R_{G,1_\infty,c}(f)\Big(\frac{1_{V(g)}}{v(g)}\Big)$$

is square-integrable as a function of h. Here we have set

$$V(g) = F_\infty^\times G(F)\backslash F_\infty^\times G(F)gK'.$$

This concludes the proof of part (i) of the proposition.

Now we will prove that, for any compact open subgroup $K' \subset K \subset G(\mathbb{A})$, there are at most finitely many $c \in C$ such that

$$(L^2_{G,1_\infty,c})^{K'} \neq (0),$$

so that $R_{G,1_\infty,c}(f) = 0$ for all but finitely many $c \in C$. As

$$R_{G,1_\infty}(f) = \sum_{c\in C} R_{G,1_\infty,c}(f)$$

part (ii) of the proposition will follow.

Let $c \in \mathcal{C}$. By definition $L^2_{G,1_\infty,c}$ is the closure in $L^2_{G,1_\infty}$ of the subspace generated by the pseudo-Eisenstein series θ_Φ with

$$\Phi : X_I^\Delta \longrightarrow \mathcal{A}_{I,\pi}^\Delta$$

a Paley–Wiener function and $(I,\pi) \in c$ (see (G.2.1) and (G.7.1)). Therefore $(L^2_{G,1_\infty,c})^{K'}$ is the closure in $L^2_{G,1_\infty}$ of the subspace generated by the pseudo-Eisenstein series θ_Φ with

$$\Phi : X_I^\Delta \longrightarrow (\mathcal{A}_{I,\pi}^\Delta)^{K'} \subset \mathcal{A}_{I,\pi}^\Delta$$

a Paley–Wiener function and $(I,\pi) \in c$. Indeed the operator

$$\varphi \mapsto \left(g \mapsto \int_{K'} \varphi(gk')dk'\right),$$

where dk' is the normalized Haar measure on the compact group K', is the orthogonal projection of $L^2_{G,1_\infty}$ onto its closed subspace $(L^2_{G,1_\infty})^{K'}$.

Now let $I \subset \Delta$. For any $\chi \in \mathcal{X}_{M_I}$ such that $\chi(Z_G(F_\infty)) = (1)$ the set of isomorphism classes of irreducible cuspidal automorphic representations π of $M_I(\mathbb{A})$ with central character χ such that

$$(\mathcal{A}_{I,\pi}^\Delta)^{K'} \neq (0)$$

is finite. Indeed we may assume that K' is a normal subgroup of K; then, if we have $(\mathcal{A}_{I,\pi}^\Delta)^{K'} \neq (0)$ for some π, we also have

$$(\mathcal{A}_{I,\pi}^I)^{M_I(\mathbb{A})\cap K'} \neq (0);$$

but by (9.2.10) the vector space $(\mathcal{A}_{M_I,\chi,\mathrm{cusp}})^{M_I(\mathbb{A})\cap K'}$ is finite dimensional over $\mathbb{C}$ if $\chi(Z_I(\mathbb{A}) \cap K') = (1)$ and is zero otherwise; our assertion follows. Therefore the set of isomorphism classes of irreducible cuspidal automorphic representations π of $F_\infty^\times \backslash M_I(\mathbb{A})$ such that

$$(\mathcal{A}_{I,\pi}^\Delta)^{K'} \neq (0)$$

is the union of finitely many orbits under X_I^Δ for the action $(\lambda,\pi) \mapsto \pi(\lambda)$. Indeed we have

$$\chi_{\pi(\lambda)} = (\lambda | Z_I(\mathbb{A}))\chi_\pi$$

and the obvious homomorphism

$$F_\infty^\times Z_I(F)\backslash Z_I(\mathbb{A})/(Z_I(\mathbb{A}) \cap K') \to F_\infty^\times M_I(\mathbb{A})^1 \backslash M_I(\mathbb{A})$$

has a finite kernel and a finite cokernel. This concludes the proof of part (ii) of the proposition. □

REMARK (11.3.3). — Part (i) of the proposition and its proof hold as well if we replace P_c by any orthogonal projection of $L^2_{G,1_\infty}$ onto a closed subspace. □

COROLLARY (11.3.4). — *If $T \in \mathfrak{a}_\emptyset^+$ we have*

$$J^T_{\mathrm{spec}}(f) = \sum_{c \in \mathcal{C}} J^T_c(f)$$

where $J^T_{\mathrm{spec}}(f)$ has been defined in (11.2.4) *and where we have set*

$$J^T_c(f) = \int_{F_\infty^\times G(F)\backslash G(\mathbb{A})} \Lambda_2^T K_c(g,g) \frac{dg}{dz_\infty d\gamma}$$

for each $c \in \mathcal{C}$. □

(11.4) Evaluation of $J^T_c(f)$ in a special case: the operators $M^T_{I,\pi}(\lambda)$
For each cuspidal datum the theory of Eisenstein series gives the spectral decomposition of $L^2_{G,1_\infty,c}$ and a corresponding expression for the distribution J^T_c (see [Ar 6] §3 for the number field case).

In the particular case where c is regular the spectral decomposition of $L^2_{G,1_\infty,c}$ and the corresponding expression for J^T_c are quite simple.

Let us recall that c is said to be **regular** (or unramified in the sense of [Ar 6] §4) if, for any $(I,\pi) \in c$, the only $w \in W$ such that $w(I) = I$ and $w(\pi) = \pi(\lambda)$ for some $\lambda \in X_I^\Delta$ is the identity element of W. Here X_I^Δ is the abelian group of complex characters of $\mathbb{A}^\times M_I(\mathbb{A})^1\backslash M_I(\mathbb{A})$.

For a regular cuspidal datum c the spectral decomposition of $L^2_{G,1_\infty,c}$ is reviewed in appendix G (see (G.10.3)). We will use freely the notations of that appendix.

PROPOSITION (11.4.1). — *Let c be a regular cuspidal datum and let $(I,\pi) \in c$ with π unitary. Let $\mathcal{B}^\Delta_{I,\pi} \subset \mathcal{A}^\Delta_{I,\pi}$ be an orthogonal Hilbert basis of the Hilbert space $\widehat{\mathcal{A}}^\Delta_{I,\pi}$ (see* (G.10.2)(i)*). Then for each $(h,g) \in (F_\infty^\times G(F)\backslash G(\mathbb{A}))^2$ we have*

$$K_c(h,g) = \widetilde{K}_c(h,g) \overset{\mathrm{dfn}}{=\!=} \frac{f d_1 \cdots d_s}{\deg(\infty)|\mathrm{Fix}_{X_I^\Delta}(\pi)|} \int_{\mathrm{Im}\, X_I^\Delta} v(h,g;\lambda) d\lambda$$

where we have set

$$v(h,g;\lambda) = \sum_{\varphi \in \mathcal{B}^\Delta_{I,\pi}} E_I^\Delta\big((i_I^\Delta(\lambda,f)(\varphi))_\lambda, \pi(\lambda)\big)(h) \overline{E_I^\Delta(\varphi_\lambda, \pi(\lambda))(g)},$$

only finitely many terms in this sum being non-zero.

Proof: For almost every $\varphi \in \mathcal{B}_{I,\pi}^{\Delta}$ we have

$$i_I^{\Delta}(\lambda, f)(\varphi) = 0.$$

Indeed, with the notations of the proof of (G.10.2), f is right K_n-invariant for some $n > 0$ and we have

$$i_I^{\Delta}(\lambda, f)(\varphi) = i_I^{\Delta}(\lambda, f)(\varphi_{K_n});$$

but we have

$$\langle \varphi_{K_n}, \varphi_{K_n} \rangle = \langle \varphi, \varphi_{K_n} \rangle \in \langle \varphi, (\mathcal{A}_{I,\pi}^{\Delta})^{K_n} \rangle,$$

so that $\varphi_{K_n} = 0$ for almost every $\varphi \in \mathcal{B}_{I,\pi}^{\Delta}$ (we have $\langle \varphi_i, (\mathcal{A}_{I,\pi}^{\Delta})^{K_n} \rangle = (0)$ for all $i > j(n)$ with the notations of loc. cit.).

For any unitary cuspidal pairs (I', π') and (I'', π'') and for any $\Phi' \in \mathcal{P}(I', \pi')$ and $\Phi'' \in \mathcal{P}(I'', \pi'')$ let us compute the integral

$$\mathcal{I} = \int_{F_\infty^\times G(F)\backslash G(\mathbb{A})} \left(\int_{F_\infty^\times G(F)\backslash G(\mathbb{A})} \overline{\theta_{\Phi''}(h)} \widetilde{K}_c(h, g) \theta_{\Phi'}(g) \frac{dh}{dz_\infty d\gamma} \right) \frac{dg}{dz_\infty d\gamma}.$$

By definition $\mathcal{I}$ is equal to

$$\frac{f d_1 \cdots d_s}{\deg(\infty) |\mathrm{Fix}_{X_I^{\Delta}}(\pi)|} \int_{\mathrm{Im}\, X_I^{\Delta}} \Big(\sum_{\varphi \in \mathcal{B}_{I,\pi}^{\Delta}} \langle \theta_{\Phi''}, E_I^{\Delta}((i_I^{\Delta}(\lambda, f)(\varphi))_\lambda, \pi(\lambda)) \rangle$$
$$\times \overline{\langle \theta_{\Phi'}, E_I^{\Delta}(\varphi_\lambda, \pi(\lambda)) \rangle} \Big) d\lambda.$$

But thanks to (G.10.1)(iii) we have

$$\begin{aligned} &\big\langle \theta_{\Phi''}, E_I^{\Delta}\big((i_I^{\Delta}(\lambda, f)(\varphi))_\lambda, \pi(\lambda)\big)\big\rangle \\ &= \frac{\deg(\infty)}{f d_1 \cdots d_s} \big\langle \Phi''(\lambda''_{w''} + w''(\lambda))_{\lambda''_{w''}}, \big(M_I(w'', \pi(\lambda))((i_I^{\Delta}(\lambda, f)(\varphi))_\lambda)\big)_{-w''(\lambda)} \big\rangle_{w''} \end{aligned}$$

if there exist $w'' \in W$ and $\lambda''_{w''} \in \mathrm{Im}\, X_{I''}^{\Delta}$ such that $w''(I) = I''$ and $w''(\pi) = \pi''(\lambda''_{w''})$ and

$$\big\langle \theta_{\Phi''}, E_I^{\Delta}\big((i_I^{\Delta}(\lambda, f)(\varphi))_\lambda, \pi(\lambda)\big)\big\rangle = 0$$

otherwise. Similarly we have

$$\begin{aligned} &\langle \theta_{\Phi'}, E_I^{\Delta}(\varphi_\lambda, \pi(\lambda)) \rangle \\ &= \frac{\deg(\infty)}{f d_1 \cdots d_s} \big\langle \Phi'(\lambda'_{w'} + w'(\lambda))_{\lambda'_{w'}}, \big(M_I(w', \pi(\lambda))(\varphi_\lambda)\big)_{-w'(\lambda)} \big\rangle_{w'} \end{aligned}$$

if there exist $w' \in W$ and $\lambda'_{w'} \in \operatorname{Im} X_{I'}^{\Delta}$ such that $w'(I) = I'$ and $w'(\pi) = \pi'(\lambda'_{w'})$ and

$$\langle \theta_{\Phi'}, E_I^{\Delta}(\varphi_\lambda, \pi(\lambda)) \rangle = 0$$

otherwise. Therefore $\mathcal{I} = 0$ unless both (I', π') and (I'', π'') are in c and, if this last condition is satisfied, we have

$$\mathcal{I} = \langle \theta_{\Phi''}, R_{G,1_\infty}(f)(\theta_{\Phi'}) \rangle.$$

Indeed, for any $\lambda \in \operatorname{Im} X_I^{\Delta}$ we have

$$\begin{aligned} \Big\langle \Phi''(\lambda''_{w''} + w''(\lambda))_{\lambda''_{w''}}, \big(M_I(w'', \pi(\lambda))((i_I^{\Delta}(\lambda, f)(\varphi))_\lambda)\big)_{-w''(\lambda)} \Big\rangle_{w''} \\ = \big\langle i_I^{\Delta}(\lambda, f^*)(\widetilde{\Phi}''(\lambda)), \varphi \big\rangle \end{aligned}$$

and

$$\Big\langle \Phi'(\lambda'_{w'} + w'(\lambda))_{\lambda'_{w'}}, \big(M_I(w', \pi(\lambda))(\varphi_\lambda)\big)_{-w'(\lambda)} \Big\rangle_{w'} = \langle \widetilde{\Phi}'(\lambda), \varphi \rangle$$

where we have set

$$\widetilde{\Phi}''(\lambda) = \Big(M_{I''}(w''^{-1}, \pi''(\lambda''_{w''} + w''(\lambda)))\big(\Phi''(\lambda''_{w''} + w''(\lambda))_{\lambda''_{w''} + w''(\lambda)}\big)\Big)_{-\lambda}$$

and

$$\widetilde{\Phi}'(\lambda) = \Big(M_{I'}(w'^{-1}, \pi'(\lambda'_{w'} + w'(\lambda)))\big(\Phi'(\lambda'_{w'} + w'(\lambda))_{\lambda'_{w'} + w'(\lambda)}\big)\Big)_{-\lambda}$$

and where f^* is defined by

$$f^*(g) = \overline{f(g^{-1})} \qquad (\forall g \in G(\mathbb{A}));$$

moreover we have

$$\begin{aligned} \sum_{\varphi \in \mathcal{B}_{I,\pi}^{\Delta}} \langle i_I^{\Delta}(\lambda, f^*)(\widetilde{\Phi}''(\lambda)), \varphi \rangle \overline{\langle \widetilde{\Phi}'(\lambda), \varphi \rangle} &= \langle i_I^{\Delta}(\lambda, f^*)(\widetilde{\Phi}''(\lambda)), \widetilde{\Phi}'(\lambda) \rangle \\ &= \langle \widetilde{\Phi}''(\lambda), i_I^{\Delta}(\lambda, f)(\widetilde{\Phi}'(\lambda)) \rangle; \end{aligned}$$

thus our assertion follows from (G.10.1)(ii).

Now the proposition follows from the denseness of the pseudo-Eisenstein series (see (G.7.4)). □

Let us fix a regular cuspidal datum c, a unitary cuspidal pair $(I, \pi) \in c$ and $T \in \mathfrak{a}_{\emptyset}^{+}$. Let $\lambda \in \operatorname{Im} X_I^{\Delta}$.

LEMMA (11.4.2). — (i) *For each* $\varphi \in \mathcal{A}^{\Delta}_{I,\pi}$ *there exists a unique vector*

$$M^T_{I,\pi}(\lambda)(\varphi) \in \mathcal{A}^{\Delta}_{I,\pi}$$

such that

$$\langle \psi, M^T_{I,\pi}(\lambda)(\varphi)\rangle = \big\langle \Lambda^T E^{\Delta}_I(\psi_\lambda, \pi(\lambda)), E^{\Delta}_I(\varphi_\lambda, \pi(\lambda))\big\rangle \qquad (\forall \psi \in \mathcal{A}^{\Delta}_{I,\pi}).$$

Moreover, if $\varphi \in \mathcal{A}^{\Delta}_{I,\pi}$ *is right invariant under some compact open subgroup* K' *of* $K \subset G(\mathbb{A})$, $M^T_{I,\pi}(\lambda)(\varphi)$ *is also right* K'*-invariant.*

(ii) *The linear operator*

$$M^T_{I,\pi}(\lambda) : \mathcal{A}^{\Delta}_{I,\pi} \to \mathcal{A}^{\Delta}_{I,\pi}$$

is selfadjoint and non-negative.

(iii) *For each* $\mu \in \mathrm{Fix}_{X^{\Delta}_I}(\pi)$ *we have*

$$M^T_{I,\pi}(\lambda + \mu)(\varphi) = (M^T_{I,\pi}(\lambda)(\varphi_\mu))_{-\mu} \qquad (\forall \varphi \in \mathcal{A}^{\Delta}_{I,\pi}).$$

Proof: If K' is a compact open subgroup of $K \subset G(\mathbb{A})$ and if we set

$$\varphi_{K'}(g) = \int_{K'} \varphi(gk')dk' \qquad (\forall g \in G(\mathbb{A}))$$

for any $\varphi \in \mathcal{A}^{\Delta}_{I,\pi}$, where dk' is the normalized Haar measure of K', it is easy to see that

$$\begin{aligned}&\big\langle \Lambda^T E^{\Delta}_I((\psi_{K'})_\lambda, \pi(\lambda)), E^{\Delta}_I(\varphi_\lambda, \pi(\lambda))\big\rangle \\ &= \big\langle \Lambda^T E^{\Delta}_I(\psi_\lambda, \pi(\lambda)), E^{\Delta}_I((\varphi_{K'})_\lambda, \pi(\lambda))\big\rangle.\end{aligned}$$

Therefore, if φ is right K'-invariant and if φ_1 is the unique vector in $\big(\mathcal{A}^{\Delta}_{I,\pi}\big)^{K'}$ such that

$$\langle \psi, \varphi_1\rangle = \big\langle \Lambda^T E^{\Delta}_I(\psi_\lambda, \pi(\lambda)), E^{\Delta}_I(\varphi_\lambda, \pi(\lambda))\big\rangle$$

for all $\psi \in \big(\mathcal{A}^{\Delta}_{I,\pi}\big)^{K'}$ (recall that $\big(\mathcal{A}^{\Delta}_{I,\pi}\big)^{K'}$ is finite dimensional over $\mathbb{C}$) we can take $M^T_{I,\pi}(\lambda)(\varphi) = \varphi_1$ and part (i) of the lemma follows (we have

$$\langle \psi, \varphi_1\rangle = \langle \psi_{K'}, \varphi_1\rangle$$

for any $\psi \in \mathcal{A}^{\Delta}_{I,\pi}$).

By (11.1.3) we have

$$\big\langle \Lambda^T E^{\Delta}_I(\psi_\lambda, \pi(\lambda)), E^{\Delta}_I(\varphi_\lambda, \pi(\lambda))\big\rangle = \big\langle \Lambda^T E^{\Delta}_I(\psi_\lambda, \pi(\lambda)), \Lambda^T E^{\Delta}_I(\varphi_\lambda, \pi(\lambda))\big\rangle.$$

Part (ii) of the lemma follows.

Part (iii) is a direct consequence of the definitions. □

For each $\lambda \in \mathrm{Im}\, X_I^\Delta$ the operator

$$M_{I,\pi}^T(\lambda) \circ i_I^\Delta(\lambda, f) \; : \; \mathcal{A}_{I,\pi}^\Delta \to \mathcal{A}_{I,\pi}^\Delta$$

has finite rank (if f is left K'-invariant for some compact open subgroup K' of $K \subset G(\mathbb{A})$ the image of this operator is contained in $(\mathcal{A}_{I,\pi}^\Delta)^{K'}$) and it has a trace. Moreover we have

$$\mathrm{tr}\big(M_{I,\pi}^T(\lambda+\mu) \circ i_I^\Delta(\lambda+\mu, f)\big) = \mathrm{tr}\big(M_{I,\pi}^T(\lambda) \circ i_I^\Delta(\lambda, f)\big)$$

for every $\mu \in \mathrm{Fix}_{X_I^\Delta}(\pi)$.

PROPOSITION (11.4.3). — *If $T \in \mathfrak{a}_\emptyset^+$ we have*

$$J_c^T(f) = \frac{f d_1 \cdots d_s}{\deg(\infty)|\mathrm{Fix}_{X_I^\Delta}(\pi)|} \int_{\mathrm{Im}\, X_I^\Delta} \mathrm{tr}\big(M_{I,\pi}^T(\lambda) \circ i_I^\Delta(\lambda, f)\big) d\lambda.$$

Proof : This is a direct consequence of (11.4.1) and of the definitions of $M_{I,\pi}^T(\lambda)$ and $J_c^T(f)$. □

Now, using formula (G.9.3)(ii) for the constant terms of the Eisenstein series, we can compute the operators $M_{I,\pi}^T(\lambda)$.

If $I \subset K \subset L \subset \Delta$ we set

$$\Phi_I^{K,L} = \sum_{\substack{J \\ I \subset J \subset K}} (-1)^{|L-J|} \widehat{\tau}_J^L : \mathfrak{a}_\emptyset \to \mathbb{C}$$

(in particular $\Phi_I^{K,K}$ is the function Φ_I^K of (10.3.3)).

LEMMA (11.4.4). — *For each $\lambda \in \rho_I + C_I^\Delta \subset X_I^\Delta$, each $\varphi \in \mathcal{A}_{I,\pi}^\Delta$ and each $g \in G(\mathbb{A})$ we have*

$$\begin{aligned}&\Lambda^T E_I^\Delta(\varphi_\lambda, \pi(\lambda))(g) \\ &= \sum_{\substack{w \in W \\ w(I) \subset \Delta}} \sum_{\delta' \in P_{w(I)}(F)\backslash G(F)} \Phi_{w(I)}^{\Delta \cap w(R^+), \Delta}(H_\emptyset(\delta' g) - T) M_I(w, \pi(\lambda))(\varphi_\lambda)(\delta' g),\end{aligned}$$

the sum over δ' converging absolutely.

Proof : By definition of Λ^T and by (G.9.3)(ii) we have

$$\begin{aligned}\Lambda^T E_I^\Delta(\varphi_\lambda, \pi(\lambda))(g) = &\sum_{J \subset \Delta} (-1)^{|\Delta - J|} \sum_{\delta \in P_J(F)\backslash G(F)} \widehat{\tau}_J^\Delta(H_\emptyset(\delta g) - T) \\ &\times \sum_{w \in \Omega(J,I)} E_{w(I)}^J\big(M_I(w, \pi(\lambda))(\varphi_\lambda), w(\pi(\lambda))\big)(\delta g)\end{aligned}$$

the sum over δ being finite (see (11.1.1)), and by (G.6.1) we have

$$\begin{aligned} &E_{w(I)}^{J}\big(M_I(w,\pi(\lambda))(\varphi_\lambda), w(\pi(\lambda))\big)(\delta g) \\ &= \sum_{\pi_J \in P_{w(I)}(F)\backslash P_J(F)} M_I(w,\pi(\lambda))(\varphi_\lambda)(\pi_J g), \end{aligned}$$

the sum over π_J converging absolutely as long as

$$w(\lambda) - \rho_{w(I)} \in C_{w(I)}^{\Delta} + \operatorname{Re} X_J^{\Delta}.$$

But we have

$$w(\lambda) - \rho_{w(I)} = w(\lambda - \rho_I) + w(\rho_I) - \rho_{w(I)}$$

and it is not difficult to check that

$$w(C_I^{\Delta}) \subset C_{w(I)}^{\Delta} + \operatorname{Re} X_{\Delta\cap w(R^+)}^{\Delta}$$

and that

$$w(\rho_I) - \rho_{w(I)} \in C_{w(I)}^{\Delta} + \operatorname{Re} X_{\Delta\cap w(R^+)}^{\Delta},$$

so that

$$w(\lambda) - \rho_{w(I)} \in C_{w(I)}^{\Delta} + \operatorname{Re} X_{\Delta\cap w(R^+)}^{\Delta} \subset C_{w(I)}^{\Delta} + \operatorname{Re} X_J^{\Delta}$$

for any $\lambda \in \rho_I + C_I^{\Delta}$ (recall that $J \subset \Delta \cap w(R^+)$). The lemma follows (set $\delta' = \pi_J\delta$). □

LEMMA (11.4.5). — *For any $I \subset K \subset \Delta$ the function $\Phi_I^{K,\Delta} : \mathfrak{a}_\emptyset \longrightarrow \mathbb{C}$ is the product of the sign*

$$(-1)^{|\Delta - K|}$$

with the characteristic function of the subset

$$\begin{aligned} \mathfrak{a}_\Delta \oplus \{H_I^{\Delta} \in \mathfrak{a}_I^{\Delta} \mid &\varpi_i(H_I^{\Delta}) > 0,\ \forall i \in \Delta - K, \\ &\textit{and}\ \ \varpi_i(H_I^{\Delta}) \leq 0,\ \forall i \in K - I\} \oplus \mathfrak{a}_\emptyset^{I}. \end{aligned}$$

□

Let us consider the lattice

$$\mathfrak{a}_{\emptyset,\mathbb{Z}} = \{H \in \mathfrak{a}_\emptyset \mid H_i \in \mathbb{Z},\ \forall i = 1,\ldots,d\}$$

in $\mathfrak{a}_\emptyset$ and for any $I \subset \Delta$ the lattice

$$\mathfrak{a}_{I,\mathbb{Z}} = \{H_I \mid H \in \mathfrak{a}_{\emptyset,\mathbb{Z}}\}$$

in $\mathfrak{a}_I$. We have $\mathfrak{a}_I \cap \mathfrak{a}_{\emptyset,\mathbb{Z}} \subset \mathfrak{a}_{I,\mathbb{Z}}$ but in general the inclusion is strict. If $d_I = (d_1, \ldots, d_s)$ is the partition of d corresponding to I and if we identify $\mathfrak{a}_I$ with $\mathbb{Q}^s$ as in (10.3) we have

$$\mathfrak{a}_I \cap \mathfrak{a}_{\emptyset,\mathbb{Z}} = \mathbb{Z}^s \subset \mathfrak{a}_{I,\mathbb{Z}} = \bigoplus_{j=1}^{s} \frac{1}{d_j}\mathbb{Z} \subset \mathfrak{a}_I = \mathbb{Q}^s.$$

We may identify the lattice $\mathfrak{a}_{I,\mathbb{Z}}$ with the abelian group

$$M_I(\mathbb{A})^1 \backslash M_I(\mathbb{A}) \cong (GL_{d_1}(\mathbb{A})^1 \backslash GL_{d_1}(\mathbb{A})) \times \cdots \times (GL_{d_s}(\mathbb{A})^1 \backslash GL_{d_s}(\mathbb{A}))$$

by

$$m_I = (g_1, \ldots, g_s) \mapsto (\frac{\deg(\det g_1)}{d_1}, \ldots, \frac{\deg(\det g_s)}{d_s}).$$

Let us denote by X_I the abelian group of characters of $M_I(\mathbb{A})^1 \backslash M_I(\mathbb{A})$. We may identify X_I with

$$\bigoplus_{j=1}^{s} \mathbb{C} \Big/ \frac{2\pi i}{\log q} d_j \mathbb{Z},$$

where $q = p^f$ is the number of elements in the field of constants of F. Then the canonical pairing

$$X_I \times \mathfrak{a}_{I,\mathbb{Z}} \to \mathbb{C}^\times$$

is given by

$$(\lambda, H_I) \mapsto q^{\langle \lambda, H_I \rangle}$$

where

$$\langle \cdot, \cdot \rangle : X_I \times \mathfrak{a}_{I,\mathbb{Z}} \to \mathbb{C} \Big/ \frac{2\pi i}{\log q} \mathbb{Z}, \ (\lambda, H_I) \mapsto \sum_{j=1}^{s} \lambda_j H_{I,j}$$

(see also (G.1)). Obviously the last pairing induces a pairing

$$\langle \cdot, \cdot \rangle : X_I^\Delta \times \mathfrak{a}_{I,\mathbb{Z}} / (\mathfrak{a}_\Delta \cap \mathfrak{a}_{\emptyset,\mathbb{Z}}) \longrightarrow \mathbb{C} \Big/ \frac{2\pi i}{\log q} \mathbb{Z}.$$

Now, at least formally, we may consider the series

$$(11.4.6) \qquad \sum_{H_I \in \mathfrak{a}_{I,\mathbb{Z}} / (\mathfrak{a}_\Delta \cap \mathfrak{a}_{\emptyset,\mathbb{Z}})} q^{-\langle \lambda, H_I \rangle} \Phi_I^{K,\Delta}(H_I)$$

for $\lambda \in X_I^\Delta$.

For each $j = 1, \ldots, s-1$ let us set

$$\alpha_{I,j}^\vee = (0, \ldots, 0, \frac{1}{d_j}, -\frac{1}{d_{j+1}}, 0, \ldots, 0) \in \mathfrak{a}_{I,\mathbb{Z}}.$$

Let e be the quotient of d by the g.c.d. of $d_1, \ldots, d_s$, i.e. the smallest positive integer such that d divides ed_j for all j. For each $n \in \mathbb{Z}/e\mathbb{Z}$ let us set

$$(n)_I = \sum_{j=1}^{s-1} (n)_I^j \alpha_{I,j}^\vee - \frac{n}{d}(1, \ldots, 1) \in \mathfrak{a}_I$$

where

$$(n)_I^j = \frac{d_1 + \cdots + d_j}{d} n + \left[- \frac{d_1 + \cdots + d_j}{d} n \right] \in \left]-1, 0\right]$$

for $j = 1, \ldots, s-1$. Finally let us set

$$C_I^{K,\Delta} = \{\lambda \in X_I^\Delta \mid \langle \mathrm{Re}\,\lambda, \alpha_{I,j}^\vee \rangle > 0 \text{ if } d_1 + \cdots + d_j \in \Delta - K \text{ and } \langle \mathrm{Re}\,\lambda, \alpha_{I,j}^\vee \rangle < 0 \text{ if } d_1 + \cdots + d_j \in K - I\}.$$

LEMMA (11.4.7). — (i) *The series (11.4.6) converges uniformly on any compact subset of $C_I^{K,\Delta}$ and for any $\lambda \in C_I^{K,\Delta}$ its sum is equal to*

$$\frac{d}{e} \Big(\sum_{n \in \mathbb{Z}/e\mathbb{Z}} q^{-\langle \lambda, (n)_I \rangle} \Big) \Big/ \prod_{j=1}^{s-1} \left(1 - q^{\langle \lambda, \alpha_{I,j}^\vee \rangle} \right)$$

if we have $\sum_{j=1}^{s} \lambda_j \in \frac{2\pi i}{\log q} \frac{d}{e} \mathbb{Z} \subset \frac{2\pi i}{\log q} \mathbb{Z}$ and is equal to 0 otherwise.

(ii) *For any $\Lambda \in \mathrm{Re}\, X_I^\Delta$ such that $\langle \Lambda, \alpha_{I,j}^\vee \rangle \neq 0$, $\forall j = 1, \ldots, s-1$, let us set*

$$\Phi_I^\Lambda(H) = \frac{d}{e} \int_{\mathrm{Im}\, X_I^\Delta} 1_{\frac{2\pi i}{\log q} \frac{d}{e} \mathbb{Z}} \Big(\sum_{j=1}^{s} \lambda_j \Big) \frac{\displaystyle\sum_{n \in \mathbb{Z}/e\mathbb{Z}} q^{\langle \Lambda + \lambda, H_I - (n)_I \rangle}}{\displaystyle\prod_{j=1}^{s-1} \left(1 - q^{\langle \Lambda + \lambda, \alpha_{I,j}^\vee \rangle} \right)} d\lambda \qquad (\forall H \in \mathfrak{a}_{\emptyset, \mathbb{Z}})$$

where $d\lambda$ is the normalized Haar measure on the compact Lie group $\mathrm{Im}\, X_I^\Delta$. Then we have

$$\Phi_I^\Lambda = \Phi_I^{K,\Delta}$$

where

$$K = \{d_1 + \cdots + d_j \in \Delta - I \mid \langle \Lambda, \alpha_{I,j}^\vee \rangle < 0\} \cup I.$$

Proof: Part (ii) follows immediately from part (i) by Fourier inversion. To prove part (i) let us first consider the exact sequence

$$0 \longrightarrow \frac{e}{d}\mathbb{Z}/\mathbb{Z} \longrightarrow \Big(\bigoplus_{j=1}^{s} \frac{1}{d_j} \mathbb{Z} \Big) \Big/ \mathbb{Z} \xrightarrow{\varpi} \mathbb{Z}^{s-1} + \Big(\frac{d_1}{d}, \ldots, \frac{d_1 + \cdots + d_{s-1}}{d} \Big) \mathbb{Z} \longrightarrow 0$$

where

$$\frac{e}{d}m \mapsto (\frac{em}{d}, \ldots, \frac{em}{d}),$$

$$\mathbb{Z}^{s-1} + (\frac{d_1}{d}, \ldots, \frac{d_1 + \cdots + d_{s-1}}{d})\mathbb{Z} \subset \mathbb{Q}^s$$

and

$$\varpi(\mu)_j = d_1\mu_1 + \cdots + d_j\mu_j - \frac{d_1 + \cdots + d_j}{d}(d_1\mu_1 + \cdots + d_s\mu_s)$$

for $j = 1, \ldots, s-1$. If

$$\nu = \nu' + (\frac{d_1}{d}, \ldots, \frac{d_1 + \cdots + d_{s-1}}{d})n \in \mathbb{Z}^{s-1} + (\frac{d_1}{d}, \ldots, \frac{d_1 + \cdots + d_{s-1}}{d})\mathbb{Z},$$

we have $\varpi(\mu) = \nu$ if and only if

$$\mu - \sum_{j=1}^{s-1} \nu_j \alpha_{I,j}^\vee + (1, \ldots, 1)\frac{n}{d} \in (1, \ldots, 1)\frac{e}{d}\mathbb{Z}.$$

Then, at least formally, it follows from lemma (11.4.5) that the series (11.4.6) is equal to

$$(-1)^{|\Delta - K|} \sum_{\nu} q^{-\langle \lambda, \sum_{j=1}^{s-1} \nu_j \alpha_{I,j}^\vee \rangle} \sum_{m \in \mathbb{Z}/\frac{d}{e}\mathbb{Z}} q^{\left(\sum_{j=1}^{s} \lambda_j\right)\frac{n-em}{d}}$$

where ν runs through the set

$$\{\nu \in \mathbb{Z}^{s-1} + (\frac{d_1}{d}, \ldots, \frac{d_1 + \cdots + d_{s-1}}{d})\mathbb{Z} \mid \nu_j > 0 \text{ if } d_1 + \cdots + d_j \in \Delta - K,$$
$$\nu_j \leq 0 \text{ if } d_1 + \cdots + d_j \in K - I\}.$$

Now on the one hand we have

$$\sum_{m \in \mathbb{Z}/\frac{d}{e}\mathbb{Z}} q^{-\left(\sum_{j=1}^{s} \lambda_j\right)\frac{em}{d}} = \begin{cases} \frac{d}{e} & \text{if } \sum_{j=1}^{s} \lambda_j \in \frac{2\pi i}{\log q}\frac{d}{e}\mathbb{Z}, \\ 0 & \text{otherwise} \end{cases}$$

(recall that $\sum_{j=1}^{s} \lambda_j \in \frac{2\pi i}{\log q}\mathbb{Z}$). On the other hand we have

$$\mathbb{Z}^{s-1} \cap (\frac{d_1}{d}, \ldots, \frac{d_1 + \cdots + d_{s-1}}{d})\mathbb{Z} = (\frac{d_1}{d}, \ldots, \frac{d_1 + \cdots + d_{s-1}}{d})e\mathbb{Z}.$$

and, for any fixed m, the above sum over ν is equal to

$$\sum_{n=0}^{e-1} q^{\left(\sum_{j=1}^{s} \lambda_j\right)\frac{n-em}{d}} \prod_{j=1}^{s-1} \sum_{\nu_j'} q^{-\langle \lambda, \alpha_{I,j}^\vee \rangle \left(\nu_j' + \frac{d_1 + \cdots + d_j}{d} n\right)}$$

where ν_j' runs through the set

$$\{\nu_j' \in \mathbb{Z} \mid \nu_j' + \frac{d_1 + \cdots + d_j}{d} n > 0\}$$

if $d_1 + \cdots + d_j \in \Delta - K$ and through the set

$$\{\nu_j' \in \mathbb{Z} \mid \nu_j' + \frac{d_1 + \cdots + d_j}{d} n \leq 0\}$$

if $d_1 + \cdots + d_j \in K - I$.

To finish the proof of the lemma we remark that, for any $\alpha \in \mathbb{R}$, the series

$$\sum_{\substack{\nu \in \mathbb{Z} \\ \nu + \alpha > 0}} q^{-z(\nu+\alpha)}$$

(resp.

$$\sum_{\substack{\nu \in \mathbb{Z} \\ \nu + \alpha \leq 0}} q^{-z(\nu+\alpha)} \;)$$

is uniformly convergent on any compact subset of $\{z \in \mathbb{C} \mid \mathrm{Re}\, z > 0\}$ (resp. $\{z \in \mathbb{C} \mid \mathrm{Re}\, z < 0\}$) and that, for any $z \in \mathbb{C}$ with $\mathrm{Re}\, z > 0$ (resp. $\mathrm{Re}\, z < 0$), its sum is equal to

$$\frac{q^{-z(\alpha+[-\alpha]+1)}}{1 - q^{-z}} = -\frac{q^{-z(\alpha+[-\alpha])}}{1 - q^{z}}$$

(resp.

$$\frac{q^{-z(\alpha+[-\alpha])}}{1 - q^{z}} \;).$$

□

From now on we assume that $T \in \mathfrak{a}_{\emptyset,\mathbb{Z}}$. Let us fix such a T and let us fix $\lambda \in \mathrm{Im}\, X_I^{\Delta}$. Let $\varphi \in \mathcal{A}_{I,\pi}^{\Delta}$ and let $\Lambda \in \mathrm{Re}\, X_I^{\Delta} \cap (\rho_I + C_I^{\Delta})$ (as $\rho_I + C_I^{\Delta} \subset C_I^{\Delta}$ we have $\langle \Lambda, \alpha_{I,j}^{\vee} \rangle > 0$ for all $j = 1, \ldots, s-1$). Then for each $w \in W$ such that $w(I) \subset \Delta$ and each $\lambda' \in \mathrm{Im}\, X_{w(I)}^{\Delta}$ we set

$$\widetilde{\Psi}_w(\lambda') = \frac{d}{e} \frac{\displaystyle\sum_{n \in \mathbb{Z}/e\mathbb{Z}} q^{-\langle \lambda' - w(\lambda+\Lambda), T_{w(I)} + (n)_{w(I)} \rangle}}{\displaystyle\prod_{j'=1}^{s-1} \left(1 - q^{\langle \lambda' - w(\lambda+\Lambda), \alpha_{w(I),j'}^{\vee} \rangle}\right)} \big(M_I(w, \pi(\lambda+\Lambda))(\varphi_{\lambda+\Lambda})\big)_{-w(\lambda+\Lambda)}$$

if

$$\sum_{j'=1}^{s} \lambda'_{j'} - \sum_{j=1}^{s} \lambda_j \in \frac{2\pi i}{\log q}\frac{d}{e}\mathbb{Z}$$

and

$$\widetilde{\Psi}_w(\lambda') = 0$$

otherwise, so that $\widetilde{\Psi}_w$ is the restriction to $\operatorname{Im} X^{\Delta}_{w(I)}$ of some element in $\mathcal{R}(X^{\Delta}_{w(I)}) \otimes \mathcal{A}^{\Delta}_{w(I),w(\pi)}$. Moreover we set

$$\Psi_w(\lambda') = \sum_{\mu' \in \mathrm{Fix}_{X^{\Delta}_{w(I)}}(w(\pi))} \widetilde{\Psi}_w(\lambda' + \mu')_{\mu'}.$$

Generalizing (G.3.1) and (G.7.1) slightly we introduce the Fourier transform

$$\Psi^{\vee}_w = \frac{1}{|\mathrm{Fix}_{X^{\Delta}_{w(I)}}(w(\pi))|} \int_{\operatorname{Im} X^{\Delta}_{w(I)}} \Psi_w(\lambda')_{\lambda'} d\lambda' = \int_{\operatorname{Im} X^{\Delta}_{w(I)}} \widetilde{\Psi}_w(\lambda')_{\lambda'} d\lambda'$$

and the pseudo-Eisenstein series

$$\theta_{\Psi_w}(g) = \sum_{\delta' \in P_{w(I)}(F)\backslash G(F)} \Psi^{\vee}_w(\delta' g) \qquad (\forall g \in G(\mathbb{A})).$$

It follows from part (ii) of the previous lemma that

$$\Psi^{\vee}_w(g) = \Phi^{\Delta\cap w(R^+),\Delta}_{w(I)}(H_\emptyset(g) - T) M_I(w, \pi(\lambda+\Lambda))(\varphi_{\lambda+\Lambda})(g) \qquad (\forall g \in G(\mathbb{A}))$$

(we have

$$\Delta \cap w(R^+) - w(I) = \{d'_1 + \cdots + d'_{j'} \in \Delta - w(I) \mid \langle -w(\Lambda), \alpha^{\vee}_{w(I),j'}\rangle < 0\})$$

and it follows from (11.4.4) that

$$\Lambda^T E^{\Delta}_I(\varphi_{\lambda+\Lambda}, \pi(\lambda+\Lambda)) = \sum_{\substack{w\in W \\ w(I)\subset\Delta}} \theta_{\Psi_w}.$$

Therefore, for any $\lambda \in \operatorname{Im} X^{\Delta}_I$, any $\varphi, \varphi' \in \mathcal{A}^{\Delta}_{I,\pi}$ and any $\Lambda, \Lambda' \in \operatorname{Re} X^{\Delta}_I \cap (\rho_I + C^{\Delta}_I)$ we have

$$\begin{aligned}
&\langle \Lambda^T E^{\Delta}_I(\varphi'_{\lambda+\Lambda'}, \pi(\lambda+\Lambda')), E^{\Delta}_I(\varphi_{\lambda+\Lambda}, \pi(\lambda+\Lambda))\rangle \\
&\quad = \langle \Lambda^T E^{\Delta}_I(\varphi'_{\lambda+\Lambda'}, \pi(\lambda+\Lambda')), \Lambda^T E^{\Delta}_I(\varphi_{\lambda+\Lambda}, \pi(\lambda+\Lambda))\rangle \\
&\quad = \sum_{\substack{w'\in W \\ w'(I)\subset\Delta}} \sum_{\substack{w\in W \\ w(I)\subset\Delta}} \langle \theta_{\Psi'_{w'}}, \theta_{\Psi_w}\rangle
\end{aligned}$$

where $\Psi'_{w'}$ is the function Ψ_w for $\varphi := \varphi'$, $w := w'$ and $\Lambda := \Lambda'$ (see (11.3.3)).

Now for any $w, w' \in W$ such that $w(I) \subset \Delta$ and $w'(I) \subset \Delta$ we may compute the scalar product of our two (generalized) pseudo-Eisenstein series $\theta_{\Psi'_{w'}}$, θ_{Ψ_w} by a slight extension of (G.8.1) (see [Mo–Wa 2] II.2.1 for more details). We have $\Omega((w'(I), w'(\pi)), (w(I), w(\pi))) = \{w'w^{-1}\}$ by regularity of (I, π). Let us consider the set of $\lambda' \in \operatorname{Re} X_I^\Delta$ such that

$$\langle \lambda' - w(\Lambda), \alpha^\vee_{w(I),j'}\rangle\langle -w(\Lambda), \alpha^\vee_{w(I),j'}\rangle > 0$$

and

$$\langle w'w^{-1}(\lambda') - w'(\Lambda'), \alpha^\vee_{w'(I),j''}\rangle\langle -w'(\Lambda'), \alpha^\vee_{w'(I),j''}\rangle > 0$$

for all $j', j'' = 1, \ldots, s-1$. The rational functions $\Psi_w(\lambda')$ and $\Psi'_{w'}(w'w^{-1}(\lambda'))$ are holomorphic on this set and, if Λ and Λ' are regular enough, i.e. if $\langle \Lambda, \alpha^\vee_{I,j}\rangle$ and $\langle \Lambda', \alpha^\vee_{I,j}\rangle$ are large enough for every $j = 1, \ldots, s-1$, this set contains a ball in $\operatorname{Re} X_{w(I)}^\Delta \cong \mathbb{R}^s/\mathbb{R}$, centered at the origin and of arbitrarily large radius. In particular, if Λ and Λ' are regular enough, we can choose λ'_0 in that set such that $\lambda'_0 \in \operatorname{Re} X_{w(I)}^\Delta \cap (\rho_{w(I)} + C_{w(I)}^\Delta)$ and we get

$$\langle \theta_{\Psi'_{w'}}, \theta_{\Psi_w}\rangle = \frac{1}{|\operatorname{Fix}_{X_{w(I)}^\Delta}(w(\pi))|}\int_{\operatorname{Im} X_{w(I)}^\Delta} u(\lambda' + \lambda'_0)d\lambda'$$

where $u(\lambda')$ is equal to

$$\frac{\deg(\infty)}{fd_1\cdots d_s}\langle \Psi'_{w'}(-w'w^{-1}(\overline{\lambda}')),$$
$$(M_{w(I)}(w'w^{-1}, w(\pi)(\lambda'))(\Psi_w(\lambda')_{\lambda'}))_{-w'w^{-1}(\lambda')}\rangle_{w'w^{-1}}$$

for every $\lambda' \in X_{w(I)}^\Delta$.

Therefore, if Λ and Λ' are regular enough, the scalar product

$$\langle \Lambda^T E_I^\Delta(\varphi'_{\lambda+\Lambda'}, \pi(\lambda + \Lambda')), E_I^\Delta(\varphi_{\lambda+\Lambda}, \pi(\lambda+\Lambda))\rangle$$

is equal to

$$\frac{d^2}{e^2}\frac{\deg(\infty)}{fd_1\cdots d_s}\sum_{\substack{w\in W\\ w(I)\subset\Delta}}\int_{S_w(\lambda)}\frac{\sum_{n\in\mathbb{Z}/e\mathbb{Z}} q^{-\langle \lambda'+\lambda'_0-w(\lambda+\Lambda), T_{w(I)}+(n)_{w(I)}\rangle}}{\prod_{j'=1}^{s-1}\left(1-q^{\langle \lambda'+\lambda'_0-w(\lambda+\Lambda), \alpha^\vee_{w(I),j'}\rangle}\right)}v_w(\lambda'+\lambda'_0)d\lambda'$$

where

$$S_w(\lambda) = \{\lambda' \in \operatorname{Im} X_{w(I)}^\Delta \mid \sum_{j'=1}^{s}\lambda'_{j'} - \sum_{j=1}^{s}\lambda_j \in \frac{2\pi i}{\log q}\frac{d}{e}\mathbb{Z}\}$$

and where, for any $\lambda' \in X^{\Delta}_{w(I)}$ such that $\sum_{j'=1}^{s} \lambda'_{j'} - \sum_{j=1}^{s} \lambda_j \in \frac{2\pi i}{\log q}\frac{d}{e}\mathbb{Z}$, $v_w(\lambda')$ is the sum over $w' \in W$ such that $w'(I) \subset \Delta$ and over $\mu'' \in \mathrm{Fix}_{X^{\Delta}_{w'(I)}}(w'(\pi))$ such that $\sum_{j''=1}^{s} \mu''_{j''} \in \frac{2\pi i}{\log q}\frac{d}{e}\mathbb{Z}$ of the product of

$$\frac{\sum_{n\in\mathbb{Z}/e\mathbb{Z}} q^{\langle w'(w^{-1}(\lambda')-\lambda+\Lambda')+\mu'', T_{w'(I)}+(n)_{w'(I)}\rangle}}{\prod_{j''=1}^{s-1}\left(1-q^{-\langle w'(w^{-1}(\lambda')-\lambda+\Lambda')+\mu'', \alpha^{\vee}_{w'(I),j''}\rangle}\right)}$$

with

$$\Big\langle \big(M_I(w',\pi(\lambda+\Lambda'))(\varphi'_{\lambda+\Lambda'})\big)_{\mu''-w'(\lambda+\Lambda')},$$
$$\Big(M_{w(I)}(w'w^{-1}, w(\pi)(\lambda'))$$
$$\big((M_I(w,\pi(\lambda+\Lambda))(\varphi_{\lambda+\Lambda}))_{\lambda'-w(\lambda+\Lambda)}\big)\Big)_{-w'w^{-1}(\lambda')}\Big\rangle_{w'w^{-1}}$$

(we have $\overline{\mu}'' = -\mu''$ and $\overline{\lambda} = -\lambda$).

LEMMA (11.4.8). — (i) *For any $w \in W$ such that $w(I) \subset \Delta$ the rational function $v_w(\lambda')$ of $\lambda' \in X^{\Delta}_{w(I)}$ is holomorphic on $\rho_I + C^{\Delta}_{w(I)}$.*

(ii) *For any $w \neq 1$ in W such that $w(I) \subset \Delta$ we have*

$$\int_{S_w(\lambda)} \frac{\sum_{n\in\mathbb{Z}/e\mathbb{Z}} q^{-\langle \lambda'+\lambda'_0-w(\lambda+\Lambda), T_{w(I)}+(n)_{w(I)}\rangle}}{\prod_{j'=1}^{s-1}\left(1-q^{\langle \lambda'+\lambda'_0-w(\lambda+\Lambda), \alpha^{\vee}_{w(I),j'}\rangle}\right)} v_w(\lambda'+\lambda'_0)d\lambda' = 0.$$

(iii) *The integral*

$$\int_{S_1(\lambda)} \frac{\sum_{n\in\mathbb{Z}/e\mathbb{Z}} q^{-\langle \lambda'+\lambda'_0-\lambda-\Lambda, T_I+(n)_I\rangle}}{\prod_{j=1}^{s-1}\left(1-q^{\langle \lambda'+\lambda'_0-\lambda-\Lambda, \alpha^{\vee}_{I,j}\rangle}\right)} v_1(\lambda'+\lambda'_0)d\lambda'$$

is equal to the sum over $w' \in W$ such that $w'(I) \subset \Delta$ and over $\mu \in \mathrm{Fix}_{X^{\Delta}_I}(\pi)$ such that $\sum_{j=1}^{s} \mu_j \in \frac{2\pi i}{\log q}\frac{d}{e}\mathbb{Z}$ of the product of

$$\frac{e}{d}\,\frac{\sum_{n\in\mathbb{Z}/e\mathbb{Z}} q^{\langle w'(\Lambda+\Lambda'+\mu), T_{w'(I)}+(n)_{w'(I)}\rangle}}{\prod_{j''=1}^{s-1}\left(1-q^{-\langle w'(\Lambda+\Lambda'+\mu), \alpha^{\vee}_{w'(I),j''}\rangle}\right)}$$

with

$$\langle (M_I(w',\pi(\lambda+\Lambda'))(\varphi'_{\lambda+\Lambda'}))_{w'(\mu-\lambda-\Lambda')}, (M_I(w',\pi(\lambda+\Lambda))(\varphi_{\lambda+\Lambda}))_{-w'(\lambda+\Lambda)}\rangle .$$

Proof : First of all let us remark that we may replace μ'' by $w'(\mu)$ with $\mu \in \mathrm{Fix}_{X_I^\Delta}(\pi)$ such that $\sum_{j=1}^{s} \mu_j \in \frac{2\pi i}{\log q}\frac{d}{e}\mathbb{Z}$ in the above expression for $v_w(\lambda')$.

By (G.5.7) the function

$$\lambda' \mapsto (M_{w(I)}(w'w^{-1}, w(\pi)(\lambda'))(\psi_{\lambda'}))_{-w'w^{-1}(\lambda')}$$

is holomorphic on $\rho_I + C_{w(I)}^\Delta$. Thus the only singularities of $v_w(\lambda')$ in $\rho_I + C_{w(I)}^\Delta$ are at most simple poles along hyperplanes

$$H_{\mu,w',j''} = \{\lambda' \in X_{w(I)}^\Delta \mid \langle w'(w^{-1}(\lambda') - \lambda - \Lambda' + \mu), \alpha_{w'(I),j''}^\vee\rangle \in \frac{2\pi i}{\log q}\mathbb{Z}\}$$

for fixed μ, w' and j''. Let us fix μ. For each w' and each j'' we have

$$H_{\mu,w',j''} = H_{\mu,w_{w'(I),\sigma}w',j''+1}$$

where σ is the transposition $(j'', j''+1)$ (with the notations of (G.1.5) and (G.5.2)) and we may consider the summands of $v_w(\lambda')$ corresponding to w' and $w_{w'(I),\sigma}w'$ and take their residues along the hyperplane $H_{\mu,w',j''} = H_{\mu,w_{w'(I),\sigma}w',j''+1}$. Clearly, to prove part (i) of the lemma it is sufficient to check that, for each w' and each j'', these two residues add up to 0.

But this easily follows from the functional equations

$$M_I(w_{w'(I),\sigma}w', \pi(\lambda+\Lambda')) = M_{w'(I)}(w_{w'(I),\sigma}, w'(\pi(\lambda+\Lambda')))M_I(w', \pi(\lambda+\Lambda')),$$

$$\begin{aligned} &M_{w(I)}(w_{w'(I),\sigma}w'w^{-1}, w(\pi)(\lambda')) \\ &\quad = M_{w'(I)}(w_{w'(I),\sigma}, w'(\pi(w^{-1}(\lambda'))))M_{w(I)}(w'w^{-1}, w(\pi)(\lambda')), \end{aligned}$$

$$\begin{aligned} M_{w_{w'(I),\sigma}w'(I)}(w_{w'(I),\sigma}, w_{w'(I),\sigma}w'(\pi(\lambda+\Lambda')))M_{w'(I)}(w_{w'(I),\sigma}, w'(\pi(\lambda+\Lambda'))) \\ = \mathrm{id} \end{aligned}$$

and

$$\begin{aligned} &M_{w'(I)}(w_{w'(I),\sigma}, w'(\pi(w^{-1}(\lambda'))))^* \\ &\quad = M_{w_{w'(I),\sigma}w'(I)}(w_{w'(I),\sigma}, w_{w'(I),\sigma}w'(\pi(w^{-1}(-\overline{\lambda}')))) \end{aligned}$$

(see (G.9.1)(ii)(c) and (G.9.3)(i)) and from the relation

$$\langle \nu', T_{w'(I)} + (n)_{w'(I)} \rangle - \langle w_{w'(I),\sigma}(\nu'), T_{w_{w'(I),\sigma}w'(I)} + (n)_{w_{w'(I),\sigma}w'(I)} \rangle \in \frac{2\pi i}{\log q}\mathbb{Z}$$

for any $\nu' \in X_{w'(I)}^{\Delta}$ such that $\langle \nu', \alpha_{w'(I),j''}^{\vee} \rangle \in \frac{2\pi i}{\log q}\mathbb{Z}$. Hence part (i) is proved.

If $w \neq 1$ we can choose $j_0' \in \{1, \dots, s-1\}$ such that

$$\langle w(\Lambda), \alpha_{w(I),j_0'}^{\vee} \rangle < 0$$

(we have $\Delta \cap w(R^+) \subsetneqq \Delta$). Let us define $\lambda_1' \in \operatorname{Re} X_{w(I)}^{\Delta}$ by

$$\begin{aligned}\lambda_1' =& (d_1', \dots, d_{j_0'}', -d_{j_0'+1}', \dots, -d_s') \\ & - \frac{d_1' + \dots + d_{j_0'}' - d_{j_0'+1}' - \dots - d_s'}{d}(d_1', \dots, d_s'),\end{aligned}$$

so that

$$\langle \lambda_1', \alpha_{w(I),j'}^{\vee} \rangle = \begin{cases} 2 & \text{if } j' = j_0', \\ 0 & \text{if } j' \neq j_0'. \end{cases}$$

Let us move the contour of integration from $S_w(\lambda) + \lambda_0'$ to $S_w(\lambda) + \lambda_0' + r\lambda_1'$ where r is a positive real number and let r go to $+\infty$.

Firstly our assumptions on λ_0' and j_0' imply that

$$\prod_{j'=1}^{s-1}\left(1 - q^{\langle \lambda' + \lambda_0' + r\lambda_1' - w(\lambda + \Lambda), \alpha_{w(I),j'}^{\vee} \rangle}\right) \neq 0$$

for every $\lambda' \in S_w(\lambda)$ and every $r > 0$. Moreover we have already proved that $v_w(\lambda')$ is holomorphic. Therefore, by applying the Cauchy formula we see that the integral does not change. Secondly we have

$$\begin{aligned}& q^{-\langle r\lambda_1', T_{w(I)} + (n)_{w(I)} \rangle + \langle rw'w^{-1}(\lambda_1'), T_{w'(I)} + (n')_{w'(I)} \rangle} \\ & \qquad < q^{2r(1 + |\{j''=1,\dots,s \mid \langle w'w^{-1}(\lambda_1'), \alpha_{w'(I),j''}^{\vee} \rangle = -2\}|)}\end{aligned}$$

for every $r > 0$, every $w' \in W$ such that $w'(I) \subset \Delta$ and all $n, n' \in \mathbb{Z}/e\mathbb{Z}$. Indeed we have

$$\langle w'w^{-1}(\lambda_1'), T_{w'(I)} \rangle \leq \langle \lambda_1', T_{w(I)} \rangle$$

as $T \in \mathfrak{a}_{\emptyset}^{+}$, we have

$$\langle w'w^{-1}(\lambda_1'), \alpha_{w'(I),j''}^{\vee} \rangle \in \{-2, 0, 2\}$$

and we have $(n)_{w(I)}^{j_0'} \in]-1,0]$ and $(n')_{w'(I)}^{j''} \in]-1,0]$ for all $j'' = 1,\ldots,s$. Thirdly, for each $w' \in W$ such that $w'(I) \subset \Delta$, the expression

$$\Big|\prod_{j'=1}^{s-1}\Big(1 - q^{\langle \lambda' + \lambda_0' + r\lambda_1' - w(\lambda+\Lambda), \alpha_{w(I),j'}^\vee\rangle}\Big)$$
$$\times \prod_{j''=1}^{s-1}\Big(1 - q^{\langle -w'w^{-1}(\overline{\lambda}' + \lambda_0' + r\lambda_1') + \mu'' - w'(\lambda+\Lambda'), \alpha_{w'(I),j''}^\vee\rangle}\Big)\Big|$$

with $\lambda' \in S_w(\lambda)$ is asymptotically equivalent to

$$Cq^{2r(1+|\{j''=1,\ldots,s \mid \langle w'w^{-1}(\lambda_1'), \alpha_{w'(I),j''}^\vee\rangle = -2\}|)}$$

for some constant C as $r \to +\infty$. Therefore part (ii) of the lemma is a consequence of the following assertion which is itself a direct consequence of (G.5.8).

For any fixed $w' \in W$ such that $w'(I) \subset \Delta$, any fixed $\psi \in \mathcal{A}_{w(I),w(\pi)}^{\Delta}$ and $\psi' \in \mathcal{A}_{w'(I),w'(\pi)}^{\Delta}$ and any fixed $\lambda' \in \operatorname{Im} X_{w(I)}^{\Delta}$,

$$\langle \psi', (M_{w(I)}(w'w^{-1}, w(\pi)(\lambda' + \lambda_0' + r\lambda_1'))(\psi_{\lambda'+r\lambda_1'}))_{-w'w^{-1}(\lambda'+\lambda_0'+r\lambda_1')}\rangle$$

is bounded as $r \to +\infty$.

To prove part (iii) let us first make the change of variables $\lambda' - \lambda \mapsto \lambda'$ and let us remark that we have the exact sequences

$$0 \longrightarrow S_1(0) \longrightarrow \operatorname{Im} X_I^{\Delta} \longrightarrow \mathbb{Z}/\frac{d}{e}\mathbb{Z} \longrightarrow 0$$

and

$$\begin{array}{ccccccc} 0 \longrightarrow \frac{1}{e}\mathbb{Z}/\mathbb{Z} \longrightarrow & S_1(0) & \longrightarrow & \Big(i\mathbb{R}\Big/\frac{2\pi i}{\log q}\mathbb{Z}\Big)^{s-1} & \longrightarrow 0, \\ & \lambda' & \longmapsto & ((\langle \lambda', \alpha_{I,j}^\vee\rangle)_j & \end{array}$$

(we have

$$\Big\langle (d_1,\ldots,d_k,0,\ldots,0) - \frac{d_1+\cdots+d_j}{d}(d_1,\ldots,d_s), \alpha_{I,j}^\vee\Big\rangle = \delta_{j,k}$$

for every $j = 1,\ldots,s-1$ and every $k = 1,\ldots,s$). Therefore the integral of our function

$$\lambda' \longmapsto \frac{\displaystyle\sum_{n\in\mathbb{Z}/e\mathbb{Z}} q^{-\langle \lambda'+\lambda_0'-\lambda-\Lambda, T_I+(n)_I\rangle}}{\displaystyle\prod_{j=1}^{s-1}\Big(1 - q^{\langle \lambda'+\lambda_0'-\lambda-\Lambda, \alpha_{I,j}^\vee\rangle}\Big)} v_1(\lambda'+\lambda_0')$$

over $S_1(\lambda)$ is equal to

$$\int_{\left(i\mathbb{R}/\frac{2\pi i}{\log q}\mathbb{Z}\right)^{s-1}} \frac{g(x)}{\prod_{j=1}^{s-1}\left(1-q^{x_j-\langle \Lambda-\lambda_0',\alpha_{I,j}^\vee\rangle}\right)}dx$$

where we have set

$$g(x)=\frac{1}{d}\sum_{m,n\in\mathbb{Z}/e\mathbb{Z}} q^{-\langle \widetilde{x}+\frac{2\pi i}{\log q}\frac{m}{e}(d_1,\ldots,d_s)+\lambda_0'-\Lambda,T_I+(n)_I\rangle}$$

$$\times v_1(\lambda+\widetilde{x}+\frac{2\pi i}{\log q}\frac{m}{e}(d_1,\ldots,d_s)+\lambda_0')$$

for any $x\in\left(\mathbb{C}/\frac{2\pi i}{\log q}\mathbb{Z}\right)^{s-1}$ and any $\widetilde{x}\in X_I^\Delta$ such that $\sum_{j=1}^s \widetilde{x}_j\in\frac{2\pi i}{\log q}\frac{d}{e}\mathbb{Z}$ and such that $x_j=\langle\widetilde{x},\alpha_{I,j}^\vee\rangle$, $\forall j=1,\ldots,s-1$. Here $dx=dx_1\cdots dx_{s_1}$ is the normalized Haar measure on the compact Lie group $\left(i\mathbb{R}/\frac{2\pi i}{\log q}\mathbb{Z}\right)^{s-1}$.

Then let us move the contour of integration by replacing x by $x+X$ for $X\in(\mathbb{R}^+)^{s-1}$ that we let go to infinity. Each time that, for some j, we have $X_j=\langle\Lambda-\lambda_0',\alpha_{I,j}^\vee\rangle$ we pick a residue. Moreover, if we let X_j go to $+\infty$ for some j, it can be seen as in the proof of the lemma that

$$\frac{g(x+X)}{\prod_{j=1}^{s-1}\left(1-q^{x_j+X_j-\langle\Lambda-\lambda_0',\alpha_{I,j}^\vee\rangle}\right)}\longrightarrow 0$$

for any $x\in\left(i\mathbb{R}/\frac{2\pi i}{\log q}\mathbb{Z}\right)^{s-1}$. Therefore, by the Cauchy formula the integral that we are computing is equal to the value at $x_j=\langle\Lambda-\lambda_0',\alpha_{I,j}^\vee\rangle$, $j=1,\ldots,s-1$, of $g(x)$. But for this particular x we can take $\widetilde{x}=\Lambda-\lambda_0'$ and we get that

$$\int_{S_1(\lambda)}\frac{\sum_{n\in\mathbb{Z}/e\mathbb{Z}} q^{-\langle\lambda'+\lambda_0'-\lambda-\Lambda,T_I+(n)_I\rangle}}{\prod_{j=1}^{s-1}\left(1-q^{\langle\lambda'+\lambda_0'-\lambda-\Lambda,\alpha_{I,j}^\vee\rangle}\right)}v_1(\lambda'+\lambda_0')d\lambda'$$

is equal to

$$\frac{1}{d}\sum_{m,n\in\mathbb{Z}/e\mathbb{Z}} q^{-\frac{2\pi i}{\log q}\frac{m}{e}(T_1+\cdots+T_d-n)}v_1(\lambda+\Lambda+\frac{2\pi i}{\log q}\frac{m}{e}(d_1,\ldots,d_s))=\frac{e}{d}v_1(\lambda+\Lambda),$$

i.e. to the sum over $w' \in W$ such that $w'(I) \subset \Delta$ and over $\mu \in \mathrm{Fix}_{X_I^\Delta}(\pi)$ such that $\sum_{j=1}^{s} \mu_j \in \frac{2\pi i}{\log q}\frac{d}{e}\mathbb{Z}$ of the product of

$$\frac{e}{d}\,\frac{\displaystyle\sum_{n\in\mathbb{Z}/e\mathbb{Z}} q^{\langle w'(\Lambda+\Lambda'+\mu),T_{w'(I)}+(n)_{w'(I)}\rangle}}{\displaystyle\prod_{j''=1}^{s-1}\left(1-q^{-\langle w'(\Lambda+\Lambda'+\mu),\alpha^\vee_{w'(I),j''}\rangle}\right)}$$

with

$$\big\langle \big(M_I(w',\pi(\lambda+\Lambda'))(\varphi'_{\lambda+\Lambda'})\big)_{w'(\mu-\lambda-\Lambda')},\\ \big(M_I(w',\pi(\lambda+\Lambda))(\varphi_{\lambda+\Lambda})\big)_{-w'(\lambda+\Lambda)}\big\rangle_{w'}.$$

This concludes the proof of the lemma. □

Proposition (11.4.9). — *For any $\lambda, \lambda' \in \overline{C}_I^\Delta$ such that $\sum_{j=1}^{s}(\lambda_j + \overline{\lambda}'_j) \in \frac{2\pi i}{\log q}\frac{d}{e}\mathbb{Z}$ and*

$$\langle w'(\lambda+\overline{\lambda}'+\mu),\alpha^\vee_{w'(I),j''}\rangle \notin \frac{2\pi i}{\log q}\mathbb{Z}$$

($\forall w' \in W$ with $w'(I) \subset \Delta$, $\forall \mu \in \mathrm{Fix}_{X_I^\Delta}(\pi)$ and $\forall j'' = 1,\dots,s$) the scalar product

$$\langle \Lambda^T E_I^\Delta(\varphi'_{\lambda'},\pi(\lambda')), E_I^\Delta(\varphi_\lambda,\pi(\lambda))\rangle$$

is equal to the sum over $\mu \in \mathrm{Fix}_{X_I^\Delta}(\pi)$ such that $\sum_{j=1}^{s}\mu_j \in \frac{2\pi i}{\log q}\frac{d}{e}\mathbb{Z}$ of

$$\frac{\deg(\infty)d}{fd_1\cdots d_s e}\sum_{\substack{w'\in W\\ w'(I)\subset\Delta}}\frac{\displaystyle\sum_{n\in\mathbb{Z}/e\mathbb{Z}} q^{\langle w'(\lambda+\overline{\lambda}'+\mu),T_{w'(I)}+(n)_{w'(I)}\rangle}}{\displaystyle\prod_{j''=1}^{s-1}\left(1-q^{-\langle w'(\lambda+\overline{\lambda}'+\mu),\alpha^\vee_{w'(I),j''}\rangle}\right)}\\ \times\big\langle \big(M_I(w',\pi(\lambda'))(\varphi'_{\lambda'})\big)_{w'(\mu-\lambda')}, \big(M_I(w',\pi(\lambda))(\varphi_\lambda)\big)_{-w'(\lambda)}\big\rangle_{w'}.$$

Proof : If $\mathrm{Im}\,\lambda = \mathrm{Im}\,\lambda'$, $\mathrm{Re}\,\lambda$ and $\mathrm{Re}\,\lambda'$ are regular enough the equality we want to prove follows from the previous lemma. But both sides of this equality are meromorphic functions on

$$\{(\lambda,\overline{\lambda}') \in \overline{C}_I^\Delta \times \overline{C}_I^\Delta \mid \sum_{j=1}^{s}(\lambda_j+\overline{\lambda}'_j) \in \frac{2\pi i}{\log q}\frac{d}{e}\mathbb{Z}\}$$

(see (G.9.4)) and each connected component of this set contains points with $\mathrm{Im}\,\lambda = \mathrm{Im}\,\lambda'$ and with $\mathrm{Re}\,\lambda$ and $\mathrm{Re}\,\lambda'$ as regular as we like. Therefore the proposition is proved. □

COROLLARY (11.4.10). — *For any regular unitary cuspidal pair* (I,π), *any* $T \in \mathfrak{a}_\emptyset^+ \cap \mathfrak{a}_{\emptyset,\mathbb{Z}}$ *and any* $\lambda \in \operatorname{Im} X_I^\Delta$ *the operator* $M_{I,\pi}^T(\lambda)$ *maps* $\varphi \in \mathcal{A}_{I,\pi}^\Delta$ *onto the sum over* $\mu \in \operatorname{Fix}_{X_I^\Delta}(\pi)$ *such that* $\sum_{j=1}^s \mu_j \in \frac{2\pi i}{\log q}\frac{d}{e}\mathbb{Z}$ *of*

$$\frac{\deg(\infty)d}{fd_1\cdots d_s e}\lim_{\zeta\to\mu}\sum_{\substack{w\in W\\ w(I)\subset\Delta}}\frac{\sum_{n\in\mathbb{Z}/e\mathbb{Z}} q^{\langle w(\zeta),T_{w(I)}+(n)_{w(I)}\rangle}}{\prod_{j'=1}^{s-1}\left(1-q^{-\langle w(\zeta),\alpha^\vee_{w(I),j'}\rangle}\right)}$$

$$\times\Big(M_I(w,\pi(\lambda))^{-1}\big((M_I(w,\pi(\lambda+\zeta))(\varphi_{\lambda+\zeta-\mu}))_{-w(\zeta)}\big)\Big)_{-\lambda}.$$

□

(11.5) Evaluation of $J_c^T(f)$ in a special case: $(G,M)^\sim$-families

Following Arthur we will replace the Weyl group elements by parabolic subgroups. Let $\mathcal{L}$ be the (finite) set of Levi subgroups of G containing the maximal torus T of diagonal matrices. Any Levi subgroup M of G over F is equal to

$$\{g\in G \mid g(V_j)=V_j,\ \forall j=1,\ldots,s\}$$

for some splitting

$$F^d = V_1\oplus\cdots\oplus V_s$$

of the F-vector space F^d into a direct sum of F-vector subspaces and $M\in\mathcal{L}$ if and only if each V_j can be generated by vectors belonging to the standard basis $(v_1,\ldots,v_d)$ of F^d. For each $I\subset\Delta$ and each $w\in W$ we have $\dot{w}M_I\dot{w}^{-1}\in\mathcal{L}$ and any $M\in\mathcal{L}$ is of this form. We have

$$\dot{w}M_I\dot{w}^{-1}=M_{I'}$$

for some $I,I'\subset\Delta$ and some $w\in W$ if and only if

$$w\in\{w\in W \mid w(I)=I'\}W_I$$

(recall that $\{w\in W \mid w(I)\subset R^+\}$ is a system of representatives of the classes in W/W_I).

For any $M\in\mathcal{L}$ let $\mathcal{P}(M)$ be the (finite) set of the parabolic subgroups P of G which admit M as a Levi component, i.e. which contain M and which satisfy $P=MN_P$ where N_P is the unipotent radical of P. For any $I\subset\Delta$ we set $\mathcal{P}(I)=\mathcal{P}(M_I)$. Then we have a map

(11.5.1) $$\{w\in W \mid w(I)\subset\Delta\}\to\mathcal{P}(I),\ w\mapsto \dot{w}^{-1}P_{w(I)}\dot{w},$$

and this map is bijective. Indeed, if $(d_1,\dots,d_s)$ is the partition of d corresponding to I any $w\in W$ such that $w(I)\subset\Delta$ is of the form $w_{I,\sigma}$ for a unique $\sigma\in\mathfrak{S}_s$ (see (G.1.5)) and $\dot{w}^{-1}P_{w(I)}\dot{w}$ is the stabilizer in G of the flag

$$(0)=V_0\subset V_1\subset\cdots\subset V_s=F^d$$

where we have set

$$V_j=\bigoplus_{k=1}^{j}\bigoplus_{m=1}^{d_{\sigma^{-1}(k)}}Fv_{d_1+\cdots+d_{\sigma^{-1}(k)-1}+m}$$

for $j=1,\dots,s$.

For any $M\subset L$ in $\mathcal{L}$ we denote by $\mathcal{P}^L(M)$ the set of parabolic subgroups of L which admit M as a Levi component. The map

$$\text{(11.5.2)},\qquad \mathcal{P}(L)\times\mathcal{P}^L(M)\to\mathcal{P}(M),\ (Q,R)\mapsto RN_Q,$$

is injective. For each $Q\in\mathcal{P}(L)$ it induces a bijection from $\mathcal{P}^L(M)$ onto the set of $P\in\mathcal{P}(M)$ such that $P\subset Q$ ($P=(P\cap L)N_Q$). For any $I\subset J\subset\Delta$ we set $\mathcal{P}^J(I)=\mathcal{P}^{M_J}(M_I)$. Then we have a bijection

$$\text{(11.5.3)}\qquad \{w_J\in W_J\mid w_J(I)\subset J\}\to\mathcal{P}^J(I),\ w_J\mapsto\dot{w}_J^{-1}P^J_{w_J(I)}\dot{w}_J,$$

which generalizes (11.5.1). The map (11.5.2),

$$\mathcal{P}(J)\times\mathcal{P}^J(I)\to\mathcal{P}(I),\ (Q,P^J)\mapsto P^JN_Q,$$

corresponds by (11.5.1) and (11.5.3) to the injective map

$$\begin{array}{rcl}\{w\in W\mid w(J)\subset\Delta\}\times\{w_J\in W_J\mid w_J(I)\subset J\} & \to & \{w\in W\mid w(I)\subset\Delta\},\\ (w,w_J) & \mapsto & ww_J.\end{array}$$

For any $M\in L$ and any $P\in\mathcal{P}(M)$ let us denote by

$$\Delta_P\subset R(G,Z_M)$$

the set of simple roots of (P,Z_M), where Z_M is the center of M. For any $I\subset\Delta$ and any $P\in\mathcal{P}(I)$ the map $\alpha\mapsto\alpha|Z_I$ is a bijection from $w^{-1}(\Delta)-I\subset R$ onto Δ_P, where $Z_I=Z_{M_I}$ and w is the unique element in W such that $w(I)\subset\Delta$ and $P=\dot{w}^{-1}P_{w(I)}\dot{w}$.

More generally, for any $M\subset L$ in $\mathcal{L}$, any $P\in\mathcal{P}(M)$ and any $Q\in\mathcal{P}(L)$ such that $P\subset Q$ we denote by

$$\Delta_P^Q\subset R(L,Z_M)\subset R(G,Z_M)$$

the set of simple roots of $(P\cap L, Z_M)$. Then, for any $I\subset J\subset\Delta$, any $P\in\mathcal{P}(I)$ and any $Q\in\mathcal{P}(J)$ such that $P\subset Q$, the map $\alpha\mapsto\alpha|Z_I$ is a bijection from $w_J^{-1}(J)-I\subset R$ onto Δ_P^Q, where w_J is the unique element in W_J such that $w_J(I)\subset J$ and $P\cap M_J=\dot{w}_J^{-1}P^J_{w_J(I)}\dot{w}_J$.

Let $I\subset\Delta$. Two parabolic subgroups $P=\dot{w}^{-1}P_{w(I)}\dot{w}$ and $P'=\dot{w}'^{-1}P_{w'(I)}\dot{w}'$ in $\mathcal{P}(I)$ ($w,w'\in W$ and $w(I),w'(I)\subset\Delta$) are said to be **adjacent** if

$$\ell_{w(I)}(w'w^{-1})=1,$$

i.e. if we have

$$w'=s_{w(I),\alpha}w$$

for some $\alpha\in\Delta-w(I)$. If this is the case it is clear that $w^{-1}(\alpha)|Z_I$ is the unique element of $\Delta_P\cap(-\Delta_{P'})$. More generally let $M\in\mathcal{L}$. Two parabolic subgroups $P,P'\in\mathcal{P}(M)$ are said to be **adjacent** if, for some (resp. any) $w\in W$ and some (resp. any) $I\subset\Delta$ such that $M=\dot{w}M_I\dot{w}^{-1}$, the two parabolic subgroups $\dot{w}^{-1}P\dot{w}$ and $\dot{w}^{-1}P'\dot{w}$ in $\mathcal{P}(I)$ are adjacent. If this is the case we denote by $\alpha_{P,P'}$ the unique element of $\Delta_P\cap(-\Delta_{P'})$ and it is clear that

$$\alpha_{P,P'}=-\alpha_{P',P}.$$

For any $M\in\mathcal{L}$ we have $M=\dot{w}M_I\dot{w}^{-1}$ for some $w\in W$ and some $I\subset\Delta$ and we have a subspace

$$\mathfrak{a}_M=w(\mathfrak{a}_I)=\{H\in\mathfrak{a}_\emptyset\mid\alpha(H)=0,\ \forall\alpha\in R(M,T)\}$$

of $\mathfrak{a}_\emptyset$, a canonical retraction

$$\mathfrak{a}_\emptyset\to\mathfrak{a}_M,\ H\mapsto H_M=w((w^{-1}H)_I),$$

and lattices

$$\mathfrak{a}_M\cap\mathfrak{a}_{\emptyset,\mathbb{Z}}\subset\mathfrak{a}_{M,\mathbb{Z}}=w(\mathfrak{a}_{I,\mathbb{Z}})=\{H_M\mid H\in\mathfrak{a}_{\emptyset,\mathbb{Z}}\}.$$

For any $\alpha\in R(G,Z_M)$ we have an element $\alpha^\vee\in\mathfrak{a}_{M,\mathbb{Z}}$: if $(d_1,\dots,d_s)$ is the partition of d corresponding to I and if $w^{-1}\alpha=(\epsilon_{d_1+\cdots+d_j}-\epsilon_{d_1+\cdots+d_k})|Z_I$ we define $(w^{-1}\alpha)^\vee\in\mathfrak{a}_{I,\mathbb{Z}}$ by

$$(w^{-1}\alpha)^\vee_\ell=\begin{cases}\frac{1}{d_j} & \text{if } \ell=j,\\ -\frac{1}{d_k} & \text{if } \ell=k,\\ 0 & \text{otherwise,}\end{cases}$$

and we set

$$\alpha^\vee=w((w^{-1}\alpha)^\vee).$$

For any $M \in \mathcal{L}$ we also have the group X_M of characters of $M(\mathbb{A})^1 \backslash M(\mathbb{A})$ and a canonical pairing

$$\langle \cdot, \cdot \rangle : X_M \times \mathfrak{a}_{M,\mathbb{Z}} \to \mathbb{C} \Big/ \frac{2\pi i}{\log q} \mathbb{Z}.$$

In fact, if we set

$$\mathfrak{a}_M^* = \mathrm{Hom}_{\mathbb{Q}}(\mathfrak{a}_M, \mathbb{C})$$

and

$$\begin{aligned} \mathfrak{a}_{M,\mathbb{Z}}^\vee &= \{\zeta \in \mathfrak{a}_M^* \mid \zeta(\mathfrak{a}_{M,\mathbb{Z}}) \subset \mathbb{Z}\} \subset (\mathfrak{a}_M \cap \mathfrak{a}_{\emptyset,\mathbb{Z}})^\vee \\ &= \{\zeta \in \mathfrak{a}_M^* \mid \zeta(\mathfrak{a}_M \cap \mathfrak{a}_{\emptyset,\mathbb{Z}}) \subset \mathbb{Z}\} \end{aligned}$$

we may identify

$$\mathfrak{a}_M^* \Big/ \frac{2\pi i}{\log q} \mathfrak{a}_{M,\mathbb{Z}}^\vee$$

with X_M by the map

$$\zeta \mapsto (m \mapsto q^{\zeta(H_M(m))})$$

where $H_M(m) = (H_\emptyset(m))_M \in \mathfrak{a}_M$ and where the above pairing is induced by the canonical pairing

$$\mathfrak{a}_M^* \times \mathfrak{a}_M \to \mathbb{C}.$$

For any $M \subset M'$ in $\mathcal{L}$, $\mathfrak{a}_{M'}$ is naturally a subspace of $\mathfrak{a}_M$. The projection $H \mapsto H_{M'}$ gives a retraction of $\mathfrak{a}_M$ onto $\mathfrak{a}_{M'}$. By duality $\mathfrak{a}_{M'}^*$ is a quotient space of $\mathfrak{a}_M^*$ and we have a section of $\mathfrak{a}_{M'}^* \hookrightarrow \mathfrak{a}_M^*$ (dual to the projection $H \mapsto H_{M'}$). Taking into account the lattices this section induces an inclusion $X_{M'} \subset X_M$ and we have a subgroup

$$\mathfrak{a}_M^{*M',\mathbb{Z}} = \{\zeta \in \mathfrak{a}_M^* \mid \zeta_{M'} \in \frac{2\pi i}{\log q} (\mathfrak{a}_{M'} \cap \mathfrak{a}_{\emptyset,\mathbb{Z}})^\vee\}$$

of $\mathfrak{a}_M^*$ where $\zeta_{M'}$ is the projection of ζ onto $\mathfrak{a}_{M'}^*$ and a subgroup

$$X_M^{M'} = \mathfrak{a}_M^{*M',\mathbb{Z}} \Big/ \frac{2\pi i}{\log q} \mathfrak{a}_{M,\mathbb{Z}}^\vee$$

of X_M. The inclusion $X_{M'} \subset X_M$ is also induced by the inclusion $M(\mathbb{A}) \subset M'(\mathbb{A})$ and $X_M^{M'}$ may be identified to the set of characters of $M(\mathbb{A})^1 \backslash M(\mathbb{A})$ which are trivial on $Z_{M'}(\mathbb{A})$.

We leave it to the reader to define the subspaces $\mathrm{Re}\, \mathfrak{a}_M^*$, $\mathrm{Im}\, \mathfrak{a}_M^*$, $\mathrm{Re}\, X_M$, $\mathrm{Im}\, X_M, \ldots$.

DEFINITION (11.5.4). — (i) (Arthur) *A (G, M)-family is a family of smooth functions*

$$c = (c_P(\zeta) \mid P \in \mathcal{P}(M))$$

on $\operatorname{Im}\mathfrak{a}_M^{*G,\mathbb{Z}}$ *which satisfies the following condition: for any pair* (P,P') *of adjacent parabolic subgroups in* $\mathcal{P}(M)$ *we have*

$$c_P(\zeta) = c_{P'}(\zeta)$$

for every $\zeta \in \operatorname{Im}\mathfrak{a}_M^{*G,\mathbb{Z}}$ *such that*

$$\langle \zeta, \alpha_{P,P'}^{\vee}\rangle = 0.$$

(ii) (Waldspurger) *A* $(G,M)^{\sim}$*-family is a family of smooth functions*

$$c = (c_P(\zeta) \mid P \in \mathcal{P}(M))$$

on $\operatorname{Im} X_M^G$ *which satisfies the following condition: for any pair* (P,P') *of adjacent parabolic subgroups in* $\mathcal{P}(M)$ *we have*

$$c_P(\zeta) = c_{P'}(\zeta)$$

for every $\zeta \in \operatorname{Im} X_M^G$ *such that*

$$\langle \zeta, \alpha_{P,P'}^{\vee}\rangle \in \frac{2\pi i}{\log q}\mathbb{Z}.$$

Obviously the rules

$$(\alpha c + \beta d)_P = \alpha c_P + \beta d_P$$

$(\alpha, \beta \in \mathbb{C})$ and

$$(cd)_P = c_P d_P$$

give to the set of (G,M)-families (resp. $(G,M)^{\sim}$-families) the structure of a $\mathbb{C}$-algebra.

LEMMA (11.5.5). — *Let us fix* $M \in \mathcal{L}$.

(i) *Let* $T \in \mathfrak{a}_{\emptyset,\mathbb{Z}}$. *For each* $P = \dot{w}P_I\dot{w}^{-1} \in \mathcal{P}(M)$ *with* $w \in W$ *and* $I \subset \Delta$ *let us set*

$$T_P = w(T_I).$$

Then $(c_P(\zeta) = q^{\langle \zeta, T_P\rangle} \mid P \in \mathcal{P}(M))$ *is a* $(G,M)^{\sim}$*-family.*

(ii) *For each* $P = \dot{w}P_I\dot{w}^{-1} \in \mathcal{P}(M)$ *with* $w \in W$ *and* $I \subset \Delta$ *and each* $\zeta \in \operatorname{Im} X_M^G$ *let us set*

$$c_P(\zeta) = \sum_{n \in \mathbb{Z}/e\mathbb{Z}} q^{\langle \zeta, (n)_P\rangle}$$

if $\sum_{j=1}^{s} \zeta_j \in \frac{2\pi i}{\log q}\frac{d}{e}\mathbb{Z}$ *and*

$$c_P(\zeta) = 0$$

otherwise. Here e *is the smallest positive integer such that* d *divides* ed_j *for every* $j = 1, \dots, s$, $d_I = (d_1, \dots, d_s)$ *being the partition of* d *corresponding to* I, *and* $(n)_P = w((n)_I)$. *Then* $(c_P(\zeta) \mid P \in \mathcal{P}(M))$ *is a* $(G,M)^{\sim}$*-family.*

Proof : Obviously we may assume that $M = M_I$ for some $I \subset \Delta$. Let $(P = \dot{w}^{-1}P_{w(I)}\dot{w}, P' = \dot{w}'^{-1}P_{w'(I)}\dot{w}')$ be a pair of adjacent parabolic subgroups in $\mathcal{P}(I)$ and let $\zeta \in \mathrm{Im}\, X_I^{\Delta}$ be such that $\langle \zeta, \alpha_{P,P'}^{\vee}\rangle \in \frac{2\pi i}{\log q}\mathbb{Z}$. We have $w' = s_{w(I),\alpha}w$ for some $\alpha \in \Delta - w(I)$, so that $\alpha_{P,P'} = w^{-1}(\alpha)|Z_I$, and we want to check that $c_P(\zeta) = c_{P'}(\zeta)$.

It is clear that

$$w^{-1}(T_{w(I)}) - w^{-1}s_{w(I),\alpha}^{-1}(T_{s_{w(I),\alpha}w(I)}) \in \mathbb{Z}\alpha_{P,P'}^{\vee}$$

and that

$$w^{-1}((n)_{w(I)}) - w^{-1}s_{w(I),\alpha}^{-1}((n)_{s_{w(I),\alpha}w(I)}) \in \mathbb{Z}\alpha_{P,P'}^{\vee} \qquad (\forall n \in \mathbb{Z}/e\mathbb{Z}).$$

The lemma follows. □

Before introducing our main example of a $(G,M)^{\sim}$-family let us generalize some definitions in appendix G.

From now on a regular unitary cuspidal pair will be a pair (M,π) where $M = \dot{w}M_I\dot{w}^{-1} \in \mathcal{L}$ and π is an irreducible admissible representation of $F_\infty^\times\backslash M(\mathbb{A})$ such that $(I, w^{-1}(\pi))$ is a regular unitary cuspidal pair in the sense of appendix G.

If (M,π) is a regular unitary cuspidal pair and if $P = \dot{w}P_I\dot{w}^{-1} \in \mathcal{P}(M)$ we set

$$\mathcal{A}_{P,\pi}^G = \{\varphi \in C^\infty(N_P(\mathbb{A})M(F)\backslash G(\mathbb{A}), \mathbb{C}) \mid w^{-1}(\varphi) \in \mathcal{A}_{I,\pi}^G\}$$

where we have set $(w^{-1}(\varphi))(g) = \varphi(\dot{w}g\dot{w}^{-1})$ for each $g \in G(\mathbb{A})$. For any locally constant function with compact support f on $F_\infty^\times\backslash G(\mathbb{A})$ and any $\lambda \in \mathrm{Im}\, X_M^G$ we have an operator $i_P^G(\lambda, f)$ on $\mathcal{A}_{P,\pi}^G$ defined by

$$i_P^G(\lambda,f)(\varphi)(h) = \int_{F_\infty^\times\backslash G(\mathbb{A})} f(g)\varphi(hg)q^{\langle \lambda, H_P(hg)-H_P(h)\rangle}\frac{dg}{dz_\infty} \qquad (\forall h \in G(\mathbb{A}))$$

for each $\varphi \in \mathcal{A}_{P,\pi}^G$, where $H_P(g)z = (H_\emptyset(m))_M$, $\forall g = n_Pmk \in N_P(\mathbb{A})M(\mathbb{A})K = G(\mathbb{A})$.

If (M,π) is a regular unitary cuspidal pair we denote by $\mathrm{Fix}_{X_M^G}(\pi)$ the finite group of $\mu \in X_M^G$ such that $\pi(\mu) \cong \pi$. For any $\mu \in \mathrm{Fix}_{X_M^G}(\pi)$ and any $P \in \mathcal{P}(M)$ we have $\mathcal{A}_{P,\pi(\mu)}^G = \mathcal{A}_{P,\pi}^G$ and we will denote by $[\mu]_P$ the automorphism $\varphi \mapsto \varphi(\mu)$ of $\mathcal{A}_{P,\pi}^G$.

For every $P = \dot{w}P_I\dot{w}^{-1}, P' = \dot{w}\dot{w}'^{-1}P_{w'(I)}\dot{w}'\dot{w}^{-1} \in \mathcal{P}(M)$ with $w, w' \in W$ such that $w'(I) \subset \Delta$ and for every $\lambda \in \mathrm{Im}\, X_M^G$ let us denote by

$$M_{P'|P}(\pi,\lambda) : \mathcal{A}_{P,\pi}^G \to \mathcal{A}_{P',\pi}^G$$

the operator

$$\varphi \mapsto ww'^{-1}\Big(\big(M_I(w', w^{-1}(\pi(\lambda)))(w^{-1}(\varphi_\lambda))\big)_{-w'w^{-1}(\lambda)}\Big).$$

For any function f as before this operator intertwines the operators $i_P^G(\lambda, f)$ and $i_{P'}^G(\lambda, f)$ and for any $P, P', P'' \in \mathcal{P}(M)$ we have

$$M_{P''|P}(\pi, \lambda) = M_{P''|P'}(\pi, \lambda) \circ M_{P'|P}(\pi, \lambda)$$

(see (G.9.1)(ii)(a) and (c)). Moreover it is clear that, for any $\mu \in \mathrm{Fix}_{X_M^G}(\pi)$, we have

$$M_{P'|P}(\pi, \lambda) \circ [\mu]_P = [\mu]_{P'} \circ M_{P'|P}(\pi, \lambda + \mu).$$

LEMMA (11.5.6). — *Let us fix a regular unitary cuspidal pair (M, π), $\lambda \in \mathrm{Im}\, X_M^G$ and $P_0 \in \mathcal{P}(M)$. For every $P \in \mathcal{P}(M)$ and every $\zeta \in \mathrm{Im}\, X_M^G$ let us set*

$$M_P(\zeta, \pi, \lambda; P_0) = M_{P|P_0}(\pi, \lambda)^{-1} \circ M_{P|P_0}(\pi, \lambda + \zeta) : \mathcal{A}_{P_0,\pi}^G \to \mathcal{A}_{P_0,\pi}^G.$$

Then for any $\mu \in \mathrm{Fix}_{X_M^G}(\pi)$ and for any locally constant function with compact support f on $F_\infty^\times \backslash G(\mathbb{A})$ the family of functions of $\zeta \in \mathrm{Im}\, X_M^G$,

$$\Big(c_P(\mu; \zeta) = \mathrm{tr}\big(M_P(\zeta, \pi, \lambda; P_0) \circ [-\mu]_{P_0} \circ i_{P_0}^G(\lambda + \zeta - \mu, f)\big) \mid P \in \mathcal{P}(M)\Big),$$

is a $(G, M)^{\sim}$-family.

Proof: Obviously we may assume that $M = M_I$ for some $I \subset \Delta$. We may moreover assume that $P_0 = P_I$ and therefore that $w_0 = 1$. By functional equation (G.9.1)(ii)(c), for each $g \in G(\mathbb{A})$,

$$\big(M_I(w', \pi(\lambda + \zeta))(\varphi_{\lambda+\zeta})\big)_{-w'(\zeta)}(g)$$

is equal to

$$\Big(M_{w(I)}(s_{w(I),\alpha}, w(\pi(\lambda + \zeta)))\big(M_I(w, \pi(\lambda + \zeta))(\varphi_{\lambda+\zeta})\big)\Big)_{-w'(\zeta)}(g)$$

and by definition (see (G.5.1)) this last expression is the integral over

$$(N_{w'(I)}(F) \cap \dot{s}_{w(I),\alpha} N_{w(I)}(F) \dot{s}_{w(I),\alpha}^{-1}) \backslash N_{w'(I)}(\mathbb{A})$$

of

$$q^{\langle w(\zeta), H_{w(I)}(\dot{s}_{w(I),\alpha}^{-1} n_{w'(I)} g)\rangle - \langle s_{w(I),\alpha} w(\zeta), H_{w'(I)}(g)\rangle} \times \big(M_I(w, \pi(\lambda + \zeta))(\varphi_{\lambda+\zeta})\big)_{-w(\zeta)}(\dot{s}_{w(I),\alpha}^{-1} n_{w'(I)} g)$$

with respect to the Haar measure $d\nu_{w'(I), s_{w(I),\alpha}} \backslash dn_{w'(I)}$. But it is not difficult to check that

$$H_{w(I)}(\dot{s}_{w(I),\alpha}^{-1} n_{w'(I)} g) - s_{w(I),\alpha}^{-1}(H_{w'(I)}(g)) \in \mathbb{Z}\alpha_{P,P'}^{\vee}$$

for every $n_{w'(I)} \in N_{w'(I)}(\mathbb{A})$ and every $g \in G(\mathbb{A})$. The lemma follows. □

For each $P \in \mathcal{P}(M)$ we set

$$\theta_P(\zeta) = (\log q)^{|\Delta_P|} \prod_{\alpha \in \Delta_P} \langle \zeta, \alpha^\vee \rangle \qquad (\zeta \in \mathrm{Im}\, \mathfrak{a}_M^{*G,\mathbb{Z}})$$

and

$$\widetilde{\theta}_P(\zeta) = \prod_{\alpha \in \Delta_P} \left(1 - q^{-\langle \zeta, \alpha^\vee \rangle}\right) \qquad (\zeta \in \mathrm{Im}\, X_M^G).$$

PROPOSITION (11.5.7). — (i) (Arthur) *If c is a (G,M)-family the function*

$$c_M(\zeta) = \sum_{P \in \mathcal{P}(M)} \frac{c_P(\zeta)}{\theta_P(\zeta)},$$

which is well-defined and smooth outside the union of the hyperplanes with equations

$$\{\zeta \in \mathrm{Im}\, \mathfrak{a}_M^{*G,\mathbb{Z}} \mid \langle \zeta, \alpha^\vee \rangle = 0\} \qquad (\alpha \in R(G, Z_M)),$$

extends to a smooth function on the whole $\mathrm{Im}\, \mathfrak{a}_M^{*G,\mathbb{Z}}$.

(ii) (Waldspurger) *If c is a $(G,M)^\sim$-family the function*

$$\widetilde{c_M}(\zeta) = \sum_{P \in \mathcal{P}(M)} \frac{c_P(\zeta)}{\widetilde{\theta}_P(\zeta)},$$

which is well-defined and smooth outside the union of the "hyperplanes" with equations

$$\{\zeta \in \mathrm{Im}\, X_M^G \mid \langle \zeta, \alpha^\vee \rangle \in \frac{2\pi i}{\log q}\mathbb{Z}\} \qquad (\alpha \in R(G, Z_M)),$$

extends to a smooth function on the whole of $\mathrm{Im}\, X_M^G$.

Proof of part (i) : Let $\ell_1(x), \ldots, \ell_n(x)$ be linear forms on $\mathbb{R}^m$ such that the hyperplanes $H_j = \{x \mid \ell_j(x) = 0\}$, $j = 1, \ldots, n$, are pairwise distinct. Let V be a neighborhood of the origin in $\mathbb{R}^m$ and let $f(x)$ be a smooth function on the complement in V of the union of the hyperplanes H_j. Let us assume that $f(x)\ell_1(x) \cdots \ell_n(x)$ extends to a smooth function on V and that its extension vanishes identically on $V \cap H_j$ for each $j = 1, \ldots, n$. Then $f(x)$ extends to a smooth function on the whole of V (Taylor formula).

Therefore, in order to prove part (i) of the proposition it is sufficient to check that the function

$$c_M(\zeta) \prod_{\alpha \in R^+(G, Z_M)} \langle \zeta, \alpha^\vee \rangle$$

extends to a smooth function on $\mathrm{Im}\,\mathfrak{a}_M^{*G,\mathbb{Z}}$, where $R^+(G, Z_M)$ is the set of positive roots for some order on $R(G, Z_M)$, and that its extension vanishes identically on the hyperplane $\langle\zeta, \alpha_0^\vee\rangle = 0$ for each $\alpha_0 \in R^+(G, Z_M)$. The first requirement is obviously fulfilled. Now for any fixed $\alpha_0 \in R^+(G, Z_M)$ the set

$$\{P \in \mathcal{P}(M) \mid \alpha_0 \in \Delta_P \text{ or } \alpha_0 \in (-\Delta_P)\}$$

can be partitioned into pairs (P, P') of adjacent parabolic subgroups. Indeed, if $M = M_I$ and if $P = \dot{w}^{-1}P_{w(I)}\dot{w} \in \mathcal{P}(I)$ is such that $\alpha_0 \in \Delta_P$ there exists a unique $P' = \dot{w}'^{-1}P_{w'(I)}\dot{w}' \in \mathcal{P}(I)$ such that (P, P') is a pair of ajacent parabolic subgroups and such that

$$\Delta_P \cap (-\Delta_{P'}) = \{\alpha_0\}$$

(take $w' = s_{w(I),w(\beta_0)}w$ where β_0 is the unique element in $w^{-1}(\Delta) - I$ the restriction to Z_I of which is α_0). Moreover, by definition of (G, M)-families, for any such pair (P, P'),

$$c_P(\zeta) - c_{P'}(\zeta)$$

vanishes identically on the hyperplane $\langle\zeta, \alpha_0^\vee\rangle = 0$. Thus the second requirement is fulfilled too. □

Following Waldspurger (see [Wa] Lemme II.7.2 and [Ar 8] Lemma 10.2) we will now deduce part (ii) of the proposition from part (i).

LEMMA (11.5.8) (Waldspurger). — *We can find a family of smooth functions with compact supports*

$$u = (u_P \mid P \in \mathcal{P}(M),\ M \in \mathcal{L})$$

on $\mathrm{Im}\,\mathfrak{a}_M^{*G,\mathbb{Z}}$ *with the following properties:*

(a)

$$\sum_{\nu \in \mathfrak{a}_{M,\mathbb{Z}}^\vee} u_P\Big(\zeta + \frac{2\pi i}{\log q}\nu\Big) = 1 \qquad (\forall \zeta \in \mathrm{Im}\,\mathfrak{a}_M^{*G,\mathbb{Z}}),$$

(b) *for each $M \in \mathcal{L}$ and each $P \in \mathcal{P}(M)$ the function*

$$\widetilde{u}_P = \frac{u_P\theta_P}{\widetilde{\theta}_P}$$

is smooth on $\mathrm{Im}\,\mathfrak{a}_M^{*G,\mathbb{Z}}$,

(c) *if $P' \in \mathcal{P}(M')$ contains $P \in \mathcal{P}(M)$ with $M \subset M'$ in $\mathcal{L}$, then $u_{P'}$ and $\widetilde{u}_{P'}$ are the restrictions of u_P and $\widetilde{u}_P$ to* $\mathrm{Im}\,\mathfrak{a}_{M'}^{*G,\mathbb{Z}}$.

Proof of part (ii) *of* (11.5.7) *assuming* (11.5.8) : Let us fix a family u as in (11.5.8). By property (a) of u we have

$$\widetilde{c_M}(\zeta) = \sum_{P\in\mathcal{P}(M)} \frac{c_P(\zeta)}{\widetilde{\theta}_P(\zeta)} \sum_{\nu\in\mathfrak{a}_{M,\mathbb{Z}}^\vee} u_P(\zeta + \frac{2\pi i}{\log q}\nu) \qquad (\forall\zeta\in \operatorname{Im}\mathfrak{a}_M^{*G,\mathbb{Z}}).$$

But c_P and $\widetilde{\theta}_P$ are invariant under $\frac{2\pi i}{\log q}\mathfrak{a}_{M,\mathbb{Z}}^\vee$ and we have

$$\widetilde{c_M}(\zeta) = \sum_{\nu\in\mathfrak{a}_{M,\mathbb{Z}}^\vee} \sum_{P\in\mathcal{P}(M)} \frac{c_P(\zeta + \frac{2\pi i}{\log q}\nu)\widetilde{u}_P(\zeta + \frac{2\pi i}{\log q}\nu)}{\theta_P(\zeta + \frac{2\pi i}{\log q}\nu)}.$$

Moreover $((c\widetilde{u})_P = c_P\widetilde{u}_P \mid P\in\mathcal{P}(M))$ is a (G,M)-family. Indeed, if (P,P') is a pair of adjacent parabolic subgroups in $\mathcal{P}(M)$ and if Q is the parabolic subgroup of G generated by P and P', for any $\zeta\in\operatorname{Im}\mathfrak{a}_M^{*G,\mathbb{Z}}$ such that $\langle\zeta,\alpha_{P,P'}^\vee\rangle = 0$ we have $\zeta\in\operatorname{Im}\mathfrak{a}_L^{*G,\mathbb{Z}}$, where L is the unique Levi subgroup of Q containing M, and by property (c) of u we have

$$\widetilde{u}_P(\zeta) = \widetilde{u}_Q(\zeta) = \widetilde{u}_{P'}(\zeta).$$

Therefore the function

$$\widetilde{c_M}(\zeta) = \sum_{\nu\in\mathfrak{a}_{M,\mathbb{Z}}^\vee} (c\widetilde{u})_M(\zeta + \frac{2\pi i}{\log q}\nu)$$

of ζ is smooth on $\operatorname{Im}\mathfrak{a}_M^{*G,\mathbb{Z}}$ as required. □

Proof of (11.5.8) : We have

$$\operatorname{Im}\mathfrak{a}_T^{*G,\mathbb{Z}} = \frac{2\pi i}{\log q}\{\zeta\in\mathbb{R}^d \mid \sum_{i=1}^d \zeta_i\in\mathbb{Z}\},$$

$$\mathfrak{a}_{T,\mathbb{Z}}^\vee = \mathbb{Z}^d$$

and, if we set

$$U = \{\zeta\in\mathbb{R}^d \mid \sum_{i=1}^d \zeta_i \in\{0,\ldots,d-1\} \text{ and } |\zeta_i - \zeta_{i+1}| < \frac{3}{4},\ \forall i = 1,\ldots,d-1\},$$

we have

$$\{\zeta\in\mathbb{R}^d \mid \sum_{i=1}^d \zeta_i\in\mathbb{Z}\} = U + \mathbb{Z}^d.$$

Thus, using a partition of unity we can find a smooth function u_B on $\mathrm{Im}\,\mathfrak{a}_T^{*G,\mathbb{Z}}$ with support in $\frac{2\pi i}{\log q}U$ such that

$$\sum_{\nu\in\mathfrak{a}_{T,\mathbb{Z}}^\vee} u_B(\zeta+\frac{2\pi i}{\log q}\nu)=1 \qquad (\forall\zeta\in\mathrm{Im}\,\mathfrak{a}_T^{*G,\mathbb{Z}}).$$

As the complex function

$$\frac{2\pi i}{\log q}x\big(1-q^{\frac{2\pi i}{\log q}x}\big)^{-1}$$

of the real variable x is smooth on the interval $]-\frac{3}{4},\frac{3}{4}[$ the function $\widetilde{u}_B=u_B\theta_B\widetilde{\theta}_B^{-1}$ is smooth on $\mathrm{Im}\,\mathfrak{a}_T^{*G,\mathbb{Z}}$.

Then for any $M\in\mathcal{L}$ and any $P=\dot{w}P_I\dot{w}^{-1}\in\mathcal{P}(T)$ with $I\in\Delta$ and $w\in W$ we set

$$u_P(\zeta)=u_B(w^{-1}(\zeta)) \qquad (\forall\zeta\in\mathrm{Im}\,\mathfrak{a}_M^{*G,\mathbb{Z}})$$

(I and the class wW_I are uniquely determined by P). Let us check that the family u satisfies conditions (a), (b) and (c).

Let us begin with (b) and (c). We may assume that $M=T$ and even that $P=B$, so that $P'=P_I$ for some $I\subset\Delta$. It is obvious from the definition of u that u_{P_I} is the restriction of u_B. Moreover the map

$$\Delta-I\to\Delta_{P_I},\ \alpha\mapsto\alpha|Z_I,$$

is bijective, we have $(\alpha|Z_I)^\vee=(\alpha^\vee)_I$ for any $\alpha\in\Delta-I$ and we have $\langle\zeta,\alpha^\vee\rangle=0$ for every $\zeta\in\mathrm{Im}\,\mathfrak{a}_I^*\subset\mathrm{Im}\,\mathfrak{a}_\emptyset^*$ and every $\alpha\in I$. Therefore it is clear that $\widetilde{u}_B|\,\mathrm{Im}\,\mathfrak{a}_{M_I}^{*G,\mathbb{Z}}$ coincides with $\widetilde{u}_{P_I}$ which is thus a smooth function.

It remains to establish condition (a). We may assume that $P\supset B$. Suppose that ζ is a point in $\mathrm{Im}\,\mathfrak{a}_M^{*G,\mathbb{Z}}$ and that $\nu\in\mathfrak{a}_{T,\mathbb{Z}}^\vee-\mathfrak{a}_{M,\mathbb{Z}}^\vee$. Then we easily check that $\zeta+\frac{2\pi i}{\log q}\nu\notin U$ (there exists $\alpha\in\Delta_B^P$ such that $\alpha(\nu)$ is a non-zero integer). Therefore we have

$$\sum_{\nu\in\mathfrak{a}_{M,\mathbb{Z}}^\vee} u_P(\zeta+\frac{2\pi i}{\log q}\nu)=\sum_{\nu\in\mathfrak{a}_{M,\mathbb{Z}}^\vee} u_B(\zeta+\frac{2\pi i}{\log q}\nu)=\sum_{\nu\in\mathfrak{a}_{T,\mathbb{Z}}^\vee} u_B(\zeta+\frac{2\pi i}{\log q}\nu)=1$$

as required. □

EXAMPLE (11.5.9). — If $c(\mu;\zeta)$ is the $(G,M_I)\widetilde{\ }$-family which is defined by

$$c_P(\mu;\zeta)=c'_P(\zeta)c''_P(\zeta)c'''_P(\mu;\zeta),$$

where $c'_P(\zeta)$ (resp. $c''_P(\zeta)$, $c'''_P(\mu;\zeta)$) is the $(G,M)\widetilde{\ }$-family of (11.5.5)(i) (resp. (11.5.5)(ii), (11.5.6)) for $M=M_I$ and $P_0=P_I$, then

$$\mathrm{tr}(M_{I,\pi}^T(\lambda)\circ i_I^\Delta(\lambda,f))=\frac{\deg(\infty)d}{fd_1\cdots d_se}\sum_\mu\widetilde{c_{M_I}}(\mu;\mu)$$

where μ runs through the subset of $\mathrm{Fix}_{X_I^\Delta}(\pi)$ defined by the condition $\sum_{j=1}^s\mu_j\in\frac{2\pi i}{\log q}\frac{d}{e}\mathbb{Z}$ (see (11.4.10)). □

For any $M \subset L$ in $\mathcal{L}$ we have an obvious notion of an (L, M)-family $(c_R(\zeta) \mid R \in \mathcal{P}^L(M))$, $\zeta \in \mathrm{Im}\, \mathfrak{a}_M^{*G,\mathbb{Z}}$, and we have an obvious extension of lemma (11.5.7)(i) to such families.

LEMMA (11.5.10). — *Let $M \subset L$ be in $\mathcal{L}$ and let $(c_P(\zeta) \mid P \in \mathcal{P}(M))$ be a (G, M)-family.*

(i) *If $Q \in \mathcal{P}(L)$ and if $\zeta \in \mathrm{Im}\, \mathfrak{a}_L^{*G,\mathbb{Z}} \subset \mathrm{Im}\, \mathfrak{a}_M^{*G,\mathbb{Z}}$, $c_{RN_Q}(\zeta)$ is independent of $R \in \mathcal{P}^L(M)$. Moreover*

$$(c_Q(\zeta) \overset{\mathrm{dfn}}{=\!=} c_{RN_Q}(\zeta) \mid Q \in \mathcal{P}(L))$$

is a (G, L)-family.

(ii) *Let us fix $Q \in \mathcal{P}(L)$. Then*

$$(c_R^Q(\zeta) \overset{\mathrm{dfn}}{=\!=} c_{RN_Q}(\zeta) \mid R \in \mathcal{P}^L(M)) \qquad (\forall \zeta \in \mathrm{Im}\, \mathfrak{a}_M^{*G,\mathbb{Z}})$$

is an (L, M)-family.

Proof: We may assume that $M = M_I$ and $L = M_J$ for some $I \subset J \subset \Delta$.

Let (R, R') be a pair of adjacent parabolic subgroups in $\mathcal{P}^J(I)$ and let $Q \in \mathcal{P}(J)$. Then $(RN_Q, R'N_Q)$ is a pair of adjacent parabolic subgroups in $\mathcal{P}(I)$ and

$$\alpha_{RN_Q, R'N_Q} = \alpha_{R,R'}.$$

Indeed there exist $w_J, w'_J \in W_J \subset W$ with $w_J(I), w'_J(I) \subset J$ and $\beta \in J - w_J(I)$ with $\alpha_{R,R'} = w_J^{-1}(\beta)|Z_I$ such that

$$R = \dot{w}_J^{-1} P^J_{w_J(I)} \dot{w}_J,$$

$$R' = \dot{w}'^{-1}_J P^J_{w'_J(I)} \dot{w}'_J$$

and

$$w'_J = s^J_{w_J(I),\beta} w_J$$

and there exists $w \in W$ with $w(J) \subset \Delta$ such that

$$Q = \dot{w}^{-1} P_{w(J)} \dot{w};$$

then we have

$$RN_Q = \dot{w}_J^{-1} \dot{w}^{-1} P_{ww_J(I)} \dot{w} \dot{w}_J,$$
$$R'N_Q = \dot{w}'^{-1}_J \dot{w}^{-1} P_{ww'_J(I)} \dot{w} \dot{w}'_J$$

and

$$s_{ww_J(I),w(\beta)} = w s^J_{w_J(I),\beta} w^{-1}.$$

It follows that
$$c_{RN_Q}(\zeta) = c_{R'N_Q}(\zeta)$$
either if $\zeta \in \mathrm{Im}\,\mathfrak{a}_{M_J}^{*G,\mathbb{Z}}$ or if $\zeta \in \mathrm{Im}\,\mathfrak{a}_{M_I}^{*G,\mathbb{Z}}$ and $\langle \zeta, \alpha_{R,R'}^{\vee}\rangle = 0$. The first assertion of part (i) follows and part (ii) too (by (G.5.2), for any pair (R,R') of parabolic subgroups in $\mathcal{P}^J(I)$ we can find a sequence $R^{(0)},\ldots,R^{(n)}$ of parabolic subgroups in $\mathcal{P}^J(I)$ such that $R^{(0)} = R$, $R^{(n)} = R'$ and such that, for each $k = 1,\ldots,n$, $(R^{(k-1)}, R^{(k)})$ is a pair of adjacent parabolic subgroups).

Now let $R \in \mathcal{P}^J(I)$ and let (Q,Q') be a pair of adjacent parabolic subgroups in $\mathcal{P}(J)$. It follows again from (G.5.2) that we can find a sequence $P^{(0)},\ldots,P^{(\ell)}$ of parabolic subgroups in $\mathcal{P}(I)$ such that $P^{(0)} = RN_Q$ and $P^{(\ell)} = RN_{Q'}$, such that, for each $i = 1,\ldots,\ell$, $(P^{(i-1)}, P^{(i)})$ is a pair of adjacent parabolic subgroups and such that, for each $i = 1,\ldots,\ell$, either $\alpha_{P^{(i-1)},P^{(i)}}|Z_J = 0$ or $\alpha_{P^{(i-1)},P^{(i)}}|Z_J = \alpha_{Q,Q'}$. Therefore, if $\zeta \in \mathrm{Im}\,\mathfrak{a}_{M_J}^{*G,\mathbb{Z}}$ satisfies $\langle \zeta, \alpha_{Q,Q'}^{\vee}\rangle = 0$,
$$c_Q(\zeta) - c_{Q'}(\zeta) = \sum_{i=1}^{\ell}\big(c_{P^{(i-1)}}(\zeta) - c_{P^{(i)}}(\zeta)\big)$$
is equal to zero. This concludes the proof of part (i) of the lemma. □

For any $M \subset L$ in $\mathcal{L}$ and any (G,M)-family c we set
$$c_L(\zeta) = \sum_{Q\in\mathcal{P}(L)} \frac{c_Q(\zeta)}{\theta_Q(\zeta)} \qquad (\forall \zeta \in \mathrm{Im}\,\mathfrak{a}_L^{*G,\mathbb{Z}})$$
and
$$c_M^Q(\zeta) = \sum_{R\in\mathcal{P}^L(M)} \frac{c_R^Q(\zeta)}{\theta_R^L(\zeta)} \qquad (\forall \zeta \in \mathrm{Im}\,\mathfrak{a}_M^{*G,\mathbb{Z}})$$
for any fixed $Q \in \mathcal{P}(L)$. Here, for any $R \in \mathcal{P}^L(M)$ we have set
$$\theta_R^L(\zeta) = (\log q)^{|\Delta_R^L|} \prod_{\alpha\in\Delta_R^L} \langle \zeta, \alpha^{\vee}\rangle$$
where $\Delta_R^L \subset R(G,Z_M)$ is the set of simple roots of (R, Z_M). If $M = M_I$ and $L = M_J$ for some $I \subset J \subset \Delta$ and if $(e_1,\ldots,e_t)$ is the partition of d corresponding to J we may identify M_J with $GL_{e_1} \times \cdots \times GL_{e_t}$, J with $\Delta_1 \amalg \cdots \amalg \Delta_t$ where $\Delta_k = \{1,\ldots,e_k - 1\}$, I with $I_1 \amalg \cdots \amalg I_t$ where $I_k \subset \Delta_k$, R with $P_1 \times \cdots \times P_t$ where $P_k \in \mathcal{P}(I_k)$ is a parabolic subgroup of GL_{e_k} and $\zeta \in \mathrm{Im}\,\mathfrak{a}_M^{*G,\mathbb{Z}} \subset \mathrm{Im}\,\mathfrak{a}_M^*$ with $(\zeta_1,\ldots,\zeta_t)$ where $\zeta_k \in \mathrm{Im}\,\mathfrak{a}_{M_{I_k}}^*$. Then we have
$$\theta_R^L(\zeta) = \prod_{k=1}^{t} (\log q)^{|\Delta_k - I_k|} \prod_{\alpha_k \in (\Delta_k)_{P_k}} \langle \zeta_k, \alpha_k^{\vee}\rangle.$$

Let us fix $M \in \mathcal{L}$. We denote by $\mathcal{L}(M) \subset \mathcal{L}$ the set of Levi subgroups in G which contain M. For any $M', M'' \in \mathcal{L}(M)$ let us consider the map

$$(11.5.11) \qquad (\mathfrak{a}_{M'}/\mathfrak{a}_G) \oplus (\mathfrak{a}_{M''}/\mathfrak{a}_G) \to \mathfrak{a}_M/\mathfrak{a}_G,\ (\zeta', \zeta'') \mapsto \zeta' + \zeta''.$$

If

$$M = \{g \in G \mid g\Big(\bigoplus_{i \in A_j} Fv_i\Big) = \bigoplus_{i \in A_j} Fv_i,\ \forall j \in J\}$$

for some partition

$$\{1, \ldots, d\} = \coprod_{j \in J} A_j,$$

we have

$$M' = \{g \in G \mid g\Big(\bigoplus_{i \in A'_{j'}} Fv_i\Big) = \bigoplus_{i \in A'_{j'}} Fv_i,\ \forall j' \in J'\}$$

(resp.

$$M'' = \{g \in G \mid g\Big(\bigoplus_{i \in A''_{j''}} Fv_i\Big) = \bigoplus_{i \in A''_{j''}} Fv_i,\ \forall j'' \in J''\})$$

for some partition

$$J = \coprod_{j' \in J'} K'_{j'} \qquad (\text{resp. } J = \coprod_{j'' \in J''} K''_{j''})$$

where we have set

$$A'_{j'} = \coprod_{j \in K'_{j'}} A_j \qquad (\text{resp. } A''_{j''} = \coprod_{j \in K''_{j''}} A_j)$$

for every $j' \in J'$ (resp. $j'' \in J''$). Then the map (11.5.11) may be identified with the map

$$(\mathbb{Q}^{J'}/\mathbb{Q}) \oplus (\mathbb{Q}^{J''}/\mathbb{Q}) \to \mathbb{Q}^J/\mathbb{Q}$$

which sends (z', z'') onto z with

$$z_j = z'_{j'} + z''_{j''} \qquad (\forall j \in J),$$

where j' and j'' are the unique elements in J' and J'' respectively such that $j \in K'_{j'} \cap K''_{j''}$ ($\mathbb{Q}$ is diagonally embedded into $\mathbb{Q}^{J'}$, $\mathbb{Q}^{J''}$ and $\mathbb{Q}^J$). The last map induces a linear map

$$\iota : \bigwedge^{\max}(\mathbb{Q}^{J'}/\mathbb{Q}) \otimes \bigwedge^{\max}(\mathbb{Q}^{J''}/\mathbb{Q}) \to \bigwedge^{\max}(\mathbb{Q}^J/\mathbb{Q}).$$

Let us consider the standard scalar product on $\mathbb{Q}^J$. It induces a scalar product on the one dimensional $\mathbb{Q}$-vector space $\bigwedge^{\max}(\mathbb{Q}^J/\mathbb{Q})$. Let e_J be an orthonormal basis of this one dimensional vector space (it is well-defined up to sign). Similarly we define $e_{J'}$ and $e_{J''}$. Then following Arthur we set

$$(11.5.12)\qquad d_M^G(M',M'') = \left|\frac{\iota(e_{J'}\otimes e_{J''})}{e_J}\right|.$$

In particular we have $d_M^G(M',M'')=0$ unless the map (11.5.11) is bijective and we have

$$d_M^G(G,M) = d_M^G(M,G) = 1.$$

For each $L\in\mathcal{L}$ and each $Q\in\mathcal{P}(L)$ we denote by $\mathfrak{a}_Q^+\subset\mathfrak{a}_L$ the corresponding Weyl chamber

$$\mathfrak{a}_Q^+ = \{\zeta\in\mathfrak{a}_L \mid \alpha(\zeta)>0,\ \forall\alpha\in\Delta_Q\}.$$

If ζ is a point in general position in $\mathfrak{a}_L$ there exists one and only one $Q\in\mathcal{P}(L)$ such that $\zeta\in\mathfrak{a}_Q^+$. Following Arthur we fix a point $\xi\in\mathfrak{a}_M$ in general position. For any pair $(M',M'')\in\mathcal{L}(M)\times\mathcal{L}(M)$ such that $d_M^G(M',M'')\neq 0$ we can find $\xi'\in\mathfrak{a}_{M'}$ and $\xi''\in\mathfrak{a}_{M''}$ such that $\xi=\frac{1}{2}(\xi'-\xi'')$ and we denote by $(P_{M'},P_{M''})$ the unique pair in $\mathcal{P}(M')\times\mathcal{P}(M'')$ such that $\xi'\in\mathfrak{a}_{P_{M'}}^+$ and $\xi''\in\mathfrak{a}_{P_{M''}}^+$. The map $(M',M'')\mapsto(P_{M'},P_{M''})$ is a section of the bundle

$$\coprod_{(M',M'')}\mathcal{P}(M')\times\mathcal{P}(M'')\to\mathcal{L}(M)\times\mathcal{L}(M)$$

over the set of pairs (M',M'') such that $d_M^G(M',M'')\neq 0$. This section depends on the choice of the point ξ but is independent of the choice of (ξ',ξ'').

PROPOSITION (11.5.13) (Arthur). — *Let us fix a point $\xi\in\mathfrak{a}_M$ in general position as before. Let c' and c'' be two (G,M)-families. Then for any $\zeta\in\operatorname{Im}\mathfrak{a}_M^{*G,\mathbb{Z}}$ we have*

$$(c'c'')_M(\zeta) = \sum_{M',M''\in\mathcal{L}(M)} d_M^G(M',M'')c_M'^{P_{M'}}(\zeta)c_M''^{P_{M''}}(\zeta).$$

Proof: See [Ar 12] Appendix. □

COROLLARY (11.5.14). — (i) *Let c and d be two (G,M)-families and let $\mu\in\operatorname{Im}\mathfrak{a}_M^{*G,\mathbb{Z}}$. We assume that $c_M^Q(\mu)=0$ for every $L\in\mathcal{L}(M)$ with $L\subsetneq G$ and every $Q\in\mathcal{P}(L)$. Then we have*

$$(cd)_M(\mu) = c_M(\mu)d_G(\mu)$$

if $\mu\in\operatorname{Im}\mathfrak{a}_G^{*G,\mathbb{Z}}\subset\operatorname{Im}\mathfrak{a}_M^{*G,\mathbb{Z}}$ and $(cd)_M(\mu)=c_M(\mu)=0$ otherwise.

(ii) *Let c, d be two $(G,M)^{\sim}$-families and let $\mu \in \mathrm{Im}\,\mathfrak{a}_M^{*G,\mathbb{Z}}$. We assume that $c_M^Q(\mu + \frac{2\pi i}{\log q}\nu) = 0$ for every $L \in \mathcal{L}(M)$ with $L \subsetneqq G$, every $Q \in \mathcal{P}(L)$ and every $\nu \in \mathfrak{a}_{M,\mathbb{Z}}^{\vee}$. Then we have*

$$\widetilde{(cd)}_M(\mu) = \widetilde{c_M}(\mu) d_G(\mu)$$

if $\mu \in \frac{2\pi i}{\log q}\mathfrak{a}_{M,\mathbb{Z}}^{\vee}$ and $\widetilde{(cd)}_M(\mu) = \widetilde{c_M}(\mu) = 0$ otherwise.

Proof : Let $P \in \mathcal{P}(M)$. We can find ξ in general position in $\mathfrak{a}_M$ such that $(P_G, P_M) = (G, P)$. For this choice of ξ we may apply the proposition to compute $(cd)_M(\mu)$. But by hypothesis we have $c_M^{P_{M'}}(\mu) = 0$ for every $M' \in \mathcal{L}(M)$ with $M' \subsetneqq G$, we have $d_M^G(G, M'') = 0$ unless $M'' = M$ and we have $d_M^G(G,M) = 1$. Therefore $(cd)_M(\mu) = c_M(\mu)d_P(\mu)$. Now if μ is such that $c_M(\mu) \neq 0$ it follows that $d_P(\mu)$ is independent of $P \in \mathcal{P}(M)$. This property holds for any (G,M)-family d. In particular we may take $d_P(\zeta) = q^{\langle \zeta, X_P \rangle}$ where $X = (X_P \mid P \in \mathcal{P}(M))$ is any family of points in $\mathrm{Re}\,\mathfrak{a}_M$ such that, for each pair of adjacent parabolic subgroups (P, P') in $\mathcal{P}(M)$, we have $X_P - X_{P'} \in \mathbb{R}\alpha_{P,P'}^{\vee}$. Letting X vary we get that $\langle \mu, \alpha^{\vee} \rangle = 0$ for every $\alpha \in R(G, Z_M)$ and part (i) of the corollary follows.

We have

$$\widetilde{(cd)}_M(\mu) = \sum_{\nu \in \mathfrak{a}_{M,\mathbb{Z}}^{\vee}} (cd\widetilde{u})_M(\mu + \frac{2\pi i}{\log q}\nu)$$

and

$$\widetilde{c_M}(\mu) = \sum_{\nu \in \mathfrak{a}_{M,\mathbb{Z}}^{\vee}} (c\widetilde{u})_M(\mu + \frac{2\pi i}{\log q}\nu)$$

(see the proof of (11.5.7)) and part (ii) of the corollary follows from part (i) which is already proved. □

For example let us consider the $(G,M)^{\sim}$-family $(c_P(\mu;\zeta) \mid P \in \mathcal{P}(M))$ of lemma (11.5.6) (for some fixed choices of (M,π), λ, P_0, μ and f).

LEMMA (11.5.15). — *Let $L \in \mathcal{L}(M)$ and $Q \in \mathcal{P}(L)$. Let us assume that*

$$f^Q(\ell) \stackrel{\text{dfn}}{=\!=} \delta_{Q(\mathbb{A})}^{1/2}(\ell) \int_{N_Q(\mathbb{A})} \int_K f(k^{-1}\ell n_Q k)\,dk\,dn_Q = 0$$

for every $\ell \in L(\mathbb{A})$. Then $c_M^Q(\mu; \mu + \frac{2\pi i}{\log q}\nu) = 0$ for all $\nu \in \mathfrak{a}_{M,\mathbb{Z}}^{\vee}$.

Proof : Firstly we remark that, for any $\nu \in \mathfrak{a}_{M,\mathbb{Z}}^{\vee}$, $c_M^Q(\mu; \mu + \frac{2\pi i}{\log q}\nu)$ does not depend on the choice of P_0. Indeed, if P_1 is another choice, for every $P \in \mathcal{P}(M)$ and for every $\zeta \in \mathfrak{a}_M^{*G,\mathbb{Z}}$,

$$M_P(\zeta, \pi, \lambda; P_1) \circ [-\mu]_{P_1} \circ i_{P_1}^G(\lambda + \zeta - \mu, f)$$

is equal to

$$M_{P_0|P_1}(\pi,\lambda)^{-1}\circ M_P(\zeta,\pi,\lambda;P_0)\circ[-\mu]_{P_0}\circ i_{P_0}^G(\lambda+\zeta-\mu,f)\circ M_{P_0|P_1}(\pi,\lambda+\zeta-\mu).$$

Therefore we may assume that $P_0\subset Q$, i.e. that $P_0=R_0N_Q$ for some $R_0\in\mathcal{P}^L(M)$.

Then, for any $\xi\in\mathfrak{a}_M^{*G,\mathbb{Z}}$, any $\varphi\in\mathcal{A}_{P_0,\pi}^G$, any $\ell\in L(\mathbb{A})$ and any $k\in K$ we have

$$M_{P|P_0}(\pi,\xi)(\varphi)(\ell k)=M_{R|R_0}(\pi,\xi)(\varphi_k)(\ell)$$

for each $P=RN_Q\subset Q$ with $R\in\mathcal{P}^L(M)$ and we have

$$i_{P_0}^G(\xi,f)(\varphi)(\ell k)=\int_K i_{R_0}^L(\xi,\delta_{Q(\mathbb{A})}^{-1/2}g_{k,k'})(\varphi'_k)(\ell)dk'$$

where $\varphi_k\in\mathcal{A}_{R_0,\pi}^L$ and $g_{k,k'}\in C_c^\infty(F_\infty^\times\backslash L(\mathbb{A}))$ are defined by

$$(\varphi_k)(\ell)=\varphi(\ell k)\qquad(\forall\ell\in L(\mathbb{A}))$$

and

$$g_{k,k'}(\ell)=\delta_{Q(\mathbb{A})}^{1/2}(\ell)\int_{N_Q(\mathbb{A})}f(k^{-1}\ell n'_Q k')dn'_Q\qquad(\forall\ell\in L(\mathbb{A}))$$

for all $k,k'\in K$ (we let the reader write the definitions of $M_{R|R_0}(\pi,\xi)$ and $i_{R_0}^L(\xi,\cdot)$). Therefore, regarding $\mathcal{A}_{P_0,\pi}^G$ as a space of functions from K to $\mathcal{A}_{R_0,\pi}^L$ (by $\varphi\mapsto(k\mapsto\varphi_k)$) we obtain that, for every $\zeta\in\mathfrak{a}_M^{*G,\mathbb{Z}}$, the operator

$$M_{P|P_0}(\pi,\lambda)^{-1}\circ M_{P|P_0}(\pi,\lambda+\zeta)\circ[-\mu]_{P_0}\circ i_{P_0}^G(\lambda+\zeta-\mu,f)$$

is an integral operator with kernel

$$M_{R|R_0}(\pi,\lambda)^{-1}\circ M_{R|R_0}(\pi,\lambda+\zeta)\circ[-\mu]_{R_0}\circ i_{R_0}^L(\lambda+\zeta-\mu,\delta_{Q(\mathbb{A})}^{-1/2}g_{k,k'})$$

and its trace is zero as

$$\int_K\delta_{Q(\mathbb{A})}^{-1/2}g_{k,k}dk=f^Q$$

is identically zero by hypothesis.

We have thus proved that, under the hypothesis $P_0\subset Q$, we have $c_P(\zeta)=0$ for every $\zeta\in\mathfrak{a}_M^{*G,\mathbb{Z}}$ and every $P\in\mathcal{P}(M)$ with $P\subset Q$. The lemma follows. □

At least formally we introduce the operator

$$\mathcal{M}_M(\pi,\lambda;P_0)=\lim_{\zeta\to0}\sum_{P\in\mathcal{P}(M)}M_P(\zeta,\pi,\lambda;P_0)\widetilde{\theta}_P(\zeta)^{-1}$$

on $\mathcal{A}_{P_0,\pi}^G$. We have

$$\mathrm{tr}\big(\mathcal{M}_M(\pi,\lambda;P_0)\circ i_{P_0}^G(\lambda,f)\big)=\widetilde{c_M}(0;0)$$

for the $(G,M)\tilde{}$-family $(c_P(\mu;\zeta)\mid P\in\mathcal{P}(M))$ of lemma (11.5.6).

Then, from (11.5.9), (11.5.14) and (11.5.15) we obtain

PROPOSITION (11.5.16). — *Let $I \subset \Delta$. Let us assume that $f^Q \equiv 0$ for every $L \in \mathcal{L}(M_I)$, $L \subsetneq G$, and every $Q \in \mathcal{P}(L)$. Then we have*

$$\operatorname{tr}\big(M_{I,\pi}^T(\lambda) \circ i_I^{\Delta}(\lambda, f)\big) = \frac{\deg(\infty)d}{fd_1 \cdots d_s} \operatorname{tr}\big(\mathcal{M}_{M_I}(\pi, \lambda; P_I) \circ i_I^{\Delta}(\lambda, f)\big).$$

In particular this trace is independent of T. □

(11.6) Evaluation of $J_c^T(f)$ in a special case: normalization of intertwining operators

Let (M, π) be a regular unitary cuspidal pair as before with $M \in \mathcal{L}$. For each $P \in \mathcal{P}(M)$ we may identify $\mathcal{A}_{P,\pi}^G$ with the induced representation $i_{M(\mathbb{A}),P(\mathbb{A})}^{G(\mathbb{A})}(\mathcal{A}_{M,\pi}^M)$ where

$$\mathcal{A}_{M,\pi}^M \cong \mathcal{V}^{m_{\mathrm{cusp}}(\pi)}$$

is the π-isotypic subrepresentation of $\mathcal{A}_{M,\chi_\pi,\mathrm{cusp}}$ (see (G.1.1) and (9.2.14)) and where we have denoted by $\mathcal{V}$ the space of the representation π. In this identification the function $\varphi \in \mathcal{A}_{P,\pi}^G$ corresponds to the map $k \mapsto \varphi_k$ from K to $\mathcal{A}_{M,\pi}^M$ where

$$\varphi_k(m) = \delta_{P(\mathbb{A})}^{-1/2}(m)\varphi(mk) \qquad (\forall m \in M(\mathbb{A})).$$

We may also decompose $(\mathcal{V}, \pi)$ into a tensor product

$$(\mathcal{V}, \pi) = (\mathcal{V}_\infty, \pi_\infty) \otimes (\mathcal{V}^\infty, \pi^\infty)$$

where $(\mathcal{V}_\infty, \pi_\infty)$ is a unitary irreducible admissible representation of $F_\infty^\times \backslash M(F_\infty)$ and $(\mathcal{V}^\infty, \pi^\infty)$ is a unitary irreducible admissible representation of $M(\mathbb{A}^\infty)$. If $\lambda \in X_M^G$ we have

$$(\mathcal{V}, \pi(\lambda)) = (\mathcal{V}_\infty, \pi_\infty(\lambda)) \otimes (\mathcal{V}^\infty, \pi^\infty(\lambda))$$

where $\pi_\infty(\lambda)$ and $\pi^\infty(\lambda)$ are defined by

$$\pi_\infty(\lambda)(m) = \pi_\infty(m)q^{\langle \lambda, H_{M,\infty}(m)\rangle} \qquad (\forall m \in M(F_\infty))$$

and

$$\pi^\infty(\lambda)(m) = \pi^\infty(m)q^{\langle \lambda, H_M^\infty(m)\rangle} \qquad (\forall m \in M(\mathbb{A}^\infty)).$$

Here $H_{M,\infty}$ and H_M^∞ are the restrictions of H_M to $M(F_\infty) \subset M(\mathbb{A})$ and to $M(\mathbb{A}^\infty) \subset M(\mathbb{A})$ respectively.

Then, for each $P \in \mathcal{P}(M)$ and each

$$f = f_\infty f^\infty \in \mathcal{C}_c^\infty(F_\infty^\times \backslash G(\mathbb{A})) = \mathcal{C}_c^\infty(F_\infty^\times \backslash G(F_\infty)) \otimes \mathcal{C}_c^\infty(G(\mathbb{A}^\infty)),$$

we have

$$(\mathcal{A}^G_{P,\pi}, i^G_P(\lambda, f)) \cong \left(i^{G(F_\infty)}_{M(F_\infty),P(F_\infty)}(\mathcal{V}_\infty), i^{G(F_\infty)}_{M(F_\infty),P(F_\infty)}(\pi_\infty(\lambda), f_\infty)\right)$$
$$\otimes \left(i^{G(\mathbb{A}^\infty)}_{M(\mathbb{A}^\infty),P(\mathbb{A}^\infty)}(\mathcal{V}^\infty), i^{G(\mathbb{A}^\infty)}_{M(\mathbb{A}^\infty),P(\mathbb{A}^\infty)}(\pi^\infty(\lambda), f^\infty)\right)^{m_{\mathrm{cusp}}(\pi)}$$

where we regard

$$i^{G(F_\infty)}_{M(F_\infty),P(F_\infty)}(\mathcal{V}_\infty) \qquad (\text{resp. } i^{G(\mathbb{A}^\infty)}_{M(\mathbb{A}^\infty),P(\mathbb{A}^\infty)}(\mathcal{V}^\infty))$$

as the space $\mathcal{V}_P(\pi_\infty)$ (resp. $\mathcal{V}_P(\pi^\infty)$) of smooth functions $\varphi : K_\infty \to \mathcal{V}_\infty$ (resp. $\varphi : K^\infty \to \mathcal{V}^\infty$) such that

$$\varphi(n_P m k) = \pi_\infty(m)(\varphi(k))$$

for every $n_P \in N_P(F_\infty) \cap K_\infty$, every $m \in M(F_\infty) \cap K_\infty$ and every $k \in K_\infty$ (resp.

$$\varphi(n_P m k) = \pi^\infty(m)(\varphi(k))$$

for every $n_P \in N_P(\mathbb{A}^\infty) \cap K^\infty$, every $m \in M(\mathbb{A}^\infty) \cap K^\infty$ and every $k \in K^\infty$) and where the operator $i^{G(F_\infty)}_{M(F_\infty),P(F_\infty)}(\pi_\infty(\lambda), f_\infty)$ (resp. $i^{G(\mathbb{A}^\infty)}_{M(\mathbb{A}^\infty),P(\mathbb{A}^\infty)}(\pi^\infty(\lambda), f^\infty)$) is defined by

$$i^{G(F_\infty)}_{M(F_\infty),P(F_\infty)}(\pi_\infty(\lambda), f_\infty)(\varphi)(k)$$
$$= \int_{F_\infty^\times \backslash G(F_\infty)} f_\infty(g)\pi_\infty(M_P(kg))(\varphi(K_P(kg)))q^{\langle \lambda+\rho_P, H_{P,\infty}(kg)\rangle} \frac{dg}{dz_\infty}$$

(resp.

$$i^{G(\mathbb{A}^\infty)}_{M(\mathbb{A}^\infty),P(\mathbb{A}^\infty)}(\pi^\infty(\lambda), f^\infty)(\varphi)(k)$$
$$= \int_{G(\mathbb{A}^\infty)} f^\infty(g)\pi^\infty(M_P(kg))(\varphi(K_P(kg)))q^{\langle \lambda+\rho_P, H_P^\infty(kg)\rangle} dg\,)$$

for every $\varphi \in \mathcal{V}_P(\pi_\infty)$ (resp. $\varphi \in \mathcal{V}^P(\pi_\infty)$). Here $h = N_P(h)M_P(h)K_P(h)$ is any Iwasawa decomposition of $h \in G(F_\infty)$ (resp. $h \in G(\mathbb{A}^\infty)$) with respect to P and K_∞ (resp. K^∞), ρ_P is half the sum of the roots of Z_M in N_P and $H_{P,\infty} = H_{M,\infty} \circ M_P$ (resp. $H_P^\infty = H_M^\infty \circ M_P$).

At least formally, for any $P, P' \in \mathcal{P}(M)$ we have a **local intertwining operator**

$$(11.6.1)_\infty \qquad M_{P'|P}(\pi_\infty, \lambda) : i^{G(F_\infty)}_{M(F_\infty),P(F_\infty)}(\mathcal{V}_\infty) \to i^{G(F_\infty)}_{M(F_\infty),P'(F_\infty)}(\mathcal{V}_\infty)$$

(resp.

$$(11.6.1)^{\infty} \qquad M_{P'|P}(\pi^{\infty},\lambda) : i_{M(\mathbb{A}^{\infty}),P(\mathbb{A}^{\infty})}^{G(\mathbb{A}^{\infty})}(\mathcal{V}^{\infty}) \to i_{M(\mathbb{A}^{\infty}),P'(\mathbb{A}^{\infty})}^{G(\mathbb{A}^{\infty})}(\mathcal{V}^{\infty}))$$

defined by

$$\begin{aligned} &M_{P'|P}(\pi_{\infty},\lambda)(\varphi)(k) \\ &= \int_{(N_{P'}(F_\infty)\cap N_P(F_\infty))\backslash N_{P'}(F_\infty)} \pi_{\infty}(M_P(n_{P'}))(\varphi(K_P(n_{P'})k)) \\ &\qquad q^{\langle \lambda+\rho_P, H_{P,\infty}(n_{P'})\rangle} dn_{P'} \end{aligned}$$

(resp.

$$\begin{aligned} &M_{P'|P}(\pi^{\infty},\lambda)(\varphi)(k) \\ &= \int_{(N_{P'}(\mathbb{A}^\infty)\cap N_P(\mathbb{A}^\infty))\backslash N_{P'}(\mathbb{A}^\infty)} \pi^{\infty}(M_P(n_{P'}))(\varphi(K_P(n_{P'})k)) \\ &\qquad q^{\langle \lambda+\rho_P, H_P^{\infty}(n_{P'})\rangle} dn_{P'} \,). \end{aligned}$$

The meaning of this integral is the following. On the one hand we can always find an increasing sequence of compact open subgroups

$$(z_n^{-1}N_{P'}(\mathcal{O}_\infty)z_n)_{n\geq 0} \qquad (\text{resp. } (z_n^{-1}N_{P'}(\mathcal{O}^\infty)z_n)_{n\geq 0})$$

of $N_{P'}(F_\infty)$ (resp. $N_{P'}(\mathbb{A}^\infty)$) with z_n in $Z_M(F_\infty)$ (resp. $Z_M(\mathbb{A}^\infty)$) such that their union is $N_{P'}(F_\infty)$ (resp. $N_{P'}(\mathbb{A}^\infty)$). On the other hand, as φ is smooth and K_∞ (resp. K^∞) is compact we can find a compact open subgroup $K'_\infty \subset K_\infty$ (resp. $K'^\infty \subset K^\infty$) such that φ is invariant under left translation by K'_∞ (resp. K'^∞). Then it is easy to see that, for each λ and each k, the integral

$$\psi_n(\lambda;k) = \int_{U_{\infty,n}} \pi_\infty(M_P(n_{P'}))(\varphi(K_P(n_{P'})k))q^{\langle \lambda+\rho_P, H_{P,\infty}(n_{P'})\rangle} dn_{P'}$$

(resp.

$$\psi_n(\lambda;k) = \int_{U_n^\infty} \pi^\infty(M_P(n_{P'}))(\varphi(K_P(n_{P'})k))q^{\langle \lambda+\rho_P, H_P^{\infty}(n_{P'})\rangle} dn_{P'} \,),$$

where

$$U_{\infty,n} = (z_n^{-1}N_{P'}(\mathcal{O}_\infty)z_n \cap N_P(F_\infty))\backslash z_n^{-1}N_{P'}(\mathcal{O}_\infty)z_n$$

(resp.

$$U_n^\infty = (z_n^{-1}N_{P'}(\mathcal{O}^\infty)z_n \cap N_P(\mathbb{A}^\infty))\backslash z_n^{-1}N_{P'}(\mathcal{O}^\infty)z_n \,),$$

is in fact a finite sum and takes its value in the finite dimensional vector space $\mathcal{V}_\infty^{M(F_\infty)\cap K'_\infty}$ (resp. $(\mathcal{V}^\infty)^{M(\mathbb{A}^\infty)\cap K'^\infty}$) which is independent of n, λ and k but which depends on φ. The integral

$$M_{P'|P}(\pi_\infty,\lambda)(\varphi)(k) \qquad (\text{resp. } M_{P'|P}(\pi^\infty,\lambda)(\varphi)(k))$$

is by definition the limit of the sequence $\psi_n(\lambda;k)$ in that finite dimensional space (if the limit exists).

LEMMA (11.6.2). — *If, for some* λ *and for every* k, *the above integral is absolutely convergent with respect to the Hermitian norm on* $\mathcal{V}_\infty$ *(resp.* $\mathcal{V}^\infty$*) (recall that* π *is unitary),*

$$M_{P'|P}(\pi_\infty, \lambda)(\varphi) \qquad (\textit{resp. } M_{P'|P}(\pi^\infty, \lambda)(\varphi))$$

is a well-defined element of $\mathcal{V}_P(\pi_\infty)$ *(resp.* $\mathcal{V}_P(\pi^\infty)$*). Moreover, if the convergence is uniform for* λ *in some open subset of* X_M^G *the map*

$$\lambda \mapsto M_{P'|P}(\pi_\infty, \lambda)(\varphi) \qquad (\textit{resp. } \lambda \mapsto M_{P'|P}(\pi^\infty, \lambda)(\varphi)),$$

which takes its values in a finite dimensional subspace of $\mathcal{V}_P(\pi_\infty)$ *(resp.* $\mathcal{V}_P(\pi^\infty)$*), is holomorphic on that open subset.* □

PROPOSITION (11.6.3) (Harish-Chandra, Langlands). — (i) *Let us assume that* $\langle \mathrm{Re}\,\lambda, \alpha^\vee\rangle > 0$ *(resp.* $\langle \mathrm{Re}\,\lambda - \rho_P, \alpha^\vee\rangle > 0$*) for every* $\alpha \in R(P, Z_M) \cap (-R(P', Z_M))$. *Then for each* k *the integral*

$$M_{P'|P}(\pi_\infty, \lambda)(\varphi)(k) \qquad (\textit{resp. } M_{P'|P}(\pi^\infty, \lambda)(\varphi)(k))$$

is absolutely convergent. Moreover the holomorphic map

$$\lambda \mapsto M_{P'|P}(\pi_\infty, \lambda)(\varphi) \qquad (\textit{resp. } \lambda \mapsto M_{P'|P}(\pi^\infty, \lambda)(\varphi))$$

from the open subset of X_M^G *defined by the above conditions to* $\mathcal{V}_P(\pi_\infty)$ *(resp.* $\mathcal{V}_P(\pi^\infty)$*) can be analytically continued as a rational function of* $\lambda \in X_M^G$ *(see* (G.9)*). The operator*

$$M_{P'|P}(\pi_\infty, \lambda) \otimes \big(M_{P'|P}(\pi^\infty, \lambda)\big)^{m_{\mathrm{cusp}}(\pi)} : \mathcal{A}_{P,\pi}^G \to \mathcal{A}_{P',\pi}^G$$

is nothing else than the global intertwining operator $M_{P'|P}(\pi, \lambda)$ *introduced in* (11.5).

(ii) *We have*

$$M_{P'|P}(\pi_\infty, \lambda) \circ i_{M(F_\infty),P(F_\infty)}^{G(F_\infty)}(\pi_\infty(\lambda), f_\infty) = i_{M(F_\infty),P'(F_\infty)}^{G(F_\infty)}(\pi_\infty(\lambda), f_\infty) \circ M_{P'|P}(\pi_\infty, \lambda)$$

(resp.

$$M_{P'|P}(\pi^\infty, \lambda) \circ i_{M(\mathbb{A}^\infty),P(\mathbb{A}^\infty)}^{G(\mathbb{A}^\infty)}(\pi^\infty(\lambda), f^\infty) = i_{M(\mathbb{A}^\infty),P'(\mathbb{A}^\infty)}^{G(\mathbb{A}^\infty)}(\pi^\infty(\lambda), f^\infty) \circ M_{P'|P}(\pi^\infty, \lambda)).$$

(iii) *For any* $P, P', P'' \in \mathcal{P}(M)$ *with*

$$d(P'', P) = d(P'', P') + d(P', P)$$

we have

$$M_{P''|P}(\pi_\infty, \lambda) = M_{P''|P'}(\pi_\infty, \lambda) \circ M_{P'|P}(\pi_\infty, \lambda)$$

(resp.

$$M_{P''|P}(\pi^\infty, \lambda) = M_{P''|P'}(\pi^\infty, \lambda) \circ M_{P'|P}(\pi^\infty, \lambda)).$$

Here $d(P', P) = \ell_I(w')$ *if* $P = \dot{w}P_I\dot{w}^{-1}$ *and* $P' = \dot{w}\dot{w}'^{-1}P_{w'(I)}\dot{w}'\dot{w}^{-1}$ *for some* $I \subset \Delta$ *and some* $w, w' \in W$ *such that* $w'(I) \subset \Delta$.

(iv) *Let* $L \supset M$ *in* $\mathcal{L}$, $Q \in \mathcal{P}(L)$ *and* $R, R' \in \mathcal{P}^L(M)$. *Then we have*

$$\left(M_{R'N_Q|RN_Q}(\pi_\infty, \lambda)(\varphi)\right)_k = M_{R'|R}(\pi_\infty, \lambda)(\varphi_k)$$

(resp.

$$\left(M_{R'N_Q|RN_Q}(\pi^\infty, \lambda)(\varphi)\right)_k = M_{R'|R}(\pi^\infty, \lambda)(\varphi_k))$$

for every $\varphi \in \mathcal{V}_{RN_Q}(\pi_\infty)$ *(resp.* $\varphi \in \mathcal{V}_{RN_Q}(\pi^\infty)$*) and every* $k \in K_\infty$ *(resp.* $k \in K^\infty$*), where* φ_k *is the function* $k_L \mapsto \varphi(k_L k)$, $k_L \in K_\infty \cap L(F_\infty)$ *(resp.* $k_L \in K^\infty \cap L(\mathbb{A}^\infty)$*) and where* $M_{R'|R}(\pi_\infty, \lambda)$ *(resp.* $M_{R'|R}(\pi^\infty, \lambda)$*) is defined in an obvious way.*

(v) *For any* $w \in W$, *we have*

$$[w] \circ M_{P'|P}(\pi_\infty, \lambda) \circ [w]^{-1} = M_{\dot{w}P'\dot{w}^{-1}|\dot{w}P\dot{w}^{-1}}(w(\pi_\infty), w(\lambda))$$

(resp.

$$[w] \circ M_{P'|P}(\pi^\infty, \lambda) \circ [w]^{-1} = M_{\dot{w}P'\dot{w}^{-1}|\dot{w}P\dot{w}^{-1}}(w(\pi^\infty), w(\lambda))$$

where $w(\pi_\infty)(\dot{w}m\dot{w}^{-1}) = \pi_\infty(m)$ *(resp.* $w(\pi^\infty)(\dot{w}m\dot{w}^{-1}) = \pi^\infty(m)$*) and the operator*

$$[w] : \mathcal{V}_P(\pi_\infty) \to \mathcal{V}_{\dot{w}P\dot{w}^{-1}}(w(\pi_\infty)) \text{ or } [w] : \mathcal{V}_{P'}(\pi_\infty) \to \mathcal{V}_{\dot{w}P'\dot{w}^{-1}}(w(\pi_\infty))$$

(resp.

$$[w] : \mathcal{V}_P(\pi^\infty) \to \mathcal{V}_{\dot{w}P\dot{w}^{-1}}(w(\pi^\infty)) \text{ or } [w] : \mathcal{V}_{P'}(\pi^\infty) \to \mathcal{V}_{\dot{w}P'\dot{w}^{-1}}(w(\pi^\infty)))$$

is defined by $[w](\varphi)(k) = \varphi(\dot{w}^{-1}k)$.

(vi) *If* $\langle \cdot, \cdot \rangle$ *is the Hermitian form of* $\mathcal{V}_\infty$ *(resp.* $\mathcal{V}^\infty$*) (let us recall that* π *is unitary), then*

$$M_{P'|P}(\pi_\infty, \lambda)^* = M_{P|P'}(\pi_\infty, -\overline{\lambda})$$

(resp.

$$M_{P'|P}(\pi^\infty, \lambda)^* = M_{P|P'}(\pi^\infty, -\overline{\lambda}))$$

where $(\cdot)^*$ *is the adjunction with respect to the Hermitian forms*

$$\langle \varphi, \psi \rangle = \int_{K_\infty} \langle \varphi(k), \psi(k) \rangle dk \qquad (\text{resp.} = \int_{K^\infty} \langle \varphi(k), \psi(k) \rangle dk)$$

on $\mathcal{V}_P(\pi_\infty)$ *and on* $\mathcal{V}_{P'}(\pi_\infty)$ *(resp. on* $\mathcal{V}_P(\pi^\infty)$ *and on* $\mathcal{V}_{P'}(\pi^\infty)$*).*

Proof : At least formally assertions (ii) to (vi) directly follow from definition (11.6.1) (for (iii) see (G.5.3) and for (vi) note that

$$M_P(n_{P'})^{-1} n_{P'} M_P(n_{P'}) M_P(n_{P'})^{-1} K_P(n_{P'})^{-1}$$

is an Iwasawa decomposition of $M_P(n_{P'})^{-1} N_P(n_{P'}) M_P(n_{P'})$). Then arguing as in the beginning of the proof of (G.5.7) we see that, in order to prove the proposition, it is sufficient to prove assertion (i) in the case where $M = M_{\Delta - \{\alpha\}}$ for some $\alpha \in \Delta$, P is the standard maximal parabolic subgroup $P_{\Delta-\{\alpha\}}$ of G and P' is the opposite parabolic subgroup, so that $N_{P'} \cap N_P = (1)$.

Now since π_∞ (resp. π^∞) is unitary the convergence assertion in (i) may be easily reduced to the convergence of the integral

$$\int_{N_{P'}(F_\infty)} q^{\langle \lambda + \rho_P, H_{P,\infty}(n_{P'})\rangle} dn_{P'}$$

(resp.

$$\int_{N_{P'}(\mathbb{A}^\infty)} q^{\langle \lambda + \rho_P, H_P^\infty(n_{P'})\rangle} dn_{P'}\,)$$

for every $\lambda \in \operatorname{Re} X_M^G$ such that $\langle \lambda, \alpha^\vee \rangle > 0$ (resp. $\langle \lambda - \rho_P, \alpha^\vee \rangle > 0$). But this integral may be explicitly computed. For example let us do the computation in the case $G = GL_2$. For each place x of F and each $u \in F_x$ we have

$$H_{P,x}\left(\begin{pmatrix} 1 & 0 \\ u & 1 \end{pmatrix}\right) = \begin{cases} \frac{\deg(x)}{f}(-n, n) & \text{if } u \in \varpi_x^{-n} \mathcal{O}_x^\times \text{ with } n > 0, \\ (0,0) & \text{if } u \in \mathcal{O}_x, \end{cases}$$

where $H_{P,x}$ is the restriction of H_P to $M(F_x) \subset M(\mathbb{A})$. Therefore our integral is equal to

$$1 + \frac{q_\infty - 1}{q_\infty} \sum_{n=1}^{+\infty} q_\infty^{-n(\lambda_1 - \lambda_2)}$$

(resp.

$$1 + \sum_{S \subset |X| - \{\infty\}} \prod_{x \in S} \frac{q_x - 1}{q_x} \sum_{n_x = 1}^{+\infty} q_x^{-n_x(\lambda_1 - \lambda_2)}$$

with $q_x = q^{\frac{\deg(x)}{f}}$) if the Haar measure $dn_{P'}$ is normalized by $\mathrm{vol}(N_{P'}(F_x), dn_{P'}) = 1$ for each place x. Its convergence is clear (resp. follows from the convergence of the zeta function of $X - \{\infty\}$). For a general G the computation is similar (note that, by the elementary divisor theorem, any $n_{P'} \in N_{P'}(F_x)$ may be written as

$$m^{-1} \begin{pmatrix} 1 & 0 \\ u & 1 \end{pmatrix} m$$

where $m \in M(\mathcal{O}_x)$, where $u_{ii} = \varpi^{n_i}$, $\forall i = 1, \ldots, r$, for some integer $r \geq 0$ and some integers $n_1 \leq \cdots \leq n_r$ and where all the other entries of u are zero).

The analytic continuation of $M_{P'|P}(\pi_\infty, \lambda)$ is due to Harich-Chandra and a proof may be found in [Shah] (Theorem (2.2.2)). The analytic continuation of $M_{P'|P}(\pi^\infty, \lambda)$ is then a consequence of the analytic continuation of $M_{P'|P}(\pi)$ proved by Langlands (see (G.9.1)) as $M_{P'|P}(\pi_\infty, \lambda)$ admits an analytic inverse (see (11.6.4)(ii) below). □

The condition $d(P'', P) = d(P'', P') + d(P', P)$ in (11.6.3)(iii) may be relaxed by normalizing the intertwining operators. We will need this normalization only at the place ∞.

THEOREM (11.6.4) (Langlands, Shahidi). — *There exist non-zero rational scalar valued functions*

$$r_\alpha(\pi_\infty, z) \qquad (\forall \alpha \in R(G, Z_M))$$

of $z \in \mathbb{C}/\frac{2\pi i}{\log q}\mathbb{Z}$ *such that, if we set*

$$r_{P'|P}(\pi_\infty, \lambda) = \prod_{\substack{\alpha \in R(P, Z_M) \\ -\alpha \in R(P', Z_M)}} r_\alpha(\pi_\infty, \langle \lambda, \alpha^\vee \rangle) \qquad (\forall P, P' \in \mathcal{P}(M)),$$

the normalized operators

$$R_{P'|P}(\pi_\infty, \lambda) = r_{P'|P}(\pi_\infty, \lambda)^{-1} M_{P'|P}(\pi_\infty, \lambda)$$

have the following properties.

(i) *We have*

$$R_{P'|P}(\pi_\infty, \lambda) \circ i^{G(F_\infty)}_{M(F_\infty), P(F_\infty)}(\pi_\infty(\lambda), f_\infty) = i^{G(F_\infty)}_{M(F_\infty), P'(F_\infty)}(\pi_\infty(\lambda), f_\infty) \circ R_{P'|P}(\pi_\infty, \lambda).$$

(ii) *For any* $P, P', P'' \in \mathcal{P}(M)$ *we have*

$$R_{P''|P}(\pi_\infty, \lambda) = R_{P''|P'}(\pi_\infty, \lambda) \circ R_{P'|P}(\pi_\infty, \lambda).$$

(iii) *Let* $L \supset M$ *in* $\mathcal{L}$, $Q \in \mathcal{P}(L)$ *and* $R, R' \in \mathcal{P}^L(M)$. *Then we have*

$$\big(R_{R'N_Q|RN_Q}(\pi_\infty, \lambda)(\varphi)\big)_k = R_{R'|R}(\pi_\infty, \lambda)(\varphi_k)$$

for every $\varphi \in \mathcal{V}_{RN_Q}(\pi_\infty)$ *and every* $k \in K_\infty$, *where* φ_k *is the function* $k_L \mapsto \varphi(k_L k)$ *in* $\mathcal{V}_R(\pi_\infty)$ $(k_L \in K_\infty \cap L(F_\infty))$ *and where* $R_{R'|R}(\pi_\infty, \lambda)$ *is defined in an obvious way.*

(iv) *For any $w \in W$ we have*

$$[w] \circ R_{P'|P}(\pi_\infty, \lambda) \circ [w]^{-1} = R_{\dot{w}P'\dot{w}^{-1}|\dot{w}P\dot{w}^{-1}}(w(\pi_\infty), w(\lambda))$$

with the notations of (11.6.3)(v).

(v) *If $\langle \cdot, \cdot \rangle$ is the Hermitian form on $\mathcal{V}_\infty$ (recall that π_∞ is unitary) then we have*

$$R_{P'|P}(\pi_\infty, \lambda)^* = R_{P|P'}(\pi_\infty, -\overline{\lambda})$$

where $(\cdot)^*$ is the adjunction with respect to the Hermitian form

$$\langle \varphi, \psi \rangle = \int_K \langle \varphi(k), \psi(k) \rangle dk$$

on $\mathcal{V}_P(\pi_\infty)$ and on $\mathcal{V}_{P'}(\pi_\infty)$.

(vi) *If π_∞ is spherical and if $\varphi \in \mathcal{V}_P(\pi_\infty)$ is fixed by K_∞, $R_{P'|P}(\pi_\infty, \lambda)(\varphi)$ is independent of λ.* □

Proof : The formulation of this theorem is due to Arthur (see [Ar 9] Theorem 2.1) and a proof of this theorem has been given by Langlands (see loc. cit.). In our case, π_∞ is non-degenerate as a local component of an irreducible cuspidal automorphic representation (see [Sha1] Corollary of theorem 5.9) and a proof of the theorem may be found in [Shah] §3. □

Remark (11.6.5). — It follows from the above theorem that the rational function $\lambda \to R_{P'|P}(\pi_\infty, \lambda)$ does not have zeros or poles on $\operatorname{Im} X_M^G$ and that, for any $\lambda \in \operatorname{Im} X_M^G$, the operator $R_{P'|P}(\pi_\infty, \lambda)$ is unitary. Indeed, by properties (i), (ii) and (v), for any λ in a dense open subset of $\operatorname{Im} X_M^G$ and for any compact open subgroup K'_∞ of K_∞, the operator

$$R_{P'|P}(\pi_\infty, \lambda) : \mathcal{V}_P(\pi_\infty)^{K'_\infty} \to \mathcal{V}_{P'}(\pi_\infty)^{K'_\infty}$$

is a unitary operator between finite dimensional Hermitian spaces and, if $F : \mathbb{C}^m \to \mathbb{C}^n$ is a meromorphic map such that $||F(iy)|| = 1$ for any $iy \in i\mathbb{R}^m \subset \mathbb{C}^m$ in the domain of holomorphy of F, the map F cannot have zeros or poles on $i\mathbb{R}^m$. □

Lemma (11.6.6). — *Let us fix $P_0 \in \mathcal{P}(M)$. For every $P \in \mathcal{P}(M)$ and every $\lambda, \zeta \in \operatorname{Im} X_M^G$ let us set*

$$R_P(\zeta, \pi_\infty, \lambda; P_0) = R_{P|P_0}(\pi_\infty, \lambda)^{-1} \circ R_{P|P_0}(\pi_\infty, \lambda+\zeta) : \mathcal{V}_{P_0}(\pi_\infty) \to \mathcal{V}_{P_0}(\pi_\infty)$$

(*resp.*

$$\begin{aligned} S_P(\zeta, \pi, \lambda; P_0) = & \, r_{P|P_0}(\pi_\infty, \lambda)^{-1} r_{P|P_0}(\pi_\infty, \lambda + \zeta) \\ & \times M_{P|P_0}(\pi^\infty, \lambda)^{-1} \circ M_{P|P_0}(\pi^\infty, \lambda + \zeta) : \mathcal{V}_{P_0}(\pi^\infty) \to \mathcal{V}_{P_0}(\pi^\infty)). \end{aligned}$$

Then, for every locally constant function with compact support f_∞ *(resp.* f^∞*) on* $F_\infty^\times\backslash G(F_\infty)$ *(resp.* $G(\mathbb{A}^\infty)$*), the family of functions of* $\zeta\in\operatorname{Im}X_M^G$*,*

$$\begin{aligned}c_\infty &= \big(c_{\infty,P}(\zeta)\\ &= \operatorname{tr}\big(R_P(\zeta,\pi_\infty,\lambda;P_0)\circ i_{M(F_\infty),P_0(F_\infty)}^{G(F_\infty)}(\pi_\infty(\lambda),f_\infty)\big)\mid P\in\mathcal{P}(M)\big)\end{aligned}$$

(resp.

$$c^\infty=\big(c_P^\infty(\zeta)=\operatorname{tr}\big(S_P(\zeta,\pi,\lambda;P_0)\circ i_{M(\mathbb{A}^\infty),P_0(\mathbb{A}^\infty)}^{G(\mathbb{A}^\infty)}(\pi^\infty(\lambda),f^\infty)\big)\mid P\in\mathcal{P}(M)\big)\big),$$

is a $(G,M)^{\sim}$*-family.*

Proof : Let (P,P') be a pair of adjacent parabolic subgroups in $\mathcal{P}(M)$. Then there exist a unique $L\in\mathcal{L}(M)$, a unique $Q\in\mathcal{P}(L)$ and a unique pair of adjacent parabolic subgroups (R,R') in $\mathcal{P}^L(M)$ such that $P=RN_Q$, $P'=R'N_Q$, $\Delta_R=\{\alpha_{P',P}\}$ and $\Delta_{R'}=\{-\alpha_{P',P}\}$.

Let $\zeta\in\operatorname{Im}\mathfrak{a}_M^{*G,\mathbb{Z}}$ be such that $\langle\zeta,\alpha_{P',P}^\vee\rangle\in\frac{2\pi i}{\log q}\mathbb{Z}$. We have $R_{R'|R}(\pi_\infty,\lambda+\zeta)=R_{R'|R}(\pi_\infty,\lambda)$ and then it follows from (11.6.4)(iii) that $R_{P'|P}(\pi_\infty,\lambda+\zeta)=R_{P'|P}(\pi_\infty,\lambda)$. Therefore, by (11.6.4)(ii) we have

$$R_P(\zeta,\pi_\infty,\lambda;P_0)=R_{P'}(\zeta,\pi_\infty,\lambda;P_0).$$

But by (11.5.6) we have

$$\begin{aligned}&R_P(\zeta,\pi_\infty,\lambda;P_0)\otimes S_P(\zeta,\pi,\lambda;P_0)^{m_{\mathrm{cusp}}(\pi)}\\ &=M_P(\zeta,\pi,\lambda;P_0)\\ &=M_{P'}(\zeta,\pi,\lambda;P_0)\\ &=R_{P'}(\zeta,\pi_\infty,\lambda;P_0)\otimes S_{P'}(\zeta,\pi,\lambda;P_0)^{m_{\mathrm{cusp}}(\pi)}.\end{aligned}$$

Therefore we also have

$$S_P(\zeta,\pi,\lambda;P_0)=S_{P'}(\zeta,\pi,\lambda;P_0)$$

and the lemma follows. □

LEMMA (11.6.7). — *Let* $L\in\mathcal{L}(M)$ *and* $Q\in\mathcal{P}(L)$*. Let us assume that*

$$f_\infty^Q(\ell)\overset{\mathrm{dfn}}{=\!=}\delta_{Q(F_\infty)}^{1/2}(\ell)\int_{N_Q(F_\infty)}\int_{K_\infty}f_\infty(k^{-1}\ell n_Qk)dkdn_Q=0$$

for every $\ell\in M(F_\infty)$*. Then* $c_{\infty,M}^Q(\frac{2\pi i}{\log q}\nu)=0$ *for all* $\nu\in\mathfrak{a}_{M,\mathbb{Z}}^\vee$*.*

Proof : As in the proof of (11.5.15) we see that $c^Q_{\infty,M}(\frac{2\pi i}{\log q}\nu)$ does not depend on the choice of P_0, so that we may assume that $P_0 = R_0N_Q \subset Q$, and we see that, for any $P = RN_Q$, the operator

$$R_{P|P_0}(\pi_\infty,\lambda)^{-1}\circ R_{P|P_0}(\pi_\infty,\lambda+\zeta)\circ i^{G(F_\infty)}_{M(F_\infty),P_0(F_\infty)}(\pi_\infty(\lambda),f_\infty)$$

has kernel

$$R_{R|R_0}(\pi_\infty,\lambda)^{-1}\circ R_{R|R_0}(\pi_\infty,\lambda+\zeta)\circ i^{L(F_\infty)}_{M(F_\infty),R_0(F_\infty)}(\pi_\infty(\lambda),\delta^{-1/2}_{Q(F_\infty)}g_{\infty,k,k'})$$

where we have set

$$g_{\infty,k,k'}(\ell)=\delta^{1/2}_{Q(F_\infty)}(\ell)\int_{N_Q(F_\infty)}f_\infty(k^{-1}\ell n_Q k')dn_Q \qquad (\forall \ell\in L(F_\infty))$$

for every $k,k'\in K_\infty$. As we have

$$\int_{K_\infty}\delta^{-1/2}_{Q(F_\infty)}g_{\infty,k,k}dk = f^Q_\infty$$

the lemma follows. □

At least formally we introduce the operator

$$\mathcal{R}_M(\pi_\infty,\lambda;P_0)=\lim_{\zeta\to 0}\sum_{P\in\mathcal{P}(M)}R_P(\zeta,\pi_\infty,\lambda;P_0)\widetilde{\theta}_P(\zeta)^{-1}$$

on $\mathcal{V}_{P_0}(\pi_\infty)$, so that

$$\mathrm{tr}\big(\mathcal{R}_M(\pi_\infty,\lambda;P_0)\circ i^{G(F_\infty)}_{M(F_\infty),P_0(F_\infty)}(\pi_\infty(\lambda),f_\infty)\big)=\widetilde{(c_\infty)_M}(0).$$

PROPOSITION (11.6.8). — *Let $I\subset\Delta$. Let us assume that $f=f_\infty f^\infty$ with $f^Q_\infty\equiv 0$ for every $L\in\mathcal{L}(M_I)$, $L\subsetneqq G$, and every $Q\in\mathcal{P}(L)$. Then*

$$\mathrm{tr}(M^T_{I,\pi}(\lambda)\circ i^\Delta_I(\lambda,f))$$

is equal to

$$m_{\mathrm{cusp}}(\pi)\frac{\deg(\infty)d}{fd_1\cdots d_s}\,\mathrm{tr}\big(\mathcal{R}_{M_I}(\pi_\infty,\lambda;P_I)\circ i^{G(F_\infty)}_{M_I(F_\infty),P_I(F_\infty)}(\pi_\infty(\lambda),f_\infty)\big)$$
$$\times\,\mathrm{tr}\big(i^{G(\mathbb{A}^\infty)}_{M_I(\mathbb{A}^\infty),P_I(\mathbb{A}^\infty)}(\pi^\infty(\lambda),f^\infty)\big).$$

Proof: The $(G,M_I)^{\frown}$-family

$$c=\big(c_P(\zeta)=\mathrm{tr}\big(M_{P|P_I}(\pi,\lambda)^{-1}\circ M_{P|P_I}(\pi,\lambda+\zeta)\circ i_I^{\Delta}(\lambda+\zeta,f)\big)\mid P\in\mathcal{P}(M_I)\big)$$

is equal to $m_{\mathrm{cusp}}(\pi)c_\infty c^\infty$ where c_∞ and c^∞ are the $(G,M_I)^{\frown}$-families defined in (11.6.6) (see (11.6.3)(i)). But by (11.6.7) and (11.5.14)(ii) we have

$$\widetilde{(c_\infty c^\infty)}_M(0)=\widetilde{(c_\infty)}_M(0)(c^\infty)_G(0).$$

Therefore the proposition follows from (11.5.16). □

Then combining this proposition and proposition (11.4.3) we obtain

COROLLARY (11.6.9). — *Let c be a regular cuspidal datum and let (I,π) be a unitary cuspidal pair in c. Let us assume that $f=f_\infty f^\infty$ with $f_\infty^Q\equiv 0$ for every $L\in\mathcal{L}(M_I)$, $L\subsetneqq G$, and every $Q\in\mathcal{P}(L)$. Then $J_c^T(f)$ is independent of $T\in\mathfrak{a}_\emptyset^+$ and is equal to*

$$\frac{m_{\mathrm{cusp}}(\pi)d}{|\mathrm{Fix}_{X_I^\Delta}(\pi)|}\int_{\mathrm{Im}\,X_I^\Delta}\mathrm{tr}\big(\mathcal{R}_{M_I}(\pi_\infty,\lambda;P_I)\circ i_{M_I(F_\infty),P_I(F_\infty)}^{G(F_\infty)}(\pi_\infty(\lambda),f_\infty)\big)$$
$$\times\,\mathrm{tr}\,i_{M_I(\mathbb{A}^\infty),P_I(\mathbb{A}^\infty)}^{G(\mathbb{A}^\infty)}(\pi^\infty(\lambda),f^\infty)d\lambda.$$

□

(11.7) Evaluation of $J_c^T(f)$ in a special case: Waldspurger's theorem

We have fixed a Haar measure dg on $G(\mathbb{A})$ with a splitting $dg=\prod_x dg_x$. In particular we have a Haar measure dg_∞ on $G(F_\infty)$. Let f_∞ be the corresponding Euler–Poincar function introduced in (5.1). Let us recall that it is very cuspidal (see (5.1.3)(ii)), so that $f_\infty^Q\equiv 0$ for every $L\in\mathcal{L}$, $L\subsetneqq G$, and every $Q\in\mathcal{P}(L)$ (there exist $w\in W$ and $J\subset\Delta$ such that $L=\dot{w}^{-1}M_J\dot{w}$ and $Q=\dot{w}^{-1}P_J\dot{w}$ and we have $f_\infty^Q(\ell)=f_\infty^{P_J}(\dot{w}\ell\dot{w}^{-1})=0$ for all $\ell\in L(F_\infty)$). Let us also recall that f_∞ is a pseudo-coefficient of the Steinberg representation, i.e. that, for any unitary admissible irreducible representation Π_∞ of $F_\infty^\times\backslash G(F_\infty)$, we have $\mathrm{tr}\,\Pi_\infty(f_\infty)=0$ unless Π_∞ is isomorphic either to the Steinberg representation $\mathrm{st}_{d,\infty}$ or to the trivial representation $1_{d,\infty}$ of $F_\infty^\times\backslash G(F_\infty)$ and that we have

$$\mathrm{tr}\,\mathrm{st}_{d,\infty}(f_\infty)=(-1)^{d-1}$$

and

$$\mathrm{tr}\,1_{d,\infty}(f_\infty)=1$$

(see (8.5.3)).

Let $I \subset \Delta$ and let $(d_1, \dots, d_s)$ be the partition of d corresponding to I. For each unitary admissible irreducible representation $\pi_\infty = \pi_{\infty,1} \otimes \cdots \otimes \pi_{\infty,s}$ of $F_\infty^\times \backslash M_I(F_\infty) \cong F_\infty^\times \backslash (GL_{d_1}(F_\infty) \times \cdots \times GL_{d_s}(F_\infty))$, each polynomial function $P(\lambda)$ on X_I^Δ, each $j = 1, \dots, s$ and each positive integer n we set

$$Z_I^\Delta(\pi_\infty; P(\lambda), j, n) = \int_{\operatorname{Im} X_I^\Delta} \operatorname{tr}\big(\mathcal{R}_{M_I}(\pi_\infty, \lambda; P_I) \circ i_{M_I(F_\infty), P_I(F_\infty)}^{G(F_\infty)}(\pi_\infty(\lambda), f_\infty)\big) P(\lambda) q^{-n\frac{\lambda_j}{d_j}} \, d\lambda.$$

Proposition (11.6.3), theorem (11.6.4) and lemma (11.6.6) have been formulated only when π_∞ is the component at ∞ of a global π such that (I, π) is a regular cuspidal pair. But they hold for a general π_∞ and $\mathcal{R}_{M_I}(\pi_\infty, \lambda; P_I)$ is well-defined. A polynomial function on X_I^Δ is a polynomial in the variables $q^{\frac{\lambda_j}{d_j}}, q^{-\frac{\lambda_j}{d_j}}$ $(j = 1, \dots, s)$ (see (G.2)). The measure $d\lambda$ is the normalized Haar measure on $\operatorname{Im} X_I^\Delta$.

Let us remark that, for every $\xi \in \operatorname{Im} X_I^\Delta$, we have

$$Z_I^\Delta(\pi_\infty(\xi); P(\lambda), j, n) = q^{n\frac{\xi_j}{d_j}} Z_I^\Delta(\pi_\infty; P(\lambda - \xi), j, n).$$

THEOREM (11.7.1) (Waldspurger). — (i) *We have*

$$Z_I^\Delta(\pi_\infty; P(\lambda), j, n) = 0$$

unless there exists $\xi \in \operatorname{Im} X_I^\Delta$ *such that, for each* $j = 1, \dots, s$, *the central character of* $\pi_{\infty,j}(\xi_j)$ *is trivial and* $\pi_{\infty,j}(\xi_j)$ *is isomorphic to the Steinberg representation or to the trivial representation of* $F_\infty^\times \backslash GL_{d_j}(F_\infty)$.

(ii) *Let us assume that, for each* $j = 1, \dots, s$, $\pi_{\infty,j}$ *has a trivial central character and is isomorphic to* $\mathrm{st}_{d_j,\infty}$ *or to* $\mathbf{1}_{d_j,\infty}$. *Let us set*

$$\epsilon_\infty(\pi_{\infty,j}) = \begin{cases} (-1)^{d-1} & \text{if } \pi_{\infty,j} \cong \mathrm{st}_{d_j,\infty}, \\ 1 & \text{if } \pi_{\infty,j} \cong \mathbf{1}_{d_j,\infty}, \end{cases}$$

as in (9.5.1) *and let us set*

$$\epsilon_\infty(\pi_\infty) = \epsilon_\infty(\pi_{\infty,1}) \cdots \epsilon_\infty(\pi_{\infty,s}).$$

Then there exists a positive integer n_0 *depending only on* $P(\lambda)$ *such that, for every* $n \geq n_0$, *we have*

$$Z_I^\Delta(\pi_\infty; P(\lambda), j, n) = \frac{\epsilon_\infty(\pi_\infty)}{d} \sum_{\substack{\sigma \in \mathfrak{S}_s \\ \sigma(j)=1}} P(w_{I,\sigma}^{-1}(\rho_{w_{I,\sigma}(I)})) q^{-n(d-d_j)/2}$$

where $w_{I,\sigma}$ *is the element of* W *such that* $w_{I,\sigma}(I) \subset \Delta$ *which is associated to* σ *in* (G.1.5) *and where, for each* $J \subset \Delta$,

$$\rho_J = \left(\frac{(d - e_1)e_1}{2}, \frac{(d - 2e_1 - e_2)e_2}{2}, \dots, \frac{(e_t - d)e_t}{2} \right) \in X_J^\Delta$$

is half the sum of the roots of Z_J *in* N_J, $(e_1, \dots, e_t)$ *being the partition of* d *corresponding to* J. □

The proof of this theorem is given in an appendix by Waldspurger at the end of this book.

Let us fix a place $o \neq \infty$ and let us assume that we have chosen the splitting of the Haar measure dg in such a way that $\mathrm{vol}(K_o, dg_o) = 1$. Let us denote by f_o the Drinfeld function of level r, i.e. the Hecke function on $GL_d(F_o)$ with Satake transform

$$p^{\deg(o)(d-1)/2}(z_1^r + \cdots + z_d^r),$$

for some positive integer r (see (4.2.5)). Let us fix an arbitrary function $f^{\infty,o} \in C_c^\infty(\mathbb{A}^{\infty,o})$ and let us set $f^\infty = f^{\infty,o} f_o$.

For each unitary regular cuspidal pair (I, π)

$$P(\lambda) = \mathrm{tr}\, i_{M_I(\mathbb{A}^{\infty,o}),P_I(\mathbb{A}^{\infty,o})}^{G(\mathbb{A}^{\infty,o})}(\pi^{\infty,o}(\lambda), f^{\infty,o})$$

is a polynomial function on X_I^Δ and, if we decompose the cuspidal automorphic irreducible representation π of $F_\infty^\times \backslash M_I(\mathbb{A}) \cong F_\infty^\times \backslash (GL_{d_1}(\mathbb{A}) \times \cdots \times GL_{d_s}(\mathbb{A}))$ into $\pi_1 \otimes \cdots \otimes \pi_s$, we have

$$\begin{aligned}\mathrm{tr}\, & i_{M_I(F_o),P_I(F_o)}^{G(F_o)}(\pi_o(\lambda), f_o) \\ &= \sum_{j=1}^{s} p^{r\deg(o)(d-1)/2}(z_1(\pi_{j,o})^r + \cdots + z_{d_j}(\pi_{j,o})^r) p^{-r\deg(o)\frac{\lambda_j}{d_j}}\end{aligned}$$

if π_o is spherical, i.e. if $\pi_{1,o}, \ldots, \pi_{s,o}$ are spherical, $(z_1(\pi_{j,o}), \ldots, z_{d_j}(\pi_{j,o}))$ being the Hecke eigenvalues of $\pi_{j,o}$, and we have

$$\mathrm{tr}\, i_{M_I(F_o),P_I(F_o)}^{G(F_o)}(\pi_o(\lambda), f_o) = 0$$

otherwise (see (7.5.7), (7.5.8.2) and (4.2.5)). Therefore we may apply Waldspurger's theorem and we get

COROLLARY (11.7.2). — *For each unitary regular cuspidal pair (I, π) the integral*

$$\begin{aligned}\int_{\mathrm{Im}\, X_I^\Delta} & \mathrm{tr}(\mathcal{R}_{M_I}(\pi_\infty, \lambda; P_I) \circ i_{M_I(F_\infty),P_I(F_\infty)}^{G(F_\infty)}(\pi_\infty(\lambda), f_\infty)) \\ & \times \mathrm{tr}\, i_{M_I(\mathbb{A}^\infty),P_I(\mathbb{A}^\infty)}^{G(\mathbb{A}^\infty)}(\pi^\infty(\lambda), f^\infty) d\lambda\end{aligned}$$

vanishes unless there exists $\xi \in \mathrm{Im}\, X_I^\Delta$ such that $\pi_\infty(\xi)$ has a trivial central character and is isomorphic to the Steinberg representation $\mathrm{st}_{I,\infty} = \mathrm{st}_{d_1,\infty} \otimes \cdots \otimes \mathrm{st}_{d_s,\infty}$ of $Z_I(F_\infty) \backslash M_I(F_\infty) \cong F_\infty^\times \backslash GL_{d_1}(F_\infty) \times \cdots \times F_\infty^\times \backslash GL_{d_s}(F_\infty)$ and unless π_o is spherical.

If π_∞ has a trivial central character and is isomorphic to $\mathrm{st}_{I,\infty}$ and if π_o is spherical there exists a positive integer r_0 depending only on o and $f^{\infty,o}$ such that, for every $r \geq r_0$, that integral is equal to

$$\frac{(-1)^{d-s}}{d} \sum_{\sigma\in\mathfrak{S}_s} \mathrm{tr}\, i_{M_I(\mathbb{A}^{\infty,o}),P_I(\mathbb{A}^{\infty,o})}^{G(\mathbb{A}^{\infty,o})}(\pi^{\infty,o}(w_{I,\sigma}^{-1}(\rho_{w_{I,\sigma}(I)})), f^{\infty,o})$$
$$\times p^{r\deg(o)(d_{\sigma^{-1}(1)}-1)/2}(z_1(\pi_{\sigma^{-1}(1),o})^r + \cdots + z_{d_{\sigma^{-1}(1)}}(\pi_{\sigma^{-1}(1),o})^r).$$

Proof : Let us remark that

(1) a cuspidal automorphic irreducible representation π_j of $F_\infty^\times\backslash GL_{d_j}(\mathbb{A})$ satisfies the condition $\pi_{j,\infty} \cong 1_{d_j,\infty}$ only if we have $d_j = 1$ and therefore $1_{d_j,\infty} = \mathrm{st}_{\infty,d_j}$ (see the proof of (9.5.5)(ii) and (9.5.9)),

(2) for a given o and a given $f^{\infty,o}$ there are only finitely many cuspidal automorphic irreducible representations π of $Z_I(F_\infty)\backslash M_I(\mathbb{A})$ such that $\pi_\infty \cong \mathrm{st}_{I,\infty}$, $\mathrm{tr}\, i_{M_I(\mathbb{A}^{\infty,o}),P_I(\mathbb{A}^{\infty,o})}^{G(\mathbb{A}^{\infty,o})}(\pi^{\infty,o}(\lambda), f^{\infty,o}) \neq 0$ for all $\lambda \in X_I^\Delta$ and π_o is spherical (see (7.5.7) and (9.2.14)).

Then the corollary is a direct consequence of Waldspurger's theorem. □

COROLLARY (11.7.3). — *Let c be a regular cuspidal automorphic datum and let $f = f_\infty f^{\infty,o} f_o$ with f_∞, $f^{\infty,o}$ and f_o as above. Then $J_c^T(f)$, which is independent of $T \in \mathfrak{a}_\emptyset^+$, vanishes unless there exists $(I,\pi) \in c$ such that π_∞ has a trivial central character and is isomorphic to the Steinberg representation of $Z_I(F_\infty)\backslash M_I(F_\infty)$ and such that π_o is spherical.*

Moreover there exists a positive integer r_0 depending only on o and $f^{\infty,o}$ and having the following property: for every $r \geq r_0$, $J_c^T(f)$ is equal to the sum of the expressions

$$\frac{(-1)^{d-s}}{s!} m_{\mathrm{cusp}}(\pi)\, \mathrm{tr}\, i_{M_I(\mathbb{A}^{\infty,o}),P_I(\mathbb{A}^{\infty,o})}^{G(\mathbb{A}^{\infty,o})}(\pi^{\infty,o}(\rho_I), f^{\infty,o})$$
$$\times p^{r\deg(o)(d_1-1)/2}(z_1(\pi_{1,o})^r + \cdots + z_{d_1}(\pi_{1,o})^r)$$

over the set of unitary cuspidal pairs $(I,\pi) \in c$ such that π_∞ has a trivial central character and is isomorphic to $\mathrm{st}_{I,\infty}$ and such that π_o is spherical.

Proof : Let us remark that the representation $\mathrm{st}_{I,\infty}(\lambda)$ is isomorphic to $\mathrm{st}_{I,\infty}$ for some $\lambda \in X_I^\Delta$ if and only if $\lambda = 0$ (for example consider the restriction $r_I^\emptyset(\mathrm{st}_{I,\infty}(\lambda))$ defined by (7.1.2) and computed in (8.1.2)(iii)). Let us also remark that any cuspidal pair (I,π) such that π_∞ has a trivial central character is automatically unitary. Therefore, if (I,π) is a cuspidal pair in c such that π_∞ has a trivial central character and is isomorphic to $\mathrm{st}_{I,\infty}$ we have $|\mathrm{Fix}_{X_I^\Delta}(\pi)| = 1$ and any other cuspidal pair (I',π') in c having the same properties is equal to $(w(I), w(\pi))$ for some $w \in W$ such that $w(I) \subset \Delta$ and is unitary.

Now it is clear that the corollary follows from (11.6.9) and (11.7.2). □

(11.8) A "simple" spectral side for the trace formula

For a general cuspidal datum c Langlands (and Morris) have established a spectral decomposition of $L^2_{G,1_\infty,c}$ in terms of the discrete spectrum of the Levi subgroups of G (see [Mo–Wa 2]VI.2).

Using this spectral decomposition Lafforgue has established the following theorem (see [Laf 2]) which is a paraphrase of a theorem proved by Arthur in the number field case (see [Ar 6], [Ar 10], [Ar 11], [Ar 12] and [Ar 13]).

THEOREM (11.8.1) (Lafforgue). — *Let $f = f_\infty f^\infty$ be a function in $\mathcal{C}_c^\infty(F_\infty^\times\backslash G(\mathbb{A})) = \mathcal{C}_c^\infty(F_\infty^\times\backslash G(F_\infty))\otimes\mathcal{C}_c^\infty(G(\mathbb{A}^\infty))$ such that $f_\infty^Q \equiv 0$ for every $L\in\mathcal{L}$, $L\subsetneq G$, and every $Q\in\mathcal{P}(L)$. Then $J^T_{\mathrm{spec}}(f)$ is independent of $T\in\mathfrak{a}_\emptyset^+$ and is equal to*

$$\sum_{I\subset\Delta}\frac{1}{s!}\sum_{\substack{\chi\in\mathcal{X}_I\\ \chi_\infty=1_{I,\infty}}}\sum_{\pi\in\Pi^2_{I,\chi}}\frac{m^2(\pi)d}{|\mathrm{Fix}_{X_I^\Delta}(\pi)|}$$
$$\times\int_{\mathrm{Im}\,X_I^\Delta}\mathrm{tr}\big(\mathcal{R}_{M_I}(\pi_\infty,\lambda;P_I)\circ i^{G(F_\infty)}_{M_I(F_\infty),P_I(F_\infty)}(\pi_\infty(\lambda),f_\infty)\big)$$
$$\times\,\mathrm{tr}\, i^{G(\mathbb{A}^\infty)}_{M_I(\mathbb{A}^\infty),P_I(\mathbb{A}^\infty)}(\pi^\infty(\lambda),f^\infty)d\lambda$$

where $\mathcal{X}_I = \mathcal{X}_{M_I}$ is the group of characters of $Z_I(F)\backslash Z_I(\mathbb{A})$, where $1_{I,\infty}$ is the trivial character of $Z_I(F_\infty)$, where $\Pi^2_{I,\chi} = \Pi^2_{M_I,\chi}$ is a system of representatives of the isomorphism classes of L^2-automorphic irreducible representations of $M_I(\mathbb{A})$ with central character χ and where $m^2(\pi)$ is the multiplicity of $\pi\in\Pi^2_{I,\chi}$ in $(\mathcal{A}^2_{M_I,\chi}, R^2_{M_I,\chi})$.

Combining this result with Waldspurger's theorem we obtain

THEOREM (11.8.2). — *Let $f = f_\infty$ be the very cuspidal Euler–Poincar function (corresponding to the Haar measure dg_∞) that we have introduced in (5.1). Let o be a place of F distinct from ∞ and let us assume that we have chosen the splitting of our Haar measure dg in such a way that* $\mathrm{vol}(K_o, dg_o) = 1$. *Let f_o be the Drinfeld function of level r for some positive integer r (see (4.2.5)). Let $f^{\infty,o}$ be an arbitrary function in $\mathcal{C}_c^\infty(G(\mathbb{A}^{\infty,o}))$ and let us set $f = f_\infty f^{\infty,o} f_o$.*

For each positive integer r which is large enough with respect to o and $f^{\infty,o}$ the truncated trace $J^T_{\mathrm{spec}}(f)$, which is independent of $T\in\mathfrak{a}_\emptyset^+$, is equal to the sum of the expressions

$$(-1)^{|\Delta-I|}m^2(\pi)\epsilon_\infty(\pi_\infty)\,\mathrm{tr}\, i^{G(\mathbb{A}^{\infty,o})}_{M_I(\mathbb{A}^{\infty,o}),P_I(\mathbb{A}^{\infty,o})}(\pi^{\infty,o}(\rho_I),f^{\infty,o})$$
$$\times p^{r\deg(o)(d_1-1)/2}(z_1(\pi_{1,o})^r+\cdots+z_{d_1}(\pi_{1,o})^r)$$

over the triples (I,χ,π), where $I\subset\Delta$, where $\chi\in\mathcal{X}_I$ with $\chi_\infty = 1_{I,\infty}$ and where $\pi\in\Pi^2_{I,\chi}$ with $\pi_{j,\infty}\cong\mathrm{st}_{d_j,\infty}$ or $\pi_{j,\infty}\cong 1_{d_j,\infty}$ for every $j=1,\ldots,s$ and with π_o spherical.

As usual $(d_1, \ldots, d_s)$ *is the partition of* d *corresponding to* I*, we have* $\pi = \pi_1 \otimes \cdots \otimes \pi_s$ *as a representation of* $Z_I(F_\infty)\backslash M_I(\mathbb{A}) \cong (F_\infty^\times\backslash GL_{d_1}(\mathbb{A})) \times \ldots \times (F_\infty^\times\backslash GL_{d_s}(\mathbb{A}))$ *and* $(z_1(\pi_{1,o}), \ldots, z_{d_1}(\pi_{1,o}))$ *are the Hecke eigenvalues of the spherical representation* $\pi_{1,o}$ *of* $GL_{d_1}(F_o)$*.*

REMARK (11.8.3). — Let $I \subset \Delta$, $\chi \in \mathcal{X}_I$ and $\pi \in \Pi^2_{I,\chi}$ occurring in the above sum. If, for some $j = 1, \ldots, s$, we have $\pi_{j,\infty} \cong \mathrm{st}_{d_j,\infty}$ (resp. $\pi_{j,\infty} \cong 1_{d_j,\infty}$) then π_j is cuspidal (resp. one dimensional) (see (9.5.5)). Therefore it follows from (9.5.6) and [Sha 1] that $m^2(\pi) = 1$. □

(11.9) Comments and references

We have closely followed Arthur's papers on the trace formula in the number field case, especially the paper [Ar 6]. The modifications due to function fields are minor. Similar modifications occur in the local trace formula for a reductive group over a p-adic field (see [Ar 8]). To prove theorem (11.8.1) Lafforgue uses much more direct arguments than Arthur's ones. Even for a regular c his proof is simpler.

Arthur has proved an analog of Waldspurger's theorem for a general real reductive group which admits discrete series representations.

12

Cohomology with compact supports of Drinfeld modular varieties

(12.0) Introduction

Let us fix a Drinfeld modular variety $M_{\mathcal{I}}^d$ with characteristic morphism

$$\Theta : M_{\mathcal{I}}^d \rightarrow \mathrm{Spec}(A) - V(\mathcal{I})$$

as in chapters 1, 2 and 3 of the first volume: F is a function field of characteristic $p > 0$, ∞ is a given place of F, A is the ring of functions in F which are regular outside ∞, $\mathcal{I}$ is a non-zero *proper* ideal of A and d is a positive integer. Let us also fix an algebraic closure $\overline{F}$ of F, a prime number $\ell \neq p$ and an algebraic closure $\overline{\mathbb{Q}}_\ell$ of the field $\mathbb{Q}_\ell$ of ℓ-adic numbers.

Grothendieck has defined the ℓ-adic cohomology with compact supports,

$$H_c^*(\overline{F} \otimes_F M_{\mathcal{I},\eta}^d, \overline{\mathbb{Q}}_\ell) = H_c^*(\overline{F} \otimes_F M_{\mathcal{I},\eta}^d, \mathbb{Q}_\ell) \otimes_{\mathbb{Q}_\ell} \overline{\mathbb{Q}}_\ell,$$

of the generic fiber $M_{\mathcal{I},\eta}^d$ of Θ. It is endowed with commuting actions of the Galois group $\mathrm{Gal}(\overline{F}/F)$ and the Hecke algebra $\mathcal{H}_{\mathcal{I}}^\infty$. The main purpose of this chapter is to describe the corresponding virtual module

$$\sum_n (-1)^n H_c^n(\overline{F} \otimes_F M_{\mathcal{I},\eta}^d, \overline{\mathbb{Q}}_\ell).$$

(12.1) Cohomological correspondences and the Deligne conjecture

Let k be a finite field of characteristic p and let $Y \xrightarrow{b} S$ be a morphism of separated schemes of finite type over k. In (1.6) we have introduced the $\mathbb{Q}$-algebra

$$\mathrm{Corr}(Y/S)$$

of correspondences of Y over S.

Let $\ell \neq p$ be a prime number. We consider the object

$$Rb_!\mathbb{Q}_\ell$$

of $D_c^b(S,\mathbb{Q}_\ell)$ (see [De] (1.1) or [Ek] for the definition of this derived category). We have a natural action of correspondences on it, i.e. a $\mathbb{Q}$-algebra homomorphism

$$\mathrm{Corr}(Y/S) \to \mathrm{End}(Rb_!\mathbb{Q}_\ell)^{\mathrm{opp}}.$$

If (Z,c) is a geometric correspondence of Y on S with $c=(c_1,c_2): Z \to Y\times_S Y$, its class $[Z,c]$ in $\mathrm{Corr}(Y/S)$ induces the endomorphism

$$Rb_!\mathbb{Q}_\ell \xrightarrow{Rb_!(\mathrm{ad})} Rb_!Rc_{1*}\mathbb{Q}_\ell = Rb_!c_{2*}\mathbb{Q}_\ell \xrightarrow{Rb_!(\mathrm{tr})} Rb_!\mathbb{Q}_\ell$$

where $\mathrm{ad}: \mathbb{Q}_\ell \to Rc_{1*}c_1^*\mathbb{Q}_\ell = Rc_{1*}\mathbb{Q}_\ell$ is the adjunction map and $\mathrm{tr}: c_{2*}\mathbb{Q}_\ell = c_{2!}\mathbb{Q}_\ell \to \mathbb{Q}_\ell$ is the trace map (let us recall that $c_1: Z\to Y$ is assumed to be proper and that $c_2 : Z \to Y$ is assumed to be finite and étale).

If $T \to S$ is a morphism between separated schemes of finite type over k we have an obvious base change morphism of $\mathbb{Q}$-algebras,

$$T\times_S(\cdot): \mathrm{Corr}(Y/S) \to \mathrm{Corr}(T\times_S Y/T),$$

and a base change isomorphism ([SGA4] (XVII, 5.2.6))

$$(Rb_!\mathbb{Q}_\ell)|T \xrightarrow{\sim} R(T\times_S b)_!\mathbb{Q}_\ell.$$

Moreover these morphisms are compatible: the diagram of $\mathbb{Q}$-algebra morphisms,

$$\begin{array}{ccc}
\mathrm{Corr}(Y/S) & \longrightarrow & \mathrm{End}(Rb_!\mathbb{Q}_\ell)^{\mathrm{opp}} \\
\big\downarrow & & \big\downarrow \\
\mathrm{Corr}(T\times_S Y/T) & \longrightarrow & \mathrm{End}(R(T\times_S b)_!\mathbb{Q}_\ell)^{\mathrm{opp}},
\end{array}$$

where the vertical arrows are base change morphisms, commutes.

Now let us assume that $S=\mathrm{Spec}(k)$ and let us fix an algebraic closure $\overline{k}$ of k. The fiber of

$$Rb_!\mathbb{Q}_\ell$$

at the geometric point $\mathrm{Spec}(\overline{k})$ of S is

$$R\Gamma_c(\overline{k}\otimes_k Y,\mathbb{Q}_\ell).$$

Therefore we have an action of the $\mathbb{Q}$-algebra $\mathrm{Corr}(Y/S)$ on $R\Gamma_c(\overline{k}\otimes_k Y,\mathbb{Q}_\ell)$ which commutes with the natural action of $\mathrm{Gal}(\overline{k}/k)$.

Let us assume that Y is smooth over $S = \mathrm{Spec}(k)$. Let (Z, c) be a geometric correspondence of Y over S and let Frob_k be the geometric Frobenius element in $\mathrm{Gal}(\overline{k}/k)$. In (3.1.2) we have introduced the Lefschetz numbers

$$\mathrm{Lef}_r(Z, c) \qquad (\forall r \in \mathbb{Z}_{>0}).$$

Let us recall that $\mathrm{Lef}_r(Z, c)$ is the number of fixed points of $\mathrm{Frob}_k^r \times (Z, c)$ in Z.

THEOREM (12.1.1) (Grothendieck). — *Let us assume moreover that Y is proper over k. Then for any positive integer r we have the trace formula*

$$\mathrm{tr}(\mathrm{Frob}_k^r \times [Z, c], R\Gamma(\overline{k} \otimes_k Y, \mathbb{Q}_\ell)) = \mathrm{Lef}_r\,(Z, c).$$

Proof: See [Kl] (1.3.6) or [SGA5] (III, 4.10). □

If we do not assume that Y is proper over k it may happen that there are local contributions to

$$\mathrm{tr}(\mathrm{Frob}_k^r \times [Z, c], R\Gamma(\overline{k} \otimes_k Y, \mathbb{Q}_\ell))$$

which come from the "fixed points at infinity of $\mathrm{Frob}_k^r \times (Z, c)$". Nevertheless we have

THEOREM (12.1.2) (Grothendieck). — *For any integer $r > 0$ the trace of* Frob_k^r *acting on* $R\Gamma_c(\overline{k} \otimes_k Y, \mathbb{Q}_\ell)$ *is equal to the number of fixed points of* Frob_k^r, *i.e. to the number of k_r-rational points in Y where k_r is the finite extension of degree r of k contained in $\overline{k}$.*

Proof: See [Gro 2] (5.1). □

COROLLARY (12.1.3). — *If $Z = Y$, $c_2 = id_Y$ and c_1 is an automorphism of finite order of Y over k, for any integer $r > 0$ we have*

$$\mathrm{tr}(\,\mathrm{Frob}_k^r \times [Z, c], R\Gamma(\overline{k} \otimes_k Y, \mathbb{Q}_\ell)) = \mathrm{Lef}_r\,(Z, c).$$

Proof: See [De–Lu] (Proof of (3.3)). □

These results have been the main motivations for the following conjecture.

CONJECTURE (12.1.4) (Deligne). — *If r is large enough with respect to (Z, c) the trace of* $\mathrm{Frob}_k^r \times [Z, c]$ *acting on* $R\Gamma_c(\overline{k} \otimes_k Y, \mathbb{Q}_\ell)$ *is equal to* $\mathrm{Lef}_r\,(Z, c)$. □

If Y is a (smooth) curve over k this conjecture follows from the Grothendieck–Verdier Lefschetz trace formula (see [Ve] and [SGA5] (III_B, 1.5)). More recently Zink has proved the conjecture for any Y of dimension 2 over k (see [Zi 2]) and the following general result has been obtained.

THEOREM (12.1.5) (Pink, Shpiz). — *Let us assume that there exits an open embedding of Y into a smooth and proper scheme $\overline{Y}$ over k and that the boundary $\overline{Y} - Y$ is a finite union of smooth divisors with at most normal crossing singularities. Then for any geometric correspondence (Z, c) of Y over k the Deligne conjecture holds.*

Proof: See [Pi 1] Theorem 7.2.2. □

Finally, using rigid analytic techniques Fujiwara has proved

THEOREM (12.1.6) (Fujiwara). — *The Deligne conjecture holds.*

Proof: See [Fu]. □

(12.2) Application of the Deligne conjecture to Drinfeld modular varieties

For each place o of F let us fix an algebraic closure $\overline{F}_o$ and an embedding $\overline{F} \hookrightarrow \overline{F}_o$ over the embedding $F \hookrightarrow F_o$. Then we get an embedding of $\mathrm{Gal}(\overline{F}_o/F_o)$ into $\mathrm{Gal}(\overline{F}/F)$ as a decomposition group at o. The residue field $\overline{\kappa(o)}$ of the integral closure $\overline{\mathcal{O}}_o$ of $\mathcal{O}_o$ in $\overline{F}_o$ is an algebraic closure of $\kappa(o)$ and we have a canonical epimorphism

$$\mathrm{Gal}(\overline{F}_o/F_o) \twoheadrightarrow \mathrm{Gal}(\overline{\kappa(o)}/\kappa(o))$$

with kernel the inertia subgroup of $\mathrm{Gal}(\overline{F}_o/F_o)$.

Let us recall that an **ℓ-adic representation** of $\mathrm{Gal}(\overline{F}/F)$ is a finite dimensional $\overline{\mathbb{Q}}_\ell$-vector space W equipped with an action of the group $\mathrm{Gal}(\overline{F}/F)$ satisfying the following properties:

(1) the action of $\mathrm{Gal}(\overline{F}/F)$ is defined over a finite extension E_λ of $\mathbb{Q}_\ell$ in $\overline{\mathbb{Q}}_\ell$,

(2) the action of $\mathrm{Gal}(\overline{F}/F)$ is continuous for the Krull topology on $\mathrm{Gal}(\overline{F}/F)$ and the λ-adic topology on the E_λ-structure of W,

(3) for almost every place o of F the inertia group of $\mathrm{Gal}(\overline{F}_o/F_o) \subset \mathrm{Gal}(\overline{F}/F)$ acts trivially on W.

By a **$(\mathrm{Gal}(\overline{F}/F) \times \mathcal{H}_\mathcal{I}^\infty)$-module** W we mean a finite dimensional $\overline{\mathbb{Q}}_\ell$-vector space W equipped with an action of the group $\mathrm{Gal}(\overline{F}/F)$ and an action of the $\mathbb{Q}$-algebra $\mathcal{H}_\mathcal{I}^\infty$ such that the actions of $\mathrm{Gal}(\overline{F}/F)$ and $\mathcal{H}_\mathcal{I}^\infty$ commute and are defined over a finite extension E_λ of $\mathbb{Q}_\ell$ in $\overline{\mathbb{Q}}_\ell$ and satisfying the above properties (2) and (3). In particular, if we forget the action of $\mathcal{H}_\mathcal{I}^\infty$ we have an ℓ-adic representation W of $\mathrm{Gal}(\overline{F}/F)$.

For any given place o (resp. any given finite set of places S) of F an ℓ-adic representation of $\mathrm{Gal}(\overline{F}/F)$ or a $(\mathrm{Gal}(\overline{F}/F)\times\mathcal{H}_{\mathcal{I}}^{\infty})$-module W is said to be **unramified at** o (resp. **outside** S) if the property (3) holds at o (resp. for every $o\notin S$).

The ℓ-adic representations of $\mathrm{Gal}(\overline{F}/F)$ (resp. the $(\mathrm{Gal}(\overline{F}/F)\times\mathcal{H}_{\mathcal{I}}^{\infty})$-modules) are the objects of an obvious abelian category. By a **virtual** ℓ-adic representation of $\mathrm{Gal}(\overline{F}/F)$ (resp. a **virtual** $(\mathrm{Gal}(\overline{F}/F)\times\mathcal{H}_{\mathcal{I}}^{\infty})$-module) we mean an element of the corresponding Grothendieck group. A virtual ℓ-adic representation of $\mathrm{Gal}(\overline{F}/F)$ or a virtual $(\mathrm{Gal}(\overline{F}/F)\times\mathcal{H}_{\mathcal{I}}^{\infty})$-module is said to be **unramified at** a place o (resp. **outside** a finite set of places S) of F if it can be written as a difference of two ℓ-adic representations of $\mathrm{Gal}(\overline{F}/F)$ or two $(\mathrm{Gal}(\overline{F}/F)\times\mathcal{H}_{\mathcal{I}}^{\infty})$-modules which are both unramified at o (resp. outside S).

We fix a system $\mathcal{M}$ of representatives of the isomorphism classes of finite dimensional irreducible $(\overline{\mathbb{Q}}_\ell\otimes_{\mathbb{Q}}\mathcal{H}_{\mathcal{I}}^{\infty})$-modules. Then we can decompose the semi-simplification W^{ss} of any $(\mathrm{Gal}(\overline{F}/F)\times\mathcal{H}_{\mathcal{I}}^{\infty})$-module W into its $\mathcal{H}_{\mathcal{I}}^{\infty}$-isotypic components

$$W^{\mathrm{ss}}=\bigoplus_{M\in\mathcal{M}}W(M)\otimes_{\overline{\mathbb{Q}}_\ell}M,$$

where $W(M)$ is a semi-simple ℓ-adic representation of $\mathrm{Gal}(\overline{F}/F)$ which is zero for all but finitely many $M\in\mathcal{M}$. Passing to the Grothendieck group we get a decomposition of any virtual $(\mathrm{Gal}(\overline{F}/F)\times\mathcal{H}_{\mathcal{I}}^{\infty})$-module W into its $\mathcal{H}_{\mathcal{I}}^{\infty}$-isotypic components

$$W=\sum_{M\in\mathcal{M}}W(M)\otimes_{\overline{\mathbb{Q}}_\ell}M,$$

where $W(M)$ is a virtual ℓ-adic representation of $\mathrm{Gal}(\overline{F}/F)$ which is zero for all but finitely many $M\in\mathcal{M}$.

A **geometric Frobenius element** at o is any element $\gamma\in\mathrm{Gal}(\overline{F}/F)$ which is conjugate to an element of $\mathrm{Gal}(\overline{F}_o/F_o)\subset\mathrm{Gal}(\overline{F}/F)$ which lifts the geometric Frobenius element $\mathrm{Frob}_o\in\mathrm{Gal}\big(\overline{\kappa(o)}/\kappa(o)\big)$. If W is a virtual ℓ-adic representation of $\mathrm{Gal}(\overline{F}/F)$ or a $(\mathrm{Gal}(\overline{F}/F)\times\mathcal{H}_{\mathcal{I}}^{\infty})$-module which is unramified at o the trace of γ or $\gamma\times h^{\infty}$ (for some $h^{\infty}\in\mathcal{H}_{\mathcal{I}}^{\infty}$) acting on W is independent of the Frobenius element γ at o and will be simply denoted by

$$\mathrm{tr}(\mathrm{Frob}_o,W)\ \text{or}\ \mathrm{tr}(\mathrm{Frob}_o\times h^{\infty},W).$$

LEMMA (12.2.1). — *Let W be a virtual ℓ-adic representation of* $\mathrm{Gal}(\overline{F}/F)$ *(resp.* $(\mathrm{Gal}(\overline{F}/F)\times\mathcal{H}_{\mathcal{I}}^{\infty})$*-module) which is unramified outside a finite set of places S of F and let us assume that, for almost every $o\notin S$, we have*

$$\mathrm{tr}(\mathrm{Frob}_o,W)=0$$

(resp. for almost every $o \notin S$, every $f^{\infty,o} \in \mathcal{H}_{\mathcal{I}}^{\infty,o}$ and every integer r which is large enough with respect to o and $f^{\infty,o}$ we have

$$\operatorname{tr}(\operatorname{Frob}_o^r \times f^{\infty,o}, W) = 0\,).$$

Then we have $W = 0$.

Proof : If W is a virtual ℓ-adic representation of $\operatorname{Gal}(\overline{F}/F)$ the assertion follows directly from the Čebotarev density theorem (see [Se 4] Chapter I, 2.2 and Appendix).

If W is a virtual $(\operatorname{Gal}(\overline{F}/F) \times \mathcal{H}_{\mathcal{I}}^{\infty})$-module let $\{M_1, \ldots, M_n\}$ be the finite subset of $M \in \mathcal{M}$ such that $W(M) \neq 0$. For each $j = 1, \ldots, n$ there exists $f_j^{\infty} \in \overline{\mathbb{Q}}_\ell \otimes_{\mathbb{Q}} \mathcal{H}_{\mathcal{I}}^{\infty}$ such that

$$\operatorname{tr}(f_j^{\infty}, M_i) = \delta_{ij} \qquad (\forall i = 1, \ldots, n).$$

Then for almost every place o we have

$$f_j^{\infty} = f_j^{\infty,o} 1_{K_o} \qquad (\forall j = 1, \ldots, n)$$

where $f_j^{\infty,o} \in \overline{\mathbb{Q}}_\ell \otimes_{\mathbb{Q}} \mathcal{H}_{\mathcal{I}}^{\infty,o}$. Therefore the hypothesis of the lemma is equivalent to the following one: for each fixed $i \in \{1, \ldots, n\}$ we have

$$\operatorname{tr}(\operatorname{Frob}_o^r, W(M_i)) = 0$$

for almost every $o \notin S$ and for every integer r which is large enough with respect to o.

But this hypothesis implies that, for each fixed $i \in \{1, \ldots, n\}$, we have

$$\operatorname{tr}(\operatorname{Frob}_o, W(M_i)) = 0$$

for almost every place $o \notin S$ and, as we have already seen, it follows that $W(M_i) = 0$, $\forall i = 1, \ldots, n$, i.e. that $n = 0$ and $W = 0$. □

For each integer n the ℓ-adic cohomology group $H_c^n(\overline{F} \otimes_F M_{\mathcal{I},\eta}^d, \overline{\mathbb{Q}}_\ell)$ is a $(\operatorname{Gal}(\overline{F}/F) \times \mathcal{H}_{\mathcal{I}}^{\infty})$-module (see [SGA4] (XVII, 5.3.6) and [SGA5] VI). As $M_{\mathcal{I},\eta}^d$ is a smooth affine scheme of pure relative dimension $d-1$ over F we have

$$H_c^n(\overline{F} \otimes_F M_{\mathcal{I},\eta}^d, \overline{\mathbb{Q}}_\ell) = (0)$$

if $n \notin [d-1, 2d-2]$ (see [SGA4] (XVIII, (3.2.6.2)) and (XIV, Corollary 3.2)).

In particular there exists a finite set of places of F outside which $H_c^n(\overline{F} \otimes_F M_{\mathcal{I},\eta}^d, \overline{\mathbb{Q}}_\ell)$ is unramified for every integer n. More precisely let o be a place of F and let $M_{\mathcal{I},\eta_o}^d$ be the fiber of the characteristic morphism Θ at the point

$\eta_o = \mathrm{Spec}(F_o)$ of X. If $o \notin \{\infty\} \cup V(\mathcal{I})$ we have a $\mathrm{Gal}(\overline{F}_o/F_o)$-equivariant isomorphism

$$(12.2.2.1) \qquad R\Gamma_c(\overline{F} \otimes_F M^d_{\mathcal{I},\eta}, \overline{\mathbb{Q}}_\ell) \xrightarrow{\sim} R\Gamma_c(\overline{F}_o \otimes_{F_o} M^d_{\mathcal{I},\eta_o}, \overline{\mathbb{Q}}_\ell)$$

($\mathrm{Gal}(\overline{F}_o/F_o) \hookrightarrow \mathrm{Gal}(\overline{F}/F)$) and a $\mathrm{Gal}(\overline{F}_o/F_o)$-equivariant specialization map

$$(12.2.2.2) \qquad R\Gamma_c(\overline{\kappa(o)} \otimes_{\kappa(o)} M^d_{\mathcal{I},o}, \overline{\mathbb{Q}}_\ell) \longrightarrow R\Gamma_c(\overline{F}_o \otimes_{F_o} M^d_{\mathcal{I},\eta_o}, \overline{\mathbb{Q}}_\ell)$$

($\mathrm{Gal}(\overline{F}_o/F_o) \twoheadrightarrow \mathrm{Gal}(\overline{\kappa(o)}/\kappa(o))$) (see [SGA4$\frac{1}{2}$] [Th. finitude] Appendix 2). Then, by [SGA4] (XVII, Theorems 5.3.6 and 5.2.6), there exists a finite set of places of F,

$$S^d_{\mathcal{I}} \supset \{\infty\} \cup V(\mathcal{I}),$$

such that the specialization map (12.2.2.2) is an isomorphism if $o \notin S^d_{\mathcal{I}}$. This general result does not give much information on $S^d_{\mathcal{I}}$: this set may be larger than $\{\infty\} \cup V(\mathcal{I})$ (nevertheless see remark (12.2.7) below).

Let us now consider the virtual $(\mathrm{Gal}(\overline{F}/F) \times \mathcal{H}^\infty_{\mathcal{I}})$-module

$$(12.2.3) \qquad W^d_{\mathcal{I}} = \sum_n (-1)^n H^n_c(\overline{F} \otimes_F M^d_{\mathcal{I},\eta}, \overline{\mathbb{Q}}_\ell).$$

It is unramified outside of $S^d_{\mathcal{I}}$ and, in order to determine $W^d_{\mathcal{I}}$, lemma (12.2.1) tells us that it is sufficient to determine the trace

$$\mathrm{tr}(\mathrm{Frob}^r_o \times f^{\infty,o}, W^d_{\mathcal{I}}) = 0$$

for almost every $o \notin S^d_{\mathcal{I}}$, every $f^{\infty,o} \in \mathcal{H}^{\infty,o}_{\mathcal{I}}$ and every integer r which is large enough with respect to o and $f^{\infty,o}$.

LEMMA (12.2.4). — *For every place $o \notin S^d_{\mathcal{I}}$, every $f^{\infty,o} \in \mathcal{H}^{\infty,o}_{\mathcal{I}}$ and every integer r we have*

$$\mathrm{tr}(\mathrm{Frob}^r_o \times f^{\infty,o}, W^d_{\mathcal{I}}) = \mathrm{tr}(\mathrm{Frob}^r_o \times f^{\infty,o}, R\Gamma_c(\overline{\kappa}(o) \otimes_{\kappa(o)} M^d_{\mathcal{I},o}, \overline{\mathbb{Q}}_\ell)).$$

Proof: It follows from (1.7.3) that the two maps (12.2.2.1) and (12.2.2.2) are $\mathcal{H}^{\infty,o}_{\mathcal{I}}$-equivariant for the actions given by (1.7.1), (1.7.2) and (1.7.4). □

To continue we need Pink's construction of toroidal compactifications of $M^d_{\mathcal{I}}$.

THEOREM (12.2.5) (Pink). — *If*

$$\dim_{\mathbb{F}_p}(A/\mathcal{I}) > \delta(A,d)$$

for some explicit non-negative integer $\delta(A,d)$ (depending only on A and d) there exists at least one embedding of $M_{\mathcal{I}}^d$ as a dense Zariski open subset of a scheme $M_{\mathcal{I},\Sigma}^d$ over $X-(\{\infty\}\cup V(\mathcal{I}))$ having the following properties:

(i) *the canonical morphism*

$$\Theta_\Sigma : M_{\mathcal{I},\Sigma}^d \to X-(\{\infty\}\cup V(\mathcal{I}))$$

is proper,

(ii) *$M_{\mathcal{I},\Sigma}^d$ is normal,*

(iii) *locally for the étale topology on $M_{\mathcal{I},\Sigma}^d$ the embedding $M_{\mathcal{I}}^d \hookrightarrow M_{\mathcal{I},\Sigma}^d$ is isomorphic to an embedding*

$$V/\Gamma \hookrightarrow U/\Gamma$$

where U is a smooth affine scheme over $X-(\{\infty\}\cup V(\mathcal{I}))$ equipped with an action of a finite group Γ and where V is the complement in U of a Γ-stable divisor in U with normal crossings relative to $X-(\{\infty\}\cup V(\mathcal{I}))$.

Proof: See [Pi 2]. □

COROLLARY (12.2.6) (Pink). — *Let us assume that*

$$\dim_{\mathbb{F}_p}(A/\mathcal{I}) > \delta(A,d).$$

Then, for every place $o \neq \infty$ of F with $o \notin V(\mathcal{I})$ and every Hecke correspondence

$$\begin{array}{ccccc} & & M_{\mathcal{I},o}^d(g^{\infty,o}) & & \\ & {\scriptstyle c_1}\swarrow & & \searrow{\scriptstyle c_2} & \\ M_{\mathcal{I},o}^d & & & & M_{\mathcal{I},o}^d \\ & \searrow & & \swarrow & \\ & & \mathrm{Spec}(\kappa(o)) & & \end{array}$$

(see (3.2.1)*), the Deligne conjecture* (12.1.4) *holds.*

Proof : As a variety with at most quotient singularities is as good as a smooth variety from the point of view of ℓ-adic cohomology the corollary is a consequence of the above theorem and of theorem (12.1.5). For more details see [Pi 2]. □

REMARK (12.2.7). — Another consequence of Pink's theorem is that we can take $S_{\mathcal{I}}^d = \{\infty\} \cup V(\mathcal{I})$ (compare to [SGA4$\frac{1}{2}$] [Th. finitude] Appendice 1.3.3). □

Summarizing the results of this section we get

THEOREM (12.2.8). — *Let us assume that*

$$\dim_{\mathbb{F}_p}(A/\mathcal{I}) > \delta(A, d).$$

Then, for every place $o \notin S_{\mathcal{I}}^d$ of F, every $f^{\infty,o}$ and every integer r which is large enough with respect to o and $f^{\infty,o}$, we have

$$\mathrm{tr}(\mathrm{Frob}_o^r \times f^{\infty,o}, W_{\mathcal{I}}^d) = \mathrm{Lef}_r(f^{\infty,o}).$$

Moreover these traces (for almost every $o \notin S_{\mathcal{I}}^d$) completely determine the virtual $(\mathrm{Gal}(\overline{F}/F) \times \mathcal{H}_{\mathcal{I}}^{\infty})$*-module $W_{\mathcal{I}}^d$.* □

(12.3) Application of the non-invariant Arthur trace formula to Drinfeld modular varieties

For every place $o \neq \infty$ of F with $o \notin V(\mathcal{I})$, every Hecke operator $f^{\infty,o} \in \mathcal{H}_{\mathcal{I}}^{\infty,o}$ and every positive integer r we have proved in (6.1) that the Lefschetz number $\mathrm{Lef}_r(f^{\infty,o})$ is equal to the elliptic part of the Arthur trace formula,

$$\sum_{\delta \in G(F)_{\natural,\mathrm{ell}}} \mathrm{vol}(F_\infty^\times G_\delta(F)\backslash G_\delta(\mathbb{A}), \frac{dg_\delta}{dz_\infty d\gamma_\delta}) \int_{G_\delta(\mathbb{A})\backslash G(\mathbb{A})} f_{\mathbb{A}}(g^{-1}\delta g)\frac{dg}{dg_\delta},$$

for some function $f_{\mathbb{A}} \in C_c^\infty(F_\infty^\times\backslash G(\mathbb{A}))$ and some Haar measure $dg = dg_\infty dg^{\infty,o} dg_o$ on $G(\mathbb{A})$. More precisely we have

$$f_{\mathbb{A}} = f_\infty f^{\infty,o} f_o,$$

where f_∞ is the very cuspidal Euler–Poincaré function (corresponding to dg_∞) which is introduced in (5.1) and where f_o is the Drinfeld function of level r introduced in (4.2.5), and the Haar measures dg_∞, $dg^{\infty,o}$ and dg_o are normalized by

$$\mathrm{vol}(K_\infty, dg_\infty) = \mathrm{vol}(K_{\mathcal{I}}^{\infty,o}, dg^{\infty,o}) = \mathrm{vol}(K_o, dg_o) = 1.$$

Moreover, after having replaced $f_{\mathbb{A}}$ by f with

$$f(g) = \int_{K_\infty} f_{\mathbb{A}}(k_\infty^{-1} g k_\infty) dk_\infty \qquad (\forall g \in G(\mathbb{A}))$$

(see (10.2.1)), we have proved in (10.7.10) that, for r large enough with respect to o and $f^{\infty,o}$, that elliptic part is in fact equal to the full geometric side $J_{\mathrm{geom}}(f_{\mathbb{A}}) = J_{\mathrm{geom}}(f)$ of the Arthur trace formula.

Therefore, by the trace formula (see (11.2.8)) we get

$$\mathrm{Lef}_r(f^{\infty,o}) = J^T_{\mathrm{spec}}(f_{\mathbb{A}})$$

for any $T \in \mathfrak{a}_\emptyset^+$ which is regular enough and for any r which is large enough with respect to o and $f^{\infty,o}$.

Finally, applying theorem (11.8.2) we obtain

THEOREM (12.3.1). — *For each positive integer r which is large enough with respect to o and $f^{\infty,o}$ the Lefschetz number* $\mathrm{Lef}_r(f^{\infty,o})$ *is equal to*

$$\sum_{I\subset\Delta} \sum_{\substack{\chi\in\mathcal{X}_I \\ \chi_\infty = 1_{I,\infty}}} \sum_{\pi} (-1)^{|\Delta - I|} m^2(\pi) \epsilon_\infty(\pi_\infty) \operatorname{tr} i^{G(\mathbb{A}^{\infty,o})}_{M_I(\mathbb{A}^{\infty,o}),P_I(\mathbb{A}^{\infty,o})}(\pi^{\infty,o}(\rho_I), f^{\infty,o})$$

$$\times p^{r\deg(o)(d_1-1)/2}(z_1(\pi_{1,o})^r + \cdots + z_{d_1}(\pi_{1,o})^r)$$

where π runs through the subset of $\Pi^2_{I,\chi}$ defined by the conditions

(1) $\pi_{j,\infty} \cong \mathrm{st}_{d_j,\infty}$ *or* $\pi_{j,\infty} \cong 1_{d_j,\infty}$ *for every* $j = 1, \ldots, s$, *if* $(d_1, \ldots, d_s)$ *is the partition of d corresponding to I and if* $\pi = \pi_1 \otimes \cdots \otimes \pi_s$,

(2) π_o *is spherical.* □

It will be convenient to extend slightly the definitions of ϵ_∞ and of the Hecke eigenvalues.

For any $d' = 1, \ldots, d$ and any unitary admissible irreducible representation π'_∞ of $F_\infty^\times \backslash GL_{d'}(F_\infty)$ we set

$$\epsilon_\infty(\pi'_\infty) = \begin{cases} (-1)^{d'-1} & \text{if } \pi'_\infty \cong \mathrm{st}_{d',\infty}, \\ 1 & \text{if } \pi'_\infty \cong 1_{d',\infty}, \\ 0 & \text{otherwise.} \end{cases}$$

For any unitary admissible irreducible representation π_∞ of $Z_I(F_\infty^\times)\backslash M_I(F_\infty)$ we set

$$\epsilon_\infty(\pi_\infty) = \prod_{j=1}^{s} \epsilon_\infty(\pi_{j,\infty})$$

if $(d_1, \ldots, d_s)$ is the partition of d corresponding to I and if $\pi = \pi_1 \otimes \cdots \otimes \pi_s$.

For any $d' = 1, \dots, d$ and any admissible irreducible representation π'_o of $GL_{d'}(F_o)$ we denote by $z_1(\pi'_o), \dots, z_{d'}(\pi'_o)$ the Hecke eigenvalues of π'_o if π'_o is spherical and we set $(z_1(\pi'_o), \dots, z_{d'}(\pi'_o)) = (0, \dots, 0)$ otherwise.

For each $d' = 1, \dots d$, we have the Grothendieck group $K_a(GL_{d'}(\mathbb{A}^\infty))$ of admissible representations of $GL_{d'}(\mathbb{A}^\infty)$ and a class map

$$[\cdot] : \mathrm{ob}\,\mathrm{Rep}_a(GL_{d'}(\mathbb{A}^\infty)) \to K_a(GL_{d'}(\mathbb{A}^\infty))$$

(see (D.11)). Obviously the trace map

$$\mathrm{ob}\,\mathrm{Rep}_a(GL_{d'}(\mathbb{A}^\infty)) \times C_c^\infty(GL_{d'}(\mathbb{A}^\infty)) \to \mathbb{C},\ (\pi, f) \mapsto \mathrm{tr}\,\pi(f),$$

induces a bilinear map

$$K_a(GL_{d'}(\mathbb{A}^\infty)) \times C_c^\infty(GL_{d'}(\mathbb{A}^\infty)) \to \mathbb{C}$$

which we will also denote by tr.

For each $d' = 1, \dots, d-1$ let us set

$$r_{d'}^\infty = \sum_{I' \subset \Delta'} (-1)^{|\Delta' - I'|+1} \sum_{\substack{\chi' \in \mathcal{X}_{I'} \\ \chi'_\infty = 1_{I',\infty}}} \sum_{\pi' \in \Pi^2_{I',\chi'}} \epsilon_\infty(\pi'_\infty) \times \left[i^{GL_{d'}(\mathbb{A}^\infty)}_{M_{I'}(\mathbb{A}^\infty), P_{I'}(\mathbb{A}^\infty)}(\pi'^\infty(\rho_{I'}))\right] \tag{12.3.2}$$

in $K_a(GL_{d'}(\mathbb{A}^\infty))$ where $\Delta' = \{1, \dots, d'-1\}$, $(d'_1, \dots, d'_{s'})$ is the partition of d' corresponding to I', ...

Let J be the subset of Δ corresponding to the partition $(d', d-d')$ of d. The additive functor $i^{GL_d(\mathbb{A}^\infty)}_{M_J(\mathbb{A}^\infty), P_J(\mathbb{A}^\infty)}$ induces a group homomorphism

$$K_a\big(GL_{d'}(\mathbb{A}^\infty) \times GL_{d-d'}(\mathbb{A}^\infty)\big) \to K_a(GL_d(\mathbb{A}^\infty))$$

and we will simply denote by

$$(r', r'') \mapsto r' \times r''$$

the composition of this group homomorphism with the map

$$\otimes : K_a(GL_{d'}(\mathbb{A}^\infty)) \times K_a(GL_{d-d'}(\mathbb{A}^\infty)) \to K_a\big(GL_{d'}(\mathbb{A}^\infty) \times GL_{d-d'}(\mathbb{A}^\infty)\big).$$

Combining theorems (12.2.8) and (12.3.1) we obtain

COROLLARY (12.3.3). — *Let us assume that*

$$\dim_{\mathbb{F}_p}(A/\mathcal{I}) > \delta(A,d).$$

Then, for every place $o \notin S_{\mathcal{I}}^d$ of F, every $f^{\infty,o} \in C_c^\infty(G(\mathbb{A}^{\infty,o})/\!\!/K_{\mathcal{I}}^{\infty,o})$ and every integer r, the geometric expression

$$\mathrm{tr}(\mathrm{Frob}_o^r \times f^{\infty,o}, W_{\mathcal{I}}^d)$$

and the spectral expression

$$\sum_{\substack{\chi\in\mathcal{X}_{GL_d}\\ \chi_\infty=1_\infty}} \sum_{\pi\in\Pi^2_{GL_d,\chi}} \epsilon_\infty(\pi_\infty)\,\mathrm{tr}\,\pi^\infty(f^{\infty,o}1_{K_o})p^{r\deg(o)(d-1)/2}(z_1(\pi_o)^r$$

$$+\cdots+z_d(\pi_o)^r)+\sum_{d'=1}^{d-1}\sum_{\substack{\chi''\in\mathcal{X}_{GL_{d-d'}}\\ \chi''_\infty=1_\infty}}\sum_{\pi''\in\Pi^2_{GL_{d-d'},\chi''}}\epsilon_\infty(\pi''_\infty)\,\mathrm{tr}\Big(r_{d'}^\infty\big(\frac{d'-d}{2}\big)$$

$$\times\Big[\pi''^\infty\big(\frac{d'}{2}\big)\Big]\Big)(f^{\infty,o}1_{K_o})\times p^{r\deg(o)(d-d'-1)/2}(z_1(\pi''_o)^r+\cdots+z_{d-d'}(\pi''_o)^r)$$

are both rational numbers and are equal.

Proof: For r large enough with respect to o and $f^{\infty,o}$ the corollary follows directly from (12.2.8) and (12.3.1) (we have

$$\rho_I=\Big(\frac{(d-d_1)d_1}{2},\frac{(d-2d_1-d_2)d_2}{2},\ldots,\frac{(d_s-d)d_s}{2}\Big),$$

i.e.

$$\rho_I=\Big(\frac{d'(d-d')}{2},\rho_{I'}+\frac{d'-d}{2}(d_2,\ldots,d_s)\Big)$$

if $d'=d-d_1$ and $d_{I'}=(d'_1,\ldots,d'_{s'})$ with $d'_{j'}=d_{j'+1}$ and $s'=s-1$).

Now in $\overline{\mathbb{Q}}_\ell$ (resp. in $\mathbb{C}$) there exist distinct non-zero numbers $\alpha_1,\ldots,\alpha_M$ (resp. $\beta_1,\ldots,\beta_N$) and non-zero numbers $c_1,\ldots,c_M$ (resp. $d_1,\ldots,d_N$) such that the above geometric expression (resp. the above spectral expression) is equal to

$$\sum_{m=1}^M c_m\alpha_m^r$$

(resp.

$$\sum_{n=1}^N d_n\beta_n^r).$$

We have already proved that these two sums are equal rational numbers for every integer $r \geq r_0$, where r_0 is a sufficiently large integer. Let us set

$$R(T) = \sum_{r \geq r_0} \sum_{m=1}^{M} c_m \alpha_m^r T^r = \sum_{m=1}^{M} \frac{c_m \alpha_m^{r_0} T^{r_0}}{1 - \alpha_m T}$$

and

$$S(T) = \sum_{r \geq r_0} \sum_{n=1}^{N} d_n \beta_n^r T^r = \sum_{n=1}^{N} \frac{d_n \beta_n^{r_0} T^{r_0}}{1 - \beta_n T}.$$

We have $R(T) \in \mathbb{Q}[[T]] \cap \overline{\mathbb{Q}}_\ell(T) \subset \mathbb{Q}(T)$, $S(T) \in \mathbb{Q}[[T]] \cap \mathbb{C}(T) \subset \mathbb{Q}(T)$ and $R(T) = S(T)$. Then, by considering the divisors of poles of these rational fractions and the corresponding residues, we see that $M = N$ and that, up to a permutation of the indices, we have $\alpha_m = \beta_m$ and $c_m = d_m$ for every $m = 1, \ldots, M$. Moreover we have

$$\prod_{m=1}^{M} (1 - \alpha_m T) \in \mathbb{Q}[T]$$

and

$$c_m = \lim_{T \to 1/\alpha_m} (1 - \alpha_m T) R(T) \in \mathbb{Q}(\alpha_m) \qquad (\forall m = 1, \ldots, M),$$

so that

$$\sum_{m=1}^{M} c_m \alpha_m^r \in \overline{\mathbb{Q}} \qquad (\forall r \in \mathbb{Z})$$

where $\overline{\mathbb{Q}}$ is the algebraic closure of $\mathbb{Q}$ in $\overline{\mathbb{Q}}_\ell$ and that

$$\sigma\Big(\sum_{m=1}^{M} c_m \alpha_m^r\Big) = \sum_{m=1}^{M} c_m \alpha_m^r \qquad (\forall r \in \mathbb{Z},\ \forall \sigma \in \mathrm{Gal}(\overline{\mathbb{Q}}/\mathbb{Q})).$$

This conclude the proof of the corollary. □

(12.4) Langlands correspondence

Let us fix a cuspidal automorphic irreducible representation $\pi \cong \pi_\infty \otimes \pi^\infty$ of $F_\infty^\times \backslash GL_d(\mathbb{A})$ such that π_∞ is isomorphic to the Steinberg representation of $F_\infty^\times \backslash GL_d(F_\infty)$. *From now on we will use Shalika's result*

$$m_{\mathrm{cusp}}(\pi) = 1$$

(*π occurs with multiplicity one in the space of cuspidal automorphic forms for GL_d*; see [Sha 1] Theorem 5.9).

Let $\mathbb{Q}(\pi)$ be the subfield of $\mathbb{C}$ generated by the complex numbers $\mathrm{tr}\,\pi(f)$ for $f \in C_c^\infty(F_\infty^\times \backslash GL_d(\mathbb{A}))$. Then $\mathbb{Q}(\pi)$ is a number field, i.e. a finite extension of $\mathbb{Q}$, and π admits a rational structure over $\mathbb{Q}(\pi)$ (see (D.10.1), (D.10.2) and (9.2.16)).

THEOREM (12.4.1). — (i) (Ramanujan–Petersson conjecture) *For almost every place o such that π_o is spherical we have*

$$p^{r\deg(o)(d-1)/2}(z_1(\pi_o)^r + \cdots + z_d(\pi_o)^r) \in \mathbb{Q}(\pi) \qquad (\forall r \in \mathbb{Z})$$

and

$$|z_i(\pi_o)| = 1 \qquad (\forall i = 1, \ldots, d)$$

where $(z_1(\pi_o), \ldots, z_d(\pi_o))$ are the Hecke eigenvalues of π_o.

(ii) (Langlands conjecture) *For each prime λ of $\mathbb{Q}(\pi)$ dividing ℓ there exists a unique (up to isomorphism) irreducible continuous representation $\sigma_\lambda(\pi)$ of* $\mathrm{Gal}(\overline{F}/F)$ *on a $\mathbb{Q}(\pi)_\lambda$-vector space of dimension d which has the following property: for almost every place o such that π_o is spherical the inertia group of* $\mathrm{Gal}(\overline{F}_o/F_o) \subset \mathrm{Gal}(\overline{F}/F)$ *acts trivially in $\sigma_\lambda(\pi)$ and we have*

$$\mathrm{tr}\big(\sigma_\lambda(\pi)(\mathrm{Frob}_o^r)\big) = p^{r\deg(o)(d-1)/2}(z_1(\pi_o)^r + \cdots + z_d(\pi_o)^r) \qquad (\forall r \in \mathbb{Z})$$

in $\mathbb{Q}(\pi)_\lambda$.

Before proving the theorem let us recall on the one hand some results of Jacquet and Shalika and on the other hand some results of Grothendieck and Deligne.

LEMMA (12.4.2) (Jacquet and Shalika). — *For any positive integer d', any irreducible cuspidal automorphic representation π' of $F_\infty^\times\backslash GL_{d'}(\mathbb{A})$ and any place o of F such that π'_o is spherical we have the estimates*

$$1/p^{\deg(o)/2} < |z_i(\pi'_o)| < p^{\deg(o)/2} \qquad (\forall i = 1, \ldots, d')$$

where $(z_1(\pi'_o), \ldots, z_{d'}(\pi'_o))$ are the Hecke eigenvalues of π'_o.

Proof: See [Ja–Sh 1] (2.5) and [Sha 1] Corollary of Theorem 5.9. □

In particular we get the following.

LEMMA (12.4.3). — *For any $I \subsetneq \Delta$, any $\chi \in \mathcal{X}_I$ with $\chi_\infty = 1_{I,\infty}$ and any $\pi_I \in \Pi^2_{I,\chi}$ with $\epsilon_\infty(\pi_{I,\infty}) \neq 0$ the irreducible admissible representation π^∞ which is fixed in this section does not occur as a subquotient of the induced representation $i_{M_I(\mathbb{A}^\infty),P_I(\mathbb{A}^\infty)}^{GL_d(\mathbb{A}^\infty)}(\pi_I^\infty(\rho_I))$.*

Proof: Let us assume that π^∞ does occur as a subquotient of the induced representation $i_{M_I(\mathbb{A}^\infty),P_I(\mathbb{A}^\infty)}^{GL_d(\mathbb{A}^\infty)}(\pi_I^\infty(\rho_I))$ and let us search for a contradiction.

Let o be a place of F such that π_o is spherical. Then, if $\pi_I \cong \pi_1 \otimes \cdots \otimes \pi_s$ as a representation of $M_I(\mathbb{A}) \cong GL_{d_1}(\mathbb{A}) \times \cdots \times GL_{d_s}(\mathbb{A})$, $\pi_{j,o}$ is also spherical for

$j = 1, \ldots, s$ and, if $(z_{d_1+\cdots+d_{j-1}+1}(\pi_{I,o}), \ldots, z_{d_1+\cdots+d_j}(\pi_{I,o}))$ are the Hecke eigenvalues of $\pi_{j,o}$, the Hecke eigenvalues of π_o are equal to

$$\begin{gathered}(z_1(\pi_{I,o})p^{\deg(o)(d-d_1)/2}, \ldots, z_{d_1}(\pi_{I,o})p^{\deg(o)(d-d_1)/2}, \ldots \\ \ldots, z_{d_1+\cdots+d_{s-1}+1}(\pi_{I,o})p^{\deg(o)(d_s-d)/2}, \ldots, z_d(\pi_{I,o})p^{\deg(o)(d_s-d)/2})\end{gathered}$$

(up to some reordering). But, for each $j = 1, \ldots, s$, either π_j is cuspidal and then the Hecke eigenvalues $(z_{d_1+\cdots+d_{j-1}+1}(\pi_{I,o}), \ldots, z_{d_1+\cdots+d_j}(\pi_{I,o}))$ satisfy the estimates of lemma (12.4.3), or π_j is one dimensional and then the absolute values of these Hecke eigenvalues are $(p^{\deg(o)(1-d_j)/2}, p^{\deg(o)(3-d_j)/2}, \ldots, p^{\deg(o)(d_j-1)/2})$ (up to some reordering). Therefore the Hecke eigenvalues of π_o cannot all satisfy the estimates of lemma (12.4.3) and we have obtained a contradiction. □

Let d' and d'' be two positive integers and let π' and π'' be cuspidal automorphic unitary irreducible representations of $F_\infty^\times \backslash GL_{d'}(\mathbb{A})$ and $F_\infty^\times \backslash GL_{d''}(\mathbb{A})$ respectively. Let $\widetilde{\pi}''$ be the contragredient representation of π''. Let S be a finite set of places of F containing ∞ such that π'_o and π''_o are spherical for every place $o \notin S$. Then for each place $o \notin S$ we may consider the Euler factor

$$L_o(\pi' \times \widetilde{\pi}'', s) = \prod_{\substack{1 \le j' \le d' \\ 1 \le j'' \le d''}} \left(1 - \frac{z_{j'}(\pi'_o)}{z_{j''}(\pi''_o)} p^{-\deg(o)s}\right)^{-1}$$

where $(z_1(\pi'_o), \ldots, z_{d'}(\pi'_o))$ and $(z_1(\pi''_o), \ldots, z_{d''}(\pi''_o))$ are the Hecke eigenvalues of π'_o and π''_o respectively and we may form the Euler product

$$L^S(\pi' \times \widetilde{\pi}'', s) = \prod_{o \notin S} L_o(\pi' \times \widetilde{\pi}'', s).$$

This Euler product is absolutely convergent for any $s \in \mathbb{C}$ with $\mathrm{Re}(s) > 1$ and defines a holomorphic function on the open half plane $\{s \in \mathbb{C} \mid \mathrm{Re}(s) > 1\}$ (see [Ja–Sh 1] (5.3) Theorem).

PROPOSITION (12.4.4) (Jacquet and Shalika). — (i) *If $d' \neq d''$, $L^S(\pi' \times \widetilde{\pi}'', s)$ admits a continuous extension to the closed half plane $\{s \in \mathbb{C} \mid \mathrm{Re}(s) \geq 1\}$.*

(ii) *If $d' = d''$ and if Σ is the set of $s \in \mathbb{C}$ on the line $\mathrm{Re}(s) = 1$ such that $\pi'(s-1)$ is isomorphic to π'', $L^S(\pi' \times \widetilde{\pi}'', s)$ admits a continuous extension to $\{s \in \mathbb{C} - \Sigma \mid \mathrm{Re}(s) \geq 1\}$ and, for each $s_0 \in \Sigma$, the limit*

$$\lim_{\substack{s \to s_0 \\ \mathrm{Re}(s) \geq 1 \\ s \notin \Sigma}} (s - s_0) L^S(\pi' \times \widetilde{\pi}'', s)$$

exists, is finite and is non-zero.

Proof: See [Ja–Sh 2] (3.3) Proposition and (3.6) Proposition. □

REMARK (12.4.5). — In fact $L^S(\pi' \times \widetilde{\pi}'', s)$ extends to a meromorphic function on the whole complex plane and the above proposition describes the poles of this meromorphic function on the line $\mathrm{Re}(s) = 1$. But we will not need this fact. □

Let W' and W'' be two irreducible ℓ-adic representations of $\mathrm{Gal}(\overline{F}/F)$. Let $W''^{\vee}$ be the dual representation of W''. Let $S \neq \emptyset$ be a finite set of places outside which W' and W'' are both unramified. We assume that W' and W'' are **ι-pure of weight** m outside S for some fixed $m \in \mathbb{R}$ and some fixed embedding $\iota : \overline{\mathbb{Q}}_\ell \hookrightarrow \mathbb{C}$, i.e. that, for each place $o \notin S$ and each eigenvalue α of Frob_o acting on W' or W'', we have $|\iota(\alpha)| = p^{\deg(o)m/2}$ (see [De] (1.2)). Then for each place $o \notin S$ we may consider the Euler factor

$$L_o(W' \otimes W''^{\vee}, s) = \det\bigl(1 - p^{-\deg(o)s}\,\mathrm{Frob}_o, \mathbb{C} \otimes_{\iota, \overline{\mathbb{Q}}_\ell} (W' \otimes_{\overline{\mathbb{Q}}_\ell} W''^{\vee})\bigr)^{-1}$$

and we may form the Euler product

$$L^S(W' \otimes W''^{\vee}, s) = \prod_{o \notin S} L_o(W' \otimes W''^{\vee}, s).$$

This Euler product is absolutely convergent for any $s \in \mathbb{C}$ with $\mathrm{Re}(s) > 1$ and defines a holomorphic function on the open half plane $\{s \in \mathbb{C} \mid \mathrm{Re}(s) > 1\}$.

Let $\mathbb{F}_q$ ($q = p^f$) be the field of constants in F, i.e. the algebraic closure of $\mathbb{F}_p$ in F, and let $\overline{\mathbb{F}}_p$ be the algebraic closure of $\mathbb{F}_q$ (or $\mathbb{F}_p$) in $\overline{F}$. The Galois group $\mathrm{Gal}(\overline{\mathbb{F}}_pF/F)$ is canonically isomorphic to

$$\mathrm{Gal}(\overline{\mathbb{F}}_p/\mathbb{F}_q) = \mathrm{Frob}_q^{\widehat{\mathbb{Z}}},$$

where $\mathrm{Frob}_q = \mathrm{Frob}_p^f$ is the geometric Frobenius element (Frob_p is the inverse of the p-th power automorphism of $\overline{\mathbb{F}}_p$). For any ℓ-adic representation W of $\mathrm{Gal}(\overline{F}/F)$ and any complex number $s \in \log_q(\iota(\overline{\mathbb{Z}}_\ell^\times))$, where $\overline{\mathbb{Z}}_\ell$ is the integral closure of $\mathbb{Z}_\ell$ in $\overline{\mathbb{Q}}_\ell$, the **Tate twist** $W(s)$ of W is defined as follows. Let us write $q^{-s} = \iota(\sigma)$ for some $\sigma \in \overline{\mathbb{Z}}_\ell^\times$. Then we may form the ℓ-adic character

$$\chi_\sigma : \mathrm{Gal}(\overline{F}/F) \twoheadrightarrow \mathrm{Gal}(\overline{\mathbb{F}}_pF/F) \to \overline{\mathbb{Z}}_\ell^\times, \qquad \mathrm{Frob}_q \mapsto \sigma,$$

and $W(s)$ is the tensor product of W by χ_σ.

THEOREM (12.4.6). — (i) (Grothendieck) *The function $L^S(W' \otimes W''^{\vee}, s)$ extends to a meromorphic function on the whole complex plane. This meromorphic function of s is in fact a rational function of q^{-s} of the form*

$$\frac{P_1(q^{-s})}{P_2(q^{-s})}$$

with $P_1(T), P_2(T) \in \mathbb{C}[T]$ and $P_1(0) = P_2(0) = 1$. The zeros of $P_2(q^{-s})$ are all simple and are exactly the complex numbers $s \in \log_q(\iota(\overline{\mathbb{Z}}_\ell^\times))$ such that W'' is isomorphic to $W'(1-s)$.

(ii) (Deligne) *The polynomials $P_1(T)$ and $P_2(T)$ do not have any root in common.*

Proof : The Euler product $L^S(W' \otimes W''^\vee, s)$ admits the following cohomological interpretation. We may view W' and W'' as smooth ℓ-adic sheaves $\mathcal{W}'$ and $\mathcal{W}''$ on the affine curve $X - S$ over $\mathbb{F}_q$ and we may consider the ℓ-adic cohomology groups

$$H_c^n = H_c^n(\overline{\mathbb{F}}_p \otimes_{\mathbb{F}_q} (X-S), \mathcal{W}' \otimes \mathcal{W}''^\vee) \qquad (n \in \mathbb{Z}).$$

Then Grothendieck has proved (see [Gro 2] and [SGA4$\frac{1}{2}$] [Rapport] 3) that

(i) $H_c^n = 0$ for every integer $n \notin \{1, 2\}$,

(ii) H_c^1 and H_c^2 are finite dimensional $\overline{\mathbb{Q}}_\ell$-vector spaces with a natural action of $\mathrm{Gal}(\overline{\mathbb{F}}_p/\mathbb{F}_q)$ and in particular with an action of the geometric Frobenius element Frob_q,

(iii) we have

$$H_c^2 = \mathrm{Hom}_{\overline{\mathbb{Q}}_\ell}(W', W'')^{\mathrm{Gal}(\overline{F}/\overline{\mathbb{F}}_p F)}(-1),$$

(iv) we have

$$L^S(W' \otimes W''^\vee, s) = \frac{P_1(q^{-s})}{P_2(q^{-s})}$$

where

$$P_n(T) = \det(1 - T\,\mathrm{Frob}_q, \mathbb{C} \otimes_{\iota, \overline{\mathbb{Q}}_\ell} H_c^n) \qquad (n = 1, 2).$$

This gives part (i) of the theorem.

Moreover Deligne has proved that the inverse roots α of $P_1(T)$ satisfy

$$|\alpha| \leq q^{1/2}$$

(see [De] Corollary (3.3.4) and (3.3.10)). This gives part (ii) of the theorem.

□

Proof of theorem (12.4.1) : To choose a place λ of $\mathbb{Q}(\pi)$ dividing ℓ is the same as to choose an embedding λ of $\mathbb{Q}(\pi)$ into our fixed algebraic closure $\overline{\mathbb{Q}}_\ell$ of $\mathbb{Q}_\ell$. Let us fix such an embedding and let us fix an embedding ι of $\overline{\mathbb{Q}}_\ell$ into $\mathbb{C}$ such that $\iota \circ \lambda : \mathbb{Q}(\pi) \hookrightarrow \mathbb{C}$ is the inclusion. Let us also fix a non-zero proper ideal $\mathcal{I}$ of A such that $(\pi^\infty)^{K_\mathcal{I}^\infty} \neq (0)$ and such that $\dim_{\mathbb{F}_p}(A/\mathcal{I}) > \delta(A, d)$ (see (12.2.5)).

For each irreducible admissible representation Π^∞ of $GL_d(\mathbb{A}^\infty)$ having non-zero fixed vectors under $K_\mathcal{I}^\infty$ let us simply denote by $H_c^n(\Pi^\infty)$, $n = d-1,\ldots,2d-2$ (resp. $W(\Pi^\infty)$), the $(\mathbb{C}\otimes_{\iota,\overline{\mathbb{Q}}_\ell}\mathcal{H}_\mathcal{I}^\infty)$-isotypic component of $(\Pi^\infty)^{K_\mathcal{I}^\infty}$ in the semi-simple (resp. virtual) $(\mathrm{Gal}(\overline{F}/F)\times(\mathbb{C}\otimes_{\iota,\overline{\mathbb{Q}}_\ell}\mathcal{H}_\mathcal{I}^\infty))$-module $\mathbb{C}\otimes_{\iota,\overline{\mathbb{Q}}_\ell} H_c^n(\overline{F}\otimes_F M_{\mathcal{I},\eta}^d,\overline{\mathbb{Q}}_\ell)^{\mathrm{ss}}$ (resp. $\mathbb{C}\otimes_{\iota,\overline{\mathbb{Q}}_\ell} W_\mathcal{I}^d$), so that

$$W(\Pi^\infty)=\sum_{n=d-1}^{2d-2}(-1)^n H_c^n(\Pi^\infty).$$

By the finite dimensionality of the ℓ-adic cohomology there are only finitely many isomorphism classes of irreducible admissible representations Π^∞ of $GL_d(\mathbb{A}^\infty)$ having non-zero fixed vectors under $K_\mathcal{I}^\infty$ and such that $H_c^n(\Pi^\infty)\neq(0)$ for at least one integer n. Let $\{\Pi_1^\infty,\ldots,\Pi_{N'}^\infty\}$ be a system of representatives of these isomorphism classes.

There are also only finitely many isomorphism classes of irreducible admissible representations Π^∞ of $GL_d(\mathbb{A}^\infty)$ having non-zero fixed vectors under $K_\mathcal{I}^\infty$ such that Π^∞ occurs as a subquotient of $i_{M_I(\mathbb{A}^\infty),P_I(\mathbb{A}^\infty)}^{GL_d(\mathbb{A}^\infty)}(\pi_I^\infty(\rho_I))$ for some $I\subset\Delta$, some $\chi\in\mathcal{X}_I$ with $\chi_\infty=1_{I,\infty}$ and some $\pi_I\in\Pi_{I,\chi}^2$ with $\epsilon_\infty(\pi_{I,\infty})\neq 0$ (see (9.2)). Let $\{\Pi_{N'+1}^\infty,\ldots,\Pi_{N'+N''}^\infty\}$ be a system of representatives of these isomorphism classes.

Let us fix $f^\infty\in\mathbb{C}\otimes\mathcal{H}_\mathcal{I}^\infty$ such that $\mathrm{tr}\,\pi^\infty(f^\infty)=1$ and such that $\mathrm{tr}\,\Pi_n^\infty(f^\infty)=0$ for any Π_n^∞, $n=1,\ldots,N'+N''$, which is not isomorphic to π^∞. We can find a finite set $S\supset S_\mathcal{I}^d$ of places of F depending on λ, ι, $\mathcal{I}$ and f^∞ such that, for every place $o\notin S$, we may split f^∞ into a product $f^{\infty,o}1_{K_o}$ and the trace formula (12.3.3) holds. By our choice of f^∞ and by lemma (12.4.3) this trace formula reduces to

$$(*)\quad \mathrm{tr}(\mathrm{Frob}_o^r,W(\pi^\infty))=(-1)^{d-1}p^{r\deg(o)(d-1)/2}(z_1(\pi_o)^r+\cdots+z_d(\pi_o)^r).$$

Now we know, thanks to Deligne (see [De] Theorem (3.3.1)), that for each integer $n=d-1,\ldots,2d-2$, the ℓ-adic representation $H_c^n(\overline{F}\otimes_F M_{\mathcal{I},\eta}^d,\overline{\mathbb{Q}}_\ell)$ of $\mathrm{Gal}(\overline{F}/F)$ admits a canonical weight filtration

$$(0)=\mathcal{W}^{-1}H_c^n\subset\mathcal{W}^0H_c^n\subset\cdots\subset\mathcal{W}^nH_c^n=H_c^n(\overline{F}\otimes_F M_{\mathcal{I},\eta}^d,\overline{\mathbb{Q}}_\ell)$$

such that, for each $m=0,\ldots,n$, $\mathrm{gr}_\mathcal{W}^m H_c^n$ is pure of weight m (outside $S_\mathcal{I}^d$). In particular, for each place $o\notin S_\mathcal{I}^d$ of F and each complex number α which occurs as an eigenvalue of Frob_o in $\mathbb{C}\otimes_{\iota,\overline{\mathbb{Q}}_\ell}\mathrm{gr}_\mathcal{W}^m H_c^n$ we have $|\alpha|=p^{\deg(o)m/2}$. Moreover this filtration is invariant under the action of $\mathcal{H}_\mathcal{I}^\infty$. It follows that the virtual $\mathrm{Gal}(\overline{F}/F)$-module $W(\pi^\infty)$ admits the decomposition

$$W(\pi^\infty)=\sum_{m=0}^{2d-2}W^m(\pi^\infty),$$

where we have set

$$W^m(\pi^\infty) \stackrel{\text{dfn}}{=\!=} \sum_{n=d-1}^{2d-2} (-1)^n (\mathbb{C} \otimes_{\iota,\overline{\mathbb{Q}}_\ell} \mathrm{gr}_{\mathcal{W}}^m H_c^n)^{\mathrm{ss}}(\pi^\infty)$$

for each integer $m \in \{0, \ldots, 2d-2\}$.

Now it follows from the trace formula $(*)$ and lemma (12.4.2) that, for each place $o \notin S$ of F and each integer r, we have

$$\mathrm{tr}(\mathrm{Frob}_o^r, W^m(\pi^\infty)) = 0 \qquad (\forall m \neq d-1)$$

and

$$(**)\ \mathrm{tr}(\mathrm{Frob}_o^r, (-1)^{d-1} W^{d-1}(\pi^\infty)) = p^{r\deg(o)(d-1)/2}(z_1(\pi_o)^r + \cdots + z_d(\pi_o)^r)$$

with

$$|z_j(\pi_o)| = 1 \qquad (\forall j = 1, \ldots, d).$$

Therefore we have

$$W^m(\pi^\infty) = 0 \qquad (\forall m \neq d-1)$$

(see lemma (12.2.1)) and part (i) of the theorem is proved.

Next we will show that

$$\sum_{n=d-1}^{2d-2} (-1)^{n-d+1} (\mathrm{gr}_{\mathcal{W}}^{d-1} H_c^n)^{\mathrm{ss}}(\pi^\infty)$$

is the virtual ℓ-adic representation associated to a true irreducible ℓ-adic representation of dimension d of $\mathrm{Gal}(\overline{F}/F)$. We may write $(-1)^{d-1} W^{d-1}(\pi^\infty)$ as a linear combination of irreducible ℓ-adic representations of $\mathrm{Gal}(\overline{F}/F)$ which are all pure of weight $d-1$ outside $S_{\mathcal{I}}^d$:

$$(-1)^{d-1} W^{d-1}(\pi^\infty) = \sum_{\alpha=1}^{a} m_\alpha W_\alpha$$

with $m_\alpha \in \mathbb{Z}$, $m_\alpha \neq 0$ and W_α not isomorphic to W_β for every $\alpha \neq \beta$. Then it follows from the trace formula $(**)$ that

$$L^S(\pi \times \widetilde{\pi}, s) = \prod_{\alpha,\beta=1}^{a} L^S(W_\alpha \otimes W_\beta^\vee, s)^{m_\alpha m_\beta}$$

for all s in the half plane $\{s \in \mathbb{C} \mid \mathrm{Re}(s) > 1\}$. Therefore the function $L^S(\pi \times \widetilde{\pi}, s)$ extends to a meromorphic function on the whole complex plane and the order of its pole at $s = 1$ is on the one hand equal to 1 (see (12.4.4)(ii)) and on the other hand equal to $\sum_{\alpha=1}^{a} m_\alpha^2$ (see (12.4.6)). Therefore we have

$$\sum_{\alpha=1}^{a} m_\alpha^2 = 1.$$

We deduce from this equality that $a = 1$ and $m_1 = \pm 1$. But it follows from the trace formula (**) that the virtual dimension of $(-1)^{d-1}W^{d-1}(\pi^\infty)$ is equal to d, so that

$$d = m_1 \dim W_1$$

and $m_1 = 1$ (with $a = 1$) as required.

Finally, the virtual ℓ-adic representation $\sum_{n=d-1}^{2d-2}(-1)^{n-d+1}(\mathrm{gr}_{\mathcal{W}}^{d-1}H_c^n)^{\mathrm{ss}}$ (π^∞) is defined over $\mathbb{Q}(\pi)_\lambda \subset \overline{\mathbb{Q}}_\ell$. Indeed the weight filtration of H_c^* is defined over $\mathbb{Q}_\ell$ and the representation π, and therefore the representation π^∞, admits a rational structure over $\mathbb{Q}(\pi)$. Let $\sigma_\lambda(\pi)$ be the "the" irreducible representation of dimension d of $\mathrm{Gal}(\overline{F}/F)$ over $\mathbb{Q}(\pi)_\lambda$ corresponding to this virtual representation. Then it is clear that $\sigma_\lambda(\pi)$ has all the required properties and part (ii) of the theorem is proved. □

LEMMA (12.4.7) (Deligne). — *Let σ be an irreducible ℓ-adic representation of* $\mathrm{Gal}(\overline{F}/F)$. *Let S_σ be the (finite) set of places of o of F such that σ is ramified at o. Let S be any finite set of places of F containing S_σ. We fix an embedding $\iota : \overline{\mathbb{Q}}_\ell \hookrightarrow \mathbb{C}$ and we assume that σ is ι-pure of some weight $m \in \mathbb{R}$ outside S. Then σ is ι-pure of weight m outside S_σ*

Proof: See [De] (1.8). □

LEMMA (12.4.8) (Henniart). — *Let σ be an irreducible ℓ-adic representation of dimension d of* $\mathrm{Gal}(\overline{F}/F)$ *and let π be a cuspidal automorphic irreducible representation of $F_\infty^\times \backslash GL_d(\mathbb{A})$. Let S_σ be the (finite) set of places of o of F such that σ is ramified at o and let S_π be the (finite) set of places o of F such that π_o is not spherical. Let S be a finite set of places of F containing $S_\sigma \cup S_\pi$. We fix an embedding $\iota : \overline{\mathbb{Q}}_\ell \hookrightarrow \mathbb{C}$ and we assume that σ is ι-pure of some weight $m \in \mathbb{R}$ outside S. Then, if the local L-factors*

$$L_o(\sigma, s) = \det(1 - p^{-\deg(o)s}\,\mathrm{Frob}_o, \mathbb{C} \otimes_{\iota, \overline{\mathbb{Q}}_\ell} \sigma)$$

and

$$L_o(\pi, s - \frac{m}{2}) = \prod_{j=1}^{d}\big(1 - z_j(\pi_o)p^{-\deg(o)(s-\frac{m}{2})}\big)^{-1}$$

coincide for every place $o \notin S$, we have $S_\sigma = S_\pi$ and these local L-factors coincide for every place $o \notin S_\sigma = S_\pi$

Proof : See [He] Corollary of Proposition 4.5 and the remark following Theorem 4.1 (the purity assumption replaces the unitarity property of Σ in the proof of loc. cit.; by lemma (12.4.7) we have it for every place $o \notin S_\sigma$). □

Applying these lemmas we immediately deduce the following corollary of theorem (12.4.1).

COROLLARY (12.4.9). — *Properties* (i) *and* (ii) *of theorem (*12.4.1*) hold for every place o of F such that π_o is spherical.* □

(12.5) The virtual module $W_{\mathcal{I}}^d$

Let $\overline{\mathbb{Q}}$ be the algebraic closure of $\mathbb{Q}$ in $\mathbb{C}$. For each $d' = 1, \ldots, d$, each $\chi' \in \mathcal{X}_{GL_{d'}}$ such that $\chi'_\infty = 1_\infty$ and each $\pi' \in \Pi_{GL_{d'},\chi',\mathrm{cusp}}$ we have already seen that the representation π' admits a rational structure over the number field $\mathbb{Q}(\pi') \subset \overline{\mathbb{Q}}$ generated by the numbers $\mathrm{tr}\,\pi'(f)$, $f \in \mathcal{C}_c^\infty(F_\infty^\times \backslash GL_{d'}(\mathbb{A}))$. Similarly, for each $d' = 1, \ldots, d$, each $\chi' \in \mathcal{X}_{GL_{d'}}$ such that $\chi'_\infty = 1_\infty$ and each $\pi' \in \Pi_{GL_{d'},\chi'}^2$ which is one dimensional the representation π' admits a rational structure over the number field $\mathbb{Q}(\pi') \subset \overline{\mathbb{Q}}$ generated by the values of the character π' of $GL_{d'}(\mathbb{A})$.

Let us fix an embedding of $\overline{\mathbb{Q}}$ into $\overline{\mathbb{Q}}_\ell$. For each number field $E \subset \overline{\mathbb{Q}}$ we will denote by E_λ the closure of E in $\overline{\mathbb{Q}}_\ell$ for the ℓ-adic topology.

For each $d' = 1, \ldots, d$, each $\chi' \in \mathcal{X}_{GL_{d'}}$ such that $\chi'_\infty = 1_\infty$ and each $\pi' \in \Pi_{GL_{d'},\chi'}^2$ such that π'_∞ is isomorphic either to the Steinberg representation or to the trivial character of $F_\infty^\times \backslash GL_{d'}(F_\infty)$ we can associate a semi-simple continuous representation $\sigma_\lambda(\pi')$ of $\mathrm{Gal}(\overline{F}/F)$ on a $\mathbb{Q}(\pi')_\lambda$-vector space of dimension d'. If π'_∞ is the Steinberg representation, so that π' is cuspidal, $\sigma_\lambda(\pi')$ is the irreducible representation associated by the Langlands correspondence to π' (see (12.4.1)(ii)). If π'_∞ is the trivial character, so that $\pi' = \xi' \circ \det$ for some character $\xi' : F_\infty^\times \backslash \mathbb{A}^\times \to \mathbb{Q}(\pi')^\times$ with $(\xi')^{d'} = \chi'$, $\sigma_\lambda(\pi')$ is the semi-simple representation

$$\sigma_\lambda(\xi') \oplus \sigma_\lambda(\xi')(-1) \oplus \cdots \oplus \sigma_\lambda(\xi')(1-d')$$

where

$$\sigma_\lambda(\xi') : \mathrm{Gal}(\overline{F}/F) \to \mathbb{Q}(\pi')^\times$$

is the continuous character corresponding to ξ' under the abelian class field theory, or equivalently under the Langlands correspondence for GL_1 (see (12.4.1)(ii) for $d = 1$). In both cases, for every place o of F such that π'_o is spherical the representation $\sigma_\lambda(\pi')$ is unramified at o and

$$\mathrm{tr}\big(\sigma_\lambda(\pi')(\mathrm{Frob}_o^r)\big) = p^{r\deg(o)(d'-1)/2}(z_1(\pi'_o)^r + \cdots + z_{d'}(\pi'_o)^r) \qquad (\forall r \in \mathbb{Z}).$$

THEOREM (12.5.1). — *For any non-zero proper ideal $\mathcal{I}$ of A the virtual $(\mathrm{Gal}(\overline{F}/F) \times \mathcal{H}_{\mathcal{I}}^{\infty})$-module $W_{\mathcal{I}}^{d}$ is equal to*

$$\sum_{\substack{\chi\in\mathcal{X}_{GL_d}\\ \chi_\infty=1_\infty}} \sum_{\pi\in\Pi^2_{GL_d,\chi}} \epsilon_\infty(\pi_\infty)\big(\overline{\mathbb{Q}}_\ell\otimes_{\mathbb{Q}(\pi)_\lambda}\sigma_\lambda(\pi)\big)\otimes_{\overline{\mathbb{Q}}}\big[(\pi^\infty)^{K_{\mathcal{I}}^\infty}\big]$$

$$+\sum_{d'=1}^{d-1} \sum_{\substack{\chi''\in\mathcal{X}_{GL_{d-d'}}\\ \chi''_\infty=1_\infty}} \sum_{\pi''\in\Pi^2_{GL_{d-d'},\chi''}} \epsilon_\infty(\pi''_\infty)\big(\overline{\mathbb{Q}}_\ell\otimes_{\mathbb{Q}(\pi'')_\lambda}\sigma_\lambda(\pi'')\big)$$

$$\otimes_{\overline{\mathbb{Q}}}\Big((r_{d'}^\infty)^{K_{\mathcal{I}}^\infty}\Big(\frac{d'-d}{2}\Big)\times\Big[(\pi''^\infty)^{K_{d',\mathcal{I}}^\infty}\Big(\frac{d'}{2}\Big)\Big]\Big)$$

where we have set

$$(r_{d'}^\infty)^{K_{d',\mathcal{I}}^\infty}=\sum_{I'\subset\Delta'}(-1)^{|\Delta'-I'|+1}\sum_{\substack{\chi'\in\mathcal{X}_{I'}\\ \chi'_\infty=1_{I',\infty}}}\sum_{\pi'\in\Pi^2_{I',\chi'}}\epsilon_\infty(\pi'_\infty)$$
$$\times\Big[i^{GL_{d'}(\mathbb{A}^\infty)}_{M_{I'}(\mathbb{A}^\infty),P_{I'}(\mathbb{A}^\infty)}\big(\pi'^\infty(\rho_{I'})\big)^{K_{d',\mathcal{I}}^\infty}\Big].$$

Proof : If $\dim_{\mathbb{F}_p}(A/\mathcal{I})>\delta(A,d)$ (see (12.2.5)) this is a direct consequence of corollary (12.3.3) and lemma (12.2.1). In general we may find a power $\mathcal{I}'$ of the ideal $\mathcal{I}$ such that the condition $\dim_{\mathbb{F}_p}(A/\mathcal{I}')>\delta(A,d)$ is satisfied. Then we have

$$H_c^n(\overline{F}\otimes_F M_{\mathcal{I},\eta}^d,\overline{\mathbb{Q}}_\ell)=\big(H_c^n(\overline{F}\otimes_F M_{\mathcal{I}',\eta}^d,\overline{\mathbb{Q}}_\ell)\big)^{K_{\mathcal{I}}^d/K_{\mathcal{I}'}^d}\qquad(\forall n\in\mathbb{Z})$$

and the theorem for $\mathcal{I}$ follows from the theorem for $\mathcal{I}'$ which is already proved. □

(12.6) Comments and references

The idea of using the Deligne conjecture together with a simple form of the trace formula is due to Kazhdan and was applied by Flicker and Kazhdan to obtain similar results to ours, but under the "stronger" hypothesis that π_∞ is a supercuspidal representation of $GL_d(F_\infty)$ (see [Fl] and [Fl–Ka]).

The method for deducing the Langlands correspondence from the equality of traces (12.3.3) is standard. It was explained to me by Kottwitz. If we forget the action of $\mathrm{Gal}(\overline{F}/F)$ in theorem (12.5.1) the formula is similar to a formula proved by Franke in the number field case (see [Fr]).

The Ramanujan–Petersson conjecture and the Langlands conjecture are expected to hold for an arbitrary (without any ramification hypothesis) unitary cuspidal automorphic representation (let us recall that the main hypothesis of theorem (12.4.1) is that π_∞ is the Steinberg representation of $GL_d(F_\infty)$). In the case $d = 2$ these conjectures have been proved in full generality by Drinfeld using the modulus spaces of rank 2 shtukas (see [Dr 3] and [Dr 4]). Recently, using the modulus spaces of rank d shtukas Lafforgue has obtained a result which is very near the full Ramanujan–Petersson conjecture (see [Laf 2]).

13

Intersection cohomology of Drinfeld modular varieties: conjectures

(13.0) Introduction

As in chapter 12 let us fix a Drinfeld modular variety $M_{\mathcal{I}}^d$ with characteristic morphism

$$\Theta : M_{\mathcal{I}}^d \to \mathrm{Spec}(A) - V(\mathcal{I})$$

($\mathcal{I}$ is a non-zero *proper* ideal of A) and let us fix an algebraic closure $\overline{F}$ of F, a prime number $\ell \neq p$ and an algebraic closure $\overline{\mathbb{Q}}_\ell$ of the field $\mathbb{Q}_\ell$ of ℓ-adic numbers.

In his paper "The L^2-Lefschetz numbers of Hecke operators" [Ar 14] Arthur has computed the Lefschetz number of a Hecke operator acting on the intersection cohomology of the Satake compactification of a Shimura variety. This Lefschetz number appears to be the geometric side of a trace formula, the spectral side of which is quite simple. It is not difficult to guess the analog for function fields of that formula. Assuming that everything goes similarly as in the number field case we may also guess what should be the ordinary cohomology sheaves of the intersection complex on the Satake compactification of our Drinfeld modular variety $M_{\mathcal{I}}^d$. This is the purpose of this speculative final chapter.

(13.1) A conjectural trace formula

For every Hecke operator $f^\infty \in \mathcal{H}_{\mathcal{I}}^\infty$ let us consider the (formal) trace

$$\text{(13.1.1)} \qquad \mathrm{ftr}\, R^2_{GL_d,1_\infty,\mathrm{disc}}(f_{\mathbb{A}}) = \sum_{\substack{\chi\in\mathcal{X}_{GL_d} \\ \chi_\infty = 1_\infty}} \mathrm{ftr}\, R^2_{GL_d,\chi,\mathrm{disc}}(f_{\mathbb{A}})$$

for the adelic function $f_{\mathbb{A}} = f_\infty f^\infty$ and some Haar measure $dg_{\mathbb{A}} = dg_\infty dg^\infty$, where f_∞ is the very cuspidal Euler–Poincaré function (corresponding to dg_∞) which is introduced in (5.1) and where the Haar measure dg^∞ gives the volume 1 to the compact open subgroup K_I^∞.

As in (9.6.2) we see that this formal trace is equal to

$$\sum_{\pi\in\Pi^2_{GL_d}} \epsilon_\infty(\pi_\infty)\operatorname{tr}\pi^\infty(f^\infty) \tag{13.1.2}$$

where

$$\pi\in\Pi^2_{GL_d} \overset{\text{dfn}}{=\!=} \coprod_{\substack{\chi\in\mathcal{X}_{GL_d}\\ \chi_\infty=1_\infty}} \Pi^2_{GL_d,\chi}$$

and where $\epsilon_\infty(\pi_\infty) = (-1)^{d-1}$ if π_∞ is isomorphic to the Steinberg representation st_∞ of $F_\infty^\times\backslash GL_d(F_\infty)$, $\epsilon_\infty(\pi_\infty) = 1$ if π_∞ is the trivial character of $F_\infty^\times\backslash GL_d(F_\infty)$ and $\epsilon_\infty(\pi_\infty) = 0$ otherwise. Here we are using Shalika's result: $m_{\mathrm{cusp}}(\pi) = 1$ for each $\chi\in\mathcal{X}_{GL_d}$ and each $\pi\in\Pi_{GL_d,\chi,\mathrm{cusp}}$.

By analogy with [Ar 14] Theorem 6.1 we can easily write in the function field case a trace formula, the spectral side of which is the above (formal) trace.

The first main term of this trace formula is a suitable modification of the character of st_∞. To define it we will need the following notations.

- We denote by $\mathcal{P}$ the set of parabolic subgroups of $G = GL_d$ (over F) containing the maximal torus T of diagonal matrices. Each $P\in\mathcal{P}$ admits a unique Levi decomposition $P = M_P N_P$ such that $M_P \supset T$. We denote by $\mathcal{L}$ the set of all the Levi subgroups M_P, $P\in\mathcal{P}$. Each $M\in\mathcal{L}$ may be identified with $GL_{d_1}\times\cdots\times GL_{d_s}$ for some partition $(d_1,\ldots,d_s)$ of d which is well-defined up to some reordering. In particular the positive integer $s_M = s$ is uniquely determined by M.

- For each $M\in\mathcal{L}$ and each closed element $\mu\in M(F_\infty)$ we denote by $M(F_\infty)_\mu$ the centralizer of μ in $M(F_\infty)$ and by

$$D^M(\mu) = \det\Bigl(1-\mathrm{Ad}(\mu), \frac{\mathrm{Lie}\, M(F_\infty)}{\mathrm{Lie}\, M(F_\infty)_\mu}\Bigr)$$

its Weyl discriminant. If $M\subset L$ are two Levi subgroups in $\mathcal{L}$ we set

$$D^L_M(\mu) = \det\Bigl(1-\mathrm{Ad}(\mu), \frac{\mathrm{Lie}\, L(F_\infty)}{\mathrm{Lie}\, M(F_\infty)}\Bigr)$$

for any closed $\mu\in M(F_\infty)$.

• We denote by $G(F_\infty)_{\mathrm{reg}}$ the set of $\gamma \in G(F_\infty)$ such that γ is closed and $D^G(\gamma) \neq 0$. We have

$$G(F_\infty)_{\mathrm{reg}} = \bigcup_{M \in \mathcal{L}} \{g^{-1}\mu g \mid g \in G(F_\infty),\ \mu \in M(F_\infty)_{\mathrm{ell}} \text{ with } D^G(\mu) \neq 0\}$$

where $M(F_\infty)_{\mathrm{ell}}$ is the set of elliptic elements in $M(F_\infty)$ (see (4.3)).

• For any $M \in \mathcal{L}$ and any elliptic element $\mu = (\gamma_1, \ldots, \gamma_s) \in M(F_\infty) \cong GL_{d_1}(F_\infty) \times \cdots \times GL_{d_s}(F_\infty)$ we denote by

$$\epsilon_{M,\infty}(\mu) = \prod_{j=1}^{s} \epsilon_{GL_{d_j},\infty}(\gamma_j)$$

the Kottwitz sign of μ. Let us recall from (5.1.3)(i) that, for any $d' = 1, \ldots, d$ and any elliptic element $\gamma' \in GL_{d'}(F_\infty)$, the Kottwitz sign $\epsilon_{GL_{d'},\infty}(\gamma')$ is $(-1)^{d''-1}$ where $d'' = d'/[F_\infty[\gamma'] : F_\infty]$.

• For any $M \in \mathcal{L}$ and any elliptic element $\mu = (\gamma_1, \ldots, \gamma_s) \in M(F_\infty) \cong GL_{d_1}(F_\infty) \times \cdots \times GL_{d_s}(F_\infty)$ we set

$$\overline{M(F_\infty)_\mu} = \prod_{j=1}^{s} D_j'^{\times}$$

where D_j' is "the" central division algebra over $F_\infty[\gamma_j]$ of invariant $1/d_j'$ with $d_j' = d/[F_\infty[\gamma_j] : F_\infty]$ (see (5.1.3)(i)). For any Haar measure $dm_{\infty,\mu}$ on $M(F_\infty)_\mu$ we denote by $d\overline{m}_{\infty,\mu}$ its transfer to the inner twist $\overline{M(F_\infty)_\mu}$ of $M(F_\infty)_\mu$ as an $(F_\infty[\gamma_1] \times \cdots \times F_\infty[\gamma_s])$-group scheme (compare with (5.1.3)(i)).

Let Θ_∞^G be the locally constant function on $G(F_\infty)_{\mathrm{reg}}$ which coincides with the character of st_∞ on this open subset of $G(F_\infty)$. Harish-Chandra has given an explicit formula for this function (see [H-C 2] 15). If $M \in \mathcal{L}$ and if $\mu \in M(F_\infty)_{\mathrm{ell}}$ with $D^G(\mu) \neq 0$ we have

$$\Theta_\infty^G(g^{-1}\mu g) = \Theta_\infty^G(\mu) \qquad (\forall g \in G(F_\infty))$$

and

$$\Theta_\infty^G(\mu) = \sum_{\substack{P \in \mathcal{P} \\ M_P \supset M}} (-1)^{d - s_{M_P}} \delta_{P(F_\infty)}^{1/2}(\mu) |D_{M_P}^G(\mu)|_\infty^{-1/2}.$$

Now, following Arthur, for any $M \in \mathcal{L}$ and any $\mu \in M(F_\infty)_{\mathrm{ell}}$ we set

$$\widetilde{\Theta}_{\infty,M}^G(\mu) = \sum_{\substack{P \in \mathcal{P} \\ M_P \supset M}} (-1)^{s_{M_P}-1} \delta_{P(F_\infty)}^{1/2}(\mu) |D_M^{M_P}(\mu)|_\infty^{1/2},$$

and, if we have moreover fixed a Haar measure $dm_{\infty,\mu}$ on $M(F_\infty)_\mu$, we set

$$\Phi^G_{\infty,M}(\mu) = \frac{\epsilon_{M,\infty}(\mu)\widetilde{\Theta}^G_{\infty,M}(\mu)}{\mathrm{vol}(Z_M(F_\infty)\backslash\overline{M(F_\infty)}_\mu, dz_{M,\infty}\backslash d\overline{m}_{\infty,\mu})} \tag{13.1.3}$$

where $dz_{M,\infty}$ is the Haar measure on $Z_M(F_\infty)$ which is normalized by

$$\mathrm{vol}(Z_M(\mathcal{O}_\infty), dz_{M,\infty}) = 1.$$

It will also be convenient to set

$$\Phi^G_{\infty,M}(\mu) = 0$$

for any non-elliptic element μ in $M(F_\infty)$.

The second main term of the trace formula is an orbital integral. Let $I \subset \Delta$ and let $\mu \in M_I(F)$. We denote by $M_{I,\mu}$ the centralizer of μ in M_I. We fix a Haar measure $dm^\infty_{I,\mu}$ on $M_{I,\mu}(\mathbb{A}^\infty)$ and we normalize the Haar measures dk^∞, dn^∞_I and dm^∞_I on $K^\infty = G(\mathcal{O}^\infty)$, $N_I(\mathbb{A}^\infty)$ and $M_I(\mathbb{A}^\infty)$ by

$$\mathrm{vol}(K^\infty_{\mathcal{I}}, dk^\infty) = 1,$$

$$\mathrm{vol}(K^\infty_{\mathcal{I}} \cap N_I(\mathbb{A}^\infty), dn^\infty_I) = 1$$

and

$$\mathrm{vol}(K^\infty_{\mathcal{I}} \cap M_I(\mathbb{A}^\infty), dm^\infty_I) = 1$$

respectively, so that

$$dg^\infty = \frac{|G(A/\mathcal{I})|}{|P_I(A/\mathcal{I})|} dm^\infty_I dn^\infty_I dk^\infty.$$

Then, following Arthur, we set

$$(f^\infty)_{M_I}(\mu) = \frac{|G(A/\mathcal{I})|}{|P_I(A/\mathcal{I})|} O^{M_I}_\mu((f^\infty)^{P_I}, dm^\infty_{I,\mu}) \tag{13.1.4}$$

where

$$(f^\infty)^{P_I}(m^\infty_I) = \delta^{1/2}_{P_I(\mathbb{A}^\infty)}(m^\infty_I) \int_{K^\infty}\int_{N_I(\mathbb{A}^\infty)} f^\infty((k^\infty)^{-1} m^\infty_I n^\infty_I k^\infty) dn^\infty_I dk^\infty$$

for every $m^\infty_I \in M_I(\mathbb{A}^\infty)$ and where

$$O^{M_I}_\mu((f^\infty)^{P_I}, dm^\infty_{I,\mu}) = \int_{M_{I,\mu}(\mathbb{A}^\infty)\backslash M_I(\mathbb{A}^\infty)} (f^\infty)^{P_I}((m^\infty_I)^{-1}\mu m^\infty_I) \frac{dm^\infty_I}{dm^\infty_{I,\mu}}.$$

DEFINITION (13.1.5). — *The L^2-Lefschetz number* $\mathrm{Lef}^2(f^\infty)$ *is the expression*

$$\sum_{I\subset\Delta}\frac{(-1)^{|\Delta-I|}}{(|\Delta-I|+1)!}\sum_{\mu\in M_I(F)_{\natural,\mathrm{ell}}}\mathrm{vol}\Big(Z_{M_I}(F_\infty)M_{I,\mu}(F)\backslash M_{I,\mu}(\mathbb{A}),$$
$$\frac{dm_{I,\mu}}{dz_{M_I,\infty}d\mu_{I,\mu}}\Big)\times\Phi^G_{\infty,M_I}(\mu)(f^\infty)_{M_I}(\mu)$$

where $M_I(F)_{\natural,\mathrm{ell}}$ is a system of representatives of the elliptic conjugacy classes in $M_I(F)$ and where, for each $I\subset\Delta$ and each $\mu\in M_I(F)_{\natural,\mathrm{ell}}$, we have fixed a Haar measure $dm_{I,\mu}=dm_{I,\mu,\infty}dm^\infty_{I,\mu}$ on $M_{I,\mu}(\mathbb{A})=M_I(F_\infty)_\mu M_{I,\mu}(\mathbb{A}^\infty)$ and we have denoted by $d\mu_{I,\mu}$ the counting measure on $M_{I,\mu}(F)$.

Obviously the product

$$\mathrm{vol}\Big(Z_{M_I}(F_\infty)M_{I,\mu}(F)\backslash M_{I,\mu}(\mathbb{A}),\frac{dm_{I,\mu}}{dz_{M_I,\infty}d\mu_{I,\mu}}\Big)\Phi^G_{\infty,M_I}(\mu)(f^\infty)_{M_I}(\mu)$$

does not depend on the choice of the Haar measure $dm_{I,\mu}$ and, for any given I, there are only finitely many $\mu\in M_I(F)_{\natural,\mathrm{ell}}$ which are elliptic in $M_I(F_\infty)$ and such that $(f^\infty)_{M_I}(\mu)\neq 0$.

CONJECTURE (13.1.6). — *With the above notations and hypotheses we have the trace formula*

$$\mathrm{Lef}^2(f^\infty)=\mathrm{ftr}\,R_{G,1_\infty,\mathrm{disc}}(f_{\mathbb{A}}).$$

(13.2) Some particular cases of conjecture (13.1.6)

Firstly, for each $d'=1,\ldots,d$ let us set

$$V_{d'}(\mathcal{I})=(-1)^{d'-1}\frac{\mathrm{vol}\big(F^\times_\infty G'(F)\backslash G'(\mathbb{A}),dz_\infty d\gamma'\backslash dg'_\infty dg'^\infty\big)}{\mathrm{vol}\big(F^\times_\infty\backslash D'^\times_\infty,dz_\infty\backslash d\overline{g}'_\infty\big)}\tag{13.2.1}$$

where $G'=GL_{d'}$, dg'_∞ is any Haar measure on $G'(F_\infty)$, dg'^∞ is the Haar measure on $G'(\mathbb{A}^\infty)$ which is normalized by $\mathrm{vol}(K^\infty_{d',\mathcal{I}},dg'^\infty)=1$ with

$$K^\infty_{d',\mathcal{I}}=\mathrm{Ker}\big(G'(\mathcal{O}^\infty)\twoheadrightarrow G'(\mathcal{O}^\infty/\mathcal{O}^\infty\mathcal{I})\big),$$

D'_∞ is the central division algebra over F_∞ of invariant $1/d'$ and $d\overline{g}'_\infty$ is the transfer of the Haar measure dg'_∞ to the inner form $D'^\times_\infty$ of $G'(F_\infty)$. For each $I\subset\Delta$ let us set

$$V_I(\mathcal{I})=\prod_{j=1}^{s}V_{d_j}(\mathcal{I})\tag{13.2.2}$$

where $(d_1,\ldots,d_s)$ is the partition of d corresponding to I. Then we have

LEMMA (13.2.3). — *Conjecture (13.1.6) for $f^\infty = 1_{K_{\mathcal{I}}^\infty}$ is equivalent to the statement*

$$\sum_{I\subset\Delta}\frac{|G(A/\mathcal{I})|}{|P_I(A/\mathcal{I})|}V_I(\mathcal{I})=\sum_{\pi\in\Pi_G^2}\epsilon_\infty(\pi_\infty)\dim(\pi^\infty)^{K_{\mathcal{I}}^\infty}.$$

We will see in section (13.4) that the above statement follows from results of Harder and Stuhler.

Proof: Let $I\subset\Delta$ and let $\mu\in M_I(F)_{\natural,\mathrm{ell}}$ be such that $\Phi_{\infty,M_I}(\mu)(1_{K_{\mathcal{I}}^\infty})_{M_I}(\mu)\neq 0$. Due to our normalization of Haar measures we have

$$(1_{K_{\mathcal{I}}^\infty})^{P_I}=1_{M_I(\mathbb{A}^\infty)\cap K_{\mathcal{I}}^\infty}.$$

Moreover, if F' is a finite field extension of F such that there exists only one place ∞' over ∞ and if $\gamma\in F'$ satisfies

$$x'(\gamma)=0$$

for every place $x'\neq\infty'$ of F' and

$$x'(\gamma-1)>0$$

for at least one place $x'\neq\infty'$ of F' we necessarily have $\gamma=1$. Therefore μ must be the unit element in $M_I(F)$ (recall that we are assuming that $\mathcal{I}$ is a *proper* ideal of A), we have

$$\Phi_{\infty,M_I}^G(\mu)=\frac{(-1)^{(d_1-1)+\cdots+(d_s-1)}(-1)^{s-1}s!}{\mathrm{vol}\big(Z_{M_I}(F_\infty)\backslash\overline{M_I(F_\infty)},dz_{M_I,\infty}\backslash d\overline{m}_{I,\infty,1}\big)}$$

(there are exactly $s!$ parabolic subgroups $P\in\mathcal{P}$ such that $M_P=M_I$) and we have

$$O_\mu^{M_I}(1_{M_I(\mathbb{A}^\infty)\cap K_{\mathcal{I}}^\infty},dm_{I,\mu}^\infty)=\frac{dm_I^\infty}{dm_{I,1}^\infty}.$$

The lemma follows. □

Secondly let us fix a place $o\neq\infty$ of F with $o\notin V(\mathcal{I})$ and let us split the Haar measure dg^∞ into $dg^{\infty,o}dg_o$ where $dg^{\infty,o}$ and dg_o are respectively the Haar measure on $G(\mathbb{A}^{\infty,o})$ giving the volume 1 to the compact open subgroup $K_{\mathcal{I}}^{\infty,o}$ and the Haar measure on $G(F_o)$ giving the volume 1 to the compact open subgroup K_o. Let us assume that $f^\infty=f^{\infty,o}f_o$ where $f^{\infty,o}\in\mathcal{H}_{\mathcal{I}}^{\infty,o}$ and where f_o is the Drinfeld function of level r for some positive integer r. We denote the L^2-Lefschetz number $\mathrm{Lef}^2(f^\infty)$ by

$$\mathrm{Lef}_r^2(f^{\infty,o})$$

to emphasize the dependence on r.

When r is large enough with respect to $f^{\infty,o}$ there is an important simplification in the expression of $\mathrm{Lef}_r^2(f^{\infty,o})$. It is based on the following result (see [Cas 3]).

LEMMA (13.2.4) (Casselman). — *Let $I \subset \Delta$, let $d_I = (d_1, \dots, d_s)$ be the corresponding partition of d and let $\mu = (\lambda_1, \dots, \lambda_s)$ be an elliptic element of $M_I(F_\infty) \cong GL_{d_1}(F_\infty) \times \cdots \times GL_{d_s}(F_\infty)$. We assume that there exists an index $j \in \{1, \dots, s\}$ such that*

$$\frac{\infty(\det(\gamma_j))}{d_j} < \frac{\infty(\det(\gamma_k))}{d_k}$$

for every $k \neq j$ in $\{1, \dots, s\}$. Then we have

$$\widetilde{\Theta}_{\infty,M_I}^{G}(\mu) = -\Bigg(\prod_{\substack{k=1\\k\neq j}}^{s}\left|\frac{\det(\gamma_k)^{d_j}}{\det(\gamma_j)^{d_k}}\right|_\infty^{1/2}\Bigg)\widetilde{\Theta}_{\infty,M'_{I'}}^{G'}(\mu')$$

where $M'_{I'}$ is the standard Levi subgroup of $G' = GL_{d'}$ corresponding to the partition

$$d'_{I'} = (d_1, \dots, d_{j-1}, d_{j+1}, \dots, d_s)$$

of $d' = d - d_j$ and where μ' is the elliptic element

$$(\gamma_1, \dots, \gamma_{j-1}, \gamma_{j+1}, \dots, \gamma_s)$$

of $M'_{I'}(F_\infty)$.

Proof: We may assume that $j = 1$. Let J be the subset of Δ corresponding to the partition $(d_1, d - d_1)$ of d, let $P_J = M_J N_J$ be the standard parabolic subgroup of G corresponding to J and let $P_J = M_J \overline{N}_J$ be its opposite parabolic subgroup. We have

$$\delta_{\overline{P}_J(F_\infty)}^{1/2}(\mu) = \prod_{k=2}^{s}\left|\frac{\det(\gamma_k)^{d_1}}{\det(\gamma_1)^{d_k}}\right|_\infty^{1/2}.$$

If $P \in \mathcal{P}$ contains $M_I \cong GL_{d_1} \times M'_{I'}$, $P \cap (GL_{d_1} \times G')$ is equal to $GL_{d_1} \times P'$ for some $P' = M'_{P'}N'_{P'} \in \mathcal{P}'$ containing $M'_{I'}$ and $GL_{d_1} \times M'_{P'}$ is a Levi subgroup of G containing M_I and contained in M. Here $\mathcal{P}'$ is the set of parabolic subgroups of G' containing the maximal torus T' of diagonal matrices in G' and, for each $P' \in \mathcal{P}'$, $M'_{P'}N'_{P'}$ is the unique Levi decomposition of P' such that $M'_{P'} \supset T'$. Moreover we have

$$\begin{aligned}\delta_{P(F_\infty)}(\mu) = \delta_{P'(F_\infty)}(\mu')\big|\det\big(\mathrm{Ad}(\mu), \mathrm{Lie}(N_P(F_\infty) \cap N_J(F_\infty))\big)\big|_\infty \\ \times \big|\det\big(\mathrm{Ad}(\mu), \mathrm{Lie}(N_P(F_\infty) \cap \overline{N}_J(F_\infty))\big)\big|_\infty\end{aligned}$$

and we have

$$|D_{M_I}^{M_P}(\mu)|_\infty = |D_{M'_{I'}}^{M'_{P'}}(\mu')|_\infty |D_{GL_{d_1} \times M'_{P'}}^{M_P}(\mu)|_\infty$$

with

$$|D^{M_P}_{GL_{d_1}\times M'_{P'}}(\mu)|_\infty = \big|\det\big(1-\mathrm{Ad}(\mu), \mathrm{Lie}(M_P(F_\infty)\cap N_J(F_\infty))\big)\big|_\infty \times\big|\det\big(1-\mathrm{Ad}(\mu), \mathrm{Lie}(M_P(F_\infty)\cap \overline{N}_J(F_\infty))\big)\big|_\infty.$$

Now, due to our hypothesis on μ, we have

$$\big|\det\big(1-\mathrm{Ad}(\mu), \mathrm{Lie}(M_P(F_\infty)\cap N_J(F_\infty))\big)\big|_\infty = \big|\det\big(\mathrm{Ad}(\mu), \mathrm{Lie}(M_P(F_\infty)\cap N_J(F_\infty))\big)\big|_\infty$$

and

$$\big|\det\big(1-\mathrm{Ad}(\mu), \mathrm{Lie}(M_P(F_\infty)\cap \overline{N}_J(F_\infty))\big)\big|_\infty = 1.$$

It follows that the expression

$$\sum_{\substack{P\in\mathcal{P}\\ M_P\supset M_I}} (-1)^{s_{M_P}-1}\delta^{1/2}_{P(F_\infty)}(\mu)|D^{M_P}_{M_I}(\mu)|_\infty^{-1/2}$$

is equal to the expression

$$\sum_{\substack{P'\in\mathcal{P}'\\ M'_{P'}\supset M'_{I'}}} (-1)^{s_{M'_{P'}}-1}\delta^{1/2}_{P'(F_\infty)}(\mu')|D^{M'_{P'}}_{M'_{I'}}(\mu')|_\infty^{-1/2}\Delta_{P'}(\gamma_1,\mu')$$

where, by definition, for each $P'\in\mathcal{P}'$, each $\gamma_1\in GL_{d_1}(F_\infty)$ and each $\mu'\in M'_{P'}(F_\infty)$, $\Delta_{P'}(\gamma_1,\mu')$ is the sum over the $P\in\mathcal{P}$ such that $P\cap(GL_{d_1}\times G') = GL_{d_1}\times P'$ of the expression

$$(*)\qquad (-1)^{s_{M_P}-s_{M'_{P'}}}\big|\det\big(\mathrm{Ad}(\gamma_1,\mu'), \mathrm{Lie}(P(F_\infty)\cap N_J(F_\infty))\big)\big|_\infty^{1/2} \times\big|\det\big(\mathrm{Ad}(\gamma_1,\mu'), \mathrm{Lie}(N_P(F_\infty)\cap \overline{N}_J(F_\infty))\big)\big|_\infty^{1/2}.$$

Therefore, to conclude the proof of the lemma it is sufficient to prove that, for every $P'\in\mathcal{P}'$, every $\gamma_1\in GL_{d_1}(F_\infty)$ and every $\mu'\in M'_{P'}(F_\infty)$, we have

$$(**)\qquad \Delta_{P'}(\gamma_1,\mu') = -\delta^{1/2}_{\overline{P}_J(F_\infty)}(\gamma_1,\mu').$$

If $w'\in W'$ (the Weyl group of (G',T')) it is clear that

$$\Delta_{\dot{w}P'\dot{w}^{-1}}(\gamma_1,\dot{w}\mu'\dot{w}^{-1}) = \Delta_{P'}(\gamma_1,\mu')$$

and that

$$\delta^{1/2}_{\overline{P}_J(F_\infty)}(\gamma_1,\dot{w}\mu\dot{w}^{-1}) = \delta^{1/2}_{\overline{P}_J(F_\infty)}(\gamma_1,\mu'),$$

so that we may assume that $P' = P'_{K'}$ for some $K' \subset \Delta'$. Let $(e'_1, \ldots, e'_{t'})$ be the partition of $d' = d - d_1$ corresponding to K'. We may identify $M'_{P'}$ with $GL_{e'_1} \times \cdots \times GL_{e'_{t'}}$ and write $\mu' = (\delta'_1, \ldots, \delta'_{t'})$. Then the set of $P \in \mathcal{P}$ such that $P \cap (GL_{d_1} \times G') = GL_{d_1} \times P'$ may be identified with the set of partitions of $L' = \{1, \ldots, t'\}$ into three disjoint subsets,

$$L' = L'_{\overline{\mathrm{u}}} \amalg L'_{\mathrm{r}} \amalg L'_{\mathrm{u}},$$

in such a way that

$$M_P \cong GL_{d_1 + \sum_{\ell' \in L'_{\mathrm{r}}} e'_{\ell'}} \times \prod_{\ell' \in L'_{\overline{\mathrm{u}}} \cup L'_{\mathrm{u}}} GL_{\delta'_{\ell'}}$$

and that $N_P \cap N_J$ (resp. $N_P \cap \overline{N}_J$) is isomorphic to the additive group of matrices of size $d_1 \times (\sum_{\ell' \in L'_{\mathrm{u}}} e'_{\ell'})$ (resp. $(\sum_{\ell' \in L'_{\overline{\mathrm{u}}}} e'_{\ell'}) \times d_1$). Moreover, by this identification, the expression $(*)$ is transformed into

$$(-1)^{|L'_{\mathrm{r}}|-1} \Bigg(\prod_{\ell' \in L'_{\mathrm{u}} \cup L'_{\mathrm{r}}} \left| \frac{\det(\gamma_1)^{e'_{\ell'}}}{\det(\delta'_{\ell'})^{d_1}} \right|_\infty^{1/2} \Bigg) \Bigg(\prod_{\ell' \in L'_{\overline{\mathrm{u}}}} \left| \frac{\det(\delta'_{\ell'})^{d_1}}{\det(\gamma_1)^{e'_{\ell'}}} \right|_\infty^{1/2} \Bigg).$$

Setting

$$a_{\ell'} = \left| \frac{\det(\gamma_1)^{e'_{\ell'}}}{\det(\delta'_{\ell'})^{d_1}} \right|_\infty^{1/2} \qquad (\forall \ell' = 1, \ldots, t')$$

we obtain that the equality $(**)$ is equivalent to

$$\sum_{(L'_{\mathrm{r}}, L'_{\mathrm{u}})} (-1)^{|L'_{\mathrm{r}}|-1} \prod_{\ell' \in L'_{\mathrm{r}} \cup L'_{\mathrm{u}}} a_{\ell'} \prod_{\ell' \in L' - (L'_{\mathrm{r}} \cup L'_{\mathrm{u}})} a_{\ell'}^{-1} = - \prod_{\ell' \in L'} a_{\ell'}^{-1}$$

where $(L'_{\mathrm{r}}, L'_{\mathrm{u}})$ runs through the set of pairs of disjoint subsets in L'. But the last equality is equivalent to

$$\sum_{L'_{\mathrm{ru}} \subset L'} \Bigg(\sum_{L'_{\mathrm{r}} \subset L'_{\mathrm{ru}}} (-1)^{|L'_{\mathrm{r}}|} \Bigg) \prod_{\ell' \in L'_{\mathrm{ru}}} a_{\ell'}^2 = 1$$

$(L'_{\mathrm{ru}} = L'_{\mathrm{r}} \cup L'_{\mathrm{u}})$ and it thus follows from the equalities

$$\sum_{L'_{\mathrm{r}} \subset L'_{\mathrm{ru}}} (-1)^{|L'_{\mathrm{r}}|} = \begin{cases} 1 & \text{if } L'_{\mathrm{ru}} = \emptyset, \\ 0 & \text{otherwise.} \end{cases}$$

□

PROPOSITION (13.2.5). — *Given the place o of F and the Hecke operator $f^{\infty,o} \in \mathcal{H}_{\mathcal{I}}^{\infty,o}$ we can find a non-negative integer $R(o, f^{\infty,o})$ having the following property. For each integer $r > R(o, f^{\infty,o})$ the L^2-Lefschetz number $\mathrm{Lef}_r^2(f^{\infty,o})$ is equal to*

$$\sum_{\gamma\in G(F)_{\natural,\mathrm{ell}}} \mathrm{vol}\Big(F_\infty^\times G_\gamma(F)\backslash G_\gamma(\mathbb{A}), \frac{dg_\gamma}{dz_\infty d\gamma_\gamma}\Big)$$
$$\times \frac{\epsilon_{G,\infty}(\gamma)}{\mathrm{vol}(F_\infty^\times\backslash\overline{G(F_\infty)}_\gamma, dz_\infty\backslash d\overline{g}_{\infty,\gamma})} O_\gamma^G(f^{\infty,o}f_o, dg_\gamma^\infty)$$

plus the sum over $I \subsetneqq \Delta$ and $\mu \in M_I(F)_{\natural,\mathrm{ell}}$ of the products of the following six expressions:

(i) $$\frac{(-1)^{|\Delta'-I'|}}{(|\Delta'-I'|+1)!},$$

(ii) $$\frac{|G(A/\mathcal{I})|}{|P_J(A/\mathcal{I})|}\frac{|G'(A/\mathcal{I})|}{|P'_{I'}(A/\mathcal{I})|},$$

(iii) $$\mathrm{vol}\Big(F_\infty^\times GL_{d_1,\gamma_1}(F)\backslash GL_{d_1,\gamma_1}(\mathbb{A}), \frac{dg_{1,\gamma_1}}{dz_\infty d\gamma_{1,\gamma_1}}\Big)$$
$$\times\, \mathrm{vol}\Big(Z_{M'_{I'}}(F_\infty)M'_{I',\mu'}(F)\backslash M'_{I',\mu'}(\mathbb{A}), \frac{dm'_{I',\mu'}}{dz_{M'_{I'},\infty}d\mu'_{I',\mu'}}\Big),$$

(iv) $$\frac{\epsilon_{GL_{d_1},\infty}(\gamma_1)}{\mathrm{vol}(F_\infty^\times\backslash\overline{GL_{d_1}(F_\infty)}_{\gamma_1}, dz_\infty\backslash d\overline{g}_{1,\gamma_1,\infty})}\Phi^{G'}_{\infty,M'_{I'}}(\mu'),$$

(v) $$O_{(\gamma_1,\mu')}^{GL_{d_1}\times M'_{I'}}\Big(\big(\delta_{P_J(\mathbb{A}^{\infty,o})}^{1/2}(f^{\infty,o})^{P_J}\big)^{GL_{d_1}\times P'_{I'}}, dg_{1,\gamma_1}^{\infty,o}\times dm'^{\infty,o}_{I',\mu'}\Big),$$

and

(vi) $$O_{\gamma_1}^{GL_{d_1}}(f_{1,o}, dg_{1,\gamma_1,o})O_{\mu'}^{M'_{I'}}(1_{K'_{I',o}}, dm'_{I',\mu',o}).$$

Here we are using the following notations:

- *f_o is the Drinfeld function of level r for $G(F_o)$,*
- *$(d_1, \ldots, d_s)$ is the partition of d corresponding to I and J is the subset of Δ corresponding to the partition $(d_1, d-d_1)$ of d,*

• *I' is the subset of $\Delta' = \{1,\dots,d-d_1-1\}$ corresponding to the partition $(d_2,\dots,d_s)$ of $d-d_1$ and $P'_{I'} = M'_{I'}N'_{I'}$ is the standard parabolic subgroup of $G' = GL_{d-d_1}$, with its standard Levi decomposition, corresponding to I',*

• *$\mu = (\gamma_1,\mu') \in M_I(F) \cong GL_{d_1}(F) \times M'_{I'}(F)$, $dm_{I,\mu} = dg_{1,\gamma_1} \times dm'_{I',\mu'}$ where $dg_{1,\gamma_1} = dg_{1,\gamma_1,\infty} dg_{1,\gamma_1}^{\infty,o} dg_{1,\gamma_1,o}$ and $dm'_{I',\mu'} = dm'_{I',\mu',\infty} dm'^{\infty,o}_{I',\mu'} dm'_{I',\mu',o}$ are Haar measures on $GL_{d_1,\gamma_1}(\mathbb{A})$ and $M'_{I'}(\mathbb{A})$ and $d\mu_{I,\mu} = d\gamma_{1,\gamma_1} \times d\mu'_{I',\mu'}$ where $d\gamma_{1,\gamma_1}$ and $d\mu'_{I',\mu'}$ are the counting measures on $GL_{d_1,\gamma_1}(F)$ and $M'_{I'}(F)$,*

• *$\overline{GL_{d_1}(F_\infty)_{\gamma_1}} = D_1'^{\times}$ where D'_1 is "the" central division algebra over $F_\infty[\gamma_1]$ of invariant $1/d'_1$ with $d'_1 = d_1/[F_\infty[\gamma_1] : F_\infty]$ and $d\overline{g}_{1,\gamma_1,\infty}$ is the transfer of the Haar measure $dg_{1,\gamma_1,\infty}$ to the inner twist $\overline{GL_{d_1}(F_\infty)_{\gamma_1}}$ of $GL_{d_1}(F_\infty)_{\gamma_1}$,*

• *$f_{1,o}$ is the Drinfeld function of level r for $GL_{d_1}(F_o)$ and $K'_{I',o} = M'_{I'}(\mathcal{O}_o)$.*

Proof: Let $I \subsetneqq \Delta$ and $\mu \in M_I(F)_{\natural,\mathrm{ell}}$. By (4.2.5) we have

$$O_\mu^{M_I}((f^\infty)^{P_I}, dm_{I,\mu}^\infty) = \sum_{j=1}^{s} O_\mu^{M_I}((f^{\infty,o})^{P_I}, dm_{I,\mu}^{\infty,o})$$
$$\times p^{r\deg(o)(d-d_j)/2} O_{\gamma_j}^{GL_{d_j}}(f_{j,o}, dg_{j,\gamma_j,o}) O_{\mu'}^{G'}(1_{K'_{I',o}}, dm'_{I',\mu',o})$$

where $G' = GL_{d-d_j}$, $d'_{I'} = (d_1,\dots,d_{j-1},d_{j+1},\dots,d_s)$, ... are as in the statement of the proposition. Let us consider an index $j = 1,\dots,s$ and let us assume that the j-th term of the above sum is non-zero. Then we have

$$O_\mu^{M_I}((f^{\infty,o})^{P_I}, dm_{I,\mu}^{\infty,o}) \neq 0$$

and in particular, for every $k = 1,\dots,s$, there exists a non-negative integer $R_{I,k}$ depending only on I, k, o and the support of $(f^{\infty,o})^{P_I}$ such that

$$\left|\deg(\infty)\infty(\det(\gamma_k)) + \deg(o)o(\det(\gamma_k))\right| = \Big|\sum_{x\neq\infty,o} \deg(x)x(\det(\gamma_k))\Big| \leq R_{I,k}.$$

We also have

$$o(\det(\gamma_j)) = r$$

and

$$o(\det(\gamma_k)) = 0 \qquad (\forall k = 1,\dots,s \text{ with } k \neq j).$$

Therefore, if we assume that

$$r > \frac{d_j}{\deg(o)}\Big(\frac{R_{I,j}}{d_j} + \frac{R_{I,k}}{d_k}\Big) \qquad (\forall k = 1,\dots,s \text{ with } k \neq j)$$

we have

$$\frac{\infty(\det(\gamma_j))}{d_j} < \frac{\infty(\det(\gamma_k))}{d_k} \qquad (\forall k = 1, \dots, s \text{ with } k \neq j)$$

and by lemma (13.2.4) it follows that

$$\Phi^G_{\infty, M_I}(\mu)$$
$$= -\Bigg(\prod_{\substack{k=1\\k\neq j}}^{s} \left|\frac{\det(\gamma_k)^{d_j}}{\det(\gamma_j)^{d_k}}\right|_\infty^{1/2}\Bigg) \frac{\epsilon_{GL_{d_j},\infty}(\gamma_j)}{\mathrm{vol}(F_\infty^\times \backslash \overline{GL_{d_j}(F_\infty)_{\gamma_j}}, dz_\infty \backslash d\overline{g}_{j,\gamma_j,\infty})} \Phi^{G'}_{\infty, M'_{I'}}(\mu').$$

Now let us set

$$R(o, f^{\infty,o}) = \sup\Big\{\frac{d_j}{\deg(o)}\Big(\frac{R_{I,j}}{d_j} + \frac{R_{I,k}}{d_k}\Big) \mid I \subset \Delta,\ j, k = 1, \dots, s \text{ with } k \neq j\Big\}.$$

Permuting the sum over μ and the sum over j we obtain that, if $r > R(o, f^{\infty,o})$, $\mathrm{Lef}^2_r(f^{\infty,o})$ is equal to

$$\sum_{s=1}^{d} \sum_{\substack{I\subset\Delta\\|\Delta - I| = s-1}} \sum_{j=1}^{s} E(I, j)$$

for certain expressions $E(I, j)$ satisfying the relations

$$E(w_{I,\sigma}(I), \sigma(j)) = E(I, j) \qquad (\forall \sigma \in \mathfrak{S}_s)$$

(see (G.1.5) for the definition of $w_{I,\sigma}$). Indeed we have

$$(f^\infty)_{M_{w(I)}}(\dot{w}\mu\dot{w}^{-1}) = (f^\infty)_{M_I}(\mu) \qquad (\forall \mu \in M_I(\mathbb{A}^\infty))$$

for any $w \in W$ such that $w(I) \subset \Delta$. Hence $\mathrm{Lef}^2_r(f^{\infty,o})$ is equal to

$$\sum_{s=1}^{d} \sum_{\substack{I\subset\Delta\\|\Delta - I| = s-1}} \sum_{j=1}^{s} E(w_{I,\sigma_j}(I), 1)$$

where

$$\sigma_j(k) = \begin{cases} k+1 & \text{if } k < j, \\ 1 & \text{if } k = j, \\ k & \text{if } k > j, \end{cases}$$

so that $\mathrm{Lef}_r^2(f^{\infty,o})$ is equal to

$$\sum_{s=1}^{d}\sum_{j=1}^{s}\Big(\sum_{\substack{I\subset\Delta\\ |\Delta-I|=s-1}} E(w_{I,\sigma_j}(I),1)\Big)=\sum_{s=1}^{d}s\Big(\sum_{\substack{I\subset\Delta\\ |\Delta-I|=s-1}} E(I,1)\Big)$$
$$=\sum_{I\subset\Delta}(|\Delta-I|+1)E(I,1).$$

Finally we have

$$\frac{(-1)^{|\Delta-I|}}{(|\Delta-I|+1)!}=-\frac{1}{(|\Delta-I|+1)}\frac{(-1)^{|\Delta'-I'|}}{(|\Delta'-I'|+1)!}$$

and

$$\frac{|G(A/\mathcal{I})|}{|P_I(A/\mathcal{I})|}=\frac{|G(A/\mathcal{I})|}{|P_J(A/\mathcal{I})|}\frac{|G'(A/\mathcal{I})|}{|P'_{I'}(A/\mathcal{I})|}.$$

The volume

$$\mathrm{vol}\Big(Z_{M_I}(F_\infty)M_{I,\mu}(F)\backslash M_{I,\mu}(\mathbb{A}),\frac{dm_{I,\mu}}{dz_{M_I,\infty}d\mu_{I,\mu}}\Big)$$

is the product of the two volumes

$$\mathrm{vol}\Big(F_\infty^\times GL_{d_1,\gamma_1}(F)\backslash GL_{d_1,\gamma_1}(\mathbb{A}),\frac{dg_{1,\gamma_1}}{dz_\infty d\gamma_{1,\gamma_1}}\Big)$$

and

$$\mathrm{vol}\Big(Z_{M'_{I'}}(F_\infty)M'_{I',\mu'}(F)\backslash M'_{I',\mu'}(\mathbb{A}),\frac{dm'_{I',\mu'}}{dz_{M'_{I'},\infty}d\mu'_{I',\mu'}}\Big).$$

We have

$$\Big(\prod_{k=2}^{s}\Big|\frac{\det(\gamma_k)^{d_1}}{\det(\gamma_1)^{d_k}}\Big|_\infty^{1/2}\Big)p^{r\deg(o)(d-d_1)/2}=\delta_{P_J(\mathbb{A}^{\infty,o})}^{1/2}(\mu)$$

and

$$\delta_{P_J(\mathbb{A}^{\infty,o})}^{1/2}(\mu)O_\mu^{M_I}((f^{\infty,o})^{P_I},dm_{I,\mu}^{\infty,o})=O_\mu^{M_I}(\delta_{P_J(\mathbb{A}^{\infty,o})}^{1/2}(f^{\infty,o})^{P_I},dm_{I,\mu}^{\infty,o})$$

for any relevant μ and we have

$$\delta_{P_J(\mathbb{A}^{\infty,o})}^{1/2}(f^{\infty,o})^{P_I}=(\delta_{P_J(\mathbb{A}^{\infty,o})}^{1/2}(f^{\infty,o})^{P_J})^{GL_{d_1}\times M'_{I'}}.$$

The proposition follows. □

REMARK (13.2.6). — If the support of $f^{\infty,o}$ is contained in $K^{\infty,o} = G(\mathcal{O}^{\infty,o})$ we can take $R(o,f^{\infty,o}) = 0$ in the above proposition. Indeed, for each $I \subset \Delta$ the support of $(f^{\infty,o})^{P_I}$ is contained in $M_I(\mathcal{O}^{\infty,o})$; therefore, if

$$O_\mu^{M_I}((f^{\infty,o})^{P_I}, dm_{I,\mu}^{\infty,o}) \neq 0$$

we have

$$\sum_{x\neq\infty,o} \deg(x)x(\det(\gamma_k)) = 0$$

and we can take $R_{I,k} = 0$, $\forall k = 1,\ldots,s$, and then $R(o,f^{\infty,o}) = 0$. □

Thirdly we restrict ourself to the case $f^{\infty,o} = 1_{K_{\mathcal{I}}^{\infty,o}}$ and we will give an expression for the L^2-Lefschetz number $\mathrm{Lef}_r^2(1_{K_{\mathcal{I}}^{\infty,o}})$ in terms of ordinary Lefschetz numbers.

For each $d' = 1,\ldots,d$ we set $G' = GL_{d'}$ and for each $I' \subset \Delta' = \{1,\ldots,d'-1\}$ we set

$$V_{I'}(\mathcal{I}) = \prod_{j'=1}^{s'} V_{d'_{j'}}(\mathcal{I})$$

where $(d'_1,\ldots,d'_{s'})$ is the partion of d' corresponding to I' (see (13.2.1) and (13.2.2)).

For each $d'' = 1,\ldots,d$ we set $G'' = GL_{d''}$ and

$$K_{d'',\mathcal{I}}^{\infty,o} = \mathrm{Ker}\big(G''(\mathcal{O}^{\infty,o}) \twoheadrightarrow G''(\mathcal{O}^{\infty,o}/\mathcal{O}^{\infty,o}\mathcal{I})\big).$$

Then for each positive integer r we can define the ordinary Lefschetz number $\mathrm{Lef}_r(1_{K_{d'',\mathcal{I}}^{\infty,o}})$ as in (3.2.2). We have proved in (6.1) that

$$\begin{aligned}\mathrm{Lef}_r(1_{K_{d'',\mathcal{I}}^{\infty,o}}) = &\sum_{\gamma''\in G''(F)_{\natural,\mathrm{ell}}} \mathrm{vol}\Big(F_\infty^\times G''_{\gamma''}(F)\backslash G''_{\gamma''}(\mathbb{A}), \frac{dg''_{\gamma''}}{dz_\infty d\gamma''_{\gamma''}}\Big)\\ &\times \frac{\epsilon_{G'',\infty}(\gamma'')}{\mathrm{vol}\big(F_\infty^\times\backslash\overline{G''(F_\infty)_{\gamma''}}, dz_\infty\backslash d\overline{g}''_{\gamma'',\infty}\big)} O_{\gamma''}^{G''}(f''^\infty, dg''^\infty_{\gamma''})\end{aligned}$$

where $f''^\infty = 1_{K_{d'',\mathcal{I}}^{\infty,o}} f''_o$ and f''_o is the Drinfeld function of level r on $G''(F_o)$ (with the obvious modifications of the notations of loc. cit.).

LEMMA (13.2.7). — *For any positive integer r we have*

$$\begin{aligned}&\mathrm{Lef}_r^2(1_{K_{d,\mathcal{I}}^{\infty,o}})\\ &= \mathrm{Lef}_r(1_{K_{d,\mathcal{I}}^{\infty,o}}) + \sum_{d'=1}^{d-1} \mathrm{Lef}_r(1_{K_{d-d',\mathcal{I}}^{\infty,o}}) \frac{|G(A/\mathcal{I})|}{|P_J(A/\mathcal{I})|} \sum_{I'\subset\Delta'} \frac{|G'(A/\mathcal{I})|}{|P'_{I'}(A/\mathcal{I})|} V_{I'}(\mathcal{I})\end{aligned}$$

where J is the subset of Δ corresponding to the partion $(d-d',d')$ of d.

Proof : Let us consider some $I \subsetneq \Delta$ and some $\mu \in M_I(F)_{\natural,\mathrm{ell}}$ which is still elliptic in $M_I(F_\infty)$ and such that expressions (v) and (vi) of proposition (13.2.5) are non-zero and let us freely use the notations of loc. cit. Due to our normalization of the Haar measures we have

$$\left(\delta_{P_J(\mathbb{A}^{\infty,o})}^{1/2}(1_{K_{\mathcal{I}}^{\infty,o}})^{P_J}\right)^{GL_{d_1}\times P'_{I'}} = \frac{|G(A/\mathcal{I})|}{|P_I(A/\mathcal{I})|}1_{K_{d_1,\mathcal{I}}^{\infty,o}}1_{K_{I',\mathcal{I}}^{\infty,o}}$$

where

$$K_{d_1,\mathcal{I}}^{\infty,o} = \mathrm{Ker}\big(GL_{d_1}(\mathcal{O}^{\infty,o}) \twoheadrightarrow GL_{d_1}(\mathcal{O}^{\infty,o}/\mathcal{O}^{\infty,o}\mathcal{I})\big)$$

and

$$K_{I',\mathcal{I}}^{\infty,o} = \mathrm{Ker}\big(M'_{I'}(\mathcal{O}^{\infty,o}) \twoheadrightarrow M'_{I'}(\mathcal{O}^{\infty,o}/\mathcal{O}^{\infty,o}\mathcal{I})\big).$$

Therefore, arguing as in the proof of lemma (13.2.3) we see that the component μ' of μ must be the unit element in $M'_{I'}(F)$ (recall once more that we are assuming that $\mathcal{I}$ is a *proper* ideal of A), that

$$\Phi_{\infty,M'_{I'}}^{G'}(\mu') = \frac{(-1)^{(d_2-1)+\cdots+(d_s-1)}(-1)^{s-2}(s-1)!}{\mathrm{vol}\big(Z_{M'_{I'}}(F_\infty)\backslash\overline{M'_{I'}(F_\infty)}, dz_{M'_{I'},\infty}\backslash d\overline{m}'_{I',1,\infty}\big)}$$

(there are exactly $(s-1)!$ parabolic subgroups P' of G' which admit $M'_{I'}$ as a Levi subgroup) and that

$$O_{\mu'}^{M'_{I'}}(1_{K_{I',\mathcal{I}}^{\infty}}) = \frac{dm'^{\infty}_{I'}}{dm'^{\infty}_{I',1}}.$$

Now it is clear that the lemma follows from proposition (13.2.5) and remark (13.2.6). □

Finally, let us check the compatibility of conjecture (13.1.6) and theorem (12.3.1) in the particular case $f^{\infty,o} = 1_{K_{\mathcal{I}}^{\infty,o}}$.

LEMMA (13.2.8). — *For each positive integer r and each $d' = 1,\ldots,d$ let us set*

$$N_{d',r}^2 = \sum_{\pi'\in\Pi_{G'}^2} \epsilon_\infty(\pi'_\infty)\dim(\pi'^{\infty,o})^{K_{d',\mathcal{I}}^{\infty,o}}p^{r\deg(o)(d'-1)/2}(z_1(\pi'_o)^r+\cdots+z_{d'}(\pi'_o)^r)$$

and

$$N_{d',r}$$

$$= \sum_{I'\subset\Delta'}\sum_{\pi'\in\Pi_{M_{I'}}^2}(-1)^{|\Delta'-I'|}\epsilon_\infty(\pi'_\infty)\dim\big(i_{M'_{I'}(\mathbb{A}^{\infty,o}),P'_{I'}(\mathbb{A}^{\infty,o})}^{G'(\mathbb{A}^{\infty,o})}(\pi'^{\infty,o}(\rho'_{I'}))\big)^{K_{d',\mathcal{I}}^{\infty,o}}$$

$$\times\, p^{r\deg(o)(d'_1-1)/2}(z_1(\pi'_{1,o})^r+\cdots+z_{d'_1}(\pi'_{1,o})^r),$$

with the obvious modification of the notations of (12.3.1) *and* (13.1.2). *Let us assume that, for each* $d' = 1, \dots, d$, *we have*

$$(*) \qquad \sum_{I' \subset \Delta'} \frac{|G'(A/\mathcal{I})|}{|P_{I'}(A/\mathcal{I})|} V_{I'}(\mathcal{I}) = \sum_{\pi' \in \Pi^2_{G'}} \epsilon_\infty(\pi'_\infty) \dim(\pi'^{\infty,o})^{K^\infty_{d',\mathcal{I}}}$$

where $G' = GL_{d'}$, $\Delta' = \{1, \dots, d'-1\}$ *and*

$$\Pi^2_{G'} \stackrel{\mathrm{dfn}}{=\!=} \coprod_{\substack{\chi' \in \mathcal{X}_{G'} \\ \chi'_\infty = 1_\infty}} \Pi^2_{G',\chi'}.$$

Then for each $d' = 1, \dots, d$ *and each positive integer* r *we have*

$$N^2_{d',r} = N_{d',r} + \sum_{d''=1}^{d'-1} N_{d-d'',r} \frac{|G'(A/\mathcal{I})|}{|P'_{J'}(A/\mathcal{I})|} \sum_{I'' \subset \Delta''} \frac{|G''(A/\mathcal{I})|}{|P''_{I''}(A/\mathcal{I})|} V_{I''}(\mathcal{I})$$

where J' *is the subset of* $\Delta' = \{1, \dots, d'\}$ *corresponding to the partition* $(d'-d'', d'')$ *of* d'.

We will see in section (13.4) that the hypothesis $(*)$ follows from results of Harder and Stuhler.

Proof: Let $d' \in \{1, \dots, d\}$. We may rewrite the expression for $N_{d',r}$ as

$$\sum_{\pi' \in \Pi^2_{GL_{d'}}} \epsilon_\infty(\pi'_\infty) \dim(\pi'^\infty)^{K^\infty_{d',\mathcal{I}}} p^{r\deg(o)(d'-1)/2} (z_1(\pi'_o)^r + \cdots + z_{d'}(\pi'_o)^r)$$

$$+ \sum_{d''=1}^{d'-1} \sum_{\pi''' \in \Pi^2_{GL_{d'-d''}}} \epsilon_\infty(\pi'''_\infty) \dim\left(r^\infty_{d''}\left(\frac{d''-d'}{2}\right) \times \left[\pi'''^\infty\left(\frac{d''}{2}\right)\right]\right)^{K^\infty_{d',\mathcal{I}}}$$

$$\times\, p^{r\deg(o)(d'-d''-1)/2} (z_1(\pi'''_o)^r + \cdots + z_{d'-d''}(\pi'''_o)^r)$$

(compare with (12.3.3)). In this formula, for each $d'' = 1, \dots, d'$ and each $I' \subset \Delta'$ we have

$$\dim\left(r^\infty_{d''}\left(\frac{d''-d'}{2}\right) \times \left[\pi'''^\infty\left(\frac{d''}{2}\right)\right]\right)^{K^\infty_{d',\mathcal{I}}}$$

$$= \frac{|G'(A/\mathcal{I})|}{|P'_{J'}(A/\mathcal{I})|} \dim(r^\infty_{d''})^{K^\infty_{d'',\mathcal{I}}} \dim(\pi'''^\infty)^{K^\infty_{d'-d'',\mathcal{I}}}$$

where $J' \subset \Delta'$ is the subset corresponding to the partition $(d'', d'-d'')$ of d' and where

$$\dim(r^\infty_{d''})^{K^\infty_{d'',\mathcal{I}}}$$

$$= \sum_{I'' \subset \Delta''} (-1)^{s''} \frac{|G''(A/\mathcal{I})|}{|P''_{I''}(A/\mathcal{I})|} \prod_{j''=1}^{s''} \left(\sum_{\pi''_{j''} \in \Pi^2_{GL_{d''_{j''}}}} \epsilon_\infty(\pi''_{j'',\infty}) \dim(\pi''^{\infty}_{j''})^{K^\infty_{d''_{j''},\mathcal{I}}} \right)$$

(see (12.3.2)). Indeed we have

$$\operatorname{tr}\Big(r_{d''}^{\infty}\big(\frac{d''-d'}{2}\big) \times \big[\pi'''^{\infty}\big(\frac{d''}{2}\big)\big]\Big)(f'^{\infty})$$
$$= \operatorname{tr}\Big(r_{d''}^{\infty}\big(\frac{d''-d'}{2}\big) \otimes \big[\pi'''^{\infty}\big(\frac{d''}{2}\big)\big]\Big)\big((f'^{\infty})^{P'_{J'}}\big)$$

for any $f'^{\infty} \in \mathcal{H}'^{\infty}_{\mathcal{I}}$ and we have

$$(1_{K^{\infty}_{d',\mathcal{I}}})^{P'_{J'}} = \frac{|G'(A/\mathcal{I})|}{|P'_{J'}(A/\mathcal{I})|} 1_{K^{\infty}_{J',\mathcal{I}}};$$

similarly we have

$$\operatorname{tr}\big(i^{G''(\mathbb{A}^{\infty})}_{M''_{I''}(\mathbb{A}^{\infty}),P''_{I''}(\mathbb{A}^{\infty})}(\pi''^{\infty}(\rho''_{I''}))(f''^{\infty}) = \operatorname{tr}(\pi''^{\infty}(\rho''_{I''}))\big((f''^{\infty})^{P''_{I''}}\big)$$

for any $f''^{\infty} \in \mathcal{H}''^{\infty}_{\mathcal{I}}$ and we have

$$(1_{K^{\infty}_{d'',\mathcal{I}}})^{P''_{I''}} = \frac{|G''(A/\mathcal{I})|}{|P''_{I''}(A/\mathcal{I})|} 1_{K^{\infty}_{I'',\mathcal{I}}}.$$

Now let

$$a, b, v, w : \{1, \ldots, d\} \to \mathbb{C}$$

be arbitrary functions and let

$$u : \{\underline{d}' = (d'_1, \ldots, d'_{s'}) \mid d'_{j'} \in \mathbb{Z}_{>0},\ \forall j' = 1, \ldots, s',$$
$$|\underline{d}'| = d'_1 + \cdots + d'_{s'} \leq d\} \to \mathbb{C}$$

be a function satisfying the property

$$u(\underline{d}'^{(1)}, \ldots, \underline{d}'^{(\sigma)}) = u(\underline{\delta})u(\underline{d}'^{(1)}) \cdots u(\underline{d}'^{(\sigma)})$$

for every $\underline{d}'^{(\iota)} = (d_1'^{(\iota)}, \ldots, d_{s'(\iota)}'^{(\iota)})$, $\iota = 1, \ldots, \sigma$, where we have set

$$\underline{\delta} = (d_1'^{(1)} + \cdots + d_{s'(1)}'^{(1)}, \ldots, d_1'^{(\sigma)} + \cdots + d_{s'(\sigma)}'^{(\sigma)}).$$

Then on the one hand the following two statements are equivalent:

(i) $$w(d') = \sum_{\substack{\underline{d}' \\ |\underline{d}'| = d'}} u(\underline{d}')v(d'_1) \cdots v(d'_{s'}) \qquad (\forall d' = 1, \ldots, d),$$

(ii) $$v(d') = \sum_{\substack{\underline{d}' \\ |\underline{d}'| = d'}} (-1)^{s'-1} u(\underline{d}')w(d'_1) \cdots w(d'_{s'}) \qquad (\forall d' = 1, \ldots, d).$$

On the other hand the following two statements are equivalent:

$$\text{(iii)}\qquad a(d') = b(d') + \sum_{d''=1}^{d'-1} b(d'-d'') \times \sum_{\substack{\underline{d}'' \\ |\underline{d}''|=d''}} u(d'-d'', \underline{d}'') v(d''_1)\cdots v(d''_{s''}) \qquad (\forall d' = 1,\ldots,d),$$

$$\text{(iv)}\ b(d') = a(d') - \sum_{d''=1}^{d'-1} a(d'-d'') u(d'-d'', d'') v(d'') \qquad (\forall d' = 1,\ldots,d).$$

Taking

$$v(d') = V_{d'}(\mathcal{I}),$$

$$w(d') = \sum_{\pi' \in \Pi^2_{GL_{d'}}} \epsilon_\infty(\pi'_\infty) \dim(\pi'^{\infty})^{K^\infty_{d',\mathcal{I}}}$$

and

$$u(\underline{d}') = \frac{|GL_{d'}(A/\mathcal{I})|}{|P'_{I'}(A/\mathcal{I})|}$$

for every $d' = 1,\ldots,d$ and every partition $\underline{d}'$ of d', where I' is the subset of Δ' corresponding to $\underline{d}'$, we get from the equivalence of (i) and (ii) that

$$-V_{d'}(\mathcal{I}) = \sum_{I' \subset \Delta'} (-1)^{s'} \frac{|GL_{d'}(A/\mathcal{I})|}{|P'_{I'}(A/\mathcal{I})|} \prod_{j'=1}^{s'} \Big(\sum_{\pi'_{j'} \in \Pi^2_{GL_{d'_{j'}}}} \epsilon_\infty(\pi'_{j',\infty}) \dim(\pi'^{\,\infty}_{j'})^{K^\infty_{d'_{j'},\mathcal{I}}} \Big)$$

and hence that

$$-V_{d'}(\mathcal{I}) = \dim(r^\infty_{d'})^{K^\infty_{d',\mathcal{I}}}.$$

Taking furthermore

$$a(d') = N^2_{d',r}$$

and

$$b(d') = N_{d',r}$$

for every $d' = 1,\ldots,d$, statement (iv) is satisfied by hypothesis and therefore statement (iii) too. But formula (iii) for d' is exactly the formula that we want to prove. $\square$

(13.3) A cohomological interpretation of the constants $V_{d'}(\mathcal{I})$
The zeta function ζ_X of X is defined by

$$\zeta_X(s) = \prod_{x\in|X|} \left(1 - p^{-\deg(x)s}\right)^{-1},$$

the product being absolutely convergent for $\mathrm{Re}(s) > 1$. Weil has proved that ζ_X admits the cohomological interpretation

$$\zeta_X(s) = \frac{P(q^{-s})}{(1-q^{-s})(1-q^{1-s})},$$

where $q = p^f$ is the number of elements of the field of constants in F and where

$$P(T) = \det(1 - T\,\mathrm{Frob}_q, H^1(\overline{\mathbb{F}}_q \otimes_{\mathbb{F}_q} X, \mathbb{Q}_\ell)) \in \mathbb{Q}[T].$$

Let us fix an integer d', $1 \leq d' \leq d$. If $\tau_{G'}$ is the Tamagawa measure of $G'(\mathbb{A})$ ($G' = GL_{d'}$) Weil has proved that

$$\tau_{G'}(G'(\mathcal{O})) = \frac{q^{1-g_X}}{\zeta_X(-1)\cdots\zeta_X(1-d')},$$

where $g_X = \dim_{\mathbb{F}_q} H^1(X, \mathcal{O}_X)$ is the genus of X, and that

$$\mathrm{vol}\big(F_\infty^\times G'(F)\backslash G'(\mathbb{A}), \frac{\tau_{G'}}{dz_\infty d\gamma'}\big) = -q^{1-g_X}\frac{d'\deg(\infty)}{f}\zeta_X'(0),$$

where

$$\zeta_X'(0) = \lim_{s\to 0}\big((1-q^{-s})\zeta_X(s)\big) = -\frac{P(1)}{q-1}$$

with

$$P(1) = |\mathrm{Pic}_X^0(\mathbb{F}_p)| = |F^\times\backslash(\mathbb{A}^\times)^1/\mathcal{O}^\times|$$

(see section (3.5) and [We 1] §3).

In definition (13.2.1) we may normalize the Haar measure dg'_∞ by $\mathrm{vol}(G'(\mathcal{O}_\infty), dg'_\infty) = 1$. Then we have

$$\mathrm{vol}(G'(\mathcal{O}), dg'_\infty dg'^\infty) = |G'(A/\mathcal{I})|,$$

$$\mathrm{vol}\Big(F_\infty^\times G'(F)\backslash G'(\mathbb{A}), \frac{dg'_\infty dg'^\infty}{dz_\infty d\gamma'}\Big) = -d'\frac{\deg(\infty)}{f}|G'(A/\mathcal{I})|\zeta_X'(0)\zeta_X(-1)\cdots\zeta_X(1-d')$$

and

$$\mathrm{vol}\big(F_\infty^\times\backslash D'^\times_\infty, \frac{d\overline{g}'_\infty}{dz_\infty}\big) = \frac{d'}{(p^{\deg(\infty)}-1)\cdots(p^{\deg(\infty)(d'-1)}-1)}$$

(see lemma (4.6.4) and [Ro] 3). It follows that

(13.3.1) $$V_{d'}(\mathcal{I}) = -|G'(A/\mathcal{I})|\zeta_{X-\{\infty\}}(0)\zeta_{X-\{\infty\}}(-1)\cdots\zeta_{X-\{\infty\}}(1-d')$$

where

$$\zeta_{X-\{\infty\}}(s) = (1-p^{-\deg(\infty)s})\zeta_X(s).$$

PROPOSITION (13.3.2). — (i) (Garland, Casselman, Harder) *The cohomology of the congruence subgroup*

$$G'(A,\mathcal{I}) \overset{\text{dfn}}{=\!=} \mathrm{Ker}\big(G'(A) \to G'(A/\mathcal{I})\big)$$

with $\mathbb{Q}$-coefficients satisfies the properties

$$H^0(G'(A,\mathcal{I}),\mathbb{Q}) = \mathbb{Q},$$

$$\dim_{\mathbb{Q}} H^{d'-1}(G'(A,\mathcal{I}),\mathbb{Q}) < +\infty$$

and

$$H^n(G'(A,\mathcal{I}),\mathbb{Q}) = (0) \qquad (\forall n \neq 0, d'-1).$$

(ii) *Let us set*

$$\chi_{d'}(\mathcal{I}) = \sum_n (-1)^n \dim_{\mathbb{Q}} H^n(G'(A,\mathcal{I}),\mathbb{Q}).$$

Then we have

(a) (Harder)

$$V_1(\mathcal{I})\chi_{d'}(\mathcal{I}) = \sum_{\pi' \in \Pi^2_{G'}} \epsilon_\infty(\pi')\dim(\pi'^\infty)^{K^\infty_{d',\mathcal{I}}},$$

(b) (Stuhler)

$$V_1(\mathcal{I})\chi_{d'}(\mathcal{I}) = \sum_{I' \subset \Delta'} \frac{|G'(A/\mathcal{I})|}{|P'_{I'}(A/\mathcal{I})|} V_{I'}(\mathcal{I}).$$

Proof : On the one hand parts (i) and (ii)(a) follow from [Har 3] (Einleitung, Satz 1, 2). On the other hand, as in the proof of (13.2.8), we see that part (ii)(b) (for varying d') is equivalent to

$$V_{d'}(\mathcal{I}) = \sum_{I' \subset \Delta'} (-1)^{s'-1} \frac{|G'(A/\mathcal{I})|}{|P'_{I'}(A/\mathcal{I})|} \prod_{j'=1}^{s'} (V_1(\mathcal{I})\chi_{d'_{j'}}(\mathcal{I}))$$

(for varying d' too; as usual $(d'_1,\ldots,d'_{s'})$ is the partition of d' corresponding to the subset I' of $\Delta' = \{1,\ldots,d'-1\}$). But this formula follows from [St] §4, Theorem 3. □

Corollary (13.3.3). — *Conjecture* (13.1.6) *holds in the following two cases:*

(i) $f^\infty = 1_{K_{\mathcal{I}}^\infty}$,

(ii) $f^\infty = 1_{K_{\mathcal{I}}^{\infty,o}} f_o$ *where* f_o *is the Drinfeld function of level* r *for some positive integer* r.

Proof: This follows directly from lemmas (13.2.3), (13.2.7), (13.2.8) and the above proposition. □

(13.4) Intersection cohomology

Let us recall that, for every $d' = 0, \dots, d-1$, $M_{\mathcal{I}}^{d-d'}$ is a smooth affine scheme over $X - (\{\infty\} \cup V(\mathcal{I}))$ of pure relative dimension $d-d'-1$ (we are assuming that $\mathcal{I}$ is a non-zero proper ideal of A) and that $GL_{d-d'}(A/\mathcal{I})$ acts on $M_{\mathcal{I}}^{d-d'}$ (via the level structure). For any smaller non-zero ideal $\mathcal{I}' \subset \mathcal{I} \subset A$ we have a finite étale Galois covering

$$M_{\mathcal{I}'}^{d-d'} \to M_{\mathcal{I}}^{d-d'} | (X - (\{\infty\} \cup V(\mathcal{I}')))$$

with Galois group $\operatorname{Ker}\big(GL_{d-d'}(A/\mathcal{I}') \twoheadrightarrow GL_{d-d'}(A/\mathcal{I})\big)$.

Conjecture (13.4.1). — *There exists a projective scheme* $\overline{M}_{\mathcal{I}}^d$ *over* $X - (\{\infty\} \cup V(\mathcal{I}))$ *containing* $M_{\mathcal{I}}^d$ *as a dense open subset and having the following properties.*

(i) $\overline{M}_{\mathcal{I}}^d$ *is normal.*

(ii) *The action of* $G(A/\mathcal{I})$ *on* $M_{\mathcal{I}}^d$ *extends to an action of* $G(A/\mathcal{I})$ *on* $\overline{M}_{\mathcal{I}}^d$.

(iii) $\overline{M}_{\mathcal{I}}^d$ *is an increasing union of Zariski open subsets which are fixed by the action of* $G(A/\mathcal{I})$,

$$\emptyset = \overline{M}_{\mathcal{I},0}^d \subset \overline{M}_{\mathcal{I},1}^d \subset \cdots \subset \overline{M}_{\mathcal{I},d}^d = \overline{M}_{\mathcal{I}}^d,$$

such that, if we set

$$M_{\mathcal{I}}^{d-d',d'} = \overline{M}_{\mathcal{I},d'+1}^d - \overline{M}_{\mathcal{I},d'}^d \qquad (\forall d' = 0, \dots, d-1),$$

we have

(a) $M_{\mathcal{I}}^{d,0} = \overline{M}_{\mathcal{I},1}^d = M_{\mathcal{I}}^d$ *and* $\overline{M}_{\mathcal{I},2}^d = M_{\mathcal{I}}^d \cup M_{\mathcal{I}}^{d-1,1}$ *is smooth over* $X - (\{\infty\} \cup V(\mathcal{I}))$,

(b) *as a* $G(A/\mathcal{I})$*-scheme* $M_{\mathcal{I}}^{d-d',d'}$ *is isomorphic to*

$$[M_{\mathcal{I}}^{d-d'} \times (F^\times\backslash(\mathbb{A}^\infty)^\times/K_{1,\mathcal{I}}^\infty)] \times_{P_J(A/\mathcal{I})} G(A/\mathcal{I})$$

for every $d' = 1, \ldots, d-1$.

Here J *is the subset of* Δ *corresponding to the partition* $(d-d', d')$ *of* d, $P_J(A/\mathcal{I})$ *acts on* $M_{\mathcal{I}}^{d-d'}$ *through its quotient*

$$\begin{aligned} &P_J(A/\mathcal{I}) \twoheadrightarrow M_J(A/\mathcal{I}) \\ &\cong GL_{d-d'}(A/\mathcal{I}) \times GL_{d'}(A/\mathcal{I}) \twoheadrightarrow GL_{d-d'}(A/\mathcal{I}), \end{aligned}$$

$P_J(A/\mathcal{I})$ *acts on* $F^\times\backslash(\mathbb{A}^\infty)^\times/K_{1,\mathcal{I}}^\infty$ *through its quotient*

$$\begin{aligned} &P_J(A/\mathcal{I}) \twoheadrightarrow M_J(A/\mathcal{I}) \\ &\cong GL_{d-d'}(A/\mathcal{I}) \times GL_{d'}(A/\mathcal{I}) \twoheadrightarrow GL_{d'}(A/\mathcal{I}) \overset{\det}{\twoheadrightarrow} (A/\mathcal{I})^\times \end{aligned}$$

and $P_J(A/\mathcal{I})$ *acts on* $G(A/\mathcal{I})$ *by right translation.*

(iv) *Let* $\mathcal{I}' \subset \mathcal{I} \subset A$ *be a smaller non-zero ideal of* A. *The finite étale Galois covering*

$$M_{\mathcal{I}'}^d \to M_{\mathcal{I}}^d|(X - (\{\infty\} \cup V(\mathcal{I}')))$$

extends to a finite Galois covering

$$\overline{M}_{\mathcal{I}'}^d \to \overline{M}_{\mathcal{I}}^d|(X - (\{\infty\} \cup V(\mathcal{I}')))$$

with the same Galois group, respecting the Zariski open subsets of (iii). *Moreover, for every* $d' = 1, \ldots, d-1$ *its restriction*

$$M_{\mathcal{I}'}^{d-d',d'} \to M_{\mathcal{I}}^{d-d',d'}|(X - (\{\infty\} \cup V(\mathcal{I}')))$$

is induced by the finite étale Galois covering

$$M_{\mathcal{I}'}^{d-d'} \to M_{\mathcal{I}}^{d-d'}|(X - (\{\infty\} \cup V(\mathcal{I}')))$$

and by the obvious homomorphism

$$F^\times\backslash(\mathbb{A}^\infty)^\times/K_{1,\mathcal{I}'}^\infty \twoheadrightarrow F^\times\backslash(\mathbb{A}^\infty)^\times/K_{1,\mathcal{I}}^\infty$$

(*we have* $K_{1,\mathcal{I}'}^\infty \subset K_{1,\mathcal{I}}^\infty$). □

REMARK (13.4.2). — We could have added a compatibility with the Hecke operators to the list of conjectural properties of $\overline{M}_{\mathcal{I}}^d$. □

By analogy with the number field case we will say that $\overline{M}_{\mathcal{I}}^d$ is the **Satake compactification** of $M_{\mathcal{I}}^d$. We denote by

$$\overline{\Theta} : \overline{M}_{\mathcal{I}}^d \to X - (\{\infty\} \cup V(\mathcal{I}))$$

its structural morphism.

It follows from (13.4.1)(iii)(b) and the strong approximation theorem (see (9.5.9)) that

$$|M_{\mathcal{I},o}^{d-d',d'}(\kappa(o)_r)| = V_1(\mathcal{I}) \frac{|G(A/\mathcal{I})|}{|P_J(A/\mathcal{I})|} \mathrm{Lef}_r(1_{K_{d-d',\mathcal{I}}^{\infty,o}})$$

for any positive integer r. Here $\kappa(o)_r$ is "the" degree r extension of the residue field $\kappa(o)$ of the place o. Therefore we may rewrite the formula of lemma (13.2.7) as

$$\text{(13.4.3)} \qquad \mathrm{Lef}_r^2(1_{K_{d-d',\mathcal{I}}^{\infty,o}}) = |M_{\mathcal{I},o}^d(\kappa(o)_r)| + \sum_{d'=1}^{d-1} \chi_{d'}(\mathcal{I}) |M_{\mathcal{I},o}^{d-d',d'}(\kappa(o)_r)|.$$

Let $j : M_{\mathcal{I}}^d \hookrightarrow \overline{M}_{\mathcal{I}}^d$ be the inclusion and let

$$\text{(13.4.4)} \qquad IC(\overline{M}_{\mathcal{I}}^d, \mathbb{Q}_\ell) = j_{!*}(\mathbb{Q}_{\ell, M_{\mathcal{I}}^d}[d])[-d] \in \mathrm{ob}\, D_c^b(\overline{M}_{\mathcal{I}}^d, \mathbb{Q}_\ell)$$

be the intersection complex of $\overline{M}_{\mathcal{I}}^d$ (see [Be–Be–De] Theorem 4.3.1) and let

$$IC^n(\overline{M}_{\mathcal{I}}^d, \mathbb{Q}_\ell) \qquad (n \in \mathbb{Z})$$

be its ordinary cohomology sheaves, so that

(a) $IC^n(\overline{M}_{\mathcal{I}}^d, \mathbb{Q}_\ell) = (0)$ for every integer $n < 0$ and for every integer $n \geq d$,

(b) $IC^0(\overline{M}_{\mathcal{I}}^d, \mathbb{Q}_\ell)$ is the constant $\mathbb{Q}_\ell$-sheaf with value $\mathbb{Q}_\ell$ on $\overline{M}_{\mathcal{I}}^d$ ($\overline{M}_{\mathcal{I}}^d$ is normal),

(c) for any positive integer $n \leq d-1$ we have

$$IC^n(\overline{M}_{\mathcal{I}}^d, \mathbb{Q}_\ell) | M_{\mathcal{I}}^{d-d',d'} = (0) \qquad (\forall d' = 0, \ldots, n).$$

Then by analogy with the number field case we are led to formulate the following

CONJECTURE (13.4.5). — (i) *For every place* $o \notin \{\infty\} \cup V(\mathcal{I})$ *of* F *the restriction of* $IC(\overline{M}_{\mathcal{I}}^{d}, \mathbb{Q}_{\ell})$ *to the fiber* $\overline{M}_{\mathcal{I},o}^{d}$ *at* o *of* $\overline{\Theta}$ *is equal to the intersection complex* $IC(\overline{M}_{\mathcal{I},o}^{d}, \mathbb{Q}_{\ell})$ *of this fiber. Moreover, for every positive integer* r, $\mathrm{Lef}_r^2(1_{K_{\mathcal{I}}^{\infty},o})$ *is the trace of* Frob_o^r *acting on the intersection cohomology*

$$R\Gamma\big(\overline{\kappa(o)} \otimes_{\kappa(o)} \overline{M}_{\mathcal{I},o}^{d}, IC(\overline{M}_{\mathcal{I},o}^{d}, \mathbb{Q}_{\ell})\big).$$

(ii) *For each* $d' = 0, \ldots, d-1$ *and each integer* n *the restriction of* $IC^n(\overline{M}_{\mathcal{I}}^{d}, \mathbb{Q}_{\ell})$ *to* $M_{\mathcal{I}}^{d-d',d'}$ *is the constant* $\mathbb{Q}_{\ell}$*-sheaf with value* $H^n(GL_{d'}(A,\mathcal{I}), \mathbb{Q}_{\ell})$. *In particular, for each* $d' = 1, \ldots, d-1$ *we have*

$$IC^n(\overline{M}_{\mathcal{I}}^{d}, \mathbb{Q}_{\ell})|M_{\mathcal{I}}^{d-d',d'} = (0) \qquad (\forall n \neq 0, d'-1).$$

REMARK (13.4.6) (Pink). — If the Picard group $\mathrm{Pic}(A)$ of A is not trivial our formulation of the first assertion of part (ii) of conjecture (13.4.5) may be slightly incorrect. For each $d' = 0, \ldots, d-1$ we have an obvious projection from

$$M_{\mathcal{I}}^{d-d',d'} \cong \big[M_{\mathcal{I}}^{d-d'} \times (F^{\times}\backslash(\mathbb{A}^{\infty})^{\times}/K_{1,\mathcal{I}}^{\infty})\big] \times_{P_J(A/\mathcal{I})} G(A/\mathcal{I})$$

onto

$$\mathrm{Pic}(A) \cong F^{\times}\backslash(\mathbb{A}^{\infty})^{\times}/K_{1,A}^{\infty}.$$

Then a more correct formulation may be

Let L *be an invertible* A*-module and let* $N = A^{d'-1} \oplus L$ *be the corresponding rank* d' *projective* A*-module (all the isomorphism classes of rank* d' *projective* A*-modules are obtained in this way and* L *may be recovered from* N *as its determinant). For each integer* n *the restriction of* $IC^n(\overline{M}_{\mathcal{I}}^{d}, \mathbb{Q}_{\ell})$ *to the fiber of the above projection at the class of* L *in* $\mathrm{Pic}(A)$ *is the constant* $\mathbb{Q}_{\ell}$*-sheaf with value*

$$H^n(\mathrm{Aut}_A(N,\mathcal{I}), \mathbb{Q}_{\ell}),$$

where we have set

$$\mathrm{Aut}_A(N,\mathcal{I}) = \mathrm{Ker}\big(\mathrm{Aut}_A(N) \to \mathrm{Aut}_{A/\mathcal{I}}(N/\mathcal{I}N)\big).$$

□

EXAMPLE (13.4.7). — Let us consider the particular case $A = \mathbb{F}_q[T]$ and $\mathcal{I} = (T)$ ($X = \mathbb{P}_{\mathbb{F}_q}^1$ with its usual point ∞ and $V(\mathcal{I}) = \{0\}$). Then we have

$$\zeta_{X-\{\infty\}} = \frac{1}{1-q^{1-s}}$$

and, for every $d' = 1, \ldots, d-1$, we have

$$\chi_{d'}(\mathcal{I}) = \sum_{I' \subset \Delta'} \frac{q^{d'(d'-1)/2}(q-1)\cdots(q^{d'}-1)}{q^{(d'^2-(d_1'^2+\cdots+d_{s'}'^2))}(q-1)^{s'}} \prod_{j'=1}^{s'} \frac{1}{(1-q^2)\cdots(1-q^{d'_{j'}})}.$$

It is elementary to check that

$$\chi_{d'}(\mathcal{I}) = 1 \qquad (\forall d' = 1, \ldots, d-1),$$

so that

$$H^n(GL_{d'}(A,\mathcal{I}), \mathbb{Q}_\ell) = \begin{cases} \mathbb{Q}_\ell & \text{if } n = 0, \\ (0) & \text{otherwise.} \end{cases}$$

Therefore, in this particular case we conjecture that

$$IC(\overline{M}^d_{\mathcal{I}}, \mathbb{Q}_\ell) = \mathbb{Q}_{\ell, \overline{M}^d_{\mathcal{I}}}[0].$$

This has been worked out geometrically by Pink (see [Pi 2]).

(13.5) Comments and references

The first part of conjecture (13.4.5) is reminiscent to Zucker's conjecture. The second part is in some sense a consequence of the first one. Zucker's conjecture has been proved by Saper and Stern (and also by Looijenga and by Looijenga and Rapoport). It says that the intersection cohomology of the Satake–Baily–Borel compactification of a Shimura variety is equal to the L^2-cohomology of this Shimura variety.

In the case $d = 2$ conjectures (13.4.1) and (13.4.5) have been proved by Drinfeld. Conjecture (13.4.1) is almost proved (see [Gek], [Kap] and [Pi 2]). In the case $d = 3$ conjecture (13.4.5) has been proved by Pink (private communication).

Appendices

D

Representations of unimodular, locally compact, totally discontinuous, separated, topological groups: addendum

(D.9) Restricted tensor products

Let $(H_y)_{y \in Y}$ be a countable family of unimodular, locally compact, totally discontinuous, separated, topological groups. For each $y \in Y$ let I_y be a compact open subgroup of H_y. We denote by

$$H = \prod_{y \in Y} (H_y, I_y)$$

the **restricted product of** the H_y with respect to the I_y, i.e.

$$H = \varinjlim_{T} (H_T \times I^T)$$

where T through runs the inductive system of all the finite subsets of Y, where

$$H_T = \prod_{y \in T} H_y,$$

where

$$I^T = \prod_{y \in Y-T} I_y$$

and where, for any $T' \subset T'' \subset Y$, the transition homomorphism is induced by the inclusion

$$I_{T''-T'} = \prod_{y \in T''-T'} I_y \subset \prod_{y \in T''-T'} H_y = H_{T''-T'}.$$

Endowed with the direct limit topology H is a locally compact, totally discontinuous, separated, topological group and

$$I = \prod_{y \in Y} I_y$$

is a compact open subgroup.

The topological group H is unimodular. If dh_y is a Haar measure on H_y for every $y \in Y$ and if we assume that

$$\mathrm{vol}(I_y, dh_y) = 1$$

for almost all $y \in Y$, there exists a unique Haar measure

$$dh = \prod_{y \in Y} dh_y$$

on H such that the restriction of dh to $H_T \times I^T$ is equal to $dh_T \times di^T$, where

$$dh_T = \prod_{y \in T} dh_y,$$

and where di^T is the Haar measure on the compact topological group I^T which is normalized by $\mathrm{vol}(I^T, di^T) = 1$. If dh_y is rational for every $y \in Y$ then dh is rational too.

Let us fix a rational Haar measure dh_y on H_y for every $y \in Y$, let us assume that $\mathrm{vol}(I_y, dh_y) = 1$ for almost all $y \in Y$ and let dh be the corresponding rational Haar measure on H. Then we have the convolution $\mathbb{Q}$-algebras $C_c^\infty(H_y//I_y)$, $y \in Y$, and $C_c^\infty(H//I)$. For each finite subset T of Y the convolution $\mathbb{Q}$-algebra $C_c^\infty(H_T//I_T)$ (with respect to dh_T) is canonically isomorphic to

$$\bigotimes_{y \in T} C_c^\infty(H_y//I_y)$$

(map $\bigotimes_{y \in T} f_y$ into f_T with

$$f_T(h_T) = \prod_{y \in T} f_y(h_y)$$

for every $h_T = (h_y)_{y \in T} \in H_T$). Moreover $C_c^\infty(H//I)$ is canonically isomorphic to

$$\varinjlim_{T} C_c^\infty(H_T//I_T)$$

where T again runs through the inductive system of all the finite subsets of Y and where, for any $T' \subset T'' \subset Y$, the transition morphism is defined by

$$f_{T'} \longmapsto f_{T'} \otimes e_{I_{T''-T'}}$$

(map $f_T \in C_c^\infty(H_T//I_T)$ into $f_T \otimes e_{I^T} \in C_c^\infty(H//I)$ with

$$(f_T \otimes e_{I^T})(h) = f_T(h_T)e_{I^T}(h^T)$$

for every $h = (h_T, h^T) \in H_T \times H^T = H$).

For each $y \in Y$ let $(\mathcal{V}_y, \pi_y)$ be a smooth representation of H_y and let us assume that for almost all $y \in Y$ we have

$$\dim_{\mathbb{C}}(\mathcal{V}_y^{I_y}) = 1.$$

Then we define (up to canonical isomorphism) a representation

$$(\mathcal{V}, \pi) = \bigotimes_{y \in Y}{}' (\mathcal{V}_y, \pi_y) \tag{D.9.1}$$

of H in the following way. We choose a finite set $T_0 \subset Y$ such that

$$\dim_{\mathbb{C}}(\mathcal{V}_y^{I_y}) = 1$$

for every $y \in Y - T_0$ and we choose a non-zero vector v_y^0 in $\mathcal{V}_y^{I_y}$ for every $y \in Y - T_0$. The space $\mathcal{V}$ is the direct limit

$$\varinjlim_T \Big(\bigotimes_{y \in T} \mathcal{V}_y\Big)$$

where T runs through the inductive system of all the finite subsets of Y containing T_0 and where, for any $T' \subset T'' \subset Y$, the transition morphism is defined by

$$\bigotimes_{y \in T'} v_y \longmapsto \Big(\bigotimes_{y \in T'} v_y\Big) \otimes \Big(\bigotimes_{y \in T'' - T'} v_y^0\Big).$$

We denote by

$$v_T \otimes v^{0,T}$$

the image of

$$v_T \in \mathcal{V}_T \overset{\text{dfn}}{=\!=} \bigotimes_{y \in T} \mathcal{V}_y$$

in $\mathcal{V}$. The representation π of H on $\mathcal{V}$ is given by

$$\pi(h_T, i^T)(v_T \otimes v^{0,T}) = \pi_T(h_T)(v_T) \otimes v^{0,T}$$

for every $h_T \in H_T$, every $i^T \in I^T$, every $v_T \in \mathcal{V}_T$ and every finite subset T of Y. Here

$$\pi_T = \bigotimes_{y \in Y} \pi_y$$

is a representation of H_T on $\mathcal{V}_T$. It is easy to check that $(\mathcal{V}, \pi)$ is a smooth representation. It is called the **restricted tensor product** of the family of smooth representations $(\mathcal{V}_y, \pi_y)_{y \in Y}$.

LEMMA (D.9.2). — (i) *Let us assume that* $(\mathcal{V}_y, \pi_y)$ *is admissible for every* $y \in Y$. *Then the restricted tensor product representation* $(\mathcal{V}, \pi)$ *is admissible too and its contragredient representation* $(\widetilde{\mathcal{V}}, \widetilde{\pi})$ *is canonically isomorphic to the restricted product of the contragredient representations* $(\widetilde{\mathcal{V}}_y, \widetilde{\pi}_y)$ *of* $(\mathcal{V}_y, \pi_y)$, $y \in Y$ *(note that*

$$\dim_{\mathbb{C}}(\widetilde{\mathcal{V}}_y^{I_y}) = \dim_{\mathbb{C}}(\mathcal{V}_y^{I_y}) = 1$$

for almost all $y \in Y$ *by* (D.1.10)).

(ii) *Let us assume moreover that* $(\mathcal{V}_y, \pi_y)$ *is irreducible for every* $y \in Y$. *Then the restricted tensor product representation* $(\mathcal{V}, \pi)$ *is irreducible too.*

Proof: For each finite subset T of Y and each compact open subgroup J_T of H_T, $J_T \times I^T \subset H$ is a compact open subgroup and in this way we obtain a basis of neighborhoods of 1 in H.

Arguing as in the proof of (D.7.1) we get a canonical isomorphism

$$(\mathcal{V}_{T'})^{J_T \times I_{T'-T}} \cong (\mathcal{V}_T)^{J_T} \otimes_{\mathbb{C}} \Big(\bigotimes_{y \in T'-T} \mathcal{V}_y^{I_y} \Big)$$

for any finite subsets $T \subset T'$ of Y and any compact open subgroup J_T of H_T. It follows that

$$\mathcal{V}^{J_T \times I^T} = (\mathcal{V}_T)^{J_T} \otimes v^{0,T}$$

for any finite subset T of Y containing T_0 and any compact open subgroup J_T of H_T.

As $(\mathcal{V}_T, \pi_T)$ is admissible for any finite subset T of Y (see (D.7.1)(i)), $(\mathcal{V}, \pi)$ is admissible.

As $((\mathcal{V}_T)^{\sim}, (\pi_T)^{\sim})$ is canonically isomorphic to

$$\bigotimes_{y \in T} (\widetilde{\mathcal{V}}_y, \widetilde{\pi}_y)$$

for any finite subset T of Y, the canonical map

$$\bigotimes_{y \in Y}{}' (\widetilde{\mathcal{V}}_y, \widetilde{\pi}_y) \longrightarrow (\widetilde{\mathcal{V}}, \widetilde{\pi})$$

is an isomorphism (for every finite subset T containing T_0 and every compact open subgroup J_T of H_T it induces an isomorphism on the spaces of $(J_T \times I^T)$-invariants). Here we have chosen for $\widetilde{v}_y^0 \in \widetilde{\mathcal{V}}_y^{I_y}$ the unique vector such that

$$\langle \widetilde{v}_y^0, v_y^0 \rangle = 1 \qquad (\forall y \in Y - T_0).$$

Under the assumptions of (D.9.2)(ii) $(\mathcal{V}_T, \pi_T)$ is irreducible for every finite subset T of Y (see (D.7.1)(ii)). Therefore $(\mathcal{V}, \pi)$ is irreducible. Indeed, for any finite subset T containing T_0 and any compact open subgroup J_T of H_T the left $(\mathbb{C} \otimes C_c^\infty(H//(J_T \times I^T)))$-module $\mathcal{V}^{J_T \times I^T}$ is irreducible (remark that $\mathbb{C}v^{0,T}$ is an irreducible left $(\mathbb{C} \otimes C_c^\infty(H^T//I^T))$-module). □

THEOREM (D.9.3). — *Let us assume that the* $\mathbb{Q}$*-algebra* $\mathcal{C}_c^\infty(H_y//I_y)$ *is commutative for almost every* $y \in Y$ *and let* $(\mathcal{V}, \pi)$ *be an admissible irreducible representation of* H*. Then for each* $y \in Y$ *there exists an admissible irreducible representation of* $(\mathcal{V}_y, \pi_y)$ *of* H_y *with*

$$\dim_{\mathbb{C}}(\mathcal{V}_y^{I_y}) = 1$$

for almost every $y \in Y$ *and there exists an isomorphism*

$$(\mathcal{V}, \pi) \cong \bigotimes_{y\in Y}{}'(\mathcal{V}_y, \pi_y)$$

in $\mathrm{Rep}_{\mathrm{s}}(H)$.

Moreover, for each $y \in Y$ *the isomorphism class of* $(\mathcal{V}_y, \pi_y)$ *in* $\mathrm{Rep}_{\mathrm{s}}(H_y)$ *is uniquely determined by* $(\mathcal{V}, \pi)$.

Proof: Let us choose a finite subset T of Y and a compact open subgroup J_T of H_T such that $\mathcal{C}_c^\infty(H_y//I_y)$ is commutative for all $y \in Y - T$ and such that the left $(\mathbb{C} \otimes \mathcal{C}_c^\infty(H//(J_T \times I^T)))$-module

$$\mathcal{V}^{J_T \times I^T}$$

is non-zero. Then this module is irreducible (see (D.1.8)). Moreover, by the admissibility of $(\mathcal{V}, \pi)$ it is finite dimensional over $\mathbb{C}$. Therefore there exist an irreducible left $(\mathbb{C} \otimes \mathcal{C}_c^\infty(H_T//J_T))$-module V_T, an irreducible left $(\mathbb{C} \otimes \mathcal{C}_c^\infty(H^T//I^T))$-module V^T and an isomorphism

$$\mathcal{V}^{J_T \times I^T} \cong V_T \otimes_{\mathbb{C}} V^T$$

of left $(\mathbb{C} \otimes \mathcal{C}_c^\infty(H//(J_T \times I^T)))$-modules (we have a canonical isomorphism of $\mathbb{Q}$-algebras

$$\mathcal{C}_c^\infty(H//(J_T \times I^T)) \cong \mathcal{C}_c^\infty(H_T//J_T) \otimes \mathcal{C}_c^\infty(H^T//I^T)$$

and we can apply [Bou] Algèbre, VIII, §7, Proposition 8).

The $\mathbb{Q}$-algebra

$$\mathcal{C}_c^\infty(H^T//I^T) = \varinjlim_{T' \cap T = \emptyset} \Big(\bigotimes_{y\in T'} \mathcal{C}_c^\infty(H_y//I_y)\Big)$$

is commutative and V^T is finite dimensional over $\mathbb{C}$ and non-zero. Therefore we have

$$\dim_{\mathbb{C}}(V^T) = 1.$$

For each $y \in Y - T$ let

$$\chi_y : \mathbb{C} \otimes \mathcal{C}_c^\infty(H_y//I_y) \to \mathbb{C}$$

be the character of the action of $\mathbb{C} \otimes \mathcal{C}_c^\infty(H_y//I_y)$ on V^T.

Thanks to (D.1.8) there exists an admissible irreducible representation $(\mathcal{V}_T, \pi_T)$ of H_T such that

$$(\mathcal{V}_T)^{J_T} \cong V_T$$

as a left $(\mathbb{C} \otimes \mathcal{C}_c^\infty(H_T//J_T))$-module and, for each $y \in Y - T$, there exists an admissible irreducible representation $(\mathcal{V}_y, \pi_y)$ of H_y such that

$$\mathcal{V}_y^{I_y} \cong (\mathbb{C}, \chi_y)$$

as a left $(\mathbb{C} \otimes \mathcal{C}_c^\infty(H_y//I_y))$-module. Thanks to (D.7.1), for each $y \in T$ there exists an admissible irreducible representation $(\mathcal{V}_y, \pi_y)$ of H_y such that $(\mathcal{V}_T, \pi_T)$ is isomorphic to

$$\bigotimes_{y \in T} (\mathcal{V}_y, \pi_y)$$

in $\mathrm{Rep}_s(H_T)$. We may form the restricted tensor product

$$\bigotimes_{y \in Y}{}' (\mathcal{V}_y, \pi_y).$$

This is an admissible irreducible representation of H (see (D.9.2)) and, by construction, its space of $(J_T \times I^T)$-invariants is isomorphic to

$$\mathcal{V}^{J_T \times I^T}$$

as a left $(\mathbb{C} \otimes \mathcal{C}_c^\infty(H//(J_T \times I^T)))$-module. Applying (D.1.8) once more we get that this restricted tensor product is isomorphic to $(\mathcal{V}, \pi)$ in $\mathrm{Rep}_s(H)$.

For each $y_0 \in Y$ we have a canonical isomorphism

$$\bigotimes_{y \in Y}{}' (\mathcal{V}_y, \pi_y) \cong (\mathcal{V}_{y_0}, \pi_{y_0}) \otimes \Big(\bigotimes_{y \in Y - \{y_0\}}{}' (\mathcal{V}_y, \pi_y) \Big)$$

of admissible irreducible representations of

$$H = H_{y_0} \times H^{y_0}.$$

Therefore $(\mathcal{V}, \pi)$ uniquely determines the isomorphism class of $(\mathcal{V}_{y_0}, \pi_{y_0})$ in $\mathrm{Rep}_s(H_{y_0})$ (see (D.7.1)(ii)). □

REMARKS (D.9.4.1). — In the decomposition

$$(\mathcal{V},\pi) \cong \bigotimes_{j=1}^{s} (\mathcal{V}_j,\pi_j)$$

of (D.7.1)(ii) (resp.

$$(\mathcal{V},\pi) \cong \bigotimes_{y\in Y}{}' (\mathcal{V}_y,\pi_y)$$

of (D.9.3)) the admissible irreducible representation $(\mathcal{V},\pi)$ of H is unitarizable if and only if each admissible irreducible representation $(\mathcal{V}_j,\pi_j)$ of H_j, $j=1,\dots,s$ (resp. $(\mathcal{V}_y,\pi_y)$ of H_y, $y\in Y$), is unitarizable (see (D.6.3) and (D.7.1)(i) (resp. (D.6.3) and (D.9.2)(i))).

(D.9.4.2). — For any

$$(\mathcal{V},\pi) \cong \bigotimes_{j=1}^{s} (\mathcal{V}_j,\pi_j)$$

as in (D.7.1)(ii) (resp.

$$(\mathcal{V},\pi) \cong \bigotimes_{y\in Y}{}' (\mathcal{V}_y,\pi_y)$$

as in (D.9.3)) and for any family $f_j \in \mathbb{C}\otimes \mathcal{C}_c^\infty(H_j)$, $j=1,\dots,s$ (resp. $f_y \in \mathbb{C}\otimes \mathcal{C}_c^\infty(H_y)$, $y\in Y$, with $f_y = 1_{I_y}$ for almost every $y\in Y$), we have

$$\operatorname{tr}\pi\Big(\prod_{j=1}^{s} f_j\Big) = \prod_{j=1}^{s} \operatorname{tr}\pi_j(f_j)$$

(resp.

$$\operatorname{tr}\pi\Big(\prod_{y\in Y} f_y\Big) = \prod_{y\in Y} \operatorname{tr}\pi_y(f_y))$$

where we have denoted by $\prod_{j=1}^{s} f_j \in \mathbb{C}\otimes \mathcal{C}_c^\infty(H)$ (resp. $\prod_{y\in Y} f_y \in \mathbb{C}\otimes \mathcal{C}_c^\infty(H)$) the function defined by

$$\Big(\prod_{j=1}^{s} f_j\Big)(h_1,\dots,h_s) = \prod_{j=1}^{s} f_j(h_j)$$

(resp.

$$\Big(\prod_{y\in Y} f_y\Big)((h_y)_{y\in Y}) = \prod_{y\in Y} f_y(h_y))$$

(recall that $\mathrm{vol}(I_y, dh_y) = 1$ for almost all $y\in Y$)). □

(D.10) Rationality properties

Let H be a unimodular, locally compact, totally discontinuous, separated, topological group and let dh be a rational Haar measure on H (see (D.1)). Let $(\mathcal{W},\rho)$ be an admissible representation of H and let E be a subfield of $\mathbb{C}$. We assume that $(\mathcal{W},\rho)$ admits a rational E-structure, i.e. that there exists an E-vector subspace $\mathcal{W}^\circ$ of $\mathcal{W}$ which is $\rho(H)$-invariant and such that the natural map $\mathbb{C}\otimes_E \mathcal{W}^\circ \to \mathcal{W}$ is an isomorphism. For each compact open subgroup I of H the natural map

$$\mathbb{C}\otimes_E (\mathcal{W}^\circ)^I \to \mathcal{W}^I$$

is an isomorphism too and, in particular, $(\mathcal{W}^\circ)^I$ is finite dimensional over E.

Let $(\mathcal{V},\pi)$ be an irreducible admissible representation of H which occurs as a subquotient of $(\mathcal{W},\rho)$. We will denote by $\mathbb{Q}(\pi)$ the subfield of $\mathbb{C}$ generated (over $\mathbb{Q}$) by the complex numbers $\operatorname{tr}\pi(f)$ for $f\in \mathcal{C}_c^\infty(H)$ and by $E(\pi)$ the subfield of $\mathbb{C}$ generated by E and $\mathbb{Q}(\pi)$.

Lemma (D.10.1). — *Under the above hypotheses $E(\pi)$ is a finite extension of E and $(\mathcal{V},\pi)$ admits a rational structure over a finite extension of $E(\pi)$ in $\mathbb{C}$.*

Proof : We fix a compact open subgroup I of H such that $\mathcal{V}^I \neq (0)$. Then $\mathcal{V}^I$ is a simple $(\mathbb{C}\otimes \mathcal{C}_c^\infty(H//I))$-module (see (D.1.8)) and there exist two $(E\otimes \mathcal{C}_c^\infty(H//I))$-submodules $W_1^\circ \subset W_2^\circ$ of $(\mathcal{W}^\circ)^I$ such that W_2°/W_1° is simple and $\mathcal{V}^I$ occurs as a subquotient of the $(\mathbb{C}\otimes \mathcal{C}_c^\infty(H//I))$-module $\mathbb{C}\otimes_E (W_2^\circ/W_1^\circ)$. But the last module is semi-simple (E is of characteristic 0 and W_2°/W_1° is finite dimensional over E, see [Bou] Algèbre, VIII, §7, n°5), so that $\mathcal{V}^I$ occurs as a quotient of $\mathbb{C}\otimes_E (W_2^\circ/W_1^\circ)$. By adjunction we get a surjective map of $(\mathbb{C}\otimes \mathcal{C}_c^\infty(H))$-modules,

$$(*)\qquad \mathcal{C}_c^\infty(H/I)\otimes_{\mathcal{C}_c^\infty(H//I)} (\mathbb{C}\otimes_E (W_2^\circ/W_1^\circ)) \twoheadrightarrow \mathcal{V}.$$

Arguing as in the proof of (D.1.8) we see that $\mathcal{C}_c^\infty(H/I)\otimes_{\mathcal{C}_c^\infty(H//I)} (W_2^\circ/W_1^\circ)$ admits a largest quotient $\overline{\mathcal{W}}^\circ$ such that

$$\operatorname{Hom}_{E\otimes \mathcal{C}_c^\infty(H)}(\mathcal{U},\mathbb{C}\otimes_E \overline{\mathcal{W}}^\circ) = (0)$$

for any non-degenerate $(E\otimes \mathcal{C}_c^\infty(H))$-module $\mathcal{U}$ with $\mathcal{U}^I = (0)$ and that this quotient is simple ($\mathbb{C}\otimes_E \overline{\mathcal{W}}^\circ$ is the quotient

$$\left(\mathcal{C}_c^\infty(H/I)\otimes_{\mathcal{C}_c^\infty(H//I)} (\mathbb{C}\otimes_E (W_2^\circ/W_1^\circ))\right)_2$$

with the notation of the proof of (D.1.8)). The kernel of the map $(*)$ does not have any non-zero fixed vector under I and, therefore, $(*)$ factors through

$\overline{\mathcal{W}} = \mathbb{C} \otimes_E \overline{\mathcal{W}}^\circ$. Replacing $\mathcal{W}$ by $\overline{\mathcal{W}}$ we see that we may assume that $\mathcal{W}^\circ$ is a simple $(E \otimes C_c^\infty(H))$-module and that, for any non-zero $(E \otimes C_c^\infty(H))$-submodule $\mathcal{U} \subset \mathcal{W}$, $\mathcal{U}^I$ is non-zero.

From now on we make these two assumptions. Then $(\mathcal{W}, \rho)$ is semi-simple and of finite length. Indeed $\mathcal{W}^I$ is a semi-simple $(\mathbb{C} \otimes C_c^\infty(H//I))$-module (same argument as above) and, if

$$\mathcal{W}^I \cong \bigoplus_{\alpha \in A} W_\alpha$$

is a decomposition into a finite sum of irreducible modules (recall that $\mathcal{W}^I$ is finite dimensional over $\mathbb{C}$), we have

$$\mathcal{W} \cong \left(C_c^\infty(H/I) \otimes_{C_c^\infty(H//I)} \mathcal{W}^I\right)_2 \cong \bigoplus_{\alpha \in A} \left(C_c^\infty(H/I) \otimes_{C_c^\infty(H//I)} W_\alpha\right)_2$$

with the notation of the proof of (D.1.8), each summand of the last sum being simple.

Moreover the canonical maps

$$\mathrm{End}_H(\mathcal{V}, \pi) \to \mathrm{End}_{\mathbb{C} \otimes C_c^\infty(H//I)}(\mathcal{V}^I) = \mathbb{C}$$

and

$$D^\circ = \mathrm{End}_H(\mathcal{W}^\circ, \rho^\circ) \to \mathrm{End}_{E \otimes C_c^\infty(H//I)}\left((\mathcal{W}^\circ)^I\right),$$

where ρ° is the restriction of ρ to $\mathcal{W}^\circ$, are bijective and the canonical map

$$D = \mathrm{End}_H(\mathcal{W}, \rho) \to \mathrm{End}_{\mathbb{C} \otimes C_c^\infty(H//I)}(\mathcal{W}^I)$$

is injective. Indeed we have $\mathcal{V} = C_c^\infty(H/I) \cdot \mathcal{V}^I$ and $\mathcal{W}^\circ = C_c^\infty(H/I) \cdot (\mathcal{W}^\circ)^I$ by the irreducibility of $(\mathcal{V}, \pi)$ and $(\mathcal{W}^\circ, \rho^\circ)$ and, thus, we have $\mathcal{W} = C_c^\infty(H/I) \cdot \mathcal{W}^I$ (even if $(\mathcal{W}, \rho)$ is not irreducible any more); the injectivity of the three maps follows; for the surjectivity of the first map we remark that any endomorphism of $\mathcal{V}^I$ induces an endomorphism of the left $C_c^\infty(H)$-module $C_c^\infty(H/I) \otimes_{C_c^\infty(H//I)} \mathcal{V}^I$ which maps the submodule $\{v \mid C_c^\infty(I\backslash H) \cdot v = 0\}$ into itself and which thus induces an endomorphism of the left $C_c^\infty(H)$-module $\mathcal{V}$ (see (D.1.8)); the same argument gives the surjectivity of the second map. In particular we have

$$\mathrm{End}_H(\mathcal{V}, \pi) = \mathbb{C},$$

D° is a finite dimensional division algebra over E ($\mathcal{V}^I$ and $(\mathcal{W}^\circ)^I$ are simple, see [Bou] Algèbre, VIII, §4, n°3), the canonical map

$$\mathbb{C} \otimes_E D^\circ \to D$$

is bijective (the canonical map

$$\mathbb{C}\otimes_E \mathrm{End}_{E\otimes C_c^\infty(H//I)}((\mathcal{W}^\circ)^I) \to \mathrm{End}_{\mathbb{C}\otimes C_c^\infty(H//I)}(\mathcal{W}^I)$$

is bijective by the finite dimensionality of $(\mathcal{W}^\circ)^I)$ and D is a semi-simple algebra over $\mathbb{C}$ (see [Bou] Algèbre, VIII, §7, n°5).

Now let F be the center of D°. It is a finite extension of E. If $(\mathcal{V}',\pi')\cong(\mathcal{V},\pi)^m$ is the isotypic component of $(\mathcal{V},\pi)$ in $(\mathcal{W},\rho)$, the $\mathbb{C}$-algebra $\mathrm{End}(\mathcal{V}',\pi')\cong gl_m(\mathbb{C})$ is a factor of D. Therefore the center $\mathbb{C}$ of $\mathrm{End}(\mathcal{V}',\pi')$ is a factor of the center $\mathbb{C}\otimes_E F$ of D, this factor determines an embedding of F into $\mathbb{C}$ over the inclusion $E\subset\mathbb{C}$ and we have

$$(\mathcal{V}',\pi')\cong(\mathcal{W},\rho)\otimes_{\mathbb{C}\otimes_E F}\mathbb{C}\cong(\mathcal{W}^\circ,\rho^\circ)\otimes_F\mathbb{C}.$$

On the one hand it follows that

$$E(\pi)=E(\{\mathrm{tr}\,\pi'(f)\mid f\in C_c^\infty(H)\})\subset F$$

is a finite extension of E. On the other hand, if $F\subset F'\subset\mathbb{C}$ is a finite extension of F which splits the finite dimensional central division algebra D° over F, so that $D^\circ\otimes_F F'\cong gl_n(F')$ for some positive integer n, we necessarily have $n=m$ and we have

$$(\mathcal{V},\pi)^m\cong((\mathcal{W}^\circ,\rho^\circ)\otimes_F F')\otimes_{F'}\mathbb{C}\cong(\mathcal{W}',\rho')^n$$

where $\mathcal{W}'$ is an admissible irreducible representation of H which admits a rational structure over F'. Hence $(\mathcal{V},\pi)$ admits a rational structure over F'.

□

LEMMA (D.10.2). — *Let* $(\mathcal{V},\pi)$ *and* $(\mathcal{W},\rho)$ *be as in lemma* (D.10.1). *Let us assume moreover that*

$$(\mathcal{W},\rho)\cong(\mathcal{V},\pi)^m\oplus(\mathcal{W}',\rho')$$

for some positive integer m *and some admissible representation* $(\mathcal{W}',\rho')$ *of* H *which does not admit* $(\mathcal{V},\pi)$ *as a subquotient. Then* $(\mathcal{V},\pi)^m$ *admits a rational structure over* $E(\pi)$.

Proof : We again fix a compact open subgroup I of H such that $\mathcal{V}^I\neq 0$. We have the splitting

$$\mathcal{W}^I\cong(\mathcal{V}^I)^m\oplus\mathcal{W}'^I$$

of $(\mathbb{C}\otimes C_c^\infty(H//I))$-modules.

Let us simply denote by Φ the functor

$$U\mapsto(C_c^\infty(H/I)\otimes_{C_c^\infty(H//I)}U)_2$$

from the category of $(\mathbb{C} \otimes C_c^\infty(H//I))$-modules to the category of non-degenerate $(\mathbb{C} \otimes C_c^\infty(H))$-modules with the notation of the proof of (D.1.8). Any rational E-structure U° on U induces a rational E-structure $\Phi(U^\circ)$ on $\Phi(U)$. We may replace $\mathcal{W}$ by $\Phi(\mathcal{W}^I)$ and therefore $\mathcal{W}'$ by $\Phi(\mathcal{W}'^I)$. Indeed we have seen in (D.1.8) that $\Phi(\mathcal{V}^I)$ is canonically isomorphic to $\mathcal{V}$.

Then $\mathcal{V}^I$ does not occur as a subquotient of $\mathcal{W}'^I$. Otherwise we could find a submodule $W \subset \mathcal{W}'^I$ and a surjective map $W \twoheadrightarrow \mathcal{V}^I$, we would have an epimorphism $\Phi(W) \twoheadrightarrow \mathcal{V}$ and a monomorphism $\Phi(W) \hookrightarrow \Phi(\mathcal{W}'^I) = \mathcal{W}'$ (it is clear that Φ maps epimorphisms into epimorphisms and monomorphisms into monomorphisms) and we would get a contradiction.

Let us consider the finite extension $E(\mathcal{V}^I) = E(\{\mathrm{tr}(f, \mathcal{V}^I) \mid f \in C_c^\infty(H//I)\})$ of E in $\mathbb{C}$. It is obvious that $E(\mathcal{V}^I) \subset E(\pi)$. Therefore, to finish the proof of the lemma it is sufficient to prove that $(\mathcal{V}^I)^m$ admits a rational structure over $E(\mathcal{V}^I)$. But we may find a finite extension F of E in $\mathbb{C}$ and rational structures V° and W'° on $\mathcal{V}^I$ and $\mathcal{W}'^I$ over F such that the splitting

$$\mathcal{W}^I \cong (\mathcal{V}^I)^m \oplus \mathcal{W}'^I$$

is induced by a splitting

$$F \otimes_E W^\circ \cong (V^\circ)^m \oplus W'^\circ$$

of $(F \otimes C_c^\infty(H//I))$-modules, where W° is the given rational structure on $\mathcal{W}^I$ over E. Clearly F contains $E(\mathcal{V}^I)$ and, enlarging F if it is necessary, we may assume that F is Galois over $E(\mathcal{V}^I)$. Now, if $\sigma \in \mathrm{Gal}(F/E(\mathcal{V}^I))$, the image under $\sigma \otimes \mathrm{id}_W$ of $(V^\circ)^m$ has the same trace as $(V^\circ)^m$ and is thus isomorphic to $(V^\circ)^m$ (see [Bou] Algèbre, VIII, §12, Proposition 3). But, as V° does not occur as a subquotient in W'°, this image is equal to $(V^\circ)^m \subset F \otimes_E W^\circ$ and the automorphisms $\sigma \otimes \mathrm{id}_W | (V^\circ)^m$, $\sigma \in \mathrm{Gal}(F/E(\mathcal{V}^I))$, define a descent datum on $(V^\circ)^m$ which is effective by the finite dimensionality of $(V^\circ)^m$. □

(D.11) The Grothendieck group of admissible representations

Let H be a unimodular, locally compact, totally discontinuous, separated, topological group. We fix a system of representatives $\mathcal{S}$ of the isomorphism classes of admissible irreducible representations of H and we denote by $K_{\mathrm{a}}(H)$ the abelian group of formal linear combinations

$$\sum_{\pi \in \mathcal{S}} m(\pi)\pi \in \mathbb{Z}^{\mathcal{S}}$$

such that, for each compact open subgroup I of H, the set

$$\{\pi \in \mathcal{S} \mid m(\pi) \neq 0 \text{ and } \mathcal{V}_\pi^I \neq (0)\}$$

is finite (we have denoted by $\mathcal{V}_\pi$ the space of the representation π). Up to isomorphism this abelian group does not depend on the choice of the system of representatives $\mathcal{S}$ and we call it the **Grothendieck group of admissible representations** of H.

Let $(\mathcal{W},\rho)$ be an admissible representation of H. For each $\pi\in\mathcal{S}$ and each compact open subgroup I of H such that $\mathcal{V}_\pi^I\neq(0)$ let us denote by $m_{\mathcal{W}^I}(\mathcal{V}_\pi^I)$ the multiplicity of the irreducible left $(\mathbb{C}\otimes\mathcal{C}_c^\infty(H//I))$-module $\mathcal{V}_\pi^I$ in the left $(\mathbb{C}\otimes\mathcal{C}_c^\infty(H//I))$-module $\mathcal{W}^I$ (we have $\dim_{\mathbb{C}}\mathcal{W}^I<+\infty$ and $\mathcal{V}_\pi^I$ is irreducible by (D.1.8)).

LEMMA (D.11.1). — *For any $\pi\in\mathcal{S}$ and any compact open subgroups I and J of H such that $\mathcal{V}_\pi^I\neq(0)$ and $\mathcal{V}_\pi^J\neq(0)$ we have*

$$m_{\mathcal{W}^J}(\mathcal{V}_\pi^J)=m_{\mathcal{W}^I}(\mathcal{V}_\pi^I).$$

Proof: We may assume that $J\subset I$. Let

$$(0)=(\mathcal{W}^J)_0\subsetneqq(\mathcal{W}^J)_1\subsetneqq\cdots\subsetneqq(\mathcal{W}^J)_N=\mathcal{W}^J$$

be a Jordan–Hölder tower for the left $(\mathbb{C}\otimes\mathcal{C}_c^\infty(H//J))$-module $\mathcal{W}^J$. By applying the exact functor $(\cdot)^I$ to the above tower we get a tower of $(\mathbb{C}\otimes\mathcal{C}_c^\infty(H//I))$-modules,

$$(0)=(\mathcal{W}^J)_0^I\subset(\mathcal{W}^J)_1^I\subset\cdots\subset(\mathcal{W}^J)_N^I=\mathcal{W}^I.$$

If $n_1,\ldots,n_m$ are the indices n such that $(\mathcal{W}^J)_n/(\mathcal{W}^J)_{n-1}$ is isomorphic to $\mathcal{V}_\pi^J$ we have

$$(\mathcal{W}^J)_n^I/(\mathcal{W}^J)_{n-1}^I\cong\mathcal{V}_\pi^I\qquad(\forall n=n_1,\ldots,n_m).$$

Therefore, to finish the proof of the lemma it is sufficient to prove that, for every other index n, $\mathcal{V}_\pi^I$ does not occur in the left $(\mathbb{C}\otimes\mathcal{C}_c^\infty(H//I))$-module $(\mathcal{W}^J)_n^I/(\mathcal{W}^J)_{n-1}^I$. But by (D.1.8) there exists a smooth irreducible representation $(\mathcal{W}',\rho')$ of H such that $\mathcal{W}'^J\cong(\mathcal{W}^J)_n/(\mathcal{W}^J)_{n-1}$ and $(\mathcal{W}^J)_n^I/(\mathcal{W}^J)_{n-1}^I\cong\mathcal{W}'^I$ is thus irreducible. Therefore, if $\mathcal{V}_\pi^I$ occurs in $(\mathcal{W}^J)_n^I/(\mathcal{W}^J)_{n-1}^I$, ρ' is isomorphic to π, so that $\mathcal{W}'^J$ is isomorphic to $\mathcal{V}_\pi^J$, and we get a contradiction. □

The class of an admissible representation $(\mathcal{W},\rho)$ of H may then be defined by

(D.11.2)
$$[\rho]=\sum_{\pi\in\mathcal{S}}m_\rho(\pi)\pi\in K_{\mathrm{a}}(H)$$

where

$$m_\rho(\pi)=m_{\mathcal{W}^I}(\mathcal{V}_\pi^I)\geq0$$

for every $\pi\in\mathcal{S}$ and every compact open subgroup I of H such that $\mathcal{V}_\pi^I\neq(0)$. Obviously the class map is additive on short exact sequences of admissible representations of H.

LEMMA (D.11.3). — *Let $(\mathcal{W}, \rho)$ be an admissible representation of H and let $\pi \in \mathcal{S}$. Then $m_\rho(\pi) \neq 0$ if and only if there exist two H-invariant subspaces*

$$\mathcal{W}'' \subset \mathcal{W}' \subset \mathcal{W}$$

such that $\mathcal{W}'/\mathcal{W}''$ with the induced representation of H is isomorphic to $(\mathcal{V}_\pi, \pi)$.

Proof: As the class map is additive and maps $\mathrm{ob}\,\mathrm{Rep}_{\mathrm{a}}(H)$ into $K_{\mathrm{a}}(H) \cap (\mathbb{Z}_{\geq 0})^{\mathcal{S}}$ the "if" part is clear.

Conversely let us assume that $m_\rho(\pi) \neq 0$ and let us choose a compact open subgroup I of H such that $\mathcal{V}_\pi^I \neq (0)$. Then there exist two left $(\mathbb{C} \otimes \mathcal{C}_c^\infty(H//I))$-submodules

$$W'' \subset W' \subset \mathcal{W}^I$$

such that W'/W'' is isomorphic to $\mathcal{V}_\pi^I$. Let us consider the natural map

$$\mathcal{C}_c^\infty(H/I) \otimes_{\mathcal{C}_c^\infty(H//I)} W' \to \mathcal{W},$$

its kernel $\mathcal{K}$ and its image $\mathcal{W}'$. The compound map

$$\mathcal{K} \to \mathcal{C}_c^\infty(H/I) \otimes_{\mathcal{C}_c^\infty(H//I)} W' \twoheadrightarrow \mathcal{C}_c^\infty(H/I) \otimes_{\mathcal{C}_c^\infty(H//I)} \mathcal{V}_\pi^I \twoheadrightarrow \mathcal{V}_\pi$$

is either zero or surjective. As $\mathcal{K}^I = (0)$ and $\mathcal{V}_\pi^I \neq (0)$ it cannot be surjective. Therefore it is zero and the compound map

$$\mathcal{C}_c^\infty(H/I) \otimes_{\mathcal{C}_c^\infty(H//I)} W' \twoheadrightarrow \mathcal{C}_c^\infty(H/I) \otimes_{\mathcal{C}_c^\infty(H//I)} \mathcal{V}_\pi^I \twoheadrightarrow \mathcal{V}_\pi$$

induces a surjective map $\mathcal{W}' \twoheadrightarrow \mathcal{V}_\pi$. Then to conclude the proof it is sufficient to take the kernel of this last map as the subspace $\mathcal{W}''$. □

(D.12) Comments and references

The standard reference for (D.9) is [Fla].

E

Reduction theory and strong approximation

(E.0) Introduction

In this appendix we will give proofs of lemmas (9.2.7), (9.5.8) and (9.5.9).

(E.1) Reduction theory

As M_I is isomorphic to $GL_{d_1} \times \cdots \times GL_{d_s}$ ($d_I = (d_1, \ldots, d_s)$) it is sufficient to prove lemmas (9.2.7) and (9.5.8) for $M = GL_d$ (d arbitrary).

Following Weil the elements in $GL_d(\mathbb{A})$ admit a geometric interpretation. Let us recall that X is the smooth projective model of the function field F. If $\mathcal{E}$ is a vector bundle of rank d over X, i.e. a locally free $\mathcal{O}_X$-Module of rank d , we denote by $\mathcal{E}_F$ the fiber of $\mathcal{E}$ at the generic point $\mathrm{Spec}(F) \hookrightarrow X$ of X ($\mathcal{E}_F$ is an F-vector space of dimension d) and, for each $x \in |X|$, we set

$$\mathcal{E}_{\mathcal{O}_x} = H^0(\mathrm{Spec}(\mathcal{O}_x), \mathcal{E})$$

($\mathcal{E}_{\mathcal{O}_x}$ is a free $\mathcal{O}_x$-module of rank d). For each $x \in |X|$, we have a canonical isomorphism

$$\mathrm{can}_x : F_x \otimes_{\mathcal{O}_x} \mathcal{E}_{\mathcal{O}_x} \xrightarrow{\sim} F_x \otimes_F \mathcal{E}_F.$$

Now, if $\mathcal{E}$ is a vector bundle of rank d over X equipped with a basis

$$\alpha_F : F^d \xrightarrow{\sim} \mathcal{E}_F$$

of its generic fiber and a basis

$$\alpha_{\mathcal{O}_x} : \mathcal{O}_x^d \xrightarrow{\sim} \mathcal{E}_{\mathcal{O}_x}$$

for every $x \in |X|$, the elements

$$g_x = (F_x \otimes_F \alpha_F)^{-1} \circ \mathrm{can}_x \circ (F_x \otimes_{\mathcal{O}_x} \alpha_{\mathcal{O}_x}) \in GL_d(F_x) \qquad (\forall x \in |X|)$$

define an element

$$g_{\mathbb{A}} = (g_x)_{x \in |X|}$$

of $GL_d(\mathbb{A})$, i.e. for almost every $x \in |X|$, we have $g_x \in GL_d(\mathcal{O}_x)$. By this construction we obtain a bijection from the set of the isomorphism classes of triples

$$(\mathcal{E}, \alpha_F, (\alpha_{\mathcal{O}_x})_{x \in |X|})$$

as above onto $GL_d(\mathbb{A})$ (the details are left to the reader). If $\gamma \in GL_d(F)$, $k \in GL_d(\mathcal{O})$ and if this bijection maps the triple $(\mathcal{E}, \alpha_F, (\alpha_{\mathcal{O}_x})_{x \in |X|})$ onto $g_{\mathbb{A}}$ it maps the triple

$$(\mathcal{E}, \alpha_F \circ \gamma^{-1}, (\alpha_{\mathcal{O}_x} \circ k_x)_{x \in |X|})$$

onto

$$\gamma g_{\mathbb{A}} k.$$

Therefore the above bijection induces a bijection between the set of the isomorphism classes of vector bundles of rank d over X and the double coset space

$$GL_d(F) \backslash GL_d(\mathbb{A}) / GL_d(\mathcal{O}).$$

More generally let I be a subset of Δ, let P_I be the corresponding standard parabolic subgroup of GL_d and let $d_I = (d_1, \ldots, d_s)$ be the corresponding partition of d (see (5.1)). Then we have a natural bijection from the set of the isomorphism classes of triples

$$(\mathcal{E}_\bullet, \alpha_{\bullet,F}, (\alpha_{\bullet,\mathcal{O}_x})_{x \in |X|})$$

onto $P_I(\mathbb{A})$, where

$$\mathcal{E}_\bullet = \big((0) = \mathcal{E}_0 \subset \mathcal{E}_1 \subset \cdots \subset \mathcal{E}_s\big)$$

is a flag of vector bundles of rank d_I over X (each $\mathcal{E}_j$ is a vector bundle of rank $d_1 + \cdots + d_j$ over X and each quotient $\mathcal{E}_j / \mathcal{E}_{j-1}$ is torsion free) which is equipped with an isomorphism of flags of F-vector spaces,

$$\alpha_{\bullet,F} : \big((0) \subset F^{d_1} \subset F^{d_1+d_2} \subset \cdots \subset F^{d_1+\cdots+d_s}\big) \xrightarrow{\sim} (\mathcal{E}_\bullet)_F,$$

and with an isomorphism of flags of free $\mathcal{O}_x$-modules,

$$\alpha_{\bullet,\mathcal{O}_x} : \big((0) \subset \mathcal{O}_x^{d_1} \subset \mathcal{O}_x^{d_1+d_2} \subset \cdots \subset \mathcal{O}_x^{d_1+\cdots+d_s}\big) \xrightarrow{\sim} (\mathcal{E}_\bullet)_{\mathcal{O}_x},$$

for every $x \in |X|$. Moreover this bijection induces a bijection between the set of isomorphism classes of the flags of vector bundles of rank d_I over X and the double coset space

$$P_I(F)\backslash P_I(\mathbb{A})/P_I(\mathcal{O}).$$

The natural embedding

$$P_I(\mathbb{A}) \hookrightarrow GL_d(\mathbb{A})$$

(resp. the canonical projection

$$P_I(\mathbb{A}) \twoheadrightarrow M_I(\mathbb{A}) \twoheadrightarrow GL_{d_j}(\mathbb{A})$$

for $j \in \{1, \ldots, s\}$) admits the modular interpretation

$$(\mathcal{E}_\bullet, \alpha_{\bullet,F}, (\alpha_{\bullet,\mathcal{O}_x})_{x\in|X|}) \mapsto (\mathcal{E}_s, \alpha_{s,F}, (\alpha_{s,\mathcal{O}_x})_{x\in|X|})$$

(resp.

$$(\mathcal{E}_\bullet, \alpha_{\bullet,F}, (\alpha_{\bullet,\mathcal{O}_x})_{x\in|X|}) \mapsto (\mathrm{gr}_j(\mathcal{E}_\bullet), \mathrm{gr}_j(\alpha_{\bullet,F}), (\mathrm{gr}_j(\alpha_{\bullet,\mathcal{O}_x}))_{x\in|X|}),$$

where

$$\mathrm{gr}_j(\mathcal{E}_\bullet) = \mathcal{E}_j/\mathcal{E}_{j-1}$$

and where

$$\mathrm{gr}_j(\alpha_{\bullet,F}) : F^{d_j} \xrightarrow{\sim} \mathrm{gr}_j(\mathcal{E}_\bullet)_F$$

and

$$\mathrm{gr}_j(\alpha_{\bullet,\mathcal{O}_x}) : \mathcal{O}_x^{d_j} \xrightarrow{\sim} \mathrm{gr}_j(\mathcal{E}_\bullet)_{\mathcal{O}_x} \qquad (\forall x \in |X|)$$

are induced by $\alpha_{\bullet,F}$ and $\alpha_{\bullet,\mathcal{O}_x}$ respectively).

If $g_{\mathbb{A}} \in GL_d(\mathbb{A})$ corresponds to $(\mathcal{E}, \alpha_F, (\alpha_{\mathcal{O}_x})_{x\in|X|})$ under the above bijection then $\det(g_{\mathbb{A}}) \in \mathbb{A}^\times$ corresponds to

$$\Big(\bigwedge^d \mathcal{E}, \bigwedge^d \alpha_F, (\bigwedge^d \alpha_{\mathcal{O}_x})_{x\in|X|}\Big)$$

and we have

$$\deg(\mathcal{E}) = -\deg(\det(g_{\mathbb{A}}))$$

(see (9.1)). Here the degree of $\mathcal{E}$ is defined by the Riemann–Roch formula

$$\dim_{\mathbb{F}_p} H^0(X, \mathcal{E}) - \dim_{\mathbb{F}_p} H^1(X, \mathcal{E}) = f(\deg(\mathcal{E}) + (1 - g_X)d)$$

($\dim_{\mathbb{F}_p} H^0(X, \mathcal{O}_X) = f$ and $\dim_{\mathbb{F}_p} H^1(X, \mathcal{O}_X) = fg_X$).

Using the above modular interpretations of $GL_d(\mathbb{A})$ and $P_I(\mathbb{A})$ lemmas (9.2.7) and (9.5.8) may be reformulated in the following way.

THEOREM (E.1.1) (Harder). — *Let c_2 be an integer with $c_2 \geq 2g_X$. Then for each vector bundle $\mathcal{E}$ of rank d over X there exists at least one flag of vector bundles,*

$$\mathcal{E}_\bullet = \big((0) = \mathcal{E}_0 \subset \mathcal{E}_1 \subset \cdots \subset \mathcal{E}_d\big),$$

of rank $d_\emptyset = (1, \ldots, 1)$ over X with

$$\mathcal{E}_d = \mathcal{E}$$

such that

$$\deg(\mathcal{E}_{j+1}/\mathcal{E}_j) - \deg(\mathcal{E}_j/\mathcal{E}_{j-1}) \leq c_2$$

for every $j = 1, \ldots, d-1$.

THEOREM (E.1.2) (Harder). — *For any integer c_2 there exists an integer $c_1' \leq c_2$ having the following property. Let $\mathcal{E}_\bullet$ and $\mathcal{E}'_\bullet$ be two flags of vector bundles of rank $d_\emptyset = (1, \ldots, 1)$ over X such that*

$$\deg(\mathcal{E}_{j+1}/\mathcal{E}_j) - \deg(\mathcal{E}_j/\mathcal{E}_{j-1}) \leq c_2$$

and

$$\deg(\mathcal{E}'_{j+1}/\mathcal{E}'_j) - \deg(\mathcal{E}'_j/\mathcal{E}'_{j-1}) \leq c_2$$

for every $j = 1, \ldots, d-1$ and let γ be an isomorphism of the vector bundle $\mathcal{E}_d$ onto the vector bundle $\mathcal{E}'_d$. Then for any $j_0 \in \{1, \ldots, d-1\}$ such that

$$\deg(\mathcal{E}_{j_0+1}/\mathcal{E}_{j_0}) - \deg(\mathcal{E}_{j_0}/\mathcal{E}_{j_0-1}) < c_1'$$

we have

$$\gamma(\mathcal{E}_{j_0}) = \mathcal{E}'_{j_0}.$$

Before proving (E.1.1) and (E.1.2) let us recall some basic facts about the Harder–Narasimhan filtration. A non-zero vector bundle $\mathcal{E}$ over X is said to be **semi-stable** if, for any non-zero subbundle $\mathcal{F}$ of $\mathcal{E}$, we have

$$\mu(\mathcal{F}) \leq \mu(\mathcal{E}).$$

Here, for any non-zero vector bundle $\mathcal{G}$ over X,

$$\mu(\mathcal{G}) = \deg(\mathcal{G})/\mathrm{rk}(\mathcal{G})$$

is the slope of $\mathcal{G}$. A filtration

$$F_\bullet\mathcal{E} = \big((0) = F_0\mathcal{E} \subset F_1\mathcal{E} \subset \cdots \subset F_s\mathcal{E} = \mathcal{E}\big)$$

of a vector bundle $\mathcal{E}$ over X by subbundles is called a **Harder–Narasimhan filtration** if the vector bundle

$$\mathrm{gr}_j^F \mathcal{E} \overset{\mathrm{dfn}}{=\!=} F_j\mathcal{E}/F_{j-1}\mathcal{E}$$

is non-zero and semi-stable for every $j = 1, \ldots, s$ and if

$$\mu(\mathrm{gr}_j^F \mathcal{E}) > \mu(\mathrm{gr}_{j+1}^F \mathcal{E})$$

for every $j = 1, \ldots, s-1$ (if $\mathcal{E} = (0)$ we take $s = 0$).

THEOREM (E.1.3) (Harder–Narasimhan). — *Any vector bundle over X admits a unique Harder–Narasimhan filtration.*

Proof : First of all let us prove the uniqueness. Let $\mathcal{F}$ be a non-zero subbundle of $\mathcal{E}$ and let

$$F_\bullet\mathcal{F} = (F_\bullet\mathcal{E}) \cap \mathcal{F}$$

be the filtration of $\mathcal{F}$ induced by $F_\bullet\mathcal{E}$. Then, for every $j = 1, \ldots, s$, either $\mathrm{gr}_j^F\mathcal{F} = (0)$ or

$$\mu(\mathrm{gr}_j^F\mathcal{F}) \leq \mu(\mathrm{gr}_j^F\mathcal{E}).$$

Therefore we have

$$\mu(\mathcal{F}) \leq \mu(F_1\mathcal{E})$$

and the equality holds only if $\mathrm{gr}_j^F\mathcal{F} = (0)$ for every $j = 2, \ldots, s$, i.e. only if

$$\mathcal{F} \subset F_1\mathcal{E}.$$

Let us denote by μ_1 the maximum of $\mu(\mathcal{F})$ when $\mathcal{F}$ runs through the set of all non-zero subbundles of $\mathcal{E}$. It is now clear that $F_1\mathcal{E}$ is the largest non-zero subbundle of $\mathcal{E}$ such that

$$\mu(F_1\mathcal{E}) = \mu_1.$$

Moreover it is clear that, for every $j = 1, \ldots, s$, $F_j\mathcal{E}$ is the inverse image under the canonical projection

$$\mathcal{E} \twoheadrightarrow \mathcal{E}/F_{j-1}\mathcal{E}$$

of the subbundle

$$F_1(\mathcal{E}/F_{j-1}\mathcal{E}) \subset \mathcal{E}/F_{j-1}\mathcal{E}.$$

This completes the proof of the uniqueness.

Now let us prove the existence. We may assume that $\mathcal{E}$ is non-zero. We define μ_1 as before ($\mu_1 \in \mathbb{Z}$). Among the non-zero subbundles $\mathcal{F}$ of $\mathcal{E}$ for which

$$\mu(\mathcal{F}) = \mu_1$$

there is one and only one which has maximal rank, say r_1. Indeed, if $\mathcal{F}'$ and $\mathcal{F}''$ are distinct subbundles of $\mathcal{E}$ such that

$$\mathrm{rk}(\mathcal{F}') = \mathrm{rk}(\mathcal{F}'') = r_1$$

and

$$\mu(\mathcal{F}') = \mu(\mathcal{F}'') = \mu_1,$$

let $\mathcal{F}$ be the subbundle of $\mathcal{E}$ generated by the subsheaf $\mathcal{F}' + \mathcal{F}''$ of $\mathcal{E}$, so that $\mathcal{F}' + \mathcal{F}'' \subset \mathcal{F} \subset \mathcal{E}$, $\mathcal{F}/(\mathcal{F}' + \mathcal{F}'')$ is a torsion module and $\mathcal{E}/\mathcal{F}$ is torsion free; by construction of $\mathcal{F}$ we have

$$\deg(\mathcal{F}) \geq 2r_1\mu_1 - \deg(\mathcal{F}' \cap \mathcal{F}'')$$

and
$$\mathrm{rk}(\mathcal{F}) = 2r_1 - \mathrm{rk}(\mathcal{F}' \cap \mathcal{F}'') > r_1,$$
by maximality of r_1 and μ_1 we have
$$\mu(\mathcal{F}) < \mu_1$$
and by definition of μ_1 we have either $\mathcal{F}' \cap \mathcal{F}'' = (0)$ or
$$\mu(\mathcal{F}' \cap \mathcal{F}'') \leq \mu_1;$$
but these inequalities lead to a contradiction. Moreover, if $F_1\mathcal{E}$ is the unique subbundle of $\mathcal{E}$ of rank r_1 for which $\mu(F_1\mathcal{E}) = \mu_1$, $F_1\mathcal{E}$ is obviously semi-stable. Then we define $F_\bullet\mathcal{E}$ by induction on the rank of $\mathcal{E}$. If
$$F_\bullet\mathcal{E}' = \big((0) = F_0\mathcal{E}' \subset F_1\mathcal{E}' \subset \cdots \subset F_{s'}\mathcal{E}' = \mathcal{E}'\big)$$
is the Harder–Narasimhan filtration of
$$\mathcal{E}' = \mathcal{E}/F_1\mathcal{E},$$
we set $s = s' + 1$ and we define $F_j\mathcal{E}$ as the inverse image of $F_{j-1}\mathcal{E}' \subset \mathcal{E}'$ under the canonical projection
$$\mathcal{E} \twoheadrightarrow \mathcal{E}/F_1\mathcal{E} = \mathcal{E}'$$
for every $j = 1, \ldots, s$. □

For an arbitrary vector bundle $\mathcal{E}$ over X let us set
$$\mu_{\max}(\mathcal{E}) = \sup\{\mu(\mathcal{G}) \mid (0) \neq \mathcal{G} \subset \mathcal{E}\}$$
and
$$\mu_{\min}(\mathcal{E}) = \inf\{\mu(\mathcal{E}/\mathcal{G}) \mid \mathcal{G} \subsetneqq \mathcal{E}\}$$
where $\mathcal{G}$ runs through the set of all subbundles of $\mathcal{E}$. For an arbitrary vector bundle $\mathcal{E}$ over X and an arbitrary subbundle $\mathcal{F}$ of $\mathcal{E}$ let us put
$$\mathrm{jump}_{\mathcal{E}}(\mathcal{F}) = \mu_{\min}(\mathcal{F}) - \mu_{\max}(\mathcal{E}/\mathcal{F}).$$

LEMMA (E.1.4). — *Let*
$$\mathcal{E}_\bullet = \big((0) = \mathcal{E}_0 \subset \mathcal{E}_1 \subset \cdots \subset \mathcal{E}_d = \mathcal{E}\big)$$
be a flag of vector bundles of rank $d_\emptyset = (1, \ldots, 1)$ over X. Then we have
$$\mu_{\max}(\mathcal{E}) \leq \sup\{\frac{a_{j_1} + \cdots + a_{j_r}}{r} \mid 0 < j_1 < \cdots < j_r \leq d,\ 1 \leq r \leq d\}$$
and
$$\mu_{\min}(\mathcal{E}) \geq \inf\{\frac{a_{j_1} + \cdots + a_{j_r}}{r} \mid 0 < j_1 < \cdots < j_r \leq d,\ 1 \leq r \leq d\}$$
where we have set
$$a_j = \deg(\mathcal{E}_j/\mathcal{E}_{j-1})$$
for every $j = 1, \ldots, d$.

Proof : By duality it is sufficient to prove one of the two inequalities. Indeed we have

$$\mu_{\max}(\mathcal{E}^{\vee}) = -\mu_{\min}(\mathcal{E})$$

and

$$\deg(\mathcal{E}_{d-j}^{\perp}/\mathcal{E}_{d-j+1}^{\perp}) = -\deg(\mathcal{E}_{d-j+1}/\mathcal{E}_{d-j})$$

for every $j = 1, \ldots, d$.

Now, if $\mathcal{G}$ is a non-zero subbundle of $\mathcal{E}$ and if we set $\mathcal{G}_j = \mathcal{G} \cap \mathcal{E}_j$ for every $j = 0, \ldots, d$, there exist exactly $r = \mathrm{rk}(\mathcal{G})$ indices $j \in \{1, \ldots, d\}$ such that $\mathcal{G}_j/\mathcal{G}_{j-1} \neq (0)$, let us say $0 < j_1 < \cdots < j_r \leq d$, and it is clear that

$$\mu(\mathcal{G}) \leq \frac{a_{j_1} + \cdots + a_{j_r}}{r}.$$

The first inequality of the lemma follows. □

COROLLARY (E.1.5). — *With the notations of* (E.1.4) *we have*

$$\mathrm{jump}_{\mathcal{E}}(\mathcal{E}_j) \geq \inf\Big\{\frac{a_{k_1} + \cdots + a_{k_r}}{r} - \frac{a_{\ell_1} + \cdots + a_{\ell_s}}{s} \ \Big| \ 0 < k_1 < \cdots < k_r \leq j < \ell_1 < \cdots < \ell_s \leq d,\ 1 \leq r \leq j,\ 1 \leq s \leq d-j\Big\}$$

for $j = 0, \ldots, d$. □

Proof of (E.1.1) : We are going to use an induction argument on d. The theorem is trivial for $d = 1$. Let us assume the theorem for rank $d-1$ and let us prove it for a vector bundle $\mathcal{E}$ of rank d over X.

Thanks to (E.1.3) we may assume that $\mathcal{E}$ is semi-stable. Indeed, let $F_{\bullet}\mathcal{E}$ be the Harder–Narasimhan filtration of $\mathcal{E}$; if $\mathcal{E}$ is not semi-stable we have $s \geq 2$ and

$$d_j = \mathrm{rk}(\mathrm{gr}_j^F \mathcal{E}) < d$$

for every $j = 1, \ldots, s$; therefore, by the induction hypothesis, for each $j = 1, \ldots, s$ there exists a flag of vector bundles of rank $d_{j,\emptyset} = (1, \ldots, 1)$ over X,

$$\mathcal{E}_{j,\bullet} = ((0) = \mathcal{E}_{j,0} \subset \mathcal{E}_{j,1} \subset \cdots \subset \mathcal{E}_{j,d_j}),$$

with

$$\mathcal{E}_{j,d_j} = \mathrm{gr}_j^F \mathcal{E}$$

such that

$$\deg(\mathcal{E}_{j,k+1}/\mathcal{E}_{j,k}) - \deg(\mathcal{E}_{j,k}/\mathcal{E}_{j,k-1}) \leq c_2$$

for every $k = 1, \ldots, d_j$; for each $j = 1, \ldots, s$ and each $k = 1, \ldots, d_j$ let

$$\mathcal{E}_{d_1+\cdots+d_{j-1}+k} \subset F_j\mathcal{E} \subset \mathcal{E}$$

be the inverse image of $\mathcal{E}_{j,k} \subset \mathrm{gr}_j^F \mathcal{E}$ under the canonical projection

$$F_j\mathcal{E} \twoheadrightarrow \mathrm{gr}_j^F \mathcal{E};$$

then the flag of vector bundles over X,

$$\mathcal{E}_\bullet = \big((0) = \mathcal{E}_0 \subset \mathcal{E}_1 \subset \cdots \subset \mathcal{E}_d\big),$$

has rank $d_\emptyset = (1, \ldots, 1)$ and satisfies the properties of (E.1.1) for $\mathcal{E}$ (if $\ell = d_1 + \cdots + d_{j-1} + k$ for some $j \in \{1, \ldots, s\}$ and some $k \in \{1, \ldots, d_j - 1\}$ we have

$$\deg(\mathcal{E}_{\ell+1}/\mathcal{E}_\ell) - \deg(\mathcal{E}_\ell/\mathcal{E}_{\ell-1}) = \deg(\mathcal{E}_{j,k+1}/\mathcal{E}_{j,k}) - \deg(\mathcal{E}_{j,k}/\mathcal{E}_{j,k-1}) \leq c_2;$$

if $\ell = d_1 + \cdots + d_j$ for some $j \in \{1, \ldots, s-1\}$ we have $\mathcal{E}_\ell = F_j\mathcal{E}$ and

$$\deg(\mathcal{E}_{\ell+1}/\mathcal{E}_\ell) \leq \mu(\mathrm{gr}_{j+1}^F\mathcal{E}) < \mu(\mathrm{gr}_j^F \mathcal{E}) \leq \deg(\mathcal{E}_\ell/\mathcal{E}_{\ell-1}),$$

so that

$$\deg(\mathcal{E}_{\ell+1}/\mathcal{E}_\ell) - \deg(\mathcal{E}_\ell/\mathcal{E}_{\ell-1}) \leq \mu(\mathrm{gr}_{j+1}^F\mathcal{E}) - \mu(\mathrm{gr}_j^F \mathcal{E}) < 0 \leq 2g_X \leq c_2\,).$$

Now, for any vector bundle $\mathcal{E}$ of rank d over X and for any line bundle $\mathcal{L}$ over X, the Riemann–Roch theorem implies that

$$\mathrm{Hom}(\mathcal{L}, \mathcal{E}) \neq (0)$$

as long as

$$\deg(\mathcal{E}) - d\deg(\mathcal{L}) + d(1 - g_X) \geq 1.$$

In particular there exists a subbundle $\mathcal{E}_1$ of $\mathcal{E}$ of rank 1 such that

$$\deg(\mathcal{E}_1) \geq \frac{\deg(\mathcal{E})}{d} - g_X$$

(choose $\mathcal{L}$ such that

$$1 \leq \deg(\mathcal{E}) - d\deg(\mathcal{L}) + d(1 - g_X) \leq d,$$

choose an embedding of $\mathcal{L}$ into $\mathcal{E}$ and take for $\mathcal{E}_1$ the subbundle of $\mathcal{E}$ generated by the sheaf image of $\mathcal{L}$). By the induction hypothesis there exists a flag of vector bundles of rank $(d-1)_\emptyset = (1, \ldots, 1)$ over X,

$$\mathcal{E}'_\bullet = \big((0) = \mathcal{E}'_0 \subset \mathcal{E}'_1 \subset \cdots \subset \mathcal{E}'_{d-1}\big),$$

with

$$\mathcal{E}'_{d-1} = \mathcal{E}/\mathcal{E}_1$$

such that

$$\deg(\mathcal{E}'_{j+1}/\mathcal{E}'_j) - \deg(\mathcal{E}'_j/\mathcal{E}'_{j-1}) \leq c_2$$

for every $j = 1, \ldots, d-2$. For each $j = 1, \ldots, d$ let $\mathcal{E}_j$ be the inverse image of $\mathcal{E}'_{j-1} \subset \mathcal{E}/\mathcal{E}_1$ under the canonical projection

$$\mathcal{E} \twoheadrightarrow \mathcal{E}/\mathcal{E}_1.$$

Then the flag of vector bundles over X,

$$\mathcal{E}_\bullet = \big((0) = \mathcal{E}_0 \subset \mathcal{E}_1 \subset \cdots \subset \mathcal{E}_d = \mathcal{E}\big),$$

has rank $d_\emptyset = (1, \ldots, 1)$ and satisfies all the properties of (E.1.1) for $\mathcal{E}$ except maybe the property

$$\deg(\mathcal{E}_2/\mathcal{E}_1) - \deg(\mathcal{E}_1) \leq c_2.$$

But if we assume that $\mathcal{E}$ is semi-stable we have

$$\frac{\deg(\mathcal{E}_2)}{2} \leq \frac{\deg(\mathcal{E})}{d}$$

and it follows that

$$\deg(\mathcal{E}_2/\mathcal{E}_1) - \deg(\mathcal{E}_1) \leq 2g_X \leq c_2$$

(recall that

$$\deg(\mathcal{E}_1) \geq \frac{\deg(\mathcal{E})}{d} - g_X\ ).$$

The proof of (E.1.1) is completed. □

Proof of (E.1.2) : If $c_2 < 0$ let us take $c'_1 = c_2$. Then

$$F_k\mathcal{E} = \mathcal{E}_k = \mathcal{E}'_k \quad (k = 0, \ldots, d)$$

is the Harder–Narasimhan filtration of $\mathcal{E}$ (see (E.1.3)).

Let us assume that $c_2 \geq 0$. Thanks to (E.1.5) we have

$$\mathrm{jump}_{\mathcal{E}}(\mathcal{E}_{j_0}) \geq \inf\{\frac{a_{k_1} + \cdots + a_{k_r}}{r} - \frac{a_{\ell_1} + \cdots + a_{\ell_s}}{s} \mid$$
$$0 < k_1 < \cdots < k_r \leq j_0 < \ell_1 < \cdots < \ell_s \leq d,\ 1 \leq r \leq j_0,\ 1 \leq s \leq d - j_0\}$$

where we have set

$$a_j = \deg(\mathcal{E}_j/\mathcal{E}_{j-1}) \qquad (\forall j = 1, \ldots, d).$$

Therefore we have

$$\mathrm{jump}_{\mathcal{E}}(\mathcal{E}_{j_0}) > -c_1' - (d-2)c_2$$

if

$$a_{j+1} - a_j \leq c_2 \qquad (j = 1, \ldots, d-1,\ j \neq j_0)$$

and

$$a_{j_0+1} - a_{j_0} < c_1'$$

(we have

$$a_k - a_\ell > -c_1' - (\ell - k - 1)c_2 \geq -c_1' - (d-2)c_2$$

for any k, ℓ with $0 < k \leq j_0 < \ell \leq d$).

Now theorem (E.1.2) follows from the following stronger statement (take $c_1' = -(2d-3)c_2$).

(*) *Let $\mathcal{F}$ be a subbundle of $\mathcal{E}$ with*

$$\mathrm{jump}_{\mathcal{E}}(\mathcal{F}) > (d-1)c_2$$

and let $\mathcal{E}'_\bullet$ be a flag of subbundles of $\mathcal{E}$ of rank $d_\emptyset = (1, \ldots, 1)$ such that

$$a'_{j+1} - a'_j \leq c_2 \qquad (\forall j = 1, \ldots, d-1)$$

where we have set

$$a'_j = \deg(\mathcal{E}'_j / \mathcal{E}'_{j-1}).$$

Then we have

$$\mathcal{F} = \mathcal{E}'_{j_0}$$

where j_0 is the rank of $\mathcal{F}$.

Let us prove the statement (*) by induction on d. For $d = 1$ there is nothing to prove. Let us assume (*) for rank $d-1$ and let us prove it in rank d. If $\mathcal{E}'_1 \subset \mathcal{F}$ (resp. $\mathcal{F} \subset \mathcal{E}'_{d-1}$) we have

$$\mathrm{jump}_{\mathcal{E}/\mathcal{E}'_1}(\mathcal{F}/\mathcal{E}'_1) \geq \mathrm{jump}_{\mathcal{E}}(\mathcal{F}) > (d-1)c_2 \geq (d-2)c_2$$

(resp.

$$\mathrm{jump}_{\mathcal{E}'_{d-1}}(\mathcal{F}) \geq \mathrm{jump}_{\mathcal{E}}(\mathcal{F}) > (d-1)c_2 \geq (d-2)c_2\,)$$

(recall that $c_2 \geq 0$) and by the induction hypothesis we get

$$\mathcal{F}/\mathcal{E}'_1 = \mathcal{E}'_{j_0}/\mathcal{E}'_1 \subset \mathcal{E}/\mathcal{E}'_1$$

(resp.

$$\mathcal{F} = \mathcal{E}'_{j_0} \subset \mathcal{E}'_{d-1}),$$

i.e.

$$\mathcal{F} = \mathcal{E}'_{j_0}$$

as required. If $\mathcal{E}'_1 \not\subset \mathcal{F}$ and $\mathcal{F} \not\subset \mathcal{E}'_{d-1}$ we have non-zero morphisms of sheaves,

$$\mathcal{E}'_1 \to \mathcal{E}/\mathcal{F}$$

and

$$\mathcal{F} \to \mathcal{E}/\mathcal{E}'_{d-1},$$

so that we have

$$a'_1 \leq \mu_{\max}(\mathcal{E}/\mathcal{F})$$

and

$$a'_d \geq \mu_{\min}(\mathcal{F});$$

in particular we have

$$a'_d - a'_1 \geq \mathrm{jump}_{\mathcal{E}}(\mathcal{F}) > (d-1)c_2$$

and we have obtained a contradiction. This completes the proof of $(*)$ and therefore of (E.1.2). □

(E.2) Strong approximation

Let us fix a place x of F and let us identify F with its image in $\mathbb{A}^x$ under the diagonal embedding. If V is a scheme of finite type over F we have a corresponding identification of $V(F)$ with a subset of $V(\mathbb{A}^x)$. We will say that V has the **strong approximation property** if $V(F)$ is dense in $V(\mathbb{A}^x)$ for the adelic topology. Let us recall that, if V is the generic fiber of a scheme of finite type $\mathcal{V}$ over X, a basis of the adelic topology is given by the open subsets

$$U_S \times \prod_{y \in |X| - (S \cup \{x\})} \mathcal{V}(\mathcal{O}_y)$$

where S runs through the set of the finite subsets of $|X| - \{x\}$ and where U_S runs through the set of the open subsets of

$$V(F_S) = \prod_{y \in S} V(F_y)$$

for the ϖ_S-adic topology (the product of the ϖ_y-adic topologies, $y \in S$). This topology is independent of the model $\mathcal{V}$ of V.

We can reformulate lemma (9.5.9) by saying that

THEOREM (E.2.1). — *SL_d (over F) has the strong approximation property.*

The proof of (E.2.1) is based on the following lemma.

LEMMA (E.2.2). — *Any scheme over F which is isomorphic to the standard affine space*

$$\mathrm{Aff}_F^n = \mathrm{Spec}(F[X_1, \ldots, X_n])$$

for some non-negative integer n has the strong approximation property.

Proof : Obviously it is sufficient to prove that F is dense in $\mathbb{A}^x$, i.e. that

$$\mathbb{A}^x = F + \prod_{y \in |X| - \{x\}} \varpi_y^{y(D)} \mathcal{O}_y$$

for any divisor

$$D = \sum_{y \in |X| - \{x\}} y(D) \cdot y$$

on $X - \{x\}$ ($y(D) \in \mathbb{Z}$ for every $y \in |X| - \{x\}$, $y(D) = 0$ for almost all $y \in |X| - \{x\}$ and ϖ_y is a uniformizer of $\mathcal{O}_y$ for every $y \in |X| - \{x\}$). Let us denote by $\mathcal{M}_{X-\{x\}}$ the sheaf for the Zariski topology on $X - \{x\}$ of germs of meromorphic functions. Then we have an obvious inclusion of abelian sheaves,

$$\mathcal{O}_{X-\{x\}}(-D) \hookrightarrow \mathcal{M}_{X-\{x\}},$$

and for any Zariski open subset U of $X - \{x\}$ we have

$$H^0(U, \mathcal{M}_{X-\{x\}}) = \begin{cases} F & \text{if } U \neq \emptyset, \\ (0) & \text{otherwise,} \end{cases}$$

$$H^0(U, \mathcal{M}_{X-\{x\}}/\mathcal{O}_{X-\{x\}}(-D)) = \bigoplus_{y \in |U|} (F_y/\varpi_y^{y(D)}\mathcal{O}_y)$$

and the canonical morphism

$$H^0(U, \mathcal{M}_{X-\{x\}}) \to H^0(U, \mathcal{M}_{X-\{x\}}/\mathcal{O}_{X-\{x\}}(-D))$$

is induced by the inclusions

$$F \subset F_y \qquad (y \in |U|)$$

where $|U|$ is the set of closed points in U. Therefore the lemma follows from the vanishing of the cohomology group

$$H^1(X - \{x\}, \mathcal{O}_{X-\{x\}}(-D))$$

($X - \{x\}$ is affine). □

Proof of (E.2.1) : For any subset A of a topological space B we denote by

$$\mathrm{cl}_B(A)$$

the closure of A in B.

Let $\alpha = \epsilon_1 - \epsilon_2 \in \Delta$, $N = N_{\Delta - \{\alpha\}}$ and $\widetilde{N} = \widetilde{N}_{\Delta - \{\alpha\}}$ (see (5.1) and (7.1) for the notations). Then N and $\widetilde{N}$ are isomorphic to Aff_F^{d-1} as F-schemes. Therefore they have the strong approximation property (see (E.2.2)). As $N(\mathbb{A}^x)$ and $\widetilde{N}(\mathbb{A}^x)$ are closed in $SL_d(\mathbb{A}^x)$ for the adelic topology this means that

$$\mathrm{cl}_{SL_d(\mathbb{A}^x)}(N(F)) = N(\mathbb{A}^x)$$

and

$$\mathrm{cl}_{SL_d(\mathbb{A}^x)}(\widetilde{N}(F)) = \widetilde{N}(\mathbb{A}^x).$$

It follows that

$$\mathrm{cl}_{SL_d(\mathbb{A}^x)}(SL_d(F)) \supset \langle N(\mathbb{A}^x), \widetilde{N}(\mathbb{A}^x)\rangle$$

where $\langle N(\mathbb{A}^x), \widetilde{N}(\mathbb{A}^x)\rangle$ is the subgroup of $SL_d(\mathbb{A}^x)$ generated by the subgroups $N(\mathbb{A}^x)$ and $\widetilde{N}(\mathbb{A}^x)$.

But, for any finite subset S of $|X| - \{x\}$, $SL_d(F_S)$ is generated by its subgroups $N(F_S)$ and $\widetilde{N}(F_S)$. Therefore, for any finite subset S of $|X| - \{x\}$ the subgroup

$$SL_d(F_S) \times \{1^{x,S}\} \subset SL_d(F_S) \times SL_d(\mathbb{A}^{x,S}) = SL_d(\mathbb{A}^x)$$

is contained in the subgroup

$$\mathrm{cl}_{SL_d(\mathbb{A}^x)}(SL_d(F)) \subset SL_d(\mathbb{A}^x)$$

(here $1^{x,S}$ is the unit element of $SL_d(\mathbb{A}^{x,S})$). The theorem follows. Indeed the union of the subgroups

$$SL_d(F_S) \times \{1^{x,S}\} \subset SL_d(\mathbb{A}^x)$$

where S runs through the set of all the finite subsets of $|X| - \{x\}$ is dense in $SL_d(\mathbb{A}^x)$. □

(E.3) Comments and references

The results in (E.1) were originally proved by Harder without using the Harder–Narasimhan filtration (see [Har 1] which was written before [Ha–Na]).

The proof of the strong approximation theorem for SL_d which is given in (E.2) is due to Moore (see [Moo] Ch. IV, Lemma (13.1)).

F

Proof of lemma (10.6.4)

(F.0) Notations

The statement of (10.6.4) is purely local. Therefore we will simply denote by F the local field F_o (or, more generally, an arbitrary non-archimedean local field) and we will delete all the indices o. We will denote by v the discrete valuation of F, by $\mathcal{O}$ its valuation ring, ..., as in chapters 4 and 5.

(F.1) Reductions

Obviously we may assume that

$$O_G(\gamma^{(n')})(F) \cap O_G(\gamma^{(n'')})(F) = \emptyset$$

for all $n', n'' \in \{1, \ldots, N\}$ with $n' \neq n''$.

By permuting $\gamma^{(1)}, \ldots, \gamma^{(N)}$ we may assume that

$$\dim O_G(\gamma^{(n)})(F) \leq \dim O_G(\gamma^{(n+1)})(F)$$

for every $n = 1, \ldots, N-1$ and by adding finitely many γ's to $\gamma^{(1)}, \ldots, \gamma^{(N)}$ we may assume that

$$\overline{O_G(\gamma^{(n)})(F)} \subset \bigcup_{n'=1}^{n} O_G(\gamma^{(n')})(F)$$

for every $n = 1, \ldots, N$ (see (4.3.2)). Then $O_G(\gamma^{(N)})(F)$ is open in

$$\bigcup_{n=1}^{N} O_G(\gamma^{(n)})(F)$$

and $\{\gamma^{(1)},\ldots,\gamma^{(N-1)}\}$ satisfies the same hypotheses as $\{\gamma^{(1)},\ldots,\gamma^{(N)}\}$, so that we can argue by induction on N.

(F.2) A geometric construction

Let us fix $\gamma\in G(F)$ and let $G_\gamma\subset G$ be its centralizer. If $\mathfrak{g}$ is the Lie algebra of G the tangent map at 1 of the morphism

$$i_\gamma : G\to G,\ g\mapsto g^{-1}\gamma g,$$

is

$$\mathfrak{g}\to\mathfrak{g},\ \xi\mapsto[\gamma,\xi].$$

Therefore the Lie algebra of G_γ is equal to

$$\mathfrak{g}_\gamma=\{\xi\in\mathfrak{g}\mid[\gamma,\xi]=0\}.$$

Morover, if we identify G with the Zariski open subset

$$\{\xi\in\mathfrak{g}\mid\det(\xi)\text{ is invertible}\}$$

of $\mathfrak{g}$, we have

$$G_\gamma=G\cap\mathfrak{g}_\gamma,$$

so that G_γ is smooth over F. The image $O_G(\gamma)$ of i_γ is a locally closed subscheme of G over F and i_γ induces an isomorphism of F-schemes from $G_\gamma\backslash G$ onto $O_G(\gamma)$. In particular $O_G(\gamma)$ is smooth over F and its tangent space at γ is

$$O_{\mathfrak{g}}(\gamma)=\{[\gamma,\xi]\mid\xi\in\mathfrak{g}\}\subset\mathfrak{g}$$

Analogous results for more general groups may be false. For example, if $G=SL_2$ with F of characteristic 2 and if

$$\gamma=\begin{pmatrix}1&1\\0&1\end{pmatrix},$$

G_γ is not smooth over F.

A slice through γ is a locally closed subscheme S of G over F passing through γ and having the following properties:

(a) S is irreducible,

(b) $S\cap O_G(\gamma)=\{\gamma\}$,

(c) the morphism

$$\pi : S\times_F G\to G,\ (s,g)\mapsto g^{-1}sg,$$

is smooth (in particular S is smooth over F and $\pi(S\times_F G)=U$ is a Zariski open subset of G),

(d) $O_G(\gamma)$ is a Zariski closed subset of U.

LEMMA (F.2.1). — *There exists at least one slice through* γ.

Proof: Let us choose an F-vector subspace V of $\mathfrak{g}$ such that

$$V \oplus O_{\mathfrak{g}}(\gamma) = \mathfrak{g}.$$

Then γ is an isolated point in

$$(\gamma + V) \cap O_G(\gamma)$$

and there exists a Zariski open neighborhood V° of 0 in V such that

$$(\gamma + V^\circ) \cap O_G(\gamma) = \{\gamma\}.$$

Shrinking V° if it is necessary we may assume that $\det(\gamma + v^\circ)$ is invertible and that

$$V + O_{\mathfrak{g}}(\gamma + v^\circ) = \mathfrak{g}$$

for every $v^\circ \in V^\circ$ (the map

$$V \to \mathrm{Hom}_F(V \oplus \mathfrak{g}, \mathfrak{g}),\ v \mapsto (v' \oplus \xi \mapsto [\gamma + v, \xi] + v'),$$

is algebraic over F and the set of epimorphisms in $\mathrm{Hom}_F(V \oplus \mathfrak{g}, \mathfrak{g})$ is Zariski open).

Then

$$S = \gamma + V^\circ \subset G$$

satisfies conditions (a), (b) and (c) (the tangent map of π at $(\gamma + v, g)$ is given by

$$v' \oplus \xi \mapsto g^{-1}([\gamma + v, g\xi g^{-1}] + v')g\,).$$

Now to complete the proof of the lemma it is sufficient to replace S by the Zariski open neighborhood

$$S - S \cap (\overline{O_G(\gamma)} - O_G(\gamma))$$

of γ in S. □

(F.3) The Harish-Chandra lemma

Let M and N be two ϖ-adic manifolds, let ω_M and ω_N be nowhere zero analytic volume forms on M and N respectively and let

$$p : M \twoheadrightarrow N$$

be a surjective analytic submersion. Let

$$\omega_p = p^*\omega_N \backslash \omega_M$$

be the corresponding relative analytic volume form. Then by a standard procedure (see (3.5)) we get measures $|\omega_M|$, $|\omega_N|$ and $|\omega_p|_n$ on M, N and the fiber $p^{-1}(n)$, $n \in N$, respectively.

If $f \in C_c^\infty(M,\mathbb{C})$, i.e. if f is a locally constant complex function with compact support on the underlying totally disconnected, locally compact, topological space, we set

$$p_*(f)(n) = \int_{p^{-1}(n)} (f|p^{-1}(n))|\omega_p|_n \qquad (\forall n \in N).$$

LEMMA (F.3.1) (Harish-Chandra). — *For each $f \in C_c^\infty(M,\mathbb{C})$ the complex function $p_*(f)$ is locally constant with compact support on N. We have*

$$\mathrm{Supp}(p_*(f)) \subset p(\mathrm{Supp}(f)).$$

The map

$$p_* : C_c^\infty(M,\mathbb{C}) \to C_c^\infty(N,\mathbb{C})$$

is $\mathbb{C}$-linear and surjective. We have the projection formula

$$p_*(fp^*(g)) = p_*(f)g$$

for every $f \in C_c^\infty(M,\mathbb{C})$ and every $g \in C_c^\infty(N,\mathbb{C})$ where we have set

$$p^*(g) = g \circ p.$$

Proof: We may assume that $M = \mathcal{O}^m$, $N = \mathcal{O}^n$ and that

$$p(x_1,\ldots,x_m) = (x_1,\ldots,x_n)$$

($m \geq n$) (use a partition of unity and the implicit function theorem). Then we have

$$\omega_M = \alpha(x_1,\ldots,x_m)dx_1 \wedge \cdots \wedge dx_m,$$
$$\omega_N = \beta(x_1,\ldots,x_n)dx_1 \wedge \cdots \wedge dx_n$$

and

$$\omega_p = \frac{\alpha(x_1,\ldots,x_m)}{\beta(x_1,\ldots,x_n)} dx_{n+1} \wedge \cdots \wedge dx_m,$$

where $\alpha(x_1,\ldots,x_m)$ and $\beta(x_1,\ldots,x_n)$ are nowhere zero analytic functions on $\mathcal{O}^m$ and $\mathcal{O}^n$ respectively.

But the absolute value $|\ |$ is a locally constant complex function on $F - \{0\}$. Therefore the same is true for the functions $|\alpha(x_1,\ldots,x_m)|$ and $|\beta(x_1,\ldots,x_n)|$ on $\mathcal{O}^m$ and $\mathcal{O}^n$ respectively. Again using a partition of unity we may assume that these functions are constant and even equal to 1. Then we have

$$p_*(f)(x_1,\ldots,x_n) = \int_{\mathcal{O}^{m-n}} f(x_1,\ldots,x_m)dx_{n+1}\cdots dx_m \qquad (\forall (x_1,\ldots,x_n) \in \mathcal{O}^n)$$

and the lemma is obvious. □

(F.4) An application of the Harish-Chandra lemma

Let us fix $\gamma \in G(F)$, a slice S through γ, a nowhere zero volume form ω_S on S (shrink S if it is necessary) and a non-zero volume form ω_G on G which is invariant under (left and right) translation. Then we have a surjective (algebraic) submersion

$$\pi : S \times_F G \twoheadrightarrow U \subset G$$

and a relative (algebraic) volume form

$$\omega_\pi = \pi^*(\omega_G|U)\backslash(\omega_S \wedge \omega_G).$$

Considering the corresponding ϖ-adic manifolds we get a surjective analytic submersion

$$\pi : S(F) \times G(F) \twoheadrightarrow \mathcal{U} \subset U(F) \subset G(F)$$

and analytic volume forms $\omega_S \wedge \omega_G$ and $\omega_{\mathcal{U}} = \omega_G|\mathcal{U}$ on $S(F) \times G(F)$ and $\mathcal{U}$ respectively, hence a relative analytic volume form ω_π on $S(F) \times G(F)$. Here $\mathcal{U}$ is an open subset of $U(F)$ which is $G(F)$-invariant under conjugation and which contains

$$O_G(\gamma)(F) = \{g^{-1}\gamma g \mid g \in G(F)\}$$

as a closed analytic subset (we have seen in the proof of (4.8.4) that $H^1_{\mathrm{fppf}}(\mathrm{Spec}(F), G_\gamma)$ is trivial). We also have the canonical projection

$$\mathrm{pr} : S(F) \times G(F) \to S(F)$$

and the relative analytic volume form $\omega_{\mathrm{pr}} = 1 \otimes \omega_G$.

Applying the Harish-Chandra lemma we obtain two surjective $\mathbb{C}$-linear maps

$$\pi_* : C_c^\infty(S(F) \times G(F), \mathbb{C}) \twoheadrightarrow C_c^\infty(\mathcal{U}, \mathbb{C})$$

and

$$\mathrm{pr}_* : C_c^\infty(S(F) \times G(F), \mathbb{C}) \twoheadrightarrow C_c^\infty(S(F), \mathbb{C}).$$

LEMMA (F.4.1). — *Let us fix a Haar measure dg_γ on $G_\gamma(F)$. Let D be an invariant distribution on $G(F)$ such that*

$$\mathcal{U} \cap \mathrm{Supp}(D) \subset O_G(\gamma)(F).$$

Then there exists a unique complex number c such that

$$D(\varphi) = c \int_{G_\gamma(F)\backslash G(F)} \varphi(g^{-1}\gamma g) \frac{dg}{dg_\gamma}$$

for every $\varphi \in C_c^\infty(\mathcal{U}, \mathbb{C}) \subset C_c^\infty(G(F), \mathbb{C})$.

Proof : Let us consider the distribution $f \mapsto D(\pi_*(f))$ on $S(F) \times G(F)$. For each $h \in G(F)$ let

$$\tau_h : G(F) \to G(F)$$

be the right translation by h. Then, for each $u \in \mathcal{U}$, $\mathrm{id}_{S(F)} \times \tau_h$ induces an isomorphism from $\pi^{-1}(u)$ onto $\pi^{-1}(h^{-1}uh)$ and the push-out of $|\omega_\pi|_u$ by this isomorphism is $|\omega_\pi|_{h^{-1}uh}$. It follows that

$$\pi_*(\tau_h^* f)(u) = \pi_*(f)(h^{-1}uh) \qquad (\forall u \in \mathcal{U})$$

and that

$$D(\pi_*(\tau_h^* f)) = D(\pi_*(f)) \qquad (\forall h \in G(F)).$$

Therefore the distribution $f \mapsto D(\pi_*(f))$ on $S(F) \times G(F)$ vanishes on the kernel of the surjective map

$$\mathrm{pr}_* : \mathcal{C}_c^\infty(S(F) \times G(F), \mathbb{C}) \twoheadrightarrow \mathcal{C}_c^\infty(S(F), \mathbb{C})$$

and induces a distribution $\widetilde{D}$ on $S(F)$ with

$$\widetilde{D}(\mathrm{pr}_*(f)) = D(\pi_*(f))$$

for every $f \in \mathcal{C}_c^\infty(S(F) \times G(F), \mathbb{C})$. Indeed, if $f \in \mathcal{C}_c^\infty(S(F) \times G(F), \mathbb{C})$ we can find a compact open subset Ω of $G(F)$ such that

$$\mathrm{Supp}(f) \subset S(F) \times \Omega$$

and a compact open subgroup $K' \subset G(F)$ such that

$$f(s, k'g) = f(s, g) \qquad (\forall s \in S(F),\ g \in G(F),\ k' \in K')$$

and

$$K'\Omega = \Omega;$$

let $(h_i)_{i \in I}$ be a system of representatives of the finite quotient $K' \backslash \Omega$; we have

$$f(s, g) = \sum_{i \in I} f(s, h_i) 1_{K'}(g h_i^{-1}) \qquad (\forall s \in S(F),\ g \in G(F))$$

and

$$\mathrm{pr}_*(f)(s) = \sum_{i \in I} f(s, h_i) \operatorname{vol}(K', |\omega_G|) \qquad (\forall s \in S(F)),$$

so that

$$f(s, g) = \frac{\mathrm{pr}_*(f)(s) 1_{K'}(g)}{\operatorname{vol}(K', |\omega_G|)} + \sum_{i \in I} f(s, h_i)(1_{K'}(g h_i^{-1}) - 1_{K'}(g))$$

$(\forall s \in S(F), g \in G(F))$.

Now the support of the distribution $\widetilde{D}$ is $\{\gamma\} \subset S(F)$. Indeed, if $\psi \in C_c^\infty(S(F))$ vanishes at γ we have

$$\mathrm{pr}_*\Big((s,g) \mapsto \psi(s)\frac{1_K(g)}{\mathrm{vol}(K, |\omega_G|)}\Big) = \psi$$

and

$$\pi_*\Big((s,g) \mapsto \psi(s)\frac{1_K(g)}{\mathrm{vol}(K, |\omega_G|)}\Big) \in C_c^\infty(\mathcal{U}, \mathbb{C})$$

vanishes along $O_G(\gamma)(F)$ (here $K = G(\mathcal{O})$). Therefore there exists a complex number c' such that

$$\widetilde{D} = c'\delta_\gamma$$

where

$$\delta_\gamma(\psi) = \psi(\gamma) \qquad (\forall \psi \in C_c^\infty(S(F), \mathbb{C}))$$

is the Dirac distribution at γ and we obtain the relation

$$D(\pi_*(f)) = c' \int_{G(F)} f(\gamma, g)|\omega_G|$$

for every $f \in C_c^\infty(S(F) \times G(F), \mathbb{C})$.

Similarly there exists a complex number c'' such that

$$\int_{G_\gamma(F)\backslash G(F)} \pi_*(f)(g)\frac{dg}{dg_\gamma} = c'' \int_{G(F)} f(\gamma, g)|\omega_G|$$

for every $f \in C_c^\infty(S(F) \times G(F), \mathbb{C})$ (the orbital integral is an invariant distribution on $G(F)$ with support in $\overline{O_G(\gamma)(F)}$ and we have

$$\mathcal{U} \cap \overline{O_G(\gamma)(F)} \subset O_G(\gamma)(F)\,).$$

From the surjectivity of π_* we get that $c'' \neq 0$ and the existence of c follows. Its uniqueness is obvious. □

(F.5) End of the proof of (10.6.4)

If we apply the results of (F.4) to $\gamma = \gamma^{(N)}$ we get an open subset $\mathcal{U}$ of $G(F)$ which contains $O_G(\gamma^{(N)})(F)$ as a closed subset, and a unique complex number c_N such that

$$D(\varphi) = c_N \int_{G_{\gamma^{(N)}}(F)\backslash G(F)} \varphi(g^{-1}\gamma^{(N)}g)\frac{dg}{dg_{\gamma^{(N)}}}$$

for every $\varphi \in C_c^\infty(G(F), \mathbb{C})$ such that $\mathrm{Supp}(\varphi) \subset \mathcal{U}$.

Let us set

$$\mathcal{U}' = \mathcal{U} \cap \Big(G(F) - \bigcup_{n=1}^{N-1} O_G(\gamma^{(n)})(F)\Big)$$

and

$$\mathcal{U}'' = G(F) - \bigcup_{n=1}^{N} O_G(\gamma^{(n)})(F).$$

Then $\mathcal{U}'$ and $\mathcal{U}''$ are open subsets of $G(F)$, $O_G(\gamma^{(N)})(F)$ is a closed subset of $\mathcal{U}'$ and we have

$$G(F) - \bigcup_{n=1}^{N-1} O_G(\gamma^{(n)})(F) = \mathcal{U}' \cup \mathcal{U}''.$$

Any $\varphi \in \mathcal{C}_c^\infty(G(F), \mathbb{C})$ such that

$$\mathrm{Supp}(\varphi) \cap \Big(\bigcup_{n=1}^{N-1} O_G(\gamma^{(n)})(F)\Big) = \emptyset$$

may be split into

$$\varphi = \varphi' + \varphi''$$

where $\varphi', \varphi'' \in \mathcal{C}_c^\infty(G(F), \mathbb{C})$ are such that

$$\mathrm{Supp}(\varphi') \subset \mathcal{U}' \subset \mathcal{U}$$

and

$$\mathrm{Supp}(\varphi'') \subset \mathcal{U}''.$$

It follows that the support of the invariant distribution

$$D_1(\varphi) = D(\varphi) - c_N \int_{G_{\gamma^{(N)}}(F)\backslash G(F)} \varphi(g^{-1}\gamma^{(N)}g)\frac{dg}{dg_{\gamma^{(N)}}}$$

($\varphi \in \mathcal{C}_c^\infty(G(F), \mathbb{C})$) is contained in

$$\bigcup_{n=1}^{N-1} O_G(\gamma^{(n)})(F).$$

Applying the induction hypothesis to D_1 we get the existence and the uniqueness of $c_1, \ldots, c_{N-1}$ and the proof of (10.6.4) is completed. □

(F.6) Comments and references

The proof of lemma (10.6.4) which is presented here is due to Harish-Chandra and Shalika (see [Sha 2]).

G

The decomposition of L_G^2 following the cuspidal data

(G.0) Introduction

In this appendix we will review Langlands' and Morris' fundamental results about the decomposition of

$$L^2_{G,1_\infty} = L^2(F_\infty^\times G(F)\backslash G(\mathbb{A}), dz_\infty d\gamma\backslash dg)$$

following the cuspidal data (see [Ar 7], [Go], [Lan 3], [Mor 1] and especially [Mo–Wa 2]). This is the first and the easiest step of the spectral decomposition of $L^2_{G,1_\infty}$.

We will use freely the notations of chapters 9 and 10.

(G.1) Cuspidal data

A **cuspidal pair** is a pair (I,π) where I is a subset of Δ and where π is an (isomorphism class of) irreducible cuspidal automorphic representation(s) of $M_I(\mathbb{A})$ with central character $\chi_\pi \in \mathcal{X}_{M_I}$ satisfying

$$\chi_\pi(Z_G(F_\infty)) = \{1\}.$$

We will say that the cuspidal pair (I,π) is **unitary** if χ_π is unitary.

For each cuspidal pair (I,π) we denote by

(G.1.1) $$\mathcal{A}^I_{I,\pi} \subset \mathcal{A}_{M_I,\chi_\pi,\mathrm{cusp}}$$

the $\mathbb{C}$-vector subspace of the vectors φ in $\mathcal{A}_{M_I,\chi_\pi,\mathrm{cusp}}$ such that the $M_I(\mathbb{A})$-submodule of $(\mathcal{A}_{M_I,\chi_\pi,\mathrm{cusp}}, R_{M_I,\chi_\pi,\mathrm{cusp}})$ generated by φ is isotypic of type π (thanks to (9.2.14) this submodule is semi-simple).

For each $I \subset \Delta$ let

$$\mathcal{C}_{M_I}^{G,\infty} = \mathcal{C}^\infty(N_I(\mathbb{A})M_I(F)\backslash G(\mathbb{A}), \mathbb{C})$$

be the $\mathbb{C}$-vector space of the complex functions φ on $G(\mathbb{A})$ which are invariant under left translation by $N_I(\mathbb{A})M_I(F)$ and which are invariant under right translation by some compact open subgroup of $G(\mathbb{A})$ (depending on φ). For each $\varphi \in \mathcal{C}_{M_I}^{G,\infty}$ and each $k \in K = G(\mathcal{O})$ we set

$$\varphi_k(m_I) = \delta_{P_I(\mathbb{A})}^{-1/2}(m_I)\varphi(m_I k) \qquad (\forall m_I \in M_I(\mathbb{A})),$$

so that

$$\varphi_k \in \mathcal{C}_{M_I}^{\infty}.$$

Then for each cuspidal pair (I, π) we denote by

$$\mathcal{A}_{I,\pi}^{\Delta} \subset \mathcal{C}_{M_I}^{G,\infty} \tag{G.1.2}$$

the $\mathbb{C}$-vector subspace of $\varphi \in \mathcal{C}_{M_I}^{G,\infty}$ such that

$$\varphi_k \in \mathcal{A}_{I,\pi}^{I} \qquad (\forall k \in K).$$

For each $I \subset \Delta$ we will denote by

$$X_{M_I} = X_I \tag{G.1.3}$$

the abelian group of complex characters of the discrete group

$$M_I(\mathbb{A})^1\backslash M_I(\mathbb{A}).$$

Let us recall that the homomorphism $H_I\big|_{M_I(\mathbb{A})} = -\deg_{M_I}$ induces an isomorphism

$$M_I(\mathbb{A})^1\backslash M_I(\mathbb{A}) \xrightarrow{\sim} \bigoplus_{j=1}^{s} \frac{1}{d_j}\mathbb{Z} \subset \mathbb{Q}^s = \mathfrak{a}_I$$

where $d_I = (d_1, \dots, d_s)$ is the partition of d corresponding to I (see (10.3)). Therefore, if we denote by

$$q = p^f$$

the number of elements of the field of constants of F and if we set

$$\langle \lambda, H \rangle = \lambda_1 H_1 + \cdots + \lambda_s H_s \in \mathbb{C}\Big/\frac{2\pi i}{\log q}\mathbb{Z}$$

for each $\lambda \in \bigoplus_{j=1}^{s}(\mathbb{C}/\frac{2\pi i}{\log q}d_j\mathbb{Z})$ and each $H \in \bigoplus_{j=1}^{s}\frac{1}{d_j}\mathbb{Z}$ we may identify X_I with

$$\bigoplus_{j=1}^{s}\Big(\mathbb{C}\Big/\frac{2\pi i}{\log q}d_j\mathbb{Z}\Big)$$

by

$$\lambda \mapsto (m_I \mapsto q^{\langle \lambda, H_I(m_I)\rangle}).$$

If $I \subset I' \subset \Delta$ we will denote by

$$X_{M_I}^{M_{I'}} = X_I^{I'} \subset X_I = \bigoplus_{j=1}^{s}\Big(\mathbb{C}\Big/\frac{2\pi i}{\log q}d_j\mathbb{Z}\Big) \tag{G.1.4}$$

the subgroup of complex characters of $M_I(\mathbb{A})^1\backslash M_I(\mathbb{A})$ which are trivial on $Z_{I'}(\mathbb{A})$. For example we have

$$X_I^{\Delta} = \Big\{\lambda \in \bigoplus_{j=1}^{s}\Big(\mathbb{C}\Big/\frac{2\pi i}{\log q}d_j\mathbb{Z}\Big) \mid \sum_{j=1}^{s}\lambda_j \in \frac{2\pi i}{\log q}\mathbb{Z}\Big\}$$

and

$$X_I^I = \bigoplus_{j=1}^{s}\Big(\frac{2\pi i}{\log q}\mathbb{Z}\Big/\frac{2\pi i}{\log q}d_j\mathbb{Z}\Big) \subset \bigoplus_{j=1}^{s}\Big(\mathbb{C}\Big/\frac{2\pi i}{\log q}d_j\mathbb{Z}\Big).$$

The decomposition

$$\mathbb{C}^\times = \mathbb{R}_{>0}\cdot\{z \in \mathbb{C}^\times \mid |z| = 1\}$$

induces a decomposition

$$X_I = \operatorname{Re} X_I \oplus \operatorname{Im} X_I$$

for every $I \subset \Delta$. We have

$$\operatorname{Re} X_I = \mathbb{R}^s \subset \bigoplus_{j=1}^{s}\Big(\mathbb{C}\Big/\frac{2\pi i}{\log q}d_j\mathbb{Z}\Big)$$

and

$$\operatorname{Im} X_I = \bigoplus_{j=1}^{s}\Big(i\mathbb{R}\Big/\frac{2\pi i}{\log q}d_j\mathbb{Z}\Big) \subset \bigoplus_{j=1}^{s}\Big(\mathbb{C}\Big/\frac{2\pi i}{\log q}d_j\mathbb{Z}\Big).$$

For any $I \subset I' \subset \Delta$ we set

$$\operatorname{Re} X_I^{I'} = (\operatorname{Re} X_I) \cap X_I^{I'}$$

and

$$\operatorname{Im} X_I^{I'} = (\operatorname{Im} X_I) \cap X_I^{I'}$$

and we have

$$X_I^{I'} = \operatorname{Re} X_I^{I'} \oplus \operatorname{Im} X_I^{I'}.$$

For any $I \subset J \subset I' \subset \Delta$ the canonical surjection

$$M_I(\mathbb{A})^1 \backslash M_I(\mathbb{A}) \twoheadrightarrow M_J(\mathbb{A})^1 \backslash M_J(\mathbb{A})$$

induces the inclusions

$$X_J \subset X_I$$

and

$$X_J^{I'} \subset X_I^{I'}.$$

If $(e_1, \ldots, e_t)$ is the partition of d corresponding to J and if

$$\begin{cases} e_1 = d_1 + \cdots + d_{s_1}, \\ \cdots \\ e_t = d_{s_1+\cdots+s_{t-1}+1} + \cdots + d_{s_1+\cdots+s_t}, \end{cases}$$

with $s_1 + \cdots + s_t = s$, we have

$$X_J = \bigoplus_{k=1}^{t} \Big(\mathbb{C} \Big/ \frac{2\pi i}{\log q} e_k \mathbb{Z}\Big) = \Big\{\lambda \in X_I \mid \frac{\lambda_1 e_1}{d_1} = \cdots = \frac{\lambda_{s_1} e_1}{d_{s_1}} = \mu_1, \\ \ldots, \\ \frac{\lambda_{s_1+\cdots+s_{t-1}+1} e_t}{d_{s_1+\cdots+s_{t-1}+1}} = \cdots = \frac{\lambda_s e_t}{d_s} = \mu_t\Big\}$$

where we have denoted by μ the elements of X_J. Moreover it is clear that we have

$$\operatorname{Re} X_I = \operatorname{Re} X_J \oplus \operatorname{Re} X_I^J$$

and

$$\operatorname{Re} X_I^{I'} = \operatorname{Re} X_J^{I'} \oplus \operatorname{Re} X_I^J.$$

If (I, π) is a cuspidal pair there exists a unique element

$$\operatorname{Re} \pi \in \operatorname{Re} X_I^{\Delta}$$

such that

$$|\chi_\pi(z_I)| = q^{-\langle \operatorname{Re} \pi, \deg_{Z_I}(z_I)\rangle} \qquad (\forall z_I \in Z_I(\mathbb{A}))$$

(see (9.1)). Then (I, π) is unitary if and only if $\operatorname{Re} \pi = 0$.

If (I, π) is a cuspidal pair and if $\lambda \in X_I^{\Delta}$ we denote by

$$\pi(\lambda)$$

the tensor product of π with the character λ of $M_I(\mathbb{A})$. Then $(I,\pi(\lambda))$ is another cuspidal pair, we have

$$\mathcal{A}^I_{I,\pi(\lambda)} = \{\varphi_\lambda \mid \varphi \in \mathcal{A}^I_{I,\pi}\}$$

and

$$\mathcal{A}^\Delta_{I,\pi(\lambda)} = \{\varphi_\lambda \mid \varphi \in \mathcal{A}^\Delta_{I,\pi}\}$$

where we have set

$$\varphi_\lambda(m_I) = q^{\langle\lambda, H_I(m_I)\rangle}\varphi(m_I) \qquad (\forall m_I \in M_I(\mathbb{A}))$$

(resp.

$$\varphi_\lambda(g) = q^{\langle\lambda, H_I(g)\rangle}\varphi(g) \qquad (\forall g \in G(\mathbb{A})))$$

for each $\varphi \in \mathcal{A}^I_{I,\pi}$ (resp. $\varphi \in \mathcal{A}^\Delta_{I,\pi}$) and we have

$$\operatorname{Re}\pi(\lambda) = \operatorname{Re}\pi + \operatorname{Re}\lambda.$$

LEMMA (G.1.5). — *Let $I \subset \Delta$ and let $d_I = (d_1, \dots, d_s)$ be the corresponding partition of d. For each $\sigma \in \mathfrak{S}_s$ (the permutation group of $\{1, \dots, s\}$) let us denote by*

$$w_{I,\sigma} \in \mathfrak{S}_d = W$$

the permutation of $\{1, \dots, d\}$ defined by

$$w_{I,\sigma}(d_1 + \cdots + d_{j-1} + k) = d_{\sigma^{-1}(1)} + \cdots + d_{\sigma^{-1}(\sigma(j)-1)} + k$$

for every $k = 1, \dots, d_j$ and every $j = 1, \dots, s$. Then the map $\sigma \mapsto w_{I,\sigma}$ is a bijection from $\mathfrak{S}_s$ onto the set of $w \in W$ such that $w(I) \subset \Delta$.

In particular, if $w \in W$ satisfies $w(I) \subset \Delta$ we have

$$\dot{w} M_I \dot{w}^{-1} = M_{w(I)}.$$

□

Let (I, π) be a cuspidal pair and let $w \in W$ satisfy $w(I) \subset \Delta$. Then $(w(I), w(\pi))$, with

$$w(\pi)(m_{w(I)}) = \pi(\dot{w}^{-1} m_{w(I)} \dot{w}) \qquad (\forall m_{w(I)} \in M_{w(I)}(\mathbb{A})),$$

is another cuspidal pair. We have

$$\mathcal{A}^{w(I)}_{w(I),w(\pi)} = \{w(\varphi) \mid \varphi \in \mathcal{A}^I_{I,\pi}\},$$

with

$$w(\varphi)(m_{w(I)}) = \varphi(\dot{w}^{-1}m_{w(I)}\dot{w}) \qquad (\forall m_{w(I)} \in M_{w(I)}(\mathbb{A})).$$

And if we set

$$w(\lambda)(m_{w(I)}) = \lambda(\dot{w}^{-1}m_{w(I)}\dot{w}) \qquad (\forall m_{w(I)} \in M_{w(I)}(\mathbb{A}))$$

for each $\lambda \in X_I$, so that $w(\lambda) \in X_{w(I)}$ and $w(X_I^{\Delta}) = X_{w(I)}^{\Delta}$, we have

$$\operatorname{Re} w(\pi) = w(\operatorname{Re} \pi)$$

and

$$w(\pi(\lambda)) = w(\pi)(w(\lambda)) \qquad (\forall \lambda \in X_I^{\Delta}).$$

In fact, for each $\sigma \in \mathfrak{S}_s$ we have

$$w_{I,\sigma}(\lambda) = (\lambda_{\sigma^{-1}(1)}, \ldots, \lambda_{\sigma^{-1}(s)}) \qquad (\forall \lambda \in X_I).$$

Definition (G.1.6). — *A cuspidal datum is an equivalence class of cuspidal pairs for the equivalence relation*

$$(I', \pi') \sim (I'', \pi'')$$

if and only if there exist $\lambda' \in X_{I'}^{\Delta}$ and $w' \in W$ such that

$$(I'', \pi'') = (w'(I'), w'(\pi'(\lambda'))).$$

We will denote by $\mathcal{C}$ the set of cuspidal data.

(G.2) Paley–Wiener functions

Let us fix a cuspidal pair (I, π). We will denote by

$$\mathrm{Fix}_{X_I^{\Delta}}(\pi)$$

the subgroup

$$\{\mu \in X_I^{\Delta} \mid \pi(\mu) \cong \pi\}$$

of X_I^{Δ}. By considering the central characters we see that

$$\mathrm{Fix}_{X_I^{\Delta}}(\pi) \subset X_I^I$$

and therefore that $\mathrm{Fix}_{X_I^{\Delta}}(\pi)$ is finite.

DEFINITION (G.2.1). — *A Paley–Wiener function for the cuspidal pair* (I,π) *is a map*

$$\Phi : X_I^\Delta \to \mathcal{A}_{I,\pi}^\Delta$$

which satisfies the following properties:

(i) *there exists a compact open subgroup* K_Φ *of* $G(\mathbb{A})$ *(depending on* Φ*) such that*

$$\Phi(\lambda)(gk_\Phi) = \Phi(\lambda)(g)$$

for every $\lambda \in X_I^\Delta$*, every* $g \in G(\mathbb{A})$ *and every* $k_\Phi \in K_\Phi$*,*

(ii) Φ *is polynomial in the variables*

$$q^{\lambda_j/d_j},\ q^{-\lambda_j/d_j} \qquad (j = 1,\dots,s),$$

(iii) *for each* $\lambda \in X_I^\Delta$ *and each* $\mu \in \mathrm{Fix}_{X_I^\Delta}(\pi)$ *we have*

$$\Phi(\lambda + \mu) = \Phi(\lambda)_{-\mu}.$$

We will denote by $\mathcal{P}(I,\pi)$ the $\mathbb{C}$-vector space of the Paley–Wiener functions for (I,π).

REMARK (G.2.2). — If K' is a compact open subgroup of $G(\mathbb{A})$ the $\mathbb{C}$-vector space

$$\{\varphi \in \mathcal{A}_{I,\pi}^\Delta \mid \varphi(gk') = \varphi(g),\ \forall g \in G(\mathbb{A}),\ \forall k' \in K'\}$$

is finite dimensional. Indeed, by shrinking K' if it is necessary we may assume that K' is an open normal subgroup of K and then the assertion follows from (9.2.10),

$$\dim_{\mathbb{C}}\big((\mathcal{A}_{I,\pi}^I)^{M_I(\mathbb{A})\cap K'}\big) < +\infty.$$

Thus property (i) of (G.2.1) implies that the range of Φ is contained in a finite dimensional $\mathbb{C}$-vector subspace of $\mathcal{A}_{I,\pi}^\Delta$. In particular property (ii) of (G.2.1) makes sense. □

Let $\mathcal{P}(X_I^\Delta)$ be the $\mathbb{C}$-vector space of complex functions f on X_I^Δ which are polynomial in the variables

$$q^{\lambda_j/d_j},\ q^{-\lambda_j/d_j} \qquad (j = 1,\dots,s).$$

We will identify

$$\mathcal{P}(X_I^\Delta) \otimes_{\mathbb{C}} \mathcal{A}_{I,\pi}^\Delta$$

with a subspace of the space of functions $\Phi : X_I^\Delta \to \mathcal{A}_{I,\pi}^\Delta$ by the map

$$\sum_\alpha f_\alpha \otimes \varphi_\alpha \mapsto \big(\lambda \mapsto \Phi(\lambda) = \sum_\alpha f_\alpha(\lambda)\varphi_\alpha\big).$$

We let $\mathrm{Fix}_{X_I^\Delta}(\pi)$ act on $\mathcal{P}(X_I^\Delta)\otimes_{\mathbb{C}}\mathcal{A}_{I,\pi}^\Delta$ by

$$(\mu, f\otimes\varphi)\mapsto (\lambda\mapsto f(\lambda+\mu))\otimes\varphi_\mu.$$

Then we have

$$\left(\mathcal{P}(X_I^\Delta)\otimes_{\mathbb{C}}\mathcal{A}_{I,\pi}^\Delta\right)^{\mathrm{Fix}_{X_I^\Delta}(\pi)}=\mathcal{P}(I,\pi).$$

Moreover we have a projection

$$\mathcal{P}(X_I^\Delta)\otimes_{\mathbb{C}}\mathcal{A}_{I,\pi}^\Delta\to\left(\mathcal{P}(X_I^\Delta)\otimes_{\mathbb{C}}\mathcal{A}_{I,\pi}^\Delta\right)^{\mathrm{Fix}_{X_I^\Delta}(\pi)},$$

$$f\otimes\varphi\mapsto\Big(\lambda\mapsto\frac{1}{|\mathrm{Fix}_{X_I^\Delta}(\pi)|}\sum_{\mu\in\mathrm{Fix}_{X_I^\Delta}(\pi)}f(\lambda+\mu)\otimes\varphi_\mu\Big).$$

(G.3) Fourier transformation

Let $d\lambda$ be the Haar measure on the compact abelian Lie group $\mathrm{Im}\,X_I^\Delta$ which is normalized by

$$\mathrm{vol}(\mathrm{Im}\,X_I^\Delta, d\lambda)=1.$$

The dual group of $\mathrm{Im}\,X_I^\Delta$ is the discrete group

$$Z_G(\mathbb{A})M_I(\mathbb{A})^1\backslash M_I(\mathbb{A})\xrightarrow{\sim}\mathbb{Z}\Big\backslash\Big(\bigoplus_{j=1}^{s}\frac{1}{d_j}\mathbb{Z}\Big)$$

($\mathbb{Z}$ is diagonally embedded and the isomorphism is induced by $H_{I|M_I(\mathbb{A}))}$). The dual Haar measure to $d\lambda$ on this discrete group is the counting measure.

If

$$\mathcal{C}_c(Z_G(\mathbb{A})M_I(\mathbb{A})^1\backslash M_I(\mathbb{A}),\mathbb{C})$$

is the $\mathbb{C}$-vector space of complex functions with finite support on $Z_G(\mathbb{A})M_I(\mathbb{A})^1\backslash M_I(\mathbb{A})$ the Fourier transformation

$$\mathcal{P}(X_I^\Delta)\to\mathcal{C}_c(Z_G(\mathbb{A})M_I(\mathbb{A})^1\backslash M_I(\mathbb{A}),\mathbb{C}),$$

$$f\mapsto\Big(m_I\mapsto\widehat{f}(m_I)=\int_{\mathrm{Im}\,X_I^\Delta}q^{\langle\lambda,H_I(m_I)\rangle}f(\lambda)d\lambda\Big),$$

is an isomorphism of $\mathbb{C}$-vector spaces (see [We 2] Ch. VII, §2 or [We 3] Ch. VI, §30).

For each cuspidal pair (I,π) and each $\Phi\in\mathcal{P}(X_I^\Delta)\otimes_{\mathbb{C}}\mathcal{A}_{I,\pi}^\Delta$ let us set

$$\text{(G.3.1)}\qquad \Phi^\vee(g)=\frac{1}{|\mathrm{Fix}_{X_I^\Delta}(\pi)|}\int_{\mathrm{Im}\,X_I^\Delta}\Phi(\lambda)_\lambda(g)d\lambda\qquad(\forall g\in G(\mathbb{A})),$$

so that $\Phi^\vee$ is a complex function on $N_I(\mathbb{A})M_I(F)\backslash G(\mathbb{A})$ which is invariant under right translation by $K \cap K_\Phi$. If

$$\Phi(\lambda) = f(\lambda)\varphi \qquad (\forall \lambda \in X_I^\Delta)$$

with $f \in \mathcal{P}(X_I^\Delta)$ and $\varphi \in \mathcal{A}_{I,\pi}^\Delta$ we have

$$\Phi^\vee(g) = \frac{1}{|\mathrm{Fix}_{X_I^\Delta}(\pi)|}\widehat{f}(m_I(g))\varphi(g) \qquad (\forall g \in G(\mathbb{A})).$$

Therefore, if we let X_I^Δ act on $\mathcal{C}_c(Z_G(\mathbb{A})M_I(\mathbb{A})^1\backslash M_I(\mathbb{A}), \mathbb{C}) \otimes_{\mathbb{C}} \mathcal{A}_{I,\pi}^\Delta$ by

$$(\lambda, f' \otimes \varphi) \mapsto q^{-\langle \lambda, H_I\rangle} f' \otimes \varphi_\lambda,$$

for any $\Phi \in \mathcal{P}(X_I^\Delta) \otimes_{\mathbb{C}} \mathcal{A}_{I,\pi}^\Delta$ the function $\Phi^\vee$ belongs to the image of the $\mathbb{C}$-linear map

$$\left(\mathcal{C}_c(Z_G(\mathbb{A})M_I(\mathbb{A})^1\backslash M_I(\mathbb{A}), \mathbb{C}) \otimes_{\mathbb{C}} \mathcal{A}_{I,\pi}^\Delta\right)^{X_I^\Delta} \to \mathcal{C}_{M_I}^{G,\infty},$$
$$\sum_\alpha f'_\alpha \otimes \varphi_\alpha \mapsto \Big(g = n_I m_I k \mapsto \sum_\alpha f'_\alpha(m_I)\varphi_\alpha(g)\Big).$$

In fact this image is equal to

$$\{\Phi^\vee \mid \Phi \in \mathcal{P}(I,\pi)\}$$

and we will denote it by $\mathcal{P}^\vee(I,\pi)$.

(G.4) Eisenstein series

For any cuspidal pair (I,π) and any $\varphi \in \mathcal{A}_{I,\pi}^\Delta$ we consider the Eisenstein series

$$E_I^\Delta(\varphi,\pi)(g) = \sum_{\gamma \in P_I(F)\backslash G(F)} \varphi(\gamma g) \qquad (\forall g \in G(\mathbb{A})). \tag{G.4.1}$$

If it converges for all $g \in G(\mathbb{A})$ then $E_I^\Delta(\varphi,\pi)$ is a complex function on $F_\infty^\times G(F)\backslash G(\mathbb{A})$.

Let

$$C_I^\Delta = \{\lambda \in X_I^\Delta \mid \mathrm{Re}\Big(\frac{\lambda_j}{d_j} - \frac{\lambda_{j+1}}{d_{j+1}}\Big) > 0,\ \forall j = 1,\ldots,s-1\}$$

(resp.

$$\overline{C}_I^\Delta = \{\lambda \in X_I^\Delta \mid \mathrm{Re}\Big(\frac{\lambda_j}{d_j} - \frac{\lambda_{j+1}}{d_{j+1}}\Big) \geq 0,\ \forall j = 1,\ldots,s-1\})$$

be the **open** (resp. **closed**) **positive cone** in X_I^Δ and let

$$\rho_I \in X_I^\Delta$$

be the **half sum of the positive roots** of (G, Z_I), so that

$$\rho_{I,j} = d_j(d_s + \cdots + d_{j+1} - d_j - \cdots - d_1)/2 \qquad (\forall j = 1,\ldots,s)$$

and

$$\delta_{P_I(\mathbb{A})}^{1/2}(m_I) = q^{\langle \rho_I, H_I(m_I)\rangle} \qquad (\forall m_I \in M_I(\mathbb{A})).$$

PROPOSITION (G.4.2). — *For each $I \subset \Delta$, each $\lambda_0 \in \rho_I + C_I^\Delta \subset X_I^\Delta$ and each compact subset Ω of $G(\mathbb{A})$ the series of functions of $(\lambda, g) \in X_I^\Delta \times G(\mathbb{A})$,*

$$\sum_{\gamma \in P_I(F)\backslash G(F)} q^{\langle \lambda + \rho_I, H_I(\gamma g)\rangle},$$

is normally convergent on $(\lambda_0 + \overline{C}_I^\Delta) \times \Omega$.

COROLLARY (G.4.3). — *For each cuspidal pair (I, π) and each $\Phi \in \mathcal{P}(I, \pi)$ the Eisenstein series*

$$E_I^\Delta(\Phi(\lambda)_\lambda, \pi(\lambda))(g),$$

viewed as series of functions of

$$(g, \lambda) \in G(\mathbb{A}) \times (\rho_I - \mathrm{Re}\,\pi + C_I^\Delta) \subset G(\mathbb{A}) \times X_I^\Delta,$$

is normally convergent when g stays in a compact subset of $G(\mathbb{A})$ and $\mathrm{Re}\,\lambda$ stays in a compact subset of $\rho_I - \mathrm{Re}\,\pi + \mathrm{Re}\,C_I^\Delta$. In particular it depends holomorphically on $\lambda \in \rho_I - \mathrm{Re}\,\pi + C_I^\Delta$.

Proof of the corollary (assuming the proposition) : Replacing π by $\pi(-\mathrm{Re}\,\pi)$ we may assume that χ_π is unitary. Then for each $\Phi \in \mathcal{P}(I, \pi)$ and each compact subset C of $\mathrm{Re}\,X_I^\Delta$ there exists a positive constant $c_{\Phi,C}$ such that

$$|\Phi(\lambda)(g)| \leq c_{\Phi,C}\, q^{\langle \rho_I, H_I(g)\rangle} \qquad (\forall \lambda \in C + \mathrm{Im}\,X_I^\Delta,\ \forall g \in G(\mathbb{A})).$$

Indeed we have

$$\Phi(\lambda) = \sum_{i=1}^{n} P_i(\lambda)\varphi_i \qquad (\forall \lambda \in X_I^\Delta)$$

for some $\varphi_1, \ldots, \varphi_n \in \mathcal{A}_{I,\pi}^\Delta$ and some polynomials $P_1(\lambda), \ldots, P_n(\lambda)$ in the variables $q^{\lambda_1/2}, q^{-\lambda_1/2}, \ldots, q^{\lambda_s/2}, q^{-\lambda_s/2}$ (see (G.2.2)); for each $i = 1, \ldots, n$ and each $k \in K$ we have $\varphi_{i,k} \in \mathcal{A}_{I,\pi}^I$, so that $|\varphi_{i,k}|$ is a locally constant function on $Z_I(\mathbb{A})M_I(F)\backslash M_I(\mathbb{A})$ with compact support (see (9.2.6)); therefore we can take

$$c_{\Phi,C} = \sup\Big\{\sum_{i=1}^{n} |P_i(\lambda)||\varphi_{i,k}(m_I)| \mid \mathrm{Re}\,\lambda \in C,\ m_I \in M_I(\mathbb{A}),\ k \in K\Big\}.$$

The corollary follows. □

To prove (G.4.2) we will need two results of reduction theory (see [Har 2] (1.4.1) and (1.4.2)).

LEMMA (G.4.4). — *There exists a non-negative integer N such that, for any g in the Siegel set*

$$\mathfrak{S} = \{g \in G(\mathbb{A}) \mid \alpha_i(H_\emptyset(g)) \geq -2g_X, \ \forall i \in \Delta\}$$

and any $\gamma \in G(F)$, we have

$$\varpi_i(H_\emptyset(\gamma g)) \leq \varpi_i(H_\emptyset(g)) + N \qquad (\forall i \in \Delta).$$

Proof: Using Weil's geometric interpretation of $GL_d(\mathbb{A})$ in terms of rank d vector bundles on X (see (E.1)) we easily see that the lemma is equivalent to the following statement.

There exists a non-negative integer N having the following property: for any pair of flags of vector bundles of rank $(1, \ldots, 1)$ over X,

$$\mathcal{E}_\bullet = ((0) = \mathcal{E}_0 \subset \mathcal{E}_1 \subset \cdots \subset \mathcal{E}_d = \mathcal{E})$$

and

$$\mathcal{E}'_\bullet = ((0) = \mathcal{E}'_0 \subset \mathcal{E}'_1 \subset \cdots \subset \mathcal{E}'_d = \mathcal{E}'),$$

such that

$$\deg(\mathcal{E}_{i+1}/\mathcal{E}_i) - \deg(\mathcal{E}_i/\mathcal{E}_{i-1}) \leq 2g_X \qquad (\forall i = 1, \ldots, d-1)$$

and such that

$$\mathcal{E}' = \mathcal{E},$$

we have

$$\deg(\mathcal{E}'_i) \leq \deg(\mathcal{E}_i) + N \qquad (\forall i = 1, \ldots, d-1).$$

Let us prove this geometric statement. We fix $i_0 \in \{1, \ldots, d-1\}$. Then there exists at least one rank i_0 subbundle $\mathcal{F}$ of $\mathcal{E}$ such that

$$\deg(\mathcal{F}') \leq \deg(\mathcal{F})$$

for any other rank i_0 subbundle $\mathcal{F}'$ of $\mathcal{E}$. Let us fix such an $\mathcal{F}$. Applying (E.1.1) to $\mathcal{F}$ and $\mathcal{E}/\mathcal{F}$ we can construct a flag of subbundles of $\mathcal{E}$ of rank $(1, \ldots, 1)$,

$$\widetilde{\mathcal{E}}_\bullet = ((0) = \widetilde{\mathcal{E}}_0 \subset \widetilde{\mathcal{E}}_1 \subset \cdots \subset \widetilde{\mathcal{E}}_d = \mathcal{E}),$$

such that

$$\deg(\widetilde{\mathcal{E}}_{i+1}/\widetilde{\mathcal{E}}_i) - \deg(\widetilde{\mathcal{E}}_i/\widetilde{\mathcal{E}}_{i-1}) \leq 2g_X \qquad (\forall i = 1, \ldots, d-1)$$

and

$$\widetilde{\mathcal{E}}_{i_0} = \mathcal{F}.$$

In particular we have

$$\deg(\mathcal{E}'_{i_0}) \leq \deg(\widetilde{\mathcal{E}}_{i_0})$$

for any flag $\mathcal{E}'_\bullet$ of subbundles of $\mathcal{E}$ of rank $(1,\ldots,1)$.

Now let us fix an integer c'_1 as in (E.1.2) (for $c_2 = 2g_X$) and let us set

$$I = \{i \in \Delta \mid c'_1 \leq \deg(\widetilde{\mathcal{E}}_{i+1}/\widetilde{\mathcal{E}}_i) - \deg(\widetilde{\mathcal{E}}_i/\widetilde{\mathcal{E}}_{i-1})\}.$$

Then for any $i \in \Delta - I$ we have

$$\widetilde{\mathcal{E}}_i = \mathcal{E}_i.$$

If $d_I = (d_1,\ldots,d_s)$ is the partition of d corresponding to I it is not difficult to check that, for any $j \in \{1,\ldots,s\}$ and any $k \in \{0,\ldots,d_j-1\}$, we have

$$\begin{aligned}\varpi_{d_1+\cdots+d_{j-1}+k} = &\frac{d_j-k}{d_j}\sum_{\ell=1}^{k}\ell\alpha_{d_1+\cdots+d_{j-1}+\ell}\\ &+\frac{k}{d_j}\sum_{\ell=k+1}^{d_j-1}(d_j-\ell)\alpha_{d_1+\cdots+d_{j-1}+\ell}\\ &+\frac{d_j-k}{d_j}\varpi_{d_1+\cdots+d_{j-1}}+\frac{k}{d_j}\varpi_{d_1+\cdots+d_j}.\end{aligned}$$

As

$$\big(\deg(\widetilde{\mathcal{E}}_i/\widetilde{\mathcal{E}}_{i-1}) - \deg(\widetilde{\mathcal{E}}_{i+1}/\widetilde{\mathcal{E}}_i)\big) - \big(\deg(\mathcal{E}_i/\mathcal{E}_{i-1}) - \deg(\mathcal{E}_{i+1}/\mathcal{E}_i)\big) \leq -c'_1 + 2g_X$$

for every $i \in I$ it follows that

$$\deg(\widetilde{\mathcal{E}}_{i_0}) - \deg(\mathcal{E}_{i_0}) \leq \frac{k(d_j-k)}{2}(2g_X - c'_1)$$

as long as $i_0 = d_1 + \cdots + d_{j-1} + k$ for some $j \in \{1,\ldots,s\}$ and some $k \in \{0,\ldots,d_j-1\}$.

To conclude the proof of the lemma it is sufficient to choose N in such a way that

$$N \geq \frac{k(d_j-k)}{2}(2g_X - c'_1)$$

for every partition $(d_1,\ldots,d_s)$ of d, every $j \in \{1,\ldots,s\}$ and every $k \in \{0,\ldots,d_j-1\}$. □

LEMMA (G.4.5). — *There exists a non-negative integer N' such that, for any g in the Siegel set $\mathfrak{S}$, we have*

$$[G(F)\cap(gKg^{-1}) : P_\emptyset(F)\cap(gKg^{-1})] \leq q^{N'}.$$

Proof : The lemma is equivalent to the following geometric statement.

There exists a non-negative integer N' *having the following property: for any flag of vector bundles of rank* $(1,\ldots,1)$ *over* X,

$$\mathcal{E}_\bullet = \big((0) = \mathcal{E}_0 \subset \mathcal{E}_1 \subset \cdots \mathcal{E}_d = \mathcal{E}\big),$$

such that

$$\deg(\mathcal{E}_{i+1}/\mathcal{E}_i) - \deg(\mathcal{E}_i/\mathcal{E}_{i-1}) \leq 2g_X \qquad (\forall i = 1,\ldots,d-1),$$

we have

$$\big[\mathrm{Aut}(\mathcal{E}) : \mathrm{Aut}(\mathcal{E}_\bullet)\big] \leq q^{N'},$$

where $\mathrm{Aut}(\mathcal{E})$ *is the (finite) group of automorphisms of the vector bundle* $\mathcal{E}$ *and* $\mathrm{Aut}(\mathcal{E}_\bullet)$ *is the group of* $\gamma \in \mathrm{Aut}(\mathcal{E})$ *which fix the flag* $\mathcal{E}_\bullet$.

Let us prove this geometric statement. We fix an integer c_1' as in (E.1.2) (for $c_2 = 2g_X$) and we set

$$I = \{i \in \Delta \mid c_1' \leq \deg(\mathcal{E}_{i+1}/\mathcal{E}_i) - \deg(\mathcal{E}_i/\mathcal{E}_{i-1})\}.$$

If $d_I = (d_1,\ldots,d_s)$ is the partition of d corresponding to I and if we set

$$\mathcal{F}_j = \mathcal{E}_{d_1+\cdots+d_j} \qquad (\forall j = 1,\ldots,s)$$

it follows from (E.1.2) that any automorphism of $\mathcal{E}$ fixes the (incomplete) flag

$$\mathcal{F}_\bullet = \big((0) = \mathcal{F}_0 \subset \mathcal{F}_1 \subset \cdots \subset \mathcal{F}_s = \mathcal{E}\big).$$

In other words we have

$$\mathrm{Aut}(\mathcal{E}) = \mathrm{Aut}(\mathcal{F}_\bullet).$$

Now if we denote by $\mathrm{Aut}^+(\mathcal{E}_\bullet)$ (resp. $\mathrm{Aut}^+(\mathcal{F}_\bullet)$) the group of $\gamma \in \mathrm{Aut}(\mathcal{E})$ such that

$$(\gamma - \mathrm{id}_{\mathcal{E}})(\mathcal{E}_i) \subset \mathcal{E}_{i-1} \qquad (\forall i = 1,\ldots,d)$$

(resp.

$$(\gamma - \mathrm{id}_{\mathcal{E}})(\mathcal{F}_j) \subset \mathcal{F}_{j-1} \qquad (\forall j = 1,\ldots,s))$$

we have

$$\begin{array}{ccccc} \mathrm{Aut}^+(\mathcal{F}_\bullet) & \subset & \mathrm{Aut}(\mathcal{F}_\bullet) & = & \mathrm{Aut}(\mathcal{E}) \\ \cap & & \cup & & \\ \mathrm{Aut}^+(\mathcal{E}_\bullet) & \subset & \mathrm{Aut}(\mathcal{E}_\bullet), & & \end{array}$$

so that

$$\big[\mathrm{Aut}(\mathcal{E}) : \mathrm{Aut}(\mathcal{E}_\bullet)\big] \leq \big[\mathrm{Aut}(\mathcal{F}_\bullet) : \mathrm{Aut}^+(\mathcal{F}_\bullet)\big],$$

and we have an exact sequence

$$1 \to \mathrm{Aut}^+(\mathcal{F}_\bullet) \to \mathrm{Aut}(\mathcal{F}_\bullet) \to \prod_{j=1}^{s} \mathrm{Aut}(\mathcal{F}_j/\mathcal{F}_{j-1}),$$

so that

$$\left[\operatorname{Aut}(\mathcal{F}_\bullet) : \operatorname{Aut}^+(\mathcal{F}_\bullet)\right] \leq \prod_{j=1}^{s} |\operatorname{Aut}(\mathcal{F}_j/\mathcal{F}_{j-1})|.$$

Therefore our previous geometric statement is a consequence of the following one (for $d := d_j$ and $\mathcal{E} := \mathcal{F}_j/\mathcal{F}_{j-1}$, $j = 1, \ldots, s$).

Let c_1' be any integer such that $c_1' \leq 2g_X$. Then there exists a non-negative integer $N'' = N''(c_1')$ having the following property: for any flag of vector bundles of rank $(1, \ldots, 1)$ over X,

$$\mathcal{E}_\bullet = \big((0) = \mathcal{E}_0 \subset \mathcal{E}_1 \subset \cdots \subset \mathcal{E}_d = \mathcal{E}\big),$$

such that

$$c_1' \leq \deg(\mathcal{E}_{i+1}/\mathcal{E}_i) - \deg(\mathcal{E}_i/\mathcal{E}_{i-1}) \leq 2g_X \qquad (\forall i = 1, \ldots, d-1),$$

we have

$$|\operatorname{Aut}(\mathcal{E})| \leq q^{N''}.$$

Let us prove the last geometric statement. We have

$$\operatorname{Aut}(\mathcal{E}) \subset \operatorname{End}(\mathcal{E}),$$

where $\operatorname{End}(\mathcal{E})$ is the $\mathbb{F}_p$-algebra of endomorphisms of the vector bundle $\mathcal{E}$. The filtration $\mathcal{E}_\bullet$ of $\mathcal{E}$ induces a filtration $F_\bullet \operatorname{End}(\mathcal{E})$ of $\operatorname{End}(\mathcal{E})$ with

$$F_n \operatorname{End}(\mathcal{E}) = \{\gamma \in \operatorname{End}(\mathcal{E}) \mid \gamma(\mathcal{E}_i) \subset \mathcal{E}_{i+n},\ \forall i \in \mathbb{Z}\}$$

for every $n \in \mathbb{Z}$ ($\mathcal{E}_i = (0)$ if $i \leq 0$ and $\mathcal{E}_i = \mathcal{E}$ if $i \geq d$). We have

$$F_n \operatorname{End}(\mathcal{E}) = \begin{cases} (0) & \text{if } n \leq -d, \\ \operatorname{End}(\mathcal{E}) & \text{if } n \geq d-1, \end{cases}$$

and we have an exact sequence of $\mathbb{F}_p$-vector spaces,

$$0 \to F_{n-1} \operatorname{End}(\mathcal{E}) \to F_n \operatorname{End}(\mathcal{E}) \to \bigoplus_{i \in \mathbb{Z}} \operatorname{Hom}\Big(\frac{\mathcal{E}_i}{\mathcal{E}_{i-1}}, \frac{\mathcal{E}_{i+n}}{\mathcal{E}_{i+n-1}}\Big),$$

so that

$$|\operatorname{gr}_n^F \operatorname{End}(\mathcal{E})| \leq \prod_{i \in \mathbb{Z}} \Big| \operatorname{Hom}\Big(\frac{\mathcal{E}_i}{\mathcal{E}_{i-1}}, \frac{\mathcal{E}_{i+n}}{\mathcal{E}_{i+n-1}}\Big)\Big| \qquad (\forall n \in \mathbb{Z}).$$

Therefore we have proved that

$$|\operatorname{Aut}(\mathcal{E})| \leq |\operatorname{End}(\mathcal{E})| \leq \prod_{i,j\in\mathbb{Z}} \left|\operatorname{Hom}\left(\frac{\mathcal{E}_i}{\mathcal{E}_{i-1}}, \frac{\mathcal{E}_j}{\mathcal{E}_{j-1}}\right)\right|.$$

But if $\mathcal{A}$ and $\mathcal{B}$ are two line bundles over X it is easy to see that

$$\dim_{\mathbb{F}_p} \operatorname{Hom}(\mathcal{A},\mathcal{B}) \leq f(\deg(\mathcal{B}) - \deg(\mathcal{A}) + 1)$$

if $\deg(\mathcal{A}) \leq \deg(\mathcal{B})$ and that

$$\operatorname{Hom}(\mathcal{A},\mathcal{B}) = (0)$$

otherwise. As we have

$$\deg\left(\frac{\mathcal{E}_j}{\mathcal{E}_{j-1}}\right) - \deg\left(\frac{\mathcal{E}_i}{\mathcal{E}_{i-1}}\right) \leq \begin{cases} (j-i)2g_X & \text{if } i \leq j, \\ (i-j)c_1' & \text{if } i > j, \end{cases}$$

for any $i, j \in \{1, \ldots, d\}$ our second geometric statement follows. □

Proof of proposition (G.4.2) : Obviously we may assume that λ is real, i.e. $\lambda \in \operatorname{Re} X_I^\Delta$. Moreover, if we fix $H_\emptyset \in \mathfrak{a}_\emptyset$ such that

$$\alpha_i(H_\emptyset) \geq -2g_X \qquad (\forall i = 1, \ldots, d-1)$$

it is sufficient to prove that the series

$$\sum_{\gamma\in P_I(F)\backslash G(F)} q^{\langle\lambda+\rho_I, H_I(\gamma g)\rangle}$$

is uniformly bounded for every $\lambda \in \lambda_0 + C_I^\Delta$ and every $g \in G(\mathbb{A})$ with $H_\emptyset(g) = H_\emptyset$.

Let us fix an integer N as in lemma (G.4.4) and let

$$\mathcal{U} \subset G(\mathbb{A})$$

be the set of $h \in G(\mathbb{A})$ such that

$$\varpi_i(H_\emptyset(h)) \leq \varpi_i(H_\emptyset) + N \qquad (\forall i \in \Delta - I)$$

and

$$H_\emptyset(g)_1 + \cdots + H_\emptyset(g)_d = H_{\emptyset,1} + \cdots + H_{\emptyset,d}.$$

We have

$$N_I(\mathbb{A})M_I(\mathbb{A})^1\mathcal{U}K = \mathcal{U}$$

and it follows from (G.4.4) that, for each $g \in G(\mathbb{A})$ with

$$H_\emptyset(g) = H_\emptyset,$$

we have

$$G(F)g \subset \mathcal{U}.$$

Let us fix an integer N' as in lemma (G.4.5) and let dh be the Haar measure on $G(\mathbb{A})$ which is normalized by $\mathrm{vol}(K, dh) = 1$. For each $g \in G(\mathbb{A})$ such that

$$H_\emptyset(g) = H_\emptyset$$

we have

$$[G(F) \cap (gKg^{-1}) : P_I(F) \cap (gKg^{-1})] \leq q^{N'}$$

and it follows that

$$\sum_{\gamma \in P_I(F)\backslash G(F)} q^{\langle \lambda + \rho_I, H_I(\gamma g)\rangle} \leq q^{N'} \int_{P_I(F)\backslash G(F)gK} q^{\langle \lambda + \rho_I, H_I(h)\rangle} \frac{dh}{d\pi_I}$$

where $d\pi_I$ is the counting measure on $P_I(F)$, so that

$$\sum_{\gamma \in P_I(F)\backslash G(F)} q^{\langle \lambda + \rho_I, H_I(\gamma g)\rangle} \leq q^{N'} \int_{P_I(F)\backslash \mathcal{U}} q^{\langle \lambda + \rho_I, H_I(h)\rangle} \frac{dh}{d\pi_I}.$$

Let dm_I (resp. dn_I) be the Haar measure on $M_I(\mathbb{A})$ (resp. $N_I(\mathbb{A})$) which is normalized by $\mathrm{vol}(M_I(\mathbb{A}) \cap K, dm_I) = 1$ (resp. $\mathrm{vol}(N_I(\mathbb{A}) \cap K, dn_I) = 1$) and let $d\mu_I$ (resp. $d\nu_I$) be the counting measure on $M_I(F)$ (resp. $N_I(F)$). By the integration formula (see (4.1.7); here the modulus character enters because we have permuted m_I and n_I) we get

$$\int_{P_I(F)\backslash \mathcal{U}} q^{\langle \lambda + \rho_I, H_I(h)\rangle} \frac{dh}{d\pi_I}$$
$$= \mathrm{vol}\Big(N_I(F)\backslash N_I(\mathbb{A}), \frac{dn_I}{d\nu_I}\Big) \int_{M_I(F)\backslash (M_I(\mathbb{A}) \cap \mathcal{U})} q^{\langle \lambda - \rho_I, H_I(m_I)\rangle} \frac{dm_I}{d\mu_I}$$

and we have

$$\int_{M_I(F)\backslash (M_I(\mathbb{A}) \cap \mathcal{U})} q^{\langle \lambda - \rho_I, H_I(m_I)\rangle} \frac{dm_I}{d\mu_I}$$
$$= \mathrm{vol}\Big(M_I(F)\backslash M_I(\mathbb{A})^1, \frac{dm_I}{d\mu_I}\Big) \sum_\mu q^{\langle \lambda - \rho_I, \mu\rangle}$$

where μ runs through the subset of $\bigoplus_{j=1}^s \frac{1}{d_j}\mathbb{Z}$ defined by the conditions

$$d_1\mu_1 + \cdots + d_j\mu_j - \frac{d_1 + \cdots + d_j}{d}(d_1\mu_1 + \cdots + d_s\mu_s) \leq \varpi_{d_1 + \cdots + d_j}(H_\emptyset) + N$$

for every $j = 1, \ldots, s-1$ and

$$d_1\mu_1 + \cdots + d_s\mu_s = H_{\emptyset,1} + \cdots + H_{\emptyset,d}$$

if $d_I = (d_1, \ldots, d_s)$ is the partition of d corresponding to I. Now it is elementary to check that the series of functions of λ,

$$\sum_{\mu} q^{\langle \lambda - \rho_I, \mu\rangle},$$

is normally convergent on $(\lambda_0 + \overline{C}_I^{\Delta}) \cap \operatorname{Re} X_I^{\Delta}$ (we have

$$\langle \lambda', \mu\rangle = \sum_{j=1}^{s-1}\Big(\frac{\lambda'_j}{d_j} - \frac{\lambda'_{j+1}}{d_{j+1}}\Big)\Big(d_1\mu_1 + \cdots + d_j\mu_j - \frac{d_1 + \cdots + d_j}{d}(d_1\mu_1 + \cdots + d_s\mu_s)\Big)$$

for every $\lambda' \in \operatorname{Re} X_I^{\Delta}$ and every μ as above). □

(G.5) Intertwining operators

Let (I, π) be a cuspidal pair and let $w \in W$ be such that $w(I) \subset \Delta$. For simplicity we set

$$(I', \pi') = (w(I), w(\pi)).$$

If $d_I = (d_1, \ldots, d_s)$ we have

$$w = w_{I,\sigma}$$

for some $\sigma \in \mathfrak{S}_s$ and $d_{I'} = (d_{\sigma^{-1}(1)}, \ldots, d_{\sigma^{-1}(s)})$ (see (G.1.5)).

Let $d\nu_{I'}$ (resp. $d\nu_{I',w}$) be the counting measure on $N_{I'}(F)$ (resp. $N_{I'}(F) \cap \dot{w}N_I(F)\dot{w}^{-1}$) and let $dn_{I'}$ (resp. $dn_{I',w}$) be the Haar measure on $N_{I'}(\mathbb{A})$ (resp. $N_{I'}(\mathbb{A}) \cap \dot{w}N_I(\mathbb{A})\dot{w}^{-1}$) which is normalized by

$$\operatorname{vol}\big(N_{I'}(F)\backslash N_{I'}(\mathbb{A}), d\nu_{I'}\backslash dn_{I'}\big) = 1$$

(resp.

$$\operatorname{vol}\big((N_{I'}(F) \cap \dot{w}N_I(F)\dot{w}^{-1})\backslash(N_{I'}(\mathbb{A}) \cap \dot{w}N_I(\mathbb{A})\dot{w}^{-1}), d\nu_{I',w}\backslash dn_{I',w}\big) = 1\,).$$

For any $\varphi \in \mathcal{C}_{M_I}^{G,\infty}$ we consider (at least formally) the integrals

$$\text{(G.5.1)}\quad M_I(w,\pi)(\varphi)(g) = \int_{(N_{I'}(F)\cap \dot{w}N_I(F)\dot{w}^{-1})\backslash N_{I'}(\mathbb{A})} \varphi(\dot{w}^{-1}n_{I'}g)\frac{dn_{I'}}{d\nu_{I',w}}$$
$$= \int_{(N_{I'}(\mathbb{A})\cap \dot{w}N_I(\mathbb{A})\dot{w}^{-1})\backslash N_{I'}(\mathbb{A})} \varphi(\dot{w}^{-1}n_{I'}g)\frac{dn_{I'}}{dn_{I',w}}.$$

If the integral $M_I(w,\pi)(\varphi)(g)$ converges for every $g \in G(\mathbb{A})$ then

$$M_I(w,\pi)(\varphi) \in \mathcal{C}^{G,\infty}_{M_{I'}}.$$

Moreover, if this happens for every $\varphi \in \mathcal{A}^{\Delta}_{I,\pi} \subset \mathcal{C}^{G,\infty}_{M_I}$ the operator

$$M_I(w,\pi) : \mathcal{A}^{\Delta}_{I,\pi} \to \mathcal{C}^{G,\infty}_{M_{I'}}$$

intertwines the representations of $G(\mathbb{A})$ by right translation on $\mathcal{A}^{\Delta}_{I,\pi}$ and $\mathcal{C}^{G,\infty}_{M_{I'}}$ and is called an **intertwining operator**.

LEMMA (G.5.2). — *Let $I \subset \Delta$ and let $d_I = (d_1, \dots, d_s)$.*

(i) *For each $\alpha \in \Delta - I$ there exists a unique element*

$$s_{I,\alpha} \in W_{I \cup \{\alpha\}} \subset W$$

such that $s_{I,\alpha}(I) \subset I \cup \{\alpha\}$ and $s_{I,\alpha} \neq 1$. In fact, if $\alpha = \epsilon_{d_1+\cdots+d_j} - \epsilon_{d_1+\cdots+d_j+1}$ for some $j \in \{1, \dots, s-1\}$ we have

$$s_{I,\alpha} = w_{I,\sigma}$$

where $\sigma \in \mathfrak{S}_s$ is the transposition $(j, j+1)$.

(ii) *For any $w \in W$ such that $w(I) \subset \Delta$ let us set*

$$\ell_I(w) = \ell(\sigma)$$

where $\sigma \in \mathfrak{S}_s$ is the unique element in $\mathfrak{S}_s$ such that $w = w_{I,\sigma}$ and where $\ell(\sigma)$ is the usual length of $\sigma \in \mathfrak{S}_s$. Then we have the following properties:

(a) *$\{s_{I,\alpha} \mid \alpha \in \Delta - I\}$ is exactly the set of $w \in W$ such that $w(I) \subset \Delta$ and $\ell_I(w) = 1$,*

(b) *for any $w, w' \in W$ such that $w(I) \subset \Delta$ and $w'w(I) \subset \Delta$ we have*

$$\ell_I(w'w) \leq \ell_{w(I)}(w') + \ell_I(w)$$

and

$$\ell_{w(I)}(w^{-1}) = \ell_I(w),$$

(c) *for any $w \in W$ with $w(I) \subset \Delta$ and any $\alpha \in \Delta - w(I)$ we have*

$$\ell_I(s_{w(I),\alpha} w) = \begin{cases} \ell_I(w) + 1 & \text{if } w^{-1}(\alpha) \in R^+, \\ \ell_I(w) - 1 & \text{otherwise,} \end{cases}$$

(d) *for any* $w \in W$ *with* $w(I) \subset \Delta$ *and with* $\ell_I(w) = \ell$ *there exist at least one sequence* $(\alpha_1, \ldots, \alpha_\ell)$ *in* Δ *and one sequence* $(s_1, \ldots, s_\ell)$ *in* W *such that*

$$\alpha_i \in \Delta - s_{i-1}s_{i-2}\cdots s_1(I)$$

and

$$s_i = s_{s_{i-1}s_{i-2}\cdots s_1(I),\alpha_i}$$

for every $i = 1, \ldots, \ell$ *and such that*

$$w = s_\ell s_{\ell-1}\cdots s_1.$$

(iii) *Let* $R(G, Z_I)$ *(resp.* $R^+(G, Z_I)$*) be the set of roots of* Z_I *in* G *(resp.* P_I*), so that* $R(G, Z_I)$ *(resp.* $R^+(G, Z_I)$*) is the set of*

$$(\epsilon_{d_1+\cdots+d_j} - \epsilon_{d_1+\cdots+d_k})|Z_I$$

for $j, k \in \{1, \ldots, s\}$ *with* $j \neq k$ *(resp.* $j < k$*) and that*

$$R(G, Z_I) = R^+(G, Z_I) \amalg R^-(G, Z_I)$$

where we have set

$$R^-(G, Z_I) = -R^+(G, Z_I).$$

Then for any $w \in W$ *with* $w(I) \subset \Delta$ *and any sequences* $(\alpha_1, \ldots, \alpha_\ell)$ *and* $(s_1, \ldots, s_\ell)$ *as in* (ii)(d) *the set*

$$R_I(w) \overset{\text{dfn}}{=\!=} \{\beta \in R^+(G, Z_I) \mid w(\beta) \in R^-(G, Z_{w(I)})\}$$

is equal to

$$\{\alpha_1|Z_I, s_1^{-1}(\alpha_2)|Z_I, \ldots, s_1^{-1}s_2^{-1}\cdots s_{\ell-1}^{-1}(\alpha_\ell)|Z_I\}.$$

Here, for any $\beta \in R(G, Z_I)$ *and any* $w \in W$ *with* $w(I) \subset \Delta$ *we have set* $w(\beta)(z_{w(I)}) = \beta(\dot{w}^{-1}z_{w(I)}\dot{w})$, $\forall z_{w(I)} \in Z_{w(I)}$.

Proof: Let us remark that, for any $\sigma, \sigma' \in \mathfrak{S}_s$, we have

$$w_{I,\sigma'\sigma} = w_{w_{I,\sigma}(I),\sigma'}w_{I,\sigma}$$

and that, for any $\sigma \in \mathfrak{S}_s$ and any

$$\alpha = \epsilon_{d_{\sigma^{-1}(1)}+\cdots+d_{\sigma^{-1}(j)}} - \epsilon_{d_{\sigma^{-1}(1)}+\cdots+d_{\sigma^{-1}(j)}+1} \in \Delta - w_{I,\sigma}(I),$$

the condition

$$w_{I,\sigma}^{-1}(\alpha) \in R^+$$

is equivalent to the condition

$$\sigma^{-1}(j) < \sigma^{-1}(j+1)$$

(recall that $(d_{\sigma^{-1}(1)}, \ldots, d_{\sigma^{-1}(s)})$ is the partition of d corresponding to $w_{I,\sigma}(I) \subset \Delta$).

Using these remarks it is easy to see that the lemma for a general I follows from its particular case $d = s$, $W = \mathfrak{S}_s$ and $I = \emptyset$. But this particular case is well known (for example, see [Sp] (10.2.2)) and the lemma is proved. □

LEMMA (G.5.3). — *Let $w_1, w_2 \in W$ be such that $w_1(I) \subset \Delta$ and $w_2w_1(I) \subset \Delta$. We set*

$$w_1(I) = I_1,$$
$$w_2w_1(I) = I'$$

and

$$w = w_2w_1.$$

If we assume that

$$\ell_I(w) = \ell_{I_1}(w_2) + \ell_I(w_1)$$

we have

$$N_{I'} \cap \dot{w}N_I\dot{w}^{-1} \subset N_{I'} \cap \dot{w}_2 N_{I_1}\dot{w}_2^{-1} \subset N_{I'}$$

and the morphism

$$N_{I'} \cap \dot{w}_2 N_{I_1}\dot{w}_2^{-1} \to N_{I_1},\ n_{I'} \mapsto \dot{w}_2^{-1} n_{I'} w_2$$

induces an isomorphism

$$(N_{I'} \cap \dot{w}N_I\dot{w}^{-1})\backslash(N_{I'} \cap \dot{w}_2 N_{I_1}\dot{w}_2^{-1}) \xrightarrow{\sim} (N_{I_1} \cap \dot{w}_1 N_I \dot{w}_1^{-1})\backslash N_{I_1}.$$

Proof: For each $\alpha = \epsilon_i - \epsilon_j \in R$ let

$$x_\alpha : \mathbb{G}_a \to G,\ t \mapsto 1 + tE_{ij},$$

be the corresponding 1-parameter subgroup (E_{ij} is the elementary matrix with all entries 0 except the entry on the i-th row and the j-th column which is equal to 1). For each $w \in W$ such that $w(I) \subset \Delta$ it is not difficult to prove that

$$N_{w(I)} \cap \dot{w}N_I\dot{w}^{-1}$$

is the subgroup of G generated by the $x_\alpha(\mathbb{G}_a)$ for all the $\alpha \in R$ such that

$$\alpha|Z_{w(I)} \in R^+(G, Z_{w(I)})$$

and

$$w^{-1}(\alpha|Z_{w(I)}) \in R^+(G, Z_I).$$

Now it follows easily from (G.5.2)(iii) that, for any w_1, w_2 as in the statement of the lemma, we have

$$R_{w(I)}(w^{-1}) = R_{w(I)}(w_2^{-1}) \amalg w_2(R_{w_1(I)}(w_1^{-1})),$$

so that

$$R^+(G, Z_{w(I)}) - R_{w(I)}(w^{-1}) \subset R^+(G, Z_{w(I)}) - R_{w(I)}(w_2^{-1})$$

and that

$$w_2^{-1}\Big((R^+(G, Z_{w(I)}) - R_{w(I)}(w_2^{-1})) - (R^+(G, Z_{w(I)}) - R_{w(I)}(w^{-1}))\Big)$$
$$= R^+(G, Z_{w_1(I)}) - (R^+(G, Z_{w_1(I)}) - R_{w_1(I)}(w_1^{-1})).$$

The lemma follows. □

Combining this lemma with the Fubini theorem we obtain

LEMMA (G.5.4). — *Let* (I,π) *be a cuspidal pair and let* $w_1, w_2 \in W$ *be such that* $w_1(I) \subset \Delta$, $w_2w_1(I) \subset \Delta$ *and*

$$\ell_I(w) = \ell_{w_1(I)}(w_2) + \ell_I(w_1).$$

Let $\varphi \in C^{G,\infty}_{M_I}$. *If the integrals*

$$M_I(w_1,\pi)(\varphi)(g) \qquad (\forall g \in G(\mathbb{A}))$$

and

$$M_{w_1(I)}(w_2, w_1(\pi))(M_I(w_1,\pi)(\varphi))(g) \qquad (\forall g \in G(\mathbb{A}))$$

are all absolutely convergent then the same is true for the integrals

$$M_I(w_2w_1,\pi)(\varphi)(g) \qquad (\forall g \in G(\mathbb{A}))$$

and we have

$$M_I(w_2w_1,\pi)(\varphi) = M_{w_1(I)}(w_2, w_1(\pi))\big(M_I(w_1,\pi)(\varphi)\big).$$

□

Let us set

$$C^{\Delta}_{I,w} = \{\lambda \in X^{\Delta}_I \mid \mathrm{Re}\Big(\frac{\lambda_j}{d_j} - \frac{\lambda_k}{d_k}\Big) > 0,\ \forall j,k \in \{1,\ldots,s\}$$
$$\text{such that } j < k \text{ and } \sigma^{-1}(j) > \sigma^{-1}(k)\}$$

if $w = w_{I,\sigma}$ for some $\sigma \in \mathfrak{S}_s$. In particular we have

$$C^{\Delta}_I = C^{\Delta}_{I,w_0} \subset C^{\Delta}_{I,1} = X^{\Delta}_I$$

where w_0 is the element of maximal length ℓ_I among the $w \in W$ such that $w(I) \subset \Delta$, i.e. $w_0 = w_{I,\sigma_0}$ with σ_0 the longest element in $\mathfrak{S}_s$. In fact, if for each

$$\beta = (\epsilon_{d_1+\cdots+d_j} - \epsilon_{d_1+\cdots+d_k})|Z_I \in R(G, Z_I)$$

$(j,k \in \{1,\ldots,s\},\ j \neq k)$ we define

$$\beta^\vee \in \bigoplus_{j=1}^{s} \frac{1}{d_j}\mathbb{Z} \subset \mathfrak{a}_I$$

by

$$\beta_\ell^\vee = \begin{cases} 1/d_j & \text{if } \ell = j, \\ -1/d_k & \text{if } \ell = k, \\ 0 & \text{otherwise,} \end{cases}$$

we have

$$\frac{\lambda_j}{d_j} - \frac{\lambda_k}{d_k} = \langle \lambda, \beta^\vee \rangle \qquad (\forall \lambda \in X_I^\Delta),$$

so that

$$C_{I,w}^\Delta = \{\lambda \in X_I^\Delta \mid \langle \operatorname{Re}\lambda, \beta^\vee\rangle > 0,\ \forall \beta \in R^+(G, Z_I)$$
$$\text{such that } w(\beta) \in R^-(G, Z_{w(I)})\}.$$

LEMMA (G.5.5). — *Let us fix $w \in W$ with $w(I) \subset \Delta$ and sequences $(\alpha_1, \dots, \alpha_\ell)$ and $(s_1, \dots, s_\ell)$ as in* (G.5.2)(ii)(d) *($\ell = \ell_I(w)$ and $w = s_\ell \cdots s_1$). Then we have*

$$C_{I,w}^\Delta = \bigcap_{m=0}^{\ell-1} (s_m \cdots s_1)^{-1}(C_{s_m\cdots s_1(I), s_{m+1}}^\Delta)$$

and

$$\langle s_m \cdots s_1(\rho_I) - \rho_{s_m\cdots s_1(I)}, (\alpha_{m+1}|Z_I)^\vee \rangle \geq 0$$

for each $m = 0, \dots, \ell - 1$.

Proof: The assertion about $C_{\Delta,w}^I$ directly follows from (G.5.2)(iii). Let us consider the embedding

$$X_I = \bigoplus_{j=1}^{s} \Big(\mathbb{C}/\frac{2\pi i}{\log q} d_j \mathbb{Z}\Big) \xrightarrow{\iota_I} \Big(\mathbb{C}/\frac{2\pi i}{\log q}\mathbb{Z}\Big)^s = X_\emptyset,$$
$$\lambda \mapsto \Big(\frac{\lambda_1}{d_1}, \dots, \frac{\lambda_1}{d_1}, \dots, \frac{\lambda_s}{d_s}, \dots, \frac{\lambda_s}{d_s}\Big)$$

(each λ_j/d_j occurring d_j times) which is induced by the inclusion $M_\emptyset(\mathbb{A}) \subset M_I(\mathbb{A})$ and which is "dual" to the projection

$$\mathfrak{a}_\emptyset \twoheadrightarrow \mathfrak{a}_I,\ H \mapsto H_I$$

(see (10.3)), so that

$$\langle \iota_I(\lambda), H \rangle = \langle \lambda, H_I \rangle$$

for every $\lambda \in X_I$ and every $H \in \mathbb{Z}^s \subset \mathfrak{a}_\emptyset$. Then it is not difficult to check that

$$\iota_{s_m\cdots s_1(I)}(s_m \cdots s_1(\rho_I) - \rho_{s_m\cdots s_1(I)}) = s_m \cdots s_1(\rho_\emptyset) - \rho_\emptyset$$

($\rho_\emptyset - \iota_I(\rho_I)$ is half the sum of the positive roots of $(M_I, M_\emptyset, P_\emptyset^I)$) and that

$$\langle \iota_{s_m\cdots s_1(I)}(X_{s_m\cdots s_1(I)}), \mathbb{Z}^d \cap \mathfrak{a}_\emptyset^{s_m\cdots s_1(I)}\rangle = (0)$$

for each $m = 0, \ldots, \ell - 1$. Therefore we have

$$\langle s_m \cdots s_1(\rho_I) - \rho_{s_m\cdots s_1(I)}, (\alpha_{m+1}|Z_I)^\vee\rangle = \langle s_m \cdots s_1(\rho_\emptyset) - \rho_\emptyset, \alpha_{m+1}^\vee\rangle$$

for each $m = 0, \ldots, \ell - 1$.

Now we have

$$\langle \rho_\emptyset, \alpha_{m+1}^\vee\rangle = 1$$

as $\alpha_{m+1} \in \Delta$ and we have

$$\langle s_m \cdots s_1(\rho_\emptyset), \alpha_{m+1}^\vee\rangle \geq 1$$

as $s_1^{-1} \cdots s_m^{-1}(\alpha_{m+1}) \in R^+$ (see (G.5.2)(ii)(c)), so that

$$\langle s_m \cdots s_1(\rho_\emptyset) - \rho_\emptyset, \alpha_{m+1}^\vee\rangle \geq 0.$$

This completes the proof of the lemma. □

If (I, π) is a cuspidal pair let us set

$$-\overline{\pi} = \pi(-2\mathrm{Re}\,\pi)$$

so that $(I, -\overline{\pi})$ is again a cuspidal pair. Then we have a natural scalar product

$$\langle\cdot,\cdot\rangle : \mathcal{A}_{I,-\overline{\pi}}^\Delta \times \mathcal{A}_{I,\pi}^\Delta \to \mathbb{C} \tag{G.5.6}$$

defined by

$$\begin{aligned}\langle \varphi', \varphi\rangle &= \int_{N_I(\mathbb{A})Z_I(\mathbb{A})M_I(F)\backslash G(\mathbb{A})} \overline{\varphi'(g)}\varphi(g)\frac{dg}{dn_I dz_I d\mu_I} \\ &= \int_K \Big(\int_{Z_I(\mathbb{A})M_I(F)\backslash M_I(\mathbb{A})} \overline{\varphi_k'(m_I)}\varphi_k(m_I)\frac{dm_I}{dz_I d\mu_I}\Big)dk\end{aligned}$$

(we have $dg = \delta_{P_I(\mathbb{A})}^{-1}(m_I)dn_I dm_I dk$ and $\mathrm{vol}(K, dk) = 1$). Here the integrals are absolutely convergent (by (9.2.6) φ_k and φ_k' have compact supports in $Z_I(\mathbb{A})M_I(F)\backslash M_I(\mathbb{A})$) and the Haar measures dn_I, dz_I $d\mu_I$ and dg are normalized in the following way:

$$\mathrm{vol}(N_I(F)\backslash N_I(\mathbb{A}), d\nu_I\backslash dn_I) = 1$$

(as usual $d\nu_I$ is the counting measure on $N_I(F)$),

$$\mathrm{vol}\big(Z_I(F)\backslash(Z_I(\mathbb{A}) \cap M_I(\mathbb{A})^1), d\zeta_I\backslash dz_I\big) = 1$$

($d\zeta_I$ is the counting measure on $Z_I(F)$), $d\mu_I$ is the counting measure on $M_I(F)$ and dg is the Haar measure on $G(\mathbb{A})$ which is fixed from the beginning of this appendix.

PROPOSITION (G.5.7). — *Let (I,π) be a cuspidal pair and let $w \in W$ with $w(I) \subset \Delta$.*

(i) *For any $\varphi \in \mathcal{A}_{I,\pi}^{\Delta}$ and any $\lambda \in \rho_I - \mathrm{Re}\,\pi + C_{I,w}^{\Delta}$ the integrals*

$$M_I(w,\pi(\lambda))(\varphi_\lambda)(g) \qquad (\forall g \in G(\mathbb{A}))$$

are all absolutely convergent. Moreover, if g stays in a compact subset of $G(\mathbb{A})$ the convergence is uniform and

$$\big(M_I(w,\pi(\lambda))(\varphi_\lambda)\big)_{-w(\lambda)} \in \mathcal{A}_{w(I),w(\pi)}^{\Delta}.$$

(ii) *For any $\Phi \in \mathcal{P}(I,\pi)$ and any $\Phi' \in \mathcal{P}(w(I),-w(\overline{\pi}))$ the function*

$$\lambda \mapsto \big\langle \Phi'(-w(\overline{\lambda})), \big(M_I(w,\pi(\lambda))(\Phi(\lambda)_\lambda)\big)_{-w(\lambda)}\big\rangle$$

is holomorphic on $\rho_I - \mathrm{Re}\,\pi + C_{I,w}^{\Delta}$.

Proof : Let $(\alpha_1,\ldots,\alpha_\ell)$ and $(s_1,\ldots,s_\ell)$ be as in (G.5.2)(ii)(d) ($\ell = \ell_I(w)$, $w = s_\ell \cdots s_1$). It follows from (G.5.5) that the condition

$$\lambda \in \rho_I - \mathrm{Re}\,\pi + C_{I,w}^{\Delta}$$

on $\lambda \in X_I^{\Delta}$ implies the conditions

$$s_m \cdots s_1(\lambda) \in \rho_{s_m\cdots s_1(I)} - \mathrm{Re}(s_m \cdots s_1(\pi)) + C_{s_m\cdots s_1(I),s_{m+1}}^{\Delta} \qquad (\forall m = 0,\ldots,\ell-1).$$

Moreover it follows from (G.5.4) that, at least formally,

$$M_I(w,\pi(\lambda))(\varphi_\lambda) = M_{s_{\ell-1}\cdots s_1(I)}(s_\ell, s_{\ell-1}\cdots s_1(\pi(\lambda)))\Big(\cdots \cdots \Big(M_{s_1(I)}(s_2, s_1(\pi(\lambda)))\big(M_I(s_1,\pi(\lambda))(\varphi_\lambda)\big)\Big)\cdots\Big).$$

Therefore, to prove part (i) of the proposition it is sufficient to do so for $\ell_I(w) = 1$, i.e. $w = s_{I,\alpha}$ for some $\alpha \in \Delta - I$ (see (G.5.2)). But in this case the partition of d associated to $w(I)$ is nothing else than

$$d_{w(I)} = (d_1,\ldots,d_{j-1},d_{j+1},d_j,d_{j+2},\ldots,d_s)$$

($d_I = (d_1,\ldots,d_s)$ and $\alpha = \epsilon_{d_1+\cdots+d_j} - \epsilon_{d_1+\cdots+d_j+1}$) and we have

$$\begin{array}{ccc} N_{I\cup\{\alpha\}} & \subset & N_{w(I)} \\ \cap & & \cap \\ \dot{w}N_I\dot{w}^{-1} & \subset & P_{I\cup\{\alpha\}}. \end{array}$$

Therefore, in order to prove part (i) of the proposition we may replace Δ by $\Delta' = \{1,\ldots,d_j+d_{j+1}-1\}$, α by $\alpha' = \epsilon_{d_j} - \epsilon_{d_j+1}$, I by $\Delta' - \{\alpha'\}$, w by

$$\begin{pmatrix} 1 & \cdots & d_j & d_j+1 & \cdots & d_j+d_{j+1} \\ 1+d_{j+1} & \cdots & d_j+d_{j+1} & 1 & \cdots & d_{j+1} \end{pmatrix},$$

π by $\pi' = \pi_j \otimes \pi_{j+1}$ (we have $\pi = \pi_1 \otimes \cdots \otimes \pi_s$ if we identify M_I with $GL_{d_1} \times \cdots \times GL_{d_s}$), λ by $\lambda' = (\lambda_j, \lambda_{j+1}) \in X_{I'}^{\Delta'}$ and φ by some $\varphi' \in \mathcal{A}_{I',\pi'}^{\Delta'}$ (we have

$$\mathcal{A}_{I,\pi}^{I} = \mathcal{A}_{\Delta_1,\pi_1}^{\Delta_1} \otimes \cdots \otimes \mathcal{A}_{\Delta_s,\pi_s}^{\Delta_s}$$

if $\Delta_j = \{1,\ldots,d_j-1\}$ for every $j = 1,\ldots,s$; see (D.7)). In other words it is sufficient to prove part (i) of the proposition in the case $I = \Delta - \{\alpha\}$ for some $\alpha \in \Delta$ and $w = s_{I,\alpha}$, so that

$$C_{I,w}^{\Delta} = C_I^{\Delta}.$$

Let us do this. For any cuspidal pair (I,π), any $w \in W$ with $I' = w(I) \subset \Delta$, any $\varphi \in \mathcal{A}_{I,\pi}^{\Delta}$ and any

$$\lambda \in \rho_I - \mathrm{Re}\,\pi + C_I^{\Delta} \subset \rho_I - \mathrm{Re}\,\pi + C_{I,w}^{\Delta}$$

(this inclusion is strict in general) we have

$$\begin{aligned} &\int_{(N_{I'}(F)\cap \dot{w}N_I(F)\dot{w}^{-1})\backslash N_{I'}(\mathbb{A})} |\varphi_\lambda(\dot{w}^{-1}n_{I'}g)| \frac{dn_I'}{d\nu_{I',w}} \\ &= \int_{N_{I'}(F)\backslash N_{I'}(\mathbb{A})} \Bigg(\sum_{\nu_{I'} \in (N_{I'}(F)\cap \dot{w}N_I(F)\dot{w}^{-1})\backslash N_{I'}(F)} |\varphi_\lambda(\dot{w}^{-1}\nu_{I'}n_{I'}g)| \Bigg) \frac{dn_{I'}}{d\nu_{I'}} \\ &\leq \int_{N_{I'}(F)\backslash N_{I'}(\mathbb{A})} \Bigg(\sum_{\gamma \in P_I(F)\backslash G(F)} |\varphi_\lambda(\gamma n_{I'}g)| \Bigg) \frac{dn_{I'}}{d\nu_{I'}} \end{aligned}$$

as

$$\begin{aligned} &(\dot{w}^{-1}N_{I'}(F)\dot{w} \cap N_I(F))\backslash \dot{w}^{-1}N_{I'}(F)\dot{w} \\ &\qquad = (\dot{w}^{-1}N_{I'}(F)\dot{w} \cap P_I(F))\backslash \dot{w}^{-1}N_{I'}(F)\dot{w} \subset P_I(F)\backslash G(F). \end{aligned}$$

Moreover the Eisenstein series

$$E_I^{\Delta}(\varphi_\lambda, \pi(\lambda))(g) \qquad (\forall g \in G(\mathbb{A}))$$

are normally convergent on compact subsets of $G(\mathbb{A})$ (see (G.4.3)) and $N_{I'}(F)\backslash N_{I'}(\mathbb{A})$ is compact (see (9.2.1)). Therefore, for any such (I,π), w, φ and λ the integrals

$$M_I(w,\pi(\lambda))(\varphi_\lambda)(g) \qquad (\forall g \in G(\mathbb{A}))$$

are all absolutely convergent (resp. uniformly convergent on compact subsets of $G(\mathbb{A})$). This concludes the proof of the first two assertions of part (i) of the proposition. The last assertion of part (i) follows from (9.2.10) once we have proved that the integral

$$\int_{N_{J'}^{I'}(F)\backslash N_{J'}^{I'}(\mathbb{A})}\bigg(\int_{(N_{I'}(F)\cap\dot{w}N_I(F)\dot{w}^{-1})\backslash N_{I'}(\mathbb{A})}\varphi_\lambda(\dot{w}^{-1}n_{I'}n_{J'}^{I'}m_{I'}k)\frac{dn_{I'}}{d\nu_{I',w}}\bigg)\frac{dn_{J'}^{I'}}{d\nu_{J'}^{I'}}$$

vanishes for every $J' \subsetneq I'$, every $m_{I'} \in M_{I'}(\mathbb{A})$ and every $k \in K$. But we have just proved that this integral is absolutely convergent ($N_{J'}^{I'}(F)\backslash N_{J'}^{I'}(\mathbb{A})$ is compact). Thus we can permute the integral over $n_{I'}$ and the integral over $n_{J'}^{I'}$ and we get

$$\int_{(N_{I'}(F)\cap\dot{w}N_I(F)\dot{w}^{-1})\backslash N_{I'}(\mathbb{A})}\bigg(\int_{N_J^I(F)\backslash N_J^I(\mathbb{A})}\varphi_\lambda(n_J^I\dot{w}^{-1}n_{I'}m_{I'}k)\frac{dn_J^I}{d\nu_J^I}\bigg)\frac{dn_{I'}}{d\nu_{I',w}}$$

where $J = w^{-1}(J') \subset I$ (make the change of variables $n_{J'}^{I'} \mapsto n_J^I = \dot{w}^{-1}n_{J'}^{I'}\dot{w}$ and $n_{I'} \mapsto (n_{J'}^{I'})^{-1}n_{I'}n_{J'}^{I'}$). Now, as $\varphi_\lambda \in \mathcal{A}_{I,\pi(\lambda)}^{\Delta}$ with $\pi(\lambda)$ cuspidal, the integral over n_J^I vanishes. This completes the proof of part (i).

In order to prove part (ii) of the proposition it is sufficient to prove that, for each $g \in G(\mathbb{A})$, the function

$$\lambda \mapsto \big(M_I(w,\pi(\lambda))(\Phi(\lambda)_\lambda)\big)_{-w(\lambda)}(g)$$

is holomorphic on $\rho_I - \operatorname{Re}\pi + C_{I,w}^{\Delta}$. Indeed, if we fix compact open subgroups K_Φ, $K_{\Phi'}$ of $G(\mathbb{A})$ as in (G.2.1)(i), if we set

$$K' = M_I(\mathbb{A}) \cap \Big(\bigcap_{k\in K} kK_{\Phi'}k^{-1}\Big)$$

and if we fix an open subset $C_{K'}$ of $M_I(\mathbb{A})$ as in (9.2.6), we have

$$\operatorname{Supp}\Big(\Phi'(-w(\overline{\lambda}))\Big) \subset N_I(\mathbb{A})C_{K'}K$$

for every $\lambda \in X_I^{\Delta}$ and it follows that we can find finitely many elements $g_1, \ldots, g_n$ in $G(\mathbb{A})$ and finitely many positive real numbers $a_1, \ldots, a_n$, depending only on K_Φ, $K_{\Phi'}$ and $C_{K'}$, such that

$$\begin{aligned}\langle \Phi'(-w(\overline{\lambda})), &\big(M_I(w,\pi(\lambda))(\Phi(\lambda)_\lambda)\big)_{-w(\lambda)}\rangle \\ &= \sum_{i=1}^{n} a_i\overline{\Phi'(-w(\overline{\lambda}))(g_i)}\big(M_I(w,\pi(\lambda))(\Phi(\lambda)_\lambda)\big)_{-w(\lambda)}(g_i).\end{aligned}$$

Now we may prove a slightly more general statement.

Let U be an open subset of $\operatorname{Re} X_I^\Delta$, *let*

$$\widetilde{U} = \{\lambda \in X_I^\Delta \mid \operatorname{Re}\lambda \in U\}$$

and let

$$\Phi : \widetilde{U} \to \mathcal{A}_{I,\pi}^\Delta$$

be a function having properties (i) *and* (ii) *of* (G.2.1) *(for $\lambda \in \widetilde{U}$) which is holomorphic on $\widetilde{U}$ (the last property makes sense by remark* (G.2.2)*). Then the function*

$$\lambda \mapsto \big(M_I(w,\pi(\lambda))(\Phi(\lambda)_\lambda)\big)_{-w(\lambda)}$$

on

$$\widetilde{U}_w = \{\lambda \in X_I^\Delta \mid \operatorname{Re}\lambda \in (\rho_I - \operatorname{Re}\pi + \operatorname{Re} C_{I,w}^\Delta) \cap U\}$$

satisfies the same properties.

Let us argue as in the proof of part (i) of the proposition. Then we see that our slightly more general statement is a consequence of the assertion

The series of functions

$$(\lambda, g) \mapsto \sum_{\gamma \in P_I(F)\backslash G(F)} |\Phi(\lambda)_\lambda(\gamma g)|$$

is normally convergent if $\operatorname{Re}\lambda$ *stays in a compact subset of*

$$(\rho_I - \operatorname{Re}\pi + \operatorname{Re} C_I^\Delta) \cap U$$

and g stays in a compact subset of $G(\mathbb{A})$.

But this assertion easily follows from (G.4.2) (see the proof of (G.4.3)). □

REMARK (G.5.8). — For any $\varphi \in \mathcal{A}_{I,\pi}^\Delta$, any $\varphi' \in \mathcal{A}_{w(I),w(\pi)}^\Delta$ and any $\lambda_0 \in \rho_I - \operatorname{Re}\pi + C_I^\Delta$, the function

$$\lambda \mapsto \left|\langle \varphi', \big(M_I(w,\pi(\lambda))(\varphi_\lambda)\big)_{-w(\lambda)}\rangle\right|$$

is bounded above on the set

$$\lambda_0 + \overline{C_I^\Delta} \subset \rho_I - \operatorname{Re}\pi + C_{I,w}^\Delta.$$

Indeed, arguing as in the proof of (G.5.7)(ii) we see that it is sufficient to prove that, for each $g \in G(\mathbb{A})$, the function

$$\lambda \mapsto \left|\big(M_I(w,\pi(\lambda))(\varphi_\lambda)\big)_{-w(\lambda)}(g)\right|$$

is bounded above on $\lambda_0+\overline{C}_I^{\Delta}$. Then arguing as in the proof of part (i) and as in the proof of (G.4.3) we see that this assertion follows from the statement

For each $g\in G(\mathbb{A})$ the function

$$\lambda\mapsto \sum_{\gamma\in P_I(F)\backslash G(F)} |q^{\langle\lambda+\rho_I,H_I(\gamma g)\rangle}|$$

is bounded above on $\lambda_0+\overline{C}_I^{\Delta}$.

But this statement is proved in (G.4.2). □

LEMMA (G.5.9). — *Let (I,π) be a cuspidal pair and let $w\in W$ with $w(I)\subset\Delta$. Let $\varphi\in\mathcal{A}_{I,\pi}^{\Delta}$ and $\varphi'\in\mathcal{A}_{w(I),-\overline{w(\pi)}}^{\Delta}$. Then if*

$$\mathrm{Re}\,\pi\in(\rho_I+C_{I,w}^{\Delta})\cap\left(-w^{-1}(\rho_{w(I)})+C_{I,w}^{\Delta}\right)$$

we have

$$\mathrm{Re}(-\overline{w(\pi)})\in\rho_{w(I)}+C_{w(I),w^{-1}}^{\Delta}$$

and

$$\langle\varphi',M_I(w,\pi)(\varphi)\rangle=\langle M_{w(I)}(w^{-1},-\overline{w(\pi)})(\varphi'),\varphi\rangle.$$

Proof: The first assertion is obvious: we have $\mathrm{Re}(-w(\pi))=-w(\mathrm{Re}\,\pi)$ and $-w(C_{I,w}^{\Delta})=C_{w(I),w^{-1}}^{\Delta}$.

Now if we set $I'=w(I)$ we have

$$\begin{aligned}&\langle\varphi',M_I(w,\pi)(\varphi)\rangle\\&=\int_{(N_{I'}(F)\cap\dot{w}N_I(F)\dot{w}^{-1})Z_{I'}(\mathbb{A})M_{I'}(F)\backslash G(\mathbb{A})}\overline{\varphi'(g)}\varphi(\dot{w}^{-1}g)\frac{dg}{d\nu_{I',w}dz_{I'}d\mu_{I'}}\end{aligned}$$

and

$$\begin{aligned}&\langle M_{w(I)}(w^{-1},-\overline{w(\pi)})(\varphi'),\varphi\rangle\\&=\int_{(N_I(F)\cap\dot{w}^{-1}N_{I'}(F)\dot{w})Z_I(\mathbb{A})M_I(F)\backslash G(\mathbb{A})}\overline{\varphi'(\dot{w}g)}\varphi(g)\frac{dg}{d\nu_{I,w^{-1}}dz_Id\mu_I},\end{aligned}$$

the two integrals being absolutely convergent by (G.5.7)(i) and (9.2.6). But these two integrals are equal (make the change of variables $g\mapsto\dot{w}g$ in the first one). This concludes the proof of the lemma. □

(G.6) The constant terms of the Eisenstein series

Let us fix a cuspidal pair (I,π) such that

$$\mathrm{Re}\,\pi\in\rho_I+C_I^{\Delta}.$$

Then for any $\varphi \in \mathcal{A}^{\Delta}_{I,\pi}$ and any $w \in W$ with $w(I) \subset \Delta$ the Eisenstein series

$$E_I^{\Delta}(\varphi,\pi)(g) \qquad (\forall g \in G(\mathbb{A}))$$

and the integrals

$$M_I(w,\pi)(\varphi)(g) \qquad (\forall g \in G(\mathbb{A}))$$

are all absolutely convergent and the convergence is uniform on each compact subset of $G(\mathbb{A})$ (see (G.4.3) for $\Phi(\lambda) = \varphi$, $\forall \lambda \in X_I^{\Delta}$, and (G.5.7)(i)).

More generally, if J is another subset of Δ such that

$$I \subset J \subset \Delta$$

and if (I,π) is a cuspidal pair such that

$$\operatorname{Re}\pi \in \rho_I + C_I^{\Delta} + \operatorname{Re} X_J^{\Delta}$$

we may consider the Eisenstein series

$$E_I^J(\varphi,\pi)(g) = \sum_{\mu_J \in P_I^J(F)\backslash M_J(F)} \varphi(\mu_J g) \qquad (\forall g \in G(\mathbb{A})). \tag{G.6.1}$$

Then an obvious modification of the proof of (G.4.3) implies that these series are normally convergent on each compact subset of $G(\mathbb{A})$.

For any pair (I,J) of subsets of Δ let us set

$$\Omega(J,I) = \{w \in W \mid w^{-1}(J) \subset R^+ \text{ and } w(I) \subset J\}. \tag{G.6.2}$$

Obviously we have

$$\Omega(J,I) = \emptyset$$

if $|I| > |J|$, we have

$$\Omega(J,I) = \{w \in W \mid w(I) = J\}$$

if $|I| = |J|$ and we have

$$\Omega(J,I) = \{w \in D_{J,I} \mid w(I) \subset J\}$$

in general where

$$D_{J,I} = \{w \in W \mid w^{-1}(J) \subset R^+ \text{ and } w(I) \subset R^+\}$$

is the system of representatives of the double classes in $W_J\backslash W/W_I$ introduced in (5.4.1).

LEMMA (G.6.3). — (i) *Let* $d_I = (d_1, \ldots, d_s)$ *and* $e_J = (e_1, \ldots, e_t)$ *be the partitions of* d *associated with* I *and* J *respectively. Let us consider the set of* $\sigma \in \mathfrak{S}_s$ *having the following property: there exists a partition* $(s_1, \ldots, s_t)$ *of* s *such that*

$$e_k = d_{\sigma^{-1}(s_1+\cdots+s_{k-1}+1)} + \cdots + d_{\sigma^{-1}(s_1+\cdots+s_k)}$$

for every $k = 1, \ldots, t$ *and such that*

$$\sigma^{-1}(j+1) > \sigma^{-1}(j)$$

for every $j \in \{1, \ldots, s\} - \{s_1, s_1 + s_2, \ldots, s_1 + \cdots + s_t\}$. *Then the map*

$$\mathfrak{S}_s \to \{w \in W \mid w(I) \subset \Delta\},\ \sigma \mapsto w_{I,\sigma}$$

(see (G.1.5)*) induces a bijection from that subset of* $\mathfrak{S}_s$ *onto* $\Omega(J, I)$.

(ii) *If* $w \in D_{J,I}$ *the two conditions*

$$w \in \Omega(J, I)$$

and

$$\dot{w} M_I \dot{w}^{-1} \subset M_J$$

are equivalent.

Proof : Let $\sigma \in \mathfrak{S}_s$. Then $(d_{\sigma^{-1}(1)}, \ldots, d_{\sigma^{-1}(s)})$ is the partition of d corresponding to $w_{I,\sigma}(I) \subset \Delta$. Therefore the condition

$$w_{I,\sigma}(I) \subset J$$

is equivalent to the existence of a partition $(s_1, \ldots, s_t)$ of s such that

$$e_k = d_{\sigma^{-1}(s_1+\cdots+s_{k-1}+1)} + \cdots + d_{\sigma^{-1}(s_1+\cdots+s_k)}$$

for every $k = 1, \ldots, t$. Moreover, if it is satisfied $J - w_{I,\sigma}(I)$ is the set of

$$\alpha_{d_{\sigma^{-1}(1)}+\cdots+d_{\sigma^{-1}(j)}}$$

for $j \in \{1, \ldots, s\} - \{s_1, s_1 + s_2, \ldots, s_1 + \cdots + s_t\}$ and $w_{I,\sigma}^{-1}(J) - I$ is the set of

$$\epsilon_{d_1+\cdots+d_{\sigma^{-1}(j)}} - \epsilon_{d_1+\cdots+d_{\sigma^{-1}(j+1)-1}+1}$$

for $j \in \{1, \ldots, s\} - \{s_1, s_1 + s_2, \ldots, s_1 + \cdots + s_t\}$, so that $w_{I,\sigma}^{-1}(J) - I$ is contained in R^+ if and only if

$$\sigma^{-1}(j+1) - 1 \geq \sigma^{-1}(j)$$

for all $j \in \{1, \ldots, s\} - \{s_1, s_1+s_2, \ldots, s_1+\cdots+s_t\}$. This completes the proof of part (i).

If $w(I) \subset \Delta$ we have

$$\dot{w}M_I\dot{w}^{-1} = M_{w(I)}$$

(see (G.1.5)). Therefore, if $w(I) \subset J$ we have

$$\dot{w}M_I\dot{w}^{-1} \subset M_J.$$

Conversely let $w \in D_{J,I}$ be such that

$$\dot{w}M_I\dot{w}^{-1} \subset M_J.$$

Then we have

$$w(R_I) \subset R_J$$

where

$$R_I = R(M_I, T) \subset R$$

(resp.

$$R_J = R(M_J, T) \subset R)$$

is the set of roots $\alpha \in R$ which are sums of simple roots in I (resp. in J), so that

$$w(I) \subset R^+ \cap R_J = R_J^+.$$

Now if $\alpha \in I$ we have

$$w(\alpha) = \beta_1 + \cdots + \beta_n$$

for a unique set $\{\beta_1, \ldots, \beta_n\}$ of simple roots in J and it follows that

$$\alpha = w^{-1}(\beta_1) + \cdots + w^{-1}(\beta_n)$$

with

$$w^{-1}(\beta_\nu) \in R^+ \qquad (\forall \nu = 1, \ldots, n).$$

But this implies that $n = 1$, i.e. that $w(I) \subset J$ as required. □

PROPOSITION (G.6.4). — *Let us fix $\varphi \in \mathcal{A}_{I,\pi}^{\Delta}$ and let J be another subset of Δ. Let us simply denote by $E_I^{\Delta}(\varphi, \pi)_J$ the constant term of $E_I^{\Delta}(\varphi, \pi)$ along P_J (see (9.2.2)). Then we have*

$$E_I^{\Delta}(\varphi, \pi)_J = \sum_{w \in \Omega(J,I)} E_{w(I)}^J\big(M_I(w, \pi)(\varphi), w(\pi)\big).$$

In particular, on the one hand we have

$$E_I^{\Delta}(\varphi, \pi)_J = 0$$

if $\Omega(J, I) = \emptyset$. On the other hand we have

$$E_I^{\Delta}(\varphi, \pi)_J = \sum_{\substack{w \in W \\ w(I) = J}} M_I(w, \pi)(\varphi)$$

if $|J| = |I|$.

Proof: By the Bruhat decomposition we have

$$G(F) = \coprod_{w \in D_{J,I}} P_I(F)\dot{w}^{-1}P_J(F)$$

(see [Ca] (2.7.3) and (2.8.1)), so that

$$E_I^{\Delta}(\varphi,\pi)(g) = \sum_{w \in D_{J,I}} \sum_{\mu_J} \sum_{\nu_J} \varphi(\dot{w}^{-1}\nu_J\mu_J g) \qquad (\forall g \in G(\mathbb{A}))$$

where μ_J (resp. ν_J) runs through the set

$$(M_J(F) \cap \dot{w}P_I(F)\dot{w}^{-1})\backslash M_J(F)$$

(resp.

$$(N_J(F) \cap \dot{w}P_I(F)\dot{w}^{-1})\backslash N_J(F)\,).$$

Here the partial sums are all absolutely convergent and the convergence is normal if g stays in a compact subset of $G(\mathbb{A})$. Therefore, for each $g \in G(\mathbb{A})$ we have

$$E_I^{\Delta}(\varphi,\pi)_J(g) = \sum_{w \in D_{J,I}} \sum_{\mu_J} \int_{(N_J(F)\cap \dot{w}P_I(F)\dot{w}^{-1})\backslash N_J(\mathbb{A})} \varphi(\dot{w}^{-1}n_J\mu_J g)\frac{dn_J}{d\nu'_J}$$

where $d\nu_J$ (resp. $d\nu'_J$) is the counting measure on $N_J(F)$ (resp. $N_J(F) \cap \dot{w}P_I(F)\dot{w}^{-1}$) and where dn_J is the Haar measure on $N_J(\mathbb{A})$ which is normalized by

$$\mathrm{vol}\big(N_J(F)\backslash N_J(\mathbb{A}), d\nu_J\backslash dn_J\big) = 1$$

(we can permute the sums and the integral as $N_J(F)\backslash N_J(\mathbb{A})$ is compact and as the series are normally convergent on compact subsets; we have $\delta_{P_J(\mathbb{A})}(\mu_J) = 1$, $\forall \mu_J \in M_J(F)$).

Now if $w(I)$ is not contained in J let us choose $\alpha \in I$ such that $w(\alpha) \notin J$. Then we have

$$N_{I-\{\alpha\}}^I \subset \dot{w}^{-1}N_J\dot{w} \cap M_I.$$

Indeed $N_{I-\{\alpha\}}^I$ is generated by the 1-parameter subgroups $x_\beta(\mathbb{G}_a)$ with $\beta \in R_I^+ - R_{I-\{\alpha\}}^+$ and for such a β we have

$$w(\beta) \in R^+$$

as $w(I) \subset R^+$ and we have

$$w(\beta) \notin R_J^+$$

as

$$\beta = \beta' + \alpha + \beta''$$

with $\beta', \beta'' \in R^+_{I-\{\alpha\}} \cup \{0\}$ (recall that R^+_I is the set of roots in R^+ which are sums of simple roots in I). But for any $h \in G(\mathbb{A})$ the constant term of

$$m_I \mapsto \varphi(m_I h)$$

along the standard parabolic subgroup $P^I_{I-\{\alpha\}}$ of M_I is identically zero (for each $k \in K$, φ_k is a cusp form on $M_I(\mathbb{A})$). Therefore, if $w(I)$ is not contained in J we have

$$\int_{(N_J(F)\cap \dot{w}P_I(F)\dot{w}^{-1})\backslash N_J(\mathbb{A})} \varphi(\dot{w}^{-1}n_J\mu_J g)\frac{dn_J}{d\nu'_J} = 0$$

for every $\mu_J \in M_I(F)$ and every $g \in G(\mathbb{A})$ (as we have

$$N_J \cap \dot{w}P_I\dot{w}^{-1} = (N_J \cap \dot{w}N_I\dot{w}^{-1})(N_J \cap \dot{w}M_I\dot{w}^{-1})$$

and

$$\dot{w}N^I_{I-\{\alpha\}}\dot{w}^{-1} \subset N_J \cap \dot{w}M_I\dot{w}^{-1}$$

we may compute the above integral by first integrating over

$$\dot{w}N^I_{I-\{\alpha\}}(F)\dot{w}^{-1}\backslash \dot{w}N^I_{I-\{\alpha\}}(\mathbb{A})\dot{w}^{-1}).$$

In other words, in our last formula for $E^{\Delta}_I(\varphi,\pi)_J(g)$ we may replace $D_{J,I}$ by $\Omega(J,I)$.

For each $w \in \Omega(J,I)$ we have

$$M_J \cap \dot{w}P_I\dot{w}^{-1} = P^J_{w(I)}$$

(we have $\dot{w}M_I\dot{w}^{-1} = M_{w(I)} \subset M_J$ by (G.6.3)(ii) and $w^{-1}(J) \subset R^+$) and we have

$$N_J \cap \dot{w}P_I\dot{w}^{-1} = N_J \cap \dot{w}N_I\dot{w}^{-1}.$$

The proposition follows. □

(G.7) Pseudo-Eisenstein series

Let (I,π) be a cuspidal pair and let $\Phi \in \mathcal{P}(I,\pi)$. We consider the **pseudo-Eisenstein series**

$$\theta_\Phi(g) = \sum_{\gamma\in P_I(F)\backslash G(F)} \Phi^\vee(\gamma g) \qquad (\forall g \in G(\mathbb{A})) \tag{G.7.1}$$

where $\Phi^\vee \in \mathcal{P}^\vee(I,\pi) \subset C^{G,\infty}_{M_I}$ has been defined in (G.3). If it converges for all $g \in G(\mathbb{A})$ then θ_Φ is a complex function on $F^\times_\infty G(F)\backslash G(\mathbb{A})$.

PROPOSITION (G.7.2). — (i) *For all $g \in G(\mathbb{A})$ the series $\theta_\Phi(g)$ are absolutely convergent. Moreover the convergence is uniform if g stays in a compact subset of $G(\mathbb{A})$ and the complex function θ_Φ on $F_\infty^\times G(F)\backslash G(\mathbb{A})$ is locally constant with compact support.*

(ii) *If*

$$\lambda_0 \in \rho_I - \mathrm{Re}\,\pi + C_I^\Delta$$

we have

$$\theta_\Phi(g) = \frac{1}{|\mathrm{Fix}_{X_I^\Delta}(\pi)|}\int_{\mathrm{Im}\,X_I^\Delta} E_I^\Delta(\Phi(\lambda+\lambda_0)_{\lambda+\lambda_0}, \pi(\lambda+\lambda_0))(g)d\lambda \quad (\forall g \in G(\mathbb{A}))$$

where $d\lambda$ is the Haar measure on the compact abelian Lie group $\mathrm{Im}\,X_I^\Delta$ of total measure 1.

Proof : By (G.3) we may assume that

$$\Phi^\vee(g) = \begin{cases} \varphi(g) & \text{if } H_I(g) \equiv H \pmod{\mathbb{Z}}, \\ 0 & \text{otherwise,} \end{cases}$$

for some $\varphi \in \mathcal{A}_{I,\pi}^\Delta$ and some $H \in \mathbb{Z}\backslash\big(\bigoplus_{j=1}^s \frac{1}{d_j}\mathbb{Z}\big)$. Then, by (9.2.1) and by (9.2.6) we can find a compact subset Ω of $G(\mathbb{A})$ such that

$$\mathrm{Supp}(\Phi^\vee) \subset Z_G(\mathbb{A})P_I(F)\Omega.$$

Indeed the set

$$\{z_I \in Z_I(\mathbb{A}) \mid H_I(z_I) \equiv H' \pmod{\mathbb{Z}}\}$$

is compact modulo $Z_G(\mathbb{A})Z_I(F)$ for any $H' \in \mathbb{Z}\backslash\big(\bigoplus_{j=1}^s \frac{1}{d_j}\mathbb{Z}\big)$. Therefore, for any compact subset C of $G(\mathbb{A})$ the set

$$\{\gamma \in P_I(F)\backslash G(F) \mid \gamma C \cap Z_G(\mathbb{A})P_I(F)\Omega \neq \emptyset\}$$

is finite and $\theta_\Phi(g)$, viewed as series of functions of $g \in C$, is in fact a finite sum. Moreover it is clear that

$$\mathrm{Supp}(\theta_\Phi) \subset Z_G(\mathbb{A})G(F)\Omega,$$

so that $\mathrm{Supp}(\theta_\Phi)$ is compact modulo $F_\infty^\times G(F)$ ($F_\infty^\times F^\times\backslash\mathbb{A}^\times$ is compact). This concludes the proof of part (i).

For any $\lambda_0 \in X_I^\Delta$ it is clear that

$$\Phi^\vee(g) = \frac{1}{|\mathrm{Fix}_{X_I^\Delta}(\pi)|}\int_{\mathrm{Im}\,X_I^\Delta} \Phi(\lambda+\lambda_0)_{\lambda+\lambda_0}(g)d\lambda \qquad (\forall g \in G(\mathbb{A}))$$

by the Cauchy theorem. Therefore we may assume that

$$\mathrm{Re}\,\pi \in \rho_I + C_I^\Delta$$

and that $\lambda_0 = 0$. Now we have

$$\int_{\mathrm{Im}\,X_I^\Delta} E_I^\Delta(\Phi(\lambda)_\lambda, \pi(\lambda))d\lambda = \int_{\mathrm{Im}\,X_I^\Delta}\Big(\sum_{\gamma\in P_I(F)\backslash G(F)} \Phi(\lambda)_\lambda(\gamma g)\Big)d\lambda$$

and by (G.4.3) we can permute the integral and the sum. □

In particular, for each cuspidal pair (I,π) and each $\Phi \in \mathcal{P}(I,\pi)$ the pseudo-Eisenstein series θ_Φ belongs to $L^2_{G,1_\infty}$. For each cuspidal datum c let

$$L^2_{G,1_\infty,c} \subset L^2(F_\infty^\times G(F)\backslash G(\mathbb{A}), dz_\infty d\gamma\backslash dg) \tag{G.7.3}$$

be the closed subspace generated by the pseudo-Eisenstein series θ_Φ with $(I,\pi) \in c$ and $\Phi \in \mathcal{P}(I,\pi)$. Clearly $L^2_{G,1_\infty,c}$ is stable under $R_{G,1_\infty}(C_c^\infty(F_\infty^\times\backslash G(\mathbb{A})))$. We will denote by $R_{G,1_\infty,c}$ the restriction of $R_{G,1_\infty}$ to $L^2_{G,1_\infty,c}$.

THEOREM (G.7.4). — *The sum of the subspaces $L^2_{G,1_\infty,c}$, $c \in \mathcal{C}$, is dense in $L^2_{G,1_\infty}$. In other words, if $\psi \in L^2_{G,1_\infty}$ satisfies*

$$\langle \psi, \theta_\Phi\rangle = 0$$

for every cuspidal pair (I,π) and every $\Phi \in \mathcal{P}(I,\pi)$ we have $\psi(g) = 0$ for almost every $g \in G(\mathbb{A})$ (let us recall that

$$\langle \varphi_1, \varphi_2\rangle = \int_{F_\infty^\times G(F)\backslash G(\mathbb{A})} \overline{\varphi_1(g)}\varphi_2(g)\frac{dg}{dz_\infty d\gamma}$$

for any $\varphi_1, \varphi_2 \in L^2_{G,1_\infty}$).

Proof : For any cuspidal pair (I,π), any $\Phi \in \mathcal{P}(I,\pi)$ and any $\psi \in C_G^\infty \cap L^2_{G,1_\infty}$ we have

$$\begin{aligned}
&\int_{F_\infty^\times G(F)\backslash G(\mathbb{A})} \overline{\psi(g)}\theta_\Phi(g)\frac{dg}{dz_\infty d\gamma} \\
&= \int_{F_\infty^\times P_I(F)\backslash G(\mathbb{A})} \overline{\psi(g)}\Phi^\vee(g)\frac{dg}{dz_\infty d\pi_I} \\
&= \int_{F_\infty^\times N_I(\mathbb{A})M_I(F)\backslash G(\mathbb{A})} \overline{\psi_I(g)}\Phi^\vee(g)\frac{dg}{dz_\infty dn_I d\mu_I}
\end{aligned}$$

where ψ_I is the constant term of ψ along P_I (see (9.2.2)).

Let $\psi \in C_G^\infty \cap L^2_{G,1_\infty}$ and let $I \subset \Delta$ be such that

$$\psi_J \equiv 0 \qquad (\forall J \subsetneq I)$$

and

$$\int_{F_\infty^\times N_I(\mathbb{A})M_I(F)\backslash G(\mathbb{A})} \overline{\psi_I(g)}\Phi^\vee(g)\frac{dg}{dz_\infty dn_I d\mu_I} = 0$$

for every $\chi \in \mathcal{X}_{M_I}$ with $\chi(F_\infty^\times) = \{1\}$, every $\pi \in \Pi_{M_I,\chi,\mathrm{cusp}}$ and every $\Phi^\vee \in \mathcal{P}^\vee(I,\pi)$. Then let us prove that

$$\psi_I \equiv 0.$$

Let K' be an open subgroup of K such that ψ is invariant under right translation by K' and such that

$$P_I(\mathbb{A}) \cap K' = (N_I(\mathbb{A}) \cap K')(M_I(\mathbb{A}) \cap K').$$

For any $\chi \in \mathcal{X}_{M_I}$ with $\chi(F_\infty^\times) = \{1\}$, any $\pi \in \Pi_{M_I,\chi,\mathrm{cusp}}$, any $\varphi \in (\mathcal{A}_{I,\pi}^I)^{M_I(\mathbb{A})\cap K'}$ and any $f' \in \mathcal{C}_c(Z_G(\mathbb{A})M_I(\mathbb{A})^1\backslash M_I(\mathbb{A}), \mathbb{C})$ our hypotheses on ψ imply that the function

$$m_I \mapsto \int_{F_\infty^\times Z_I(F)\backslash Z_I(\mathbb{A})} \overline{\psi_I(z_I m_I)} f'(z_I m_I)\chi(z_I) \frac{dz_I}{dz_\infty d\zeta_I}$$

is a cusp form with central character χ^{-1} on $M_I(\mathbb{A})$ (it belongs to $\mathcal{C}_{M_I}^\infty$, it admits χ^{-1} as central character and it is cuspidal so that we can apply (9.2.10)) and that

$$\int_{Z_I(\mathbb{A})M_I(F)\backslash M_I(\mathbb{A})} \Big(\int_{F_\infty^\times Z_I(F)\backslash Z_I(\mathbb{A})} \overline{\psi_I(z_I m_I)} f'(z_I m_I)\chi(z_I) \frac{dz_I}{dz_\infty d\zeta_I}\Big) \varphi(m_I) \frac{dm_I}{dz_I d\pi_I} = 0$$

(if, for each $g \in G(\mathbb{A})$, we set

$$\widetilde{\varphi}(g) = \begin{cases} \delta_{P_I(\mathbb{A})}^{1/2}(m_I)\varphi(m_I) & \text{if } g = n_I m_I k' \in N_I(\mathbb{A})M_I(\mathbb{A})K', \\ 0 & \text{otherwise,} \end{cases}$$

we have $\widetilde{\varphi} \in \mathcal{A}_{I,\pi}^\Delta$ and $f' \otimes \widetilde{\varphi}$ defines a function $\Phi^\vee$ in $\mathcal{P}^\vee(I, \pi)$, see (G.3)). It follows that, for any $\chi \in \mathcal{X}_{M_I}$ with $\chi(F_\infty^\times) = \{1\}$ and any $f' \in \mathcal{C}_c(Z_G(\mathbb{A})M_I(\mathbb{A})^1\backslash M_I(\mathbb{A}), \mathbb{C})$, we have

$$\int_{F_\infty^\times Z_I(F)\backslash Z_I(\mathbb{A})} \overline{\psi_I(z_I m_I)} f'(z_I m_I)\chi(z_I) \frac{dz_I}{dz_\infty d\zeta_I} = 0 \qquad (\forall m_I \in M_I(\mathbb{A}))$$

(apply (9.2.14)). Let us fix $m_I \in M_I(\mathbb{A})$ and let f' be the characteristic function of $Z_G(\mathbb{A})M_I(\mathbb{A})^1 m_I$ in $M_I(\mathbb{A})$. For any $\chi \in \mathcal{X}_{M_I}$ with $\chi(F_\infty^\times) = \{1\}$ we have

$$\int_{F_\infty^\times Z_I(F)\backslash Z_G(\mathbb{A})Z_I(\mathbb{A})^1} \overline{\psi_I(z_I m_I)}\chi(z_I) \frac{dz_I}{dz_\infty d\zeta_I} = 0.$$

But $F_\infty^\times Z_I(F)\backslash Z_G(\mathbb{A})Z_I(\mathbb{A})^1$ is a compact group and any character of this compact group is the restriction of some $\chi \in \mathcal{X}_{M_I}$ with $\chi(F_\infty^\times) = \{1\}$. Therefore the locally constant function

$$z_I \mapsto \psi_I(z_I m_I)$$

on $F_\infty^\times Z_I(F)\backslash Z_G(\mathbb{A})Z_I(\mathbb{A})^1$ vanishes identically and in particular we have

$$\psi_I(m_I) = 0$$

as required.

Now by induction on $|I|$ we get the following result. If $\psi \in \mathcal{C}_G^\infty \cap L^2_{G,1_\infty}$ satisfies

$$\langle \psi, \theta_\Phi \rangle = 0$$

for every cuspidal pair (I, π) and every $\Phi \in \mathcal{P}(I, \pi)$ then ψ is identically zero.

Let us finish the proof of the theorem. Let $\psi \in L^2_{G,1_\infty}$ be such that

$$\langle \psi, \theta_\Phi \rangle = 0$$

for every cuspidal pair (I, π) and every $\Phi \in \mathcal{P}(I, \pi)$. Let us fix a sequence of open subgroups of K,

$$\cdots \subset K_{n+1} \subset K_n \subset \cdots \subset K_1 \subset K,$$

which form a fundamental system of neighborhoods of the unit element in $G(\mathbb{A})$. For each positive integer n let us denote by ψ_n the function

$$g \mapsto \int_{K_n} \psi(gk_n)dk_n$$

(dk_n being the Haar measure on K_n with total volume 1). Then it is clear that

$$\psi_n \in \mathcal{C}_G^\infty \cap L^2_{G,1_\infty}$$

with

$$\|\psi_n\|_2^2 \leq \|\psi\|_2^2 \operatorname{vol}(K_n, dk_n) = \|\psi\|_2^2$$

and that

$$\lim_{n \to +\infty} \|\psi - \psi_n\|_2 = 0.$$

Moreover, for any $\Phi \in \mathcal{P}(I, \pi)$ we have

$$\langle \psi_n, \theta_\Phi \rangle = \langle \psi, \theta_{\Phi_n} \rangle = 0$$

where we have set

$$\Phi_n(\lambda)(g) = \int_{K_n} \Phi(\lambda)(gk_n^{-1})dk_n \qquad (\forall \lambda \in X_I^\Delta,\ \forall g \in G(\mathbb{A})),$$

so that $\Phi_n \in \mathcal{P}(I, \pi)$. Therefore it follows that

$$\psi_n \equiv 0$$

for every positive integer n and that

$$\psi(g) = 0$$

for almost all $g \in G(\mathbb{A})$ as required. □

(G.8) The scalar product of two pseudo-Eisenstein series
Let (I,π) and (I',π') be two cuspidal pairs. If $\Phi \in \mathcal{P}(I,\pi)$ and $\Phi' \in \mathcal{P}(I',\pi')$ we may form the inner product

$$\langle \theta_{\Phi'}, \theta_{\Phi}\rangle = \int_{F_\infty^\times G(F)\backslash G(\mathbb{A})} \overline{\theta_{\Phi'}(g)}\theta_{\Phi}(g)\frac{dg}{dz_\infty d\gamma}$$

(both θ_Φ and $\theta_{\Phi'}$ are locally constant complex functions with compact supports on $F_\infty^\times G(F)\backslash G(\mathbb{A})$).

If (I,π) and (I',π') are equivalent, i.e. define the same cuspidal datum (see (G.1.6)), we denote by

$$\Omega((I',\pi'),(I,\pi)) \subset \Omega(I',I) \subset W$$

the set of $w \in W$ such that $w(I) = I'$ and such that there exists $\lambda' \in X_{I'}^{\Delta}$ with $w(\pi) = \pi'(\lambda')$. For each $w \in \Omega((I',\pi'),(I,\pi))$ we choose $\lambda'_w \in X_{I'}^{\Delta}$ such that

$$w(\pi) = \pi'(-2\mathrm{Re}\,\pi' - \overline{\lambda}'_w) = \overline{-\pi'(\lambda'_w)}$$

and we denote by

$$\langle\cdot,\cdot\rangle_w : \mathcal{A}^{\Delta}_{I',\pi'(\lambda'_w)} \times \mathcal{A}^{\Delta}_{w(I),w(\pi)} \to \mathbb{C}$$

the scalar product defined in (G.5.6).

THEOREM (G.8.1). — *Let $\Phi \in \mathcal{P}(I,\pi)$ and $\Phi' \in \mathcal{P}(I',\pi')$.*

(i) *If (I,π) and (I',π') are not equivalent we have*

$$\langle \theta_{\Phi'}, \theta_{\Phi}\rangle = 0.$$

(ii) *If (I,π) and (I',π') are equivalent then, for any*

$$\lambda_0 \in \rho_I - \mathrm{Re}\,\pi + C_I^{\Delta}$$

we have

$$\langle \theta_{\Phi'}, \theta_{\Phi}\rangle = \frac{1}{|\mathrm{Fix}_{X_I^{\Delta}}(\pi)|}\int_{\mathrm{Im}\,X_I^{\Delta}}\Big(\sum_{w\in\Omega((I',\pi'),(I,\pi))} u(w,\lambda)\Big)d\lambda$$

where $u(w,\lambda)$ is equal to

$$\frac{\deg(\infty)}{fd_1\cdots d_s}\langle \Phi'(\lambda'_w - w(\overline{\lambda}+\lambda_0))_{\lambda'_w},$$
$$\Big(M_I(w,\pi(\lambda+\lambda_0))\big(\Phi(\lambda+\lambda_0)_{\lambda+\lambda_0}\big)\Big)_{-w(\lambda+\lambda_0)}\rangle_w$$

(*as usual f is the degree over $\mathbb{F}_p$ of the field of constants in F and $(d_1,\ldots,d_s)$ is the partition of d corresponding to I*).

Proof: Permuting (I,π,Φ) and (I',π',Φ') if it is necessary we may assume that

$$|I'|\leq |I|.$$

By definition we have

$$\begin{aligned}\langle\theta_{\Phi'},\theta_\Phi\rangle &= \int_{F_\infty^\times G(F)\backslash G(\mathbb{A})}\overline{\theta_{\Phi'}(g)}\theta_\Phi(g)\frac{dg}{dz_\infty d\gamma}\\ &= \int_{F_\infty^\times P_{I'}(F)\backslash G(\mathbb{A})}\overline{\Phi'^\vee(g)}\theta_\Phi(g)\frac{dg}{dz_\infty d\pi_{I'}}\\ &= \int_{F_\infty^\times N_{I'}(\mathbb{A})M_{I'}(F)\backslash G(\mathbb{A})}\overline{\Phi'^\vee(g)}\theta_{\Phi,I'}(g)\frac{dg}{dz_\infty dn_{I'}d\mu_{I'}}\end{aligned}$$

where $\theta_{\Phi,I'}$ is the constant term of θ_Φ along $P_{I'}$ (see (9.2.2)), where $d\pi_{I'}$, $d\mu_{I'}$ and $d\nu_{I'}$ are the counting measures on $P_{I'}(F)$, $M_{I'}(F)$ and $N_{I'}(F)$ and where the Haar measure $dn_{I'}$ on $N_{I'}(\mathbb{A})$ is normalized by $\mathrm{vol}(N_{I'}(F)\backslash N_{I'}(\mathbb{A}), d\nu_{I'}\backslash dn_{I'})=1$).

Now it follows from (G.7.2) and (G.6.4) that

$$\theta_{\Phi,I'}=0$$

if $|I'|<|I|$ (so that $\Omega(I',I)=\emptyset$) and that

$$\theta_{\Phi,I'}=\frac{1}{|\mathrm{Fix}_{X_I^\Delta}(\pi)|}\int_{\mathrm{Im}\,X_I^\Delta}\Big(\sum_{w\in\Omega(I',I)}M_I(w,\pi(\lambda+\lambda_0))(\Phi(\lambda+\lambda_0)_{\lambda+\lambda_0})\Big)d\lambda$$

if $|I'|=|I|$. Therefore, to finish the proof of the theorem we may assume that $|I'|=|I|$ and it is sufficient to check that, for each $w\in\Omega(I',I)$, i.e. for each $w\in W$ such that $w(I)=I'$, the integral

$$\widetilde{u}(w,\lambda)=\int_{F_\infty^\times N_{I'}(\mathbb{A})M_{I'}(F)\backslash G(\mathbb{A})}\overline{\Phi'^\vee(g)}M_I(w,\pi(\lambda+\lambda_0))$$
$$(\Phi(\lambda+\lambda_0)_{\lambda+\lambda_0})(g)\frac{dg}{dz_\infty dn_{I'}d\mu_{I'}}$$

is equal to $u(w,\lambda)$ if $w\in\Omega((I',\pi'),(I,\pi))$ and is identically zero otherwise (recall that

$$\{m_{I'}\in M_{I'}(\mathbb{A})\mid \Phi'^\vee(n_{I'}m_{I'}k)\neq 0 \text{ for some } n_{I'}\in N_{I'}(\mathbb{A}) \text{ and } k\in K\}$$

is compact modulo $F_\infty^\times M_{I'}(\mathbb{A})^1$; see (G.3)).

But to compute $\widetilde{u}(w,\lambda)$ we may first integrate over $F_\infty^\times Z_{I'}(F)\backslash Z_{I'}(\mathbb{A})$ for some Haar measure $dz_\infty d\zeta_{I'}\backslash dz_{I'}$ ($d\zeta_{I'}$ is the counting measure on $Z_{I'}(F)$)

and then integrate over $N_{I'}(\mathbb{A})Z_{I'}(\mathbb{A})M_{I'}(F)\backslash G(\mathbb{A})$ for the Haar measure $dn_{I'}dz_{I'}d\mu_{I'}\backslash dg$. As, for any $z_{I'} \in Z_{I'}(\mathbb{A})$ and any $g \in G(\mathbb{A})$, we have

$$\begin{aligned} M_I(w,\pi(\lambda+\lambda_0))\big(\Phi(\lambda+\lambda_0)_{\lambda+\lambda_0}\big)(z_{I'}g) \\ = \chi_{w(\pi(\lambda+\lambda_0))}(z_{I'})M_I(w,\pi(\lambda+\lambda_0))\big(\Phi(\lambda+\lambda_0)_{\lambda+\lambda_0}\big)(g) \end{aligned}$$

the integral over $F_\infty^\times Z_{I'}(F)\backslash Z_{I'}(\mathbb{A})$ is the product of

$$M_I(w,\pi(\lambda+\lambda_0))\big(\Phi(\lambda+\lambda_0)_{\lambda+\lambda_0}\big)(g)$$

and of the integral

$$\int_{F_\infty^\times Z_{I'}(F)\backslash Z_{I'}(\mathbb{A})} \overline{\Phi'^\vee(z_{I'}g)}\chi_{w(\pi(\lambda+\lambda_0))}(z_{I'})\frac{dz_{I'}}{dz_\infty d\zeta_{I'}}.$$

Let us compute the last integral. We may assume that

$$\Phi'(\lambda')(g) = f'(\lambda')\varphi'(g) \qquad (\forall \lambda' \in X_{I'}^{\Delta},\ \forall g \in G(\mathbb{A}))$$

for some $f' \in \mathcal{P}(X_{I'}^{\Delta})$ and some $\varphi' \in \mathcal{A}_{I',\pi'}^{\Delta}$ (see (G.2)). Then we have

$$\Phi'^\vee(z_{I'}g) = \frac{1}{|\mathrm{Fix}_{X_{I'}^{\Delta}}(\pi')|}\widehat{f'}(z_{I'}m_{I'}(g))\chi_{\pi'}(z_{I'})\varphi'(g)$$

(see (G.3)). Therefore, on the one hand, if the restriction of the character $\overline{\chi}_{\pi'}\chi_{w(\pi(\lambda+\lambda_0))}$ to $F_\infty^\times Z_{I'}(F)\backslash \mathbb{A}^\times Z_{I'}(\mathbb{A})^1$ is non-trivial the above integral is zero. On the other hand, if this restriction is trivial, i.e. if

$$\overline{\chi}_{\pi'}\chi_{w(\pi(\lambda+\lambda_0))}(z_{I'}) = q^{\langle -\overline{\mu}', H_{I'}(z_{I'})\rangle}$$

for some $\mu' \in X_{I'}^{\Delta}$, the above integral is equal to the product of

$$\frac{1}{|\mathrm{Fix}_{X_{I'}^{\Delta}}(\pi')|}\overline{\varphi'(g)}\,\mathrm{vol}\Big(F_\infty^\times Z_{I'}(F)\backslash \mathbb{A}^\times Z_{I'}(\mathbb{A})^1, \frac{dz_{I'}}{dz_\infty d\zeta_{I'}}\Big)$$

and of

$$\frac{1}{d_1'\cdots d_{s'}'}\sum_{\nu'}\overline{f'(\mu'+\nu')q^{\langle\overline{\mu}'+\overline{\nu}',H_{I'}(m_{I'}(g))\rangle}}$$

(Fourier inversion formula) where ν' runs through

$$X_{I'}^{I'} = \bigoplus_{j'=1}^{s'}\Big(\frac{2\pi i}{\log q}\mathbb{Z}\Big/\frac{2\pi i}{\log q}d_{j'}'\mathbb{Z}\Big) \subset \mathrm{Im}\, X_{I'}^{\Delta}$$

and where $(d'_1, \ldots, d'_{s'})$ is the partition of d corresponding to I'. It follows that $\widetilde{u}(w, \lambda)$ is equal to the product of

$$\frac{1}{d'_1 \cdots d'_{s'} |\mathrm{Fix}_{X^{\Delta}_{I'}}(\pi')|} \mathrm{vol}\Big(F^{\times}_{\infty} Z_{I'}(F) \backslash \mathbb{A}^{\times} Z_{I'}(\mathbb{A})^1, \frac{dz_{I'}}{dz_{\infty} d\zeta_{I'}}\Big)$$

and

$$\sum_{\nu' \in X^{I'}_{I'}} \int_{N_{I'}(\mathbb{A}) Z_{I'}(\mathbb{A}) M_{I'}(F) \backslash G(\mathbb{A})} \overline{(\Phi'(\mu' + \nu')_{\mu'+\nu'})(g)}$$
$$\times M_I(w, \pi(\lambda + \lambda_0)) (\Phi(\lambda + \lambda_0)_{\lambda+\lambda_0})(g) \frac{dg}{dn_{I'} dz_{I'} d\mu_{I'}}$$

if

$$\chi_{w(\pi(\lambda+\lambda_0))} = \chi_{-\overline{\pi'(\mu')}}$$

for some $\mu' \in X^{\Delta}_{I'}$ and that

$$\widetilde{u}(w, \lambda) = 0$$

otherwise.

Now, if σ' and σ'' are two irreducible cuspidal automorphic representations of $M_{I'}(\mathbb{A})$ such that

$$\chi_{\sigma'} = \chi_{-\overline{\sigma}''}$$

and if $\psi' \in \mathcal{A}^{\Delta}_{I',\sigma'}$ and $\psi'' \in \mathcal{A}^{\Delta}_{I',\sigma''}$, we have

$$\int_{N_{I'}(\mathbb{A}) Z_{I'}(\mathbb{A}) M_{I'}(F) \backslash G(\mathbb{A})} \overline{\psi''(g)} \psi'(g) \frac{dg}{dn_{I'} dz_{I'} d\mu_{I'}} = 0$$

unless σ' is isomorphic to $-\overline{\sigma}'' = \sigma''(-2\mathrm{Re}\, \sigma'')$. Therefore we have

$$\widetilde{u}(w, \lambda) = 0$$

unless there exists at least one $\mu'_w \in X^{\Delta}_{I'}$ such that

$$w(\pi(\lambda + \lambda_0)) = -\overline{\pi'(\mu'_w)},$$

i.e. unless there exists at least one $\lambda'_w \in X^{\Delta}_{I'}$ ($\lambda'_w = \mu'_w + w(\overline{\lambda} + \lambda_0)$) such that

$$w(\pi) = -\overline{\pi'(\lambda'_w)}.$$

Moreover, if we have

$$w(\pi) = -\overline{\pi'(\lambda'_w)}$$

for some $\lambda'_w \in X^{\Delta}_{I'}$ then $\widetilde{u}(w, \lambda)$ is the product of

$$\frac{1}{d'_1 \cdots d'_{s'}} \mathrm{vol}\Big(F^{\times}_{\infty} Z_{I'}(F) \backslash \mathbb{A}^{\times} Z_{I'}(\mathbb{A})^1, \frac{dz_{I'}}{dz_{\infty} d\zeta_{I'}}\Big)$$

and the integral

$$\int_{N_{I'}(\mathbb{A})Z_{I'}(\mathbb{A})M_{I'}(F)\backslash G(\mathbb{A})} \overline{(\Phi'(\mu'_w)_{\mu'_w})(g)} \times M_I(w,\pi(\lambda+\lambda_0))(\Phi(\lambda+\lambda_0)_{\lambda+\lambda_0})(g)\frac{dg}{dn_{I'}dz_{I'}d\mu_{I'}}$$

where

$$\mu'_w = \lambda'_w - w(\overline{\lambda}+\lambda_0)$$

(we have

$$\Phi'(\mu'_w+\nu')_{\mu'_w+\nu'} = \Phi'(\mu'_w)_{\mu'_w}$$

for every $\nu' \in \mathrm{Fix}_{X_{I'}^{\Delta}}(\pi')$) (see (G.2.1)(iii)).

Now we have

$$d'_1\cdots d'_{s'} = d_1\cdots d_s$$

and we have the exact sequences

$$1\to \mathcal{O}_\infty^\times Z_{I'}(F)\backslash Z_{I'}(\mathbb{A})^1 \to F_\infty^\times Z_{I'}(F)\backslash \mathbb{A}^\times Z_{I'}(\mathbb{A})^1 \to \mathbb{Z}f/\mathbb{Z}\deg(\infty)\to 0$$

and

$$1\to \mathcal{O}_\infty^\times \to Z_{I'}(F)\backslash Z_{I'}(\mathbb{A})^1 \to \mathcal{O}_\infty^\times Z_{I'}(F)\backslash Z_{I'}(\mathbb{A})^1 \to 1.$$

Therefore we have

$$\mathrm{vol}\big(F_\infty^\times Z_{I'}(F)\backslash\mathbb{A}^\times Z_{I'}(\mathbb{A})^1, dz_\infty d\zeta_{I'}\backslash dz_{I'}\big) = \frac{\deg(\infty)}{f}$$

if we normalize $dz_{I'}$ and dz_∞ by

$$\mathrm{vol}(Z_{I'}(F)\backslash Z_{I'}(\mathbb{A})^1, d\zeta_{I'}\backslash dz_{I'}) = 1$$

and

$$\mathrm{vol}(\mathcal{O}_\infty^\times, dz_\infty) = 1.$$

The theorem follows. □

COROLLARY (G.8.2). — *If c' and c'' are two distinct cuspidal data the subspaces $L^2_{G,1_\infty,c'}$ and $L^2_{G,1_\infty,c''}$ of $L^2_{G,1_\infty}$ are orthogonal. We have*

$$L^2_{G,1_\infty} = \widehat{\bigoplus_{c\in C}} L^2_{G,1_\infty,c},\quad R_{G,1_\infty} = \widehat{\bigoplus_{c\in C}} R_{G,1_\infty,c}$$

(Hilbert sum of unitary representations of $F_\infty^\times\backslash G(\mathbb{A})$).

Proof: Apply (G.8.1)(i) and (G.7.4). □

(G.9) Analytic continuation of Eisenstein series

In this section we will review (without proofs) the main results of Langlands, Harder and Morris about the analytic continuation of the Eisenstein series and of the intertwining operators. We restrict ourself to the easiest case: Eisenstein series arising from cusp forms. Complete proofs may be found in [Mor 1] and [Mo–Wa 2] (Ch. IV).

For each $I \subset \Delta$, X_I^Δ is a complex affine variety with affine ring the $\mathbb{C}$-algebra $\mathcal{P}(X_I^\Delta)$ of polynomial functions on X_I^Δ (see (G.2)). In general X_I^Δ is not irreducible. Nevertheless it is reduced and we can form the total ring of fractions $\mathcal{R}(X_I^\Delta)$ of $\mathcal{P}(X_I^\Delta)$, i.e. the ring of fractions $D(\lambda)^{-1}P(\lambda)$ with $P(\lambda) \in \mathcal{P}(X_I^\Delta)$ and $D(\lambda) \in \mathcal{P}(X_I^\Delta)$ such that $D(\lambda)$ does not vanish identically on any irreducible component of X_I^Δ. If V is a finite dimensional complex vector space we will set

$$\mathcal{P}(X_I^\Delta; V) = \mathcal{P}(X_I^\Delta) \otimes_{\mathbb{C}} V$$

and

$$\mathcal{R}(X_I^\Delta; V) = \mathcal{R}(X_I^\Delta) \otimes_{\mathbb{C}} V$$

and we will view an element of $\mathcal{P}(X_I^\Delta; V)$ (resp. $\mathcal{R}(X_I^\Delta; V)$) as a polynomial (resp. rational) function on X_I^Δ with values in V. If Ω is an open subset (for the classical topology) of X_I^Δ which meets each irreducible component of X_I^Δ and if

$$F : \Omega \to V$$

is a holomorphic function we will say that F can be analytically continued as a rational function to X_I^Δ if there exists $D(\lambda) \in \mathcal{P}(X_I^\Delta)$ which does not vanish on Ω and $P(\lambda) \in \mathcal{P}(X_I^\Delta; V)$ such that

$$D(\lambda)F(\lambda) = P(\lambda) \qquad (\forall \lambda \in \Omega).$$

Then $D(\lambda)^{-1}P(\lambda) \in \mathcal{R}(X_I^\Delta; V)$ is uniquely determined by F and will be simply denoted by $F(\lambda)$.

THEOREM (G.9.1) (Analytic continuation and functional equation of Eisenstein series). — *Let us fix a compact open subgroup $K' \subset K$ of $G(\mathbb{A})$.*

(i) *For each cuspidal pair (I, π) the functions*

$$\begin{aligned}\rho_I - \operatorname{Re}\pi + C_I^\Delta &\to \operatorname{Hom}_{\mathbb{C}}((\mathcal{A}_{I,\pi}^\Delta)^{K'}, \mathbb{C}),\\ \lambda &\mapsto (\varphi \mapsto E_I^\Delta(\varphi_\lambda, \pi(\lambda))(g))\end{aligned}$$

($g \in G(\mathbb{A})$) *and the functions*

$$\begin{aligned}\rho_I - \operatorname{Re}\pi + C_I^\Delta &\to \operatorname{Hom}_{\mathbb{C}}\big((\mathcal{A}_{I,\pi}^\Delta)^{K'}, (\mathcal{A}_{w(I),w(\pi)}^\Delta)^{K'}\big),\\ \lambda &\mapsto \Big(\varphi \mapsto \big(M_I(w, \pi(\lambda))(\varphi_\lambda)\big)_{-w(\lambda)}\Big)\end{aligned}$$

($w \in W$ *with* $w(I) \subset \Delta$) *can be analytically continued as rational functions to* X_I^Δ.

(ii) *Let (I,π) be a cuspidal pair. Then the following functional equations hold.*

(a) *Let $f \in C_c^\infty(F_\infty^\times \backslash G(\mathbb{A}))$ be K'-bi-invariant. We have*

$$E_I^\Delta\big((i_I^\Delta(\lambda,f)(\varphi))_\lambda,\pi(\lambda)\big)(h) = \int_{F_\infty^\times\backslash G(\mathbb{A})} f(g)E_I^\Delta(\varphi_\lambda,\pi(\lambda))(hg)\frac{dg}{dz_\infty}$$

and

$$\Big(M_I(w,\pi(\lambda))\big((i_I^\Delta(\lambda,f)(\varphi))_\lambda\big)\Big)_{-w(\lambda)} = i_{w(I)}^\Delta(w(\lambda),f)\big((M_I(w,\pi(\lambda))(\varphi_\lambda))_{-w(\lambda)}\big)$$

for every $\varphi \in (\mathcal{A}_{I,\pi}^\Delta)^{K'}$, every $h \in G(\mathbb{A})$ and every $w \in W$ with $w(I) \subset \Delta$. Here we have set

$$i_{I'}^\Delta(\lambda',f)(\varphi')(h) = \int_{F_\infty^\times\backslash G(\mathbb{A})} f(g)\varphi'(hg)q^{\langle\lambda',H_{I'}(hg)-H_{I'}(h)\rangle}\frac{dg}{dz_\infty} \qquad (\forall h \in G(\mathbb{A}))$$

for each cuspidal pair (I',π') and each $\varphi' \in \mathcal{A}_{I',\pi'}^\Delta$.

(b) *Let $w \in W$ with $w(I) \subset \Delta$. We have*

$$E_{w(I)}^\Delta\big(M_I(w,\pi(\lambda))(\varphi_\lambda),w(\pi(\lambda))\big) = E_I^\Delta(\varphi_\lambda,\pi(\lambda))$$

for every $\varphi \in (\mathcal{A}_{I,\pi}^\Delta)^{K'}$.

(c) *Let $w,w' \in W$ with $w(I) \subset \Delta$ and $w'(w(I)) \subset \Delta$. We have*

$$\Big(M_{w(I)}(w',w(\pi(\lambda)))\big(M_I(w,\pi(\lambda))(\varphi_\lambda)\big)\Big)_{-w'w(\lambda)} = \big(M_I(w'w,\pi(\lambda))(\varphi_\lambda)\big)_{-w'w(\lambda)}$$

for every $\varphi \in (\mathcal{A}_{I,\pi}^\Delta)^{K'}$.

Proof: See [Mo–Wa 2] (IV.1.8), (IV.1.9)(a), (IV.1.10) and (IV.1.12). □

Let (I,π) be a cuspidal pair and let $(d_1,\ldots,d_s)$ be the partition of d corresponding to I. For each pair (j,k) with $j,k \in \{1,\ldots,s\}$ and $j < k$ we will denote by $n_{jk}(\pi)$ the smallest positive integer n such that

$$n\Big(\frac{\lambda_j}{d_j}-\frac{\lambda_k}{d_k}\Big) \in \frac{2\pi i}{\log q}\mathbb{Z}$$

for every $\lambda \in \mathrm{Fix}_{X_I^\Delta}(\pi)$ and we will denote by $h_{jk}(\pi)$ the polynomial function on X_I^Δ defined by

$$h_{jk}(\pi)(\lambda) = q^{n_{jk}(\pi)\left(\frac{\lambda_j}{d_j} - \frac{\lambda_k}{d_k}\right)} - 1 \qquad (\forall \lambda \in X_I^\Delta).$$

A **radicial hyperplane** for (I, π) is an algebraic subset H of X_I^Δ of the form

$$H = \lambda_0 + \{\lambda \in X_I^\Delta \mid h_{jk}(\pi)(\lambda) = 0\}$$

for some $\lambda_0 \in X_I^\Delta$ and some pair (j, k) as above. We will denote by $\mathcal{H}(I, \pi)$ the set of all radicial hyperplanes for (I, π). For each $H \in \mathcal{H}(I, \pi)$ we have

$$\mathrm{Fix}_{X_I^\Delta}(\pi) + H = H$$

and the function

$$h_H(\lambda) = h_{jk}(\pi)(\lambda - \lambda_0)$$

defining H is uniquely determined by H.

THEOREM (G.9.2) (Singularities of Eisenstein series). — *Let us fix a compact open subgroup $K' \subset K$ of $G(\mathbb{A})$ and a cuspidal pair (I, π).*

(i) *There exists a function*

$$m : \mathcal{H}(I, \pi) \to \mathbb{Z}_{\geq 0}$$

with finite support (depending on K' and (I, π)) such that

$$\Big(\prod_{H \in \mathcal{H}(I,\pi)} h_H(\lambda)^{m(H)}\Big)\Big(\varphi \mapsto \big(M_I(w, \pi(\lambda))(\varphi_\lambda)\big)_{-w(\lambda)}\Big)$$
$$\in \mathcal{P}\Big(X_I^\Delta; \mathrm{Hom}_{\mathbb{C}}\big((\mathcal{A}_{I,\pi}^\Delta)^{K'}, (\mathcal{A}_{w(I),w(\pi)}^\Delta)^{K'}\big)\Big)$$

for every $w \in W$ with $w(I) \subset \Delta$.

Moreover for any such a function $m : \mathcal{H}(I, \pi) \to \mathbb{Z}_{\geq 0}$ we automatically have

$$\Big(\prod_{H \in \mathcal{H}(I,\pi)} h_H(\lambda)^{m(H)}\Big)\big(\varphi \mapsto E_I^\Delta(\varphi_\lambda, \pi(\lambda))(g)\big) \in \mathcal{P}\big(X_I^\Delta; \mathrm{Hom}_{\mathbb{C}}((\mathcal{A}_{I,\pi}^\Delta)^{K'}, \mathbb{C})\big).$$

In other words the singularities of the Eisenstein series are at most those of the intertwining operators.

(ii) *We can choose the function $m : \mathcal{H}(I, \pi) \to \mathbb{Z}_{\geq 0}$ in* (i) *with the following extra properties:*

(a) $m(H) \in \{0,1\}$ *for each* $H \in \mathcal{H}(I,\pi)$ *such that*

$$(-\mathrm{Re}\,\pi + \overline{C}_I^{\Delta}) \cap H \neq \emptyset,$$

where

$$\overline{C}_I^{\Delta} = \{\lambda \in X_I^{\Delta} \mid \mathrm{Re}\Big(\frac{\lambda_j}{d_j} - \frac{\lambda_{j+1}}{d_{j+1}}\Big) \geq 0,\ \forall j = 1, \ldots, s-1\},$$

(b) $m(H) = 0$ *for each* $H \in \mathcal{H}(I,\pi)$ *such that*

$$(-\mathrm{Re}\,\pi + \mathrm{Im}\,X_I^{\Delta}) \cap H \neq \emptyset,$$

(c) $m(H) = 0$ *for each* $H \in \mathcal{H}(I,\pi)$ *such that*

$$(\rho_I - \mathrm{Re}\,\pi + C_I^{\Delta}) \cap H \neq \emptyset.$$

Proof : See [Mo–Wa 2] (IV.1.11). □

REMARK (G.9.3). — Let $K' \subset K$ be a compact open subgroup of $G(\mathbb{A})$ and let (I,π) be a cuspidal pair. By analytic continuation we get the following equalities of rational functions on X_I^{Δ}.

(i) For each $w \in W$ with $w(I) \subset \Delta$, each $\varphi \in \mathcal{A}_{I,\pi}^{\Delta}$ and each $\varphi' \in \mathcal{A}_{w(I),-\overline{w(\pi)}}^{\Delta}$ we have

$$\langle \varphi', (M_I(w,\pi(\lambda))(\varphi_\lambda))_{-w(\lambda)}\rangle = \langle (M_{w(I)}(w^{-1}, -\overline{w(\pi(\lambda))})(\varphi'_{-\overline{w(\lambda)}}))_{\overline{\lambda}}, \varphi\rangle$$

(see (G.5.9)).

(ii) For each $\varphi \in \mathcal{A}_{I,\pi}^{\Delta}$ and each $J \subset \Delta$ we have

$$E_I^{\Delta}(\varphi_\lambda, \pi(\lambda))_J = \sum_{w \in \Omega(J,I)} E_{w(I)}^{J}(M_I(w,\pi(\lambda))(\varphi_\lambda), w(\pi(\lambda)))$$

(see (G.6.4)). □

We will say that a cuspidal datum c is **regular** (Arthur uses the terminology "unramified") if, for some (and therefore for every) $(I,\pi) \in c$, the only $w \in W$ such that

$$w(I) = I$$

and

$$w(\pi) = \pi(\lambda)$$

for some $\lambda \in X_I^{\Delta}$ is the identity element of W. We will say that a cuspidal pair is **regular** if it belongs to a regular cuspidal datum.

PROPOSITION (G.9.4). — *Let $K' \subset K$ be a compact open subgroup of $G(\mathbb{A})$ and let (I,π) be a cuspidal pair. If (I,π) is regular the rational functions*

$$\lambda \mapsto (\varphi \mapsto E_I^\Delta(\varphi_\lambda, \pi(\lambda))(g))$$

$(g \in G(\mathbb{A}))$ and

$$\lambda \mapsto \left(\varphi \mapsto (M_I(w,\pi(\lambda))(\varphi_\lambda))_{-w(\lambda)}\right)$$

$(w \in W$ with $w(I) \subset \Delta)$ (see (G.9.1)(i)) do not have poles in

$$-\mathrm{Re}\,\pi + \overline{C}_I^\Delta \subset X_I^\Delta.$$

Proof : By (G.9.2)(i) the statement for the Eisenstein series follows from the statement for the intertwining operators.

Now let us fix $w \in W$ with $w(I) \subset \Delta$. We will prove more precisely that the rational function

$$\lambda \mapsto \left(\varphi \mapsto (M_I(w,\pi(\lambda))(\varphi_\lambda))_{-w(\lambda)}\right)$$

does not have poles in

$$-\mathrm{Re}\,\pi + \overline{C}_{I,w}^\Delta \subset X_I^\Delta$$

where $\overline{C}_{I,w}^\Delta$ is the closure of $C_{I,w}^\Delta$ in X_I^Δ (see (G.5)). Using (G.5.2)(ii)(d), (G.5.5) and the functional equation (G.9.1)(ii)(c) we easily see that it is sufficient to prove the following statement.

For each cuspidal pair (I',π') in the equivalence class of (I,π) and for each $s' \in W$ with $s'(I') \subset \Delta$ and $\ell_{I'}(s') = 1$ (see (9.5.2)(ii)) the rational function

$$\lambda' \mapsto \left(\varphi' \mapsto (M_{I'}(s',\pi'))(\varphi'_{\lambda'}))_{-s'(\lambda')}\right)$$

does not have poles in

$$-\mathrm{Re}\,\pi' + \overline{C}_{I',s'}^\Delta \subset X_{I'}^\Delta.$$

But as (I',π') is regular the last statement immediately follows from [Mo–Wa 2] (Remark at the end of (IV.3.12)). □

(G.10) The spectral decomposition of $L^2_{G,1_\infty,c}$ for regular cuspidal data c

Let c be a cuspidal datum. For all unitary cuspidal pairs (I,π) and (I',π') in c and for all $\Phi \in \mathcal{P}(I,\pi)$ and $\Phi' \in \mathcal{P}(I',\pi')$ we have seen that the scalar product of the two pseudo-Eisenstein series θ_Φ and $\theta_{\Phi'}$ is the integral of some rational function on X_I^Δ with contour of integration

$$\lambda_0 + \operatorname{Im} X_I^\Delta \subset X_I^\Delta$$

for any fixed $\lambda_0 \in \rho_I + C_I^\Delta$. In order to obtain the spectral decomposition of $L^2_{G,1_\infty,c}$ Langlands chooses a path from λ_0 to 0 in $\overline{C}_I^\Delta$ and lets the contour of integration move along this path. Each time the path crosses a singular hyperplane a residue occurs (Cauchy formula). In general the result of this process is rather complicated (see [Mo–Wa 1] (II.3) and [Mo–Wa 2] Ch. V and VI). However, in the case of a regular cuspidal datum c no residue occurs and the result is very simple. In particular we have

PROPOSITION (G.10.1). — *Let c be a regular cuspidal datum.*

(i) *For each unitary cuspidal pair $(I,\pi) \in c$ and for each $\Phi \in \mathcal{P}(I,\pi)$ we have*

$$\theta_\Phi(g) = \frac{1}{|\mathrm{Fix}_{X_I^\Delta}(\pi)|} \int_{\operatorname{Im} X_I^\Delta} E_I^\Delta(\Phi(\lambda)_\lambda, \pi(\lambda))(g)d\lambda \qquad (\forall g \in G(\mathbb{A})).$$

(ii) *For any unitary cuspidal pairs (I,π) and (I',π') in c and for any $\Phi \in \mathcal{P}(I,\pi)$ and $\Phi' \in \mathcal{P}(I',\pi')$ we have*

$$\langle \theta_{\Phi'}, \theta_\Phi \rangle = \frac{\deg(\infty)}{fd_1 \cdots d_s |\mathrm{Fix}_{X_I^\Delta}(\pi)|}$$
$$\times \int_{\operatorname{Im} X_I^\Delta} \langle \Phi'(\lambda'_w + w(\lambda))_{\lambda'_w}, \big(M_I(w,\pi(\lambda))(\Phi(\lambda)_\lambda)\big)_{-w(\lambda)} \rangle_w d\lambda$$

where w is the unique element in W with $w(I) = I'$ and $w(\pi) = \pi'(\lambda'_w)$ for some $\lambda'_w \in \operatorname{Im} X_{I'}^\Delta$.

(iii) *For any unitary cuspidal pairs (I,π) and (I',π') in c and for any $\varphi \in \mathcal{A}_{I,\pi}^\Delta$ and $\Phi' \in \mathcal{P}(I',\pi')$ we have*

$$\langle \theta_{\Phi'}, E_I^\Delta(\varphi,\pi) \rangle = \frac{\deg(\infty)}{fd_1 \cdots d_s} \langle \Phi'(\lambda'_w)_{\lambda'_w}, M_I(w,\pi)(\varphi) \rangle_w$$

where w is the unique element in W with $w(I) = I'$ and $w(\pi) = \pi'(\lambda'_w)$ for some $\lambda'_w \in \operatorname{Im} X_{I'}^\Delta$.

Proof: Parts (i) and (ii) of the proposition directly follow from (G.7.2)(ii), (G.8.1), (G.9.4) and the Cauchy formula. Part (iii) follows from parts (i) and (ii) where we have taken

$$\Phi(\lambda) = \frac{1}{|\mathrm{Fix}_{X_I^\Delta}(\pi)|} \sum_{\mu \in \mathrm{Fix}_{X_I^\Delta}(\pi)} f(\lambda + \mu) \otimes \varphi_\mu$$

for some arbitrary $f \in \mathcal{P}(X_I^\Delta)$. □

For each unitary cuspidal pair (I,π) let $\widehat{\mathcal{A}}_{I,\pi}^{\Delta}$ be the Hilbert completion of $\mathcal{A}_{I,\pi}^{\Delta}$ with respect to the scalar product

$$\langle\varphi',\varphi\rangle=\int_{N_I(\mathbb{A})Z_I(\mathbb{A})M_I(F)\backslash G(\mathbb{A})}\overline{\varphi'(g)}\varphi(g)\frac{dg}{dn_I dz_I d\mu_I}$$

(recall that, for any φ in $\mathcal{A}_{I,\pi}^{\Delta}$, we have $\varphi(z_I g)=\chi_\pi(z_I)\varphi(g)$, $\forall z_I\in Z_I(\mathbb{A})$, $\forall g\in G(\mathbb{A})$, and that the support of φ in $N_I(\mathbb{A})M_I(F)\backslash G(\mathbb{A})$ is compact modulo $Z_I(\mathbb{A})$; see (9.2.6)). If $\widehat{\mathcal{A}}_{I,\pi}^{I}$ is the Hilbert completion of $\mathcal{A}_{I,\pi}^{I}$ with respect to its obvious scalar product ($\widehat{\mathcal{A}}_{I,\pi}^{I}$ is the π-isotypic component of the space of L^2-functions ψ on $M_I(F)\backslash M_I(\mathbb{A})$ such that

$$\psi(z_I m_I)=\chi_\pi(z_I)\psi(m_I)$$

for every $z_I\in Z_I(\mathbb{A})$ and almost every $m_I\in M_I(\mathbb{A})$) we may identify $\widehat{\mathcal{A}}_{I,\pi}^{\Delta}$ with the space of measurable functions

$$\varphi: F_\infty^\times N_I(\mathbb{A})M_I(F)\backslash G(\mathbb{A})\to\mathbb{C}$$

such that

$$\varphi(z_I g)=\chi_\pi(z_I)\varphi(g)\qquad(\forall z_I\in Z_I(\mathbb{A}),\ g\in G(\mathbb{A})),$$

$$\int_{N_I(\mathbb{A})Z_I(\mathbb{A})M_I(F)\backslash G(\mathbb{A})}|\varphi(g)|^2\frac{dg}{dn_I dz_I d\mu_I}<+\infty$$

and

$$\varphi_k\in\widehat{\mathcal{A}}_{I,\pi}^{I}\qquad(\forall k\in K),$$

where

$$\varphi_k(m_I)=\delta_{P_I(\mathbb{A})}^{-1/2}(m_I)\varphi(m_I k)$$

for almost every $m_I\in M_I(\mathbb{A})$.

LEMMA (G.10.2). — (i) *Let $K'\subset K$ be a compact open subgroup of $G(\mathbb{A})$. Any $\varphi\in\widehat{\mathcal{A}}_{I,\pi}^{\Delta}$ which is right K'-invariant belongs to $\mathcal{A}_{I,\pi}^{\Delta}\subset\widehat{\mathcal{A}}_{I,\pi}^{\Delta}$.*

(ii) *The Hilbert space $\widehat{\mathcal{A}}_{I,\pi}^{\Delta}$ admits an orthonormal Hilbert basis $(\varphi_i)_{i\in\mathbb{Z}_{>0}}$ with $\varphi_i\in\mathcal{A}_{I,\pi}^{\Delta}$ for every i.*

Proof : For each $\varphi\in\widehat{\mathcal{A}}_{I,\pi}^{\Delta}$ let us set

$$\varphi_{K'}(g)=\int_{K'}\varphi(gk')dk'\qquad(\forall g\in G(\mathbb{A}))$$

where dk' is the normalized Haar measure on K', so that $\varphi_{K'}$ is a locally constant function on $N_I(\mathbb{A})F_\infty^\times M_I(F)\backslash G(\mathbb{A})$ which is right K'-invariant and that

$$\varphi_{K'} \in \widehat{\mathcal{A}}_{I,\pi}^\Delta$$

with

$$\langle \varphi_{K'}, \varphi_{K'} \rangle \leq \langle \varphi, \varphi \rangle \operatorname{vol}(K', dk') = \langle \varphi, \varphi \rangle$$

(Cauchy–Schwarz inequality). It is easy to see that

$$\langle \psi_{K'}, \varphi \rangle = \langle \psi, \varphi_{K'} \rangle$$

for all $\varphi, \psi \in \widehat{\mathcal{A}}_{I,\pi}^\Delta$.

In particular, if $\psi \in \widehat{\mathcal{A}}_{I,\pi}^\Delta$ is right K'-invariant and is orthogonal to $(\mathcal{A}_{I,\pi}^\Delta)^{K'}$ we have

$$\langle \psi, \varphi \rangle = \langle \psi_{K'}, \varphi \rangle = \langle \psi, \varphi_{K'} \rangle = 0$$

for every $\varphi \in \widehat{\mathcal{A}}_{I,\pi}^\Delta$, so that $\psi = 0$. It follows that $(\mathcal{A}_{I,\pi}^\Delta)^{K'}$ is dense in $(\widehat{\mathcal{A}}_{I,\pi}^\Delta)^{K'}$ and, as $(\mathcal{A}_{I,\pi}^\Delta)^{K'}$ is finite dimensional over $\mathbb{C}$, part (i) of the lemma follows.

Let

$$\cdots \subset K_{n+1} \subset K_n \subset \cdots \subset K_1 \subset K$$

be a sequence of compact open subgroups of $G(\mathbb{A})$ which form a fundamental system of neighborhoods of the unit element in $G(\mathbb{A})$. Then $\mathcal{A}_{I,\pi}^\Delta$ is the increasing union of the finite dimensional $\mathbb{C}$-vector spaces $(\mathcal{A}_{I,\pi}^\Delta)^{K_n}$, $n \in \mathbb{Z}_{>0}$. Therefore we can find an algebraic basis $(\varphi_i)_{i \in \mathbb{Z}_{>0}}$ of $\mathcal{A}_{I,\pi}^\Delta$ and a map

$$j : \mathbb{Z}_{>0} \to \mathbb{Z}_{>0}$$

such that

$$j(n) \leq j(n+1) \qquad (\forall n \in \mathbb{Z}_{>0}),$$
$$\lim_{n \to +\infty} j(n) = +\infty$$

and $(\varphi_1, \ldots, \varphi_{j(n)})$ is an orthonormal basis of $(\mathcal{A}_{I,\pi}^\Delta)^{K_n}$ with respect to the scalar product induced by $\langle \cdot, \cdot \rangle$, $\forall n \in \mathbb{Z}_{>0}$. It is clear that $(\varphi_i)_{i \in \mathbb{Z}_{>0}}$ is a Hilbert basis of $\widehat{\mathcal{A}}_{I,\pi}^\Delta$ and part (ii) of the lemma is proved too. □

Let $\widehat{\mathcal{P}}(I, \pi)$ be the Hilbert space of measurable functions

$$F : \operatorname{Im} X_I^\Delta \to \widehat{\mathcal{A}}_{I,\pi}^\Delta$$

such that, for each $\mu \in \operatorname{Fix}_{X_I^\Delta}(\pi)$, we have

$$F(\lambda + \mu) = F(\lambda)_{-\mu}$$

for almost every $\lambda \in \mathrm{Im}\, X_I^{\Delta}$ and such that

$$\|F\|^2 \stackrel{\mathrm{dfn}}{=} \frac{\deg(\infty)}{f d_1 \cdots d_s |\mathrm{Fix}_{X_I^{\Delta}}(\pi)|} \int_{\mathrm{Im}\, X_I^{\Delta}} \langle F(\lambda), F(\lambda)\rangle d\lambda < +\infty.$$

We can embed $\mathcal{P}(I,\pi)$ in $\widehat{\mathcal{P}}(I,\pi)$ by restricting the polynomial functions $\Phi \in \mathcal{P}(I,\pi)$ to $\mathrm{Im}\, X_I^{\Delta}$ and it is not difficult to see that $\widehat{\mathcal{P}}(I,\pi)$ is the Hilbert completion of $\mathcal{P}(I,\pi)$ with respect to the above norm ($L^2(\mathrm{Im}\, X_I^{\Delta}, d\lambda)$ is the Hilbert completion of $\mathcal{P}(I,\pi)|\mathrm{Im}\, X_I^{\Delta}$ and

$$\widehat{\mathcal{P}}(I,\pi) \cong \left(L^2(\mathrm{Im}\, X_I^{\Delta}, d\lambda)\widehat{\otimes}\widehat{\mathcal{A}}_{I,\pi}^{\Delta}\right)^{\mathrm{Fix}_{X_I^{\Delta}}(\pi)}).$$

THEOREM (G.10.3). — *Let c be a regular cuspidal datum and lets (I,π) be a unitary cuspidal pair in c. Then the map*

$$\mathcal{P}(I,\pi) \to L^2_{G,1_\infty,c}, \ \Phi \mapsto \theta_\Phi,$$

extends in a unique way to an isometry of $\widehat{\mathcal{P}}(I,\pi)$ onto $L^2_{G,1_\infty,c}$.

Moreover, if we let $F_\infty^\times\backslash G(\mathbb{A})$ act on $L^2_{G,1_\infty,c}$ and $\widehat{\mathcal{A}}_{I,\pi}^{\Delta}$ (and consequently on $\widehat{\mathcal{P}}(I,\pi)$) by right translation this isometry is $(F_\infty^\times\backslash G(\mathbb{A}))$-equivariant.

Proof: The theorem follows from (G.10.1). The only assertion which is not completely obvious is the surjectivity of the isometry of $\widehat{\mathcal{P}}(I,\pi)$ into $L^2_{G,1_\infty,c}$.

But let (I',π') be another unitary cuspidal pair in c and let $\Phi' \in \mathcal{P}(I',\pi')$ be such that

$$\langle \theta_\Phi, \theta_{\Phi'}\rangle = 0 \qquad (\forall \Phi \in \mathcal{P}(I,\pi)).$$

Then we can apply (G.10.1)(ii) and we get that

$$\int_{\mathrm{Im}\, X_{I'}^{\Delta}} \langle \Phi(\lambda_{w'} + w'(\lambda'))_{\lambda_{w'}}, (M_{I'}(w',\pi'(\lambda'))(\Phi'(\lambda')_{\lambda'}))_{-w'(\lambda')}\rangle_{w'} d\lambda' = 0$$

$(\forall \Phi \in \mathcal{P}(I,\pi))$ where w' is the unique element in W with $w'(I') = I$ and $w'(\pi') = \pi(\lambda_{w'})$ for some $\lambda_{w'} \in \mathrm{Im}\, X_I^{\Delta}$. It easily follows that

$$(M_{I'}(w',\pi'(\lambda'))(\Phi'(\lambda')_{\lambda'}))_{-w'(\lambda')} = 0 \qquad (\forall \lambda' \in \mathrm{Im}\, X_{I'}^{\Delta}),$$

i.e.

$$M_{I'}(w',\pi'(\lambda'))(\Phi'(\lambda')_{\lambda'}) = 0 \qquad (\forall \lambda' \in \mathrm{Im}\, X_{I'}^{\Delta}).$$

Therefore, thanks to the functional equation

$$M_I(w'^{-1}, \pi(\lambda_{w'} + w'(\lambda'))) \circ M_{I'}(w',\pi'(\lambda')) = \mathrm{id}$$

(see (G.9.1)(ii)(c) and (G.9.2)(ii)(b)), we obtain that

$$\Phi'(\lambda')_{\lambda'} = 0 \qquad (\forall \lambda' \in \mathrm{Im}\, X_{I'}^{\Delta}),$$

i.e.

$$\Phi'(\lambda') = 0 \qquad (\forall \lambda' \in \mathrm{Im}\, X_{I'}^{\Delta}),$$

so that $\Phi' = 0$. Thus $\mathcal{P}(I,\pi)$ is dense in $L^2_{G,1_\infty,c}$ as required. □

(G.11) Comments and references

For a general reductive group G over a global field F Langlands, in the number field case, and Morris, in the function field case, have given the spectral decomposition of $L^2(G(F)\backslash G(\mathbb{A}))$ in terms of the discrete spectrum of the Levi subgroups. In the case where G is the general linear group, Moeglin and Waldspurger have explicitly described the discrete spectrum in terms of the cuspidal spectrum. It follows that the spectral decomposition of $L^2(GL_d(F)\backslash GL_d(\mathbb{A}))$ can be given in terms of the cuspidal spectrum of the Levi subgroups $GL_{d_1} \times \cdots \times GL_{d_s}$ (see [Mo–Wa 1] (III.3)).

References

[Ar 1] J.G. ARTHUR. — A trace formula for reductive groups I: terms associated to classes in $G(\mathbb{Q})$, *Duke Math. J.* **45**, (1978), 911–953.

[Ar 2] J. ARTHUR. — The trace formula in invariant form, *Ann. of Math.* **114**, (1981), 1–74.

[Ar 3] J. ARTHUR. — A measure on the unipotent variety, *Can. J. Math.* **37**, (1985), 1237–1274.

[Ar 4] J. ARTHUR. — On a family of distributions obtained from orbits, *Can. J. Math.* **38**, (1986), 179–214.

[Ar 5] J. ARTHUR. — The local behaviour of weighted orbital integrals, *Duke Math. J.* **56**, (1988), 223–293.

[Ar 6] J. ARTHUR. — A trace formula for reductive groups II: applications of a truncation operator, *Comp. Math.* **40**, (1980), 87–121.

[Ar 7] J. ARTHUR. — Eisenstein series and the trace formula, *Proc. Sym. Pure Math.* **33, Part 1**, (1979), 253–274.

[Ar 8] J. ARTHUR. — A local trace formula, *Publ. Math. IHES* **73**, (1991), 5–96.

[Ar 9] J. ARTHUR. — Intertwining operators and residues I. Weighted characters, *J. Funct. Anal.* **84**, (1989), 19–84.

[Ar 10] J. ARTHUR. — On a family of distribution obtained from Eisenstein series, I: Application of the Paley–Wiener theorem, *Amer. J. Math.* **104**, (1982), 1243–1288.

[Ar 11] J. ARTHUR. — On a family of distribution obtained from Eisenstein series, II: Explicit formulas, *Amer. J. Math.* **104**, (1982), 1289–1336.

[Ar 12] J. ARTHUR. — The invariant trace formula. I. Local theory, *J. Amer. Math.* **1**, (1988), 323–383.

[Ar 13] J. ARTHUR. — The invariant trace formula. II. Global theory, *J. Amer. Math.* **1**, (1988), 501–554.

[Ar 14] J. ARTHUR. — The L^2-Lefschetz numbers for Hecke operators, *Invent. Math.* **97**, (1981), 257–290.

[Be–Be–De] A.A. BEILINSON, J. BERNSTEIN, P. DELIGNE. — Faisceaux pervers, in *Analyse et topologie sur les espaces singuliers*, Astérisque, 100, (1982).

[Bo–Ja] A. BOREL, H. JACQUET. — Automorphic forms and automorphic representations, *Proc. Sym. in Pure Math.* **33, Part 1**, (1979), 189–202.

[Bo 2] A. BOREL. — *Linear algebraic groups*, Benjamin, New York, 1969.

[Cas 3] W. CASSELMAN. — Characters and Jacquet modules, *Math. Ann.* **230**, (1977), 101–105.

[Cl] L. CLOZEL. — On the cohomology of Kottwitz's arithmetic varieties, *Duke Math. J.* **72**, (1993), 757–795.

[De] P. DELIGNE. — La conjecture de Weil. I, *Publ. Math. IHES* **52**, (1980), 137–252.

[De–Lu] P. DELIGNE, G. LUSZTIG. — Representations of reductive groups over finite fields, *Ann. of Math.* **103**, (1976), 103–161.

[Dr 4] V.G. DRINFEL'D. — Cohomology of compactified manifolds of modules of F-sheaves of rank 2, *J. Soviet Math.* **46**, (1989), 1789–1821.

[Ek] T. EKEDAHL. — On the adic formalism, in *The Grothendieck Festschrift, Vol. II*, Birkhäuser, (1990), 197–218.

[Fr] J. FRANKE. — *Harmonic analysis in weighted L_2-spaces*,Preprint, 1990.

[Fu] K. FUJIWARA. — *Rigid geometry, Lefschetz–Verdier trace formula and Deligne's conjecture*, Preprint, 1993.

[Ge–PS] I.M. GELFAND, I. PIATETSKI-SHAPIRO. — Automorphic functions and representation theory, *Trans. Moscow Math. Soc.* **12**, (1963), 438-464.

[Gek] E.U. GEKELER. — Compactification du schéma des modules de Drinfel'd, in *Séminaire de théorie des nombres* 1985-86, Univesité de Bordeaux I, Talence.

[Gro 2] A. GROTHENDIECK. — Formule des traces de Lefschetz et rationalité des fonctions L, Séminaire Bourbaki 1964/65, in *Dix exposés sur la cohomologie des schémas*, North-Holland, (1968), 31–45.

[Har 1] G. HARDER. — Minkowskische Reduktionstheorie über Funktionenkörpern, *Invent. Math.* **7**, (1969), 33–54.

[Har 2] G. HARDER. — Chevalley groups over function fields and automorphic forms, *Ann. of Math.* **100**, (1974), 249–306.

[Har 3] G. HARDER. — Die Kohomologie S-arithmetischer Gruppen über Funktionenkörpern, *Invent. math.* **42**, (1977), 135–175.

[Ha-Na] G. HARDER, M.S. NARASIMHAN. — On the cohomology groups of moduli spaces of vector bundles on curves, *Math. Ann.* **212**, (1975), 215–248.

[H-C 1] HARISH-CHANDRA. — *Automorphic forms on semi-simple Lie groups*, Lecture Notes in Math. 62, Springer-Verlag, 1968.

[H-C 2] HARISH-CHANDRA. — Harmonic analysis on reductive p-adic groups, in *Harmonic analysis on homogeneous spaces, Proc. Sym. Pure Math.* **26**, (1973), 167–192.

[He] G. HENNIART. — On the local Langlands conjecture for $GL(n)$: the cyclic case, *Ann. of Math.* **123**, (1986), 145–203.

[Ja–Sh 1] H. JACQUET, J.A. SHALIKA. — On the Euler product and the classification of automorphic representations I, *Amer. J. Math.* **103**, (1981), 499–558.

[Ja–Sh 2] H. JACQUET, J.A. SHALIKA. — On the Euler product and the classification of automorphic representations II, *Amer. J. Math.* **103**, (1981), 777–815.

[Kap] M.M. KAPRANOV. — On cuspidal divisors on the modular varieties of elliptic modules, *Math. USSR Izvestiya* **30**, (1988), 533–547.

[Kl] S.L. KLEIMAN. — Algebraic cycles and the Weil conjectures, in *Dix exposés sur la cohomologie des schémas*, North-Holland, (1968), 359–386.

[La 2] J.-P. LABESSE. — La formule des traces d'Arthur–Selberg, in *Séminaire Bourbaki* 1984-85, Astérisque, 133–134, (1986), 73–88.

[Laf 1] L. LAFFORGUE. — *$\mathcal{D}$-stukas de Drinfeld*, Thèse, Université de Paris-Sud, 1994.

[Laf 2] L. LAFFORGUE. — *Formule des traces d'Arthur–Selberg sur les corps de fonctions et conjecture de Ramanujan–Petersson*, Prépublication Université de Paris-Sud, 1995.

[Lan 3] R. LANGLANDS. — *On the functional equations satisfied by Eisenstein series*, Lecture Notes in Math. 544, Springer-Verlag, 1976.

[Mo–Wa 1] C. MOEGLIN, J.-L. WALDSPURGER. — Le spectre résiduel de $Gl(n)$, *Ann. Scient. Ec. Norm. Sup.* **22**, (1989), 605–674.

[Mo–Wa 2] C. MOEGLIN, J.-L. WALDSPURGER. — *Décomposition spectrale et séries d'Eisenstein, Une paraphrase de l'Ecriture*, Birkhäuser, 1994.

[Moo] C.C. MOORE. — Group extensions of p-adic and adelic linear groups, *Publ. Math. IHES* **35**, (1968), 5–70.

[Mor 1] L.E. MORRIS. — Eisenstein series for reductive groups over global function fields I: the cusp form case, *Can. J. Math.* **34**, (1982), 91–168.

[Mor 2] L.E. MORRIS. — Eisenstein series for reductive groups over global function fields II: the general case, *Can. J. Math.* **34**, (1982), 1112–1182.

[Pi 1] R. PINK. — On the calculation of local terms in the Lefschetz–Verdier trace formula and its application to a conjecture of Deligne, *Ann. of Math.* **135**, (1992), 483–525.

[Pi 2] R. PINK. — *Smooth compactifications of Drinfeld moduli schemes*, Preprint, 1995.

[Se 4] J.-P. SERRE. — *Abelian l-adic representations and elliptic curves*, Benjamin, New York, 1968.

[Shah] F. SHAHIDI. — On certain L-functions, *Amer. J. Math.* **103**, (1980), 297–355.

[Sha 1] J.A. SHALIKA. — The multiplicity one theorem for Gl_n, *Ann. of Math.* **100**, (1974), 171–193.

[Sha 2] J.A. SHALIKA. — A theorem on semi-simple $\mathcal{P}$-adic groups, *Ann. of Math.* **95**, (1972), 226–242.

[St] U. STUHLER. — On the cohomology of SL_n over rings of algebraic functions, in *Algebraic K-theory, Part II*, Lecture Notes in Math. 967, Springer-Verlag, (1982), 316–359.

[Ta] J.T. TATE. — Fourier analysis in number fields and Hecke's zeta functions, in *Algebraic number theory* (J.W.S. Cassels and A. Fröhlich Editors), Academic Press, (1967), 305–347.

[Ve] J.-L. VERDIER. — The Lefschetz fixed point formula in étale cohomology, in *Proceedings of a Conference on Local Fields*, Springer-Verlag, (1967), 199–214.

[Wa] J.-L. WALDSPURGER. — Intégrales orbitales sphériques pour $GL(N)$ sur un corps p-adique, in *Orbites unipotentes et représentations II. Groupes p-adiques et réels*, Astérisque, 171–172, (1989), 279–337.

[Wal] N. Wallach. — On the constant term of a square-integrable automorphic form, in *Operator algebras and group representations, Vol. II*, (1984), 227–237.

[We 4] A. Weil. — *Courbes algébriques et variétés abéliennes*, Hermann, 1971.

[Zi 2] T. Zink. — The Lefschetz trace formula for an open algebraic surface, in *Automorphic forms, Shimura varieties, and L-functions, Vol. II*, Academic Press, (1990), 337–376.

[SGA4] M. Artin, A. Grothendieck, J.-L. Verdier. — *Théorie des topos et cohomologie étale des schémas*, Lecture Notes in Math. 269, 270, 305, Springer-Verlag, 1972/73.

[SGA4$\frac{1}{2}$] P. Deligne. — *Cohomologie étale*, Lecture Notes in Math. 569, Springer-Verlag, 1977.

[SGA5] A. Grothendieck. — *Cohomologie l-adique et fonctions L*, Lecture Notes in Math. 589, Springer-Verlag, 1977.

Some residue computations

J.-L. Waldspurger
Université Paris 7, CNRS

written by G. Laumon

1. Notations

Let F be a non-archimedean local field and d be a positive integer. We denote by $\mathcal{O}$ the ring of integers of F, by q the number of elements of the residue field of F and by $|\ |$ the normalized absolute value of F. We consider the group $G = GL_d(F)$, its maximal torus T of diagonal matrices and its maximal compact subgroup $K = GL_d(\mathcal{O})$. We normalize the Haar measures in such a way that $\mathrm{vol}(K) = 1$ and $\mathrm{vol}(N \cap K) = 1$ for all the unipotent subgroups N which will occur. We denote by $\mathcal{L}$ the set of the Levi subgroups containing T. For each $M \in \mathcal{L}$ we denote by $\mathcal{P}(M)$ the set of parabolic subgroups $P = MN$ which admit M as a Levi component. We will freely use the standard notations A_M, $R(G, A_M)$, Δ_P, $\alpha >_P 0$, $\alpha^\vee$, ρ_P, $\mathcal{P}^L(M)$, ...

Let $B \in \mathcal{P}(T)$ be the standard Borel subgroup of upper triangular matrices. We denote by $\mathcal{B}$ the corresponding Iwahori subgroup, i.e. $\mathcal{B} = K_1(B \cap K)$ where K_1 is the kernel of the projection $K \twoheadrightarrow GL_d(\mathbb{F}_q)$.

2. Very cuspidal functions

Let $f \in C_c^\infty(G)$. We say that f is very cuspidal if, for every $M \in \mathcal{L} - \{G\}$, every $P \in \mathcal{P}(M)$ and every $m \in M$, we have

$$\int_{K \times N} f(k^{-1} mnk) dndk = 0.$$

REMARK. — As any parabolic subgroup is conjugate to a parabolic subgroup containing B by an element of K we may restrict ourself in the above definition to the pairs (M, P) such that $P \supset B$.

3. Intertwining operators

For any reductive group H over F let us set $\mathfrak{a}_H^* = X^*(H) \otimes \mathbb{C}$, $\mathrm{Re}\,\mathfrak{a}_H^* = X^*(H) \otimes \mathbb{R}$ and $\mathrm{Im}\,\mathfrak{a}_H^* = i\mathrm{Re}\,\mathfrak{a}_H^* \subset \mathfrak{a}_H^*$ (these notations are different from Arthur's ones: he sets $\mathfrak{a}_H^* = X^*(H) \otimes \mathbb{R}$). For $\lambda \in \mathfrak{a}_H^*$ we also denote by λ the corresponding character of the group H: if $\lambda = x \otimes s$ with $x \in X^*(H)$ and $s \in \mathbb{C}$ this is the character $h \mapsto h^\lambda := |x(h)|^s$. Let us set $\mathfrak{a}_H^\vee = \{\lambda \in \mathfrak{a}_H^* \mid h^\lambda = 1, \forall h \in H\}$. By definition a polynomial function on $\mathfrak{a}_H^*/\mathfrak{a}_H^\vee$ is a linear combination of functions $\lambda \mapsto h^\lambda$ with $h \in H$. A rational function is a quotient of two polynomial functions.

Let $M \in \mathcal{L}$ and let (π, V) be an admissible representation of M of finite length. For $\lambda \in \mathfrak{a}_M^*/\mathfrak{a}_M^\vee$ and $P = MN \in \mathcal{P}(M)$ we may realize the induced representation $I_P^G(\pi \otimes \lambda)$ on the space

$$I_P^G V = \{\varphi : K \to V \mid f \text{ is locally constant and } f(mnk) = \pi(m)f(k), \\ \forall m \in M \cap K,\ n \in N \cap K,\ k \in K\}.$$

Following Harish-Chandra, for any $P, P' \in \mathcal{P}(M)$, we define the intertwining operator

$$M_{P'|P}(\pi \otimes \lambda) : I_P^G V \to I_{P'}^G V$$

by

$$M_{P'|P}(\pi \otimes \lambda)(\varphi)(g) = \int_{N \cap N' \backslash N'} \varphi(n'g)dn'.$$

This integral is convergent as long as $\langle \mathrm{Re}\,\lambda, \alpha^\vee \rangle > r$, $\forall \alpha \in R(G, A_M)$ such that $\alpha >_P 0$ and $\alpha <_{P'} 0$, where r is a sufficiently large positive real number depending only on π. Moreover it extends to a rational function of λ on the whole of $\mathfrak{a}_M^*/\mathfrak{a}_M^\vee$.

Let us assume moreover that π is irreducible. Then, following Langlands (see [Ar]), it is possible to define normalized intertwining operators

$$R_{P'|P}(\pi \otimes \lambda) = r_{P'|P}(\pi \otimes \lambda)^{-1} M_{P'|P}(\pi \otimes \lambda),$$

which are also rational functions of λ, in such a way that

$$R_{P''|P}(\pi \otimes \lambda) = R_{P''|P'}(\pi \otimes \lambda) \circ R_{P'|P}(\pi \otimes \lambda)$$

for any $P, P', P'' \in \mathcal{P}(M)$.

4. Logarithmic derivatives of intertwining operators

Let $M \in \mathcal{L}$, $P_0 \in \mathcal{P}(M)$, $\lambda \in \mathfrak{a}_M^*/\mathfrak{a}_M^\vee$ and let π be an admissible irreducible representation of M. For any $P \in \mathcal{P}(M)$ and any $\zeta \in \mathrm{Im}\,\mathfrak{a}_M^*$ we set

$$\mathcal{R}_P(\pi \otimes \lambda, \zeta; P_0) = R_{P_0|P}(\pi \otimes \lambda) \circ R_{P|P_0}(\pi \otimes (\lambda + \zeta)).$$

This expression is well-defined for any λ in a dense open subset of $\mathfrak{a}_M^*$ and the family

$$\mathcal{R}(\pi \otimes \lambda, \zeta; P_0) = \big(\mathcal{R}_P(\pi \otimes \lambda, \zeta; P_0) \mid P \in \mathcal{P}(M)\big)$$

is a (G, M)-family in the sense of Arthur. For any $P \in \mathcal{P}(M)$ and any $\zeta \in \operatorname{Im} \mathfrak{a}_M^*$ we set

$$\theta_P(\zeta) = (\log q)^{|\Delta_P|} \prod_{\alpha \in \Delta_P} \langle \zeta, \alpha^\vee \rangle.$$

Then we set

$$\mathcal{R}_M(\pi \otimes \lambda; P_0) = \lim_{\zeta \to 0} \sum_{P \in \mathcal{P}(M)} \frac{\mathcal{R}_P(\pi \otimes \lambda, \zeta; P_0)}{\theta_P(\zeta)}.$$

For any $f \in C_c^\infty(G)$ it is known that $\operatorname{tr}(\mathcal{R}_M(\pi \otimes \lambda; P_0) \circ I_{P_0}^G(\pi \otimes \lambda, f))$ is independent of P_0.

LEMMA. — *Let L be a Levi supgroup in $\mathcal{L}$ which contains M, let R be a parabolic subgroup of G which contains L and let M_R be the unique Levi subgroup of R which contains T. Following Arthur, from the (G, M)-family $\big(\mathcal{R}_P(\pi \otimes \lambda, \zeta; P_0) \mid P \in \mathcal{P}(M)\big)$ we get an (M_R, L)-family $\big(\mathcal{R}_Q^R(\pi \otimes \lambda, \zeta; P_0) \mid Q \in \mathcal{P}(L),\ Q \subset R\big)$. Let f be a very cuspidal function in C_c^∞. Then if $(R, L) \neq (G, M)$ we have*

$$\operatorname{tr}\big(\mathcal{R}_L^R(\pi \otimes \lambda; P_0) \circ I_{P_0}^G(\pi \otimes \lambda, f)\big) = 0.$$

Proof : This trace is independent of P_0, so that we may assume that $P_0 \subset R$.

Let us first consider the case $L = M$, $R \subsetneqq G$. If V is the space of π the operator $\mathcal{R}_M^R(\pi \otimes \lambda; P_0) \in \operatorname{End} I_{P_0}^G V$ is induced from the operator $\mathcal{R}_M^{M_R}(\pi \otimes \lambda; P_0 \cap M_R) \in \operatorname{End} I_{P_0 \cap M_R}^{M_R} V$ and if we set

$$g_{k,k'}(m_R) = \int_{N_R} f(k^{-1} m_R n_R k') dn_R$$

the operator $\mathcal{R}_M^R(\pi \otimes \lambda; P_0) \circ I_{P_0}^G(\pi \otimes \lambda, f)$ has kernel

$$(k, k') \mapsto \mathcal{R}_M^{M_R}(\pi \otimes \lambda; P_0 \cap M_R) \circ I_{P_0 \cap M_R}^{M_R}(\pi \otimes \lambda, g_{k,k'}).$$

But the integral of $g_{k,k}$ over K vanishes identically by the very-cuspidality of f and the lemma is proved in that case.

Now let us consider the case $L \supsetneqq M$. Following Arthur, for any (G, M)-family c, we have the descent formula

$$c_L^R(\zeta) = \sum_{\substack{L' \in \mathcal{L} \\ M \subset L' \subset M_R}} d_M^R(L, L') c_M^{Q_{L'}}(\zeta),$$

where $L' \mapsto Q_{L'} \in \mathcal{P}(L')$ is a certain section and the $d_M^R(L, L')$ are certain constants. Moreover we have

$$d_M^R(L, L') \neq 0 \Rightarrow \mathfrak{a}_M^{M_R} = \mathfrak{a}_L^{M_R} \oplus \mathfrak{a}_{L'}^{M_R}.$$

In our case, as $L \neq M$ we have $d_M^R(L, L') \neq 0$ only if $Q_{L'} \subsetneqq R \subset G$. But we have already proved that

$$\operatorname{tr}\big(\mathcal{R}_M^{Q_{L'}}(\pi \otimes \lambda; P_0) \circ I_{P_0}^G(\pi \otimes \lambda, f)\big) = 0$$

for any $Q_{L'} \subsetneqq G$. Therefore we have

$$d_M^R(L, L') \operatorname{tr}\big(\mathcal{R}_M^{Q_{L'}}(\pi \otimes \lambda; P_0) \circ I_{P_0}^G(\pi \otimes \lambda, f)\big) = 0$$

for every $L' \in \mathcal{L}$ such that $M \subset L' \subset M_R$ and Arthur's descent formula implies the required vanishing. □

Corollary. — *Let $(c_P \mid P \in \mathcal{P}(M))$ be a (G, M)-family. Then under the hypotheses of the lemma we have*

$$\operatorname{tr}\big((c\mathcal{R}(\pi \otimes \lambda; P_0))_M \circ I_{P_0}^G(\pi \otimes \lambda, f)\big) = c_{P_0}(0) \operatorname{tr}\big(\mathcal{R}_M(\pi \otimes \lambda; P_0) \circ I_{P_0}^G(\pi \otimes \lambda, f)\big).$$

Proof: Following Arthur, if c and d are two (G, M)-families we have the splitting formula

$$(cd)_M(\zeta) = \sum_{L, L' \in \mathcal{L}(M)} d_M^G(L, L') c_M^{Q_L}(\zeta) d_M^{Q_{L'}}(\zeta)$$

where $(L, L') \mapsto (Q_L, Q_{L'}) \in \mathcal{P}(L) \times \mathcal{P}(L')$ is a certain section and the $d_M^G(L, L')$ are certain constants. □

Remark. — It follows from this corollary that, for any very cuspidal function $f \in \mathcal{C}_c^\infty(G)$, $\operatorname{tr}\big(\mathcal{R}_M(\pi \otimes \lambda; P_0) \circ I_{P_0}^G(\pi \otimes \lambda, f)\big)$ does not depend on the way that we have normalized the intertwining operators.

If π is unitary it is known that $\mathcal{R}_P(\pi \otimes \lambda, \zeta; P_0)$ and $\mathcal{R}_M(\pi \otimes \lambda; P_0)$ are well-defined for every $\lambda \in \operatorname{Im} \mathfrak{a}_M^*$.

5. The theorem

Let $M \in \mathcal{L}$. Let us denote by B_M^* the basis of $X^*(M)$, and therefore of $\mathfrak{a}_M^*$, given by the determinants of the factors of $M \cong GL_{d_1}(F) \times \cdots \times GL_{d_s}(F)$ and let us denote by B_M the dual basis of $\mathfrak{a}_M$.

For each unitary admissible irreducible representation π of M (resp. $F^\times \backslash M$), each $f \in C_c^\infty(G)$ (resp. $f \in C_c^\infty(F^\times \backslash G)$), each holomorphic function $X : \mathfrak{a}_M^*/\mathfrak{a}_M^\vee \to \mathbb{C}$, each $b \in B_M$ and each integer n we set

$$Z_M(\pi, f; X(\lambda), b, n)$$
$$= \int_{\mathrm{Im}\,\mathfrak{a}_M^*/\mathfrak{a}_M^\vee} \mathrm{tr}\big(\mathcal{R}_M(\pi \otimes \lambda; P_0) \circ I_{P_0}^G(\pi \otimes \lambda, f)\big) X(\lambda) q^{-n\langle \lambda, b\rangle} d\lambda$$

(resp.

$$Z_M^G(\pi, f; X(\lambda), b, n)$$
$$= \int_{\mathrm{Im}\,\mathfrak{a}_M^{G*}/(\mathfrak{a}_M^\vee \cap \mathrm{Im}\,\mathfrak{a}_M^{G*})} \mathrm{tr}\big(\mathcal{R}_M(\pi \otimes \lambda; P_0) \circ I_{P_0}^G(\pi \otimes \lambda, f)\big) X(\lambda) q^{-n\langle \lambda, b\rangle} d\lambda\,)$$

where the Haar measure $d\lambda$ is normalized by

$$\mathrm{vol}(\mathrm{Im}\,\mathfrak{a}_M^*/\mathfrak{a}_M^\vee, d\lambda) = 1 \quad (\text{resp. } \mathrm{vol}(\mathrm{Im}\,\mathfrak{a}_M^{G*}/(\mathfrak{a}_M^\vee \cap \mathrm{Im}\,\mathfrak{a}_M^{G*}), d\lambda) = 1\,).$$

We say that a function $X : \mathfrak{a}_M^*/\mathfrak{a}_M^\vee \to \mathbb{C}$ has polynomial growth if there exists a finite set $\{X_1, \ldots, X_n\}$ of polynomial functions on $\mathfrak{a}_M^*/\mathfrak{a}_M^\vee$ such that

$$|X(\lambda)| \leq \sup\{|X_i(\lambda)|,\ i = 1, \ldots, n\}.$$

For each $P \in \mathcal{P}(M)$ we denote by $\overline{\mathcal{C}}_P$ the cone in $\mathrm{Re}\,\mathfrak{a}_M$ generated by $\mathrm{Re}\,\mathfrak{a}_G$ and the coroots $\alpha^\vee$ with $\alpha \in \Delta_P$.

We denote by St^M the Steinberg representation of M. If $M \cong GL_{d_1}(F) \times \cdots \times GL_{d_s}(F)$ we have $\mathrm{St}^M = \mathrm{St}^{GL_{d_1}(F)} \otimes \cdots \otimes \mathrm{St}^{GL_{d_s}(F)}$.

THEOREM. — *Let $M \in \mathcal{L}$, let f be a very cuspidal function in $C_c^\infty(G)$ (resp. $C_c^\infty(F^\times \backslash G)$), let $X : \mathfrak{a}_M^*/\mathfrak{a}_M^\vee \to \mathbb{C}$ be a holomorphic function with polynomial growth and let $b \in B_M$.*

(i) *There exists an integer n_0 such that, for every $n \geq n_0$, we have the equality*

$$Z_M(\mathrm{St}^M, f; X(\lambda), b, n) = \sum_{\substack{P \in \mathcal{P}(M) \\ b \in \overline{\mathcal{C}}_P}} Z_G(\mathrm{St}^G, f; X(\lambda + \rho_P), b, n) q^{-n\langle \rho_P, b\rangle}$$

with

$$Z_G(\mathrm{St}^G, f; X(\lambda), b, n) = \int_{\mathrm{Im}\,\mathfrak{a}_G^*/\mathfrak{a}_G^\vee} \mathrm{tr}(\mathrm{St}^G \otimes \lambda, f) X(\lambda) q^{-n\langle \lambda, b\rangle} d\lambda$$

(*resp.*

$$\begin{aligned} &Z_M^G(\mathrm{St}^M, f; X(\lambda), b, n) \\ &\quad = \frac{d_0}{d} \sum_{\lambda \in (\frac{d_0}{d})\mathfrak{a}_G^\vee/\mathfrak{a}_G^\vee} \mathrm{tr}(\mathrm{St}^G \otimes \lambda, f) \sum_{\substack{P \in \mathcal{P}(M) \\ b \in \overline{\mathcal{C}}_P}} X(\lambda + \rho_P) q^{-n\langle \lambda + \rho_P, b\rangle} \end{aligned}$$

where d_0 *is the g.c.d. of* $(d_1, \ldots, d_s)$ *if* $M \cong GL_{d_1}(F) \times \cdots \times GL_{d_s}(F)$).

(ii) *Let* $\pi = \pi_1 \otimes \cdots \otimes \pi_s$ *be a unitary admissible representation of* $M \cong GL_{d_1}(F) \times \cdots \times GL_{d_s}(F)$ (*resp.* $F^\times \backslash M \cong F^\times \backslash (GL_{d_1}(F) \times \cdots \times GL_{d_s}(F))$). *Let us assume that, for each* $j = 1, \ldots, s$, π_j *is isomorphic to the Steinberg representation or to the trivial representation of* GL_{d_j}. *Then, for every integer* n, *we have*

$$\epsilon(\pi) Z_M(\pi, f; X(\lambda), b, n) = \epsilon(\mathrm{St}^M) Z_M(\mathrm{St}^M, f; X(\lambda), b, n)$$

(*resp.*

$$\epsilon(\pi) Z_M^G(\pi, f; X(\lambda), b, n) = \epsilon(\mathrm{St}^M) Z_M^G(\mathrm{St}^M, f; X(\lambda), b, n))$$

where $\epsilon(\pi) = \prod_{j=1}^s \epsilon(\pi_j)$ *and* $\epsilon(\pi_j) = (-1)^{d_j - 1}$ *if* π_j *is the Steinberg representation of* $GL_{d_j}(F)$ *and* $\epsilon(\pi_j) = 1$ *if* π_j *is the trivial representation of* $GL_{d_j}(F)$.

(iii) *Let us assume that our very cuspidal function* $f \in C_c^\infty(F^\times \backslash G)$ *is bi-ivariant under the Iwahori subgroup* $\mathcal{B}$. *Let* $\pi = \pi_1 \otimes \cdots \otimes \pi_s$ *be a unitary admissible irreducible representation of* $F^\times \backslash M \cong F^\times \backslash (GL_{d_1}(F) \times \cdots \times GL_{d_s}(F))$. *Let us assume that there exists at least one* $j = 1, \ldots, s$ *such that, for any* $\mu_j \in \mathrm{Im}\,\mathfrak{a}_{GL_{d_j}(F)}^*$, $\pi_j \otimes \mu_j$ *is isomorphic neither to the Steinberg representation nor to the trivial representation of* GL_{d_j}. *Then for every integer* n *we have*

$$Z_M^G(\pi, f; X(\lambda), b, n) = 0.$$

REMARKS. — (i) If b corresponds to the block $GL_{d_j}(F)$ of M we have $\langle \rho_P, b \rangle = (d - d_j)/2$ for each $P \in \mathcal{P}(M)$ such that $b \in \overline{\mathcal{C}}_P$.

(ii) For any $\mu \in \mathrm{Im}\,\mathfrak{a}_M^*$ we have

$$Z_M(\pi \otimes \mu, f; X(\lambda), b, n) = q^{n\langle \mu, b\rangle} Z_M(\pi, f; X(\lambda - \mu), b, n).$$

(iii) The assertions "resp." of parts (i) and (ii) of the theorem easily follow from the others. Indeed we have the splittings

$$\int_{\mathrm{Im}\,\mathfrak{a}_M^*/\mathfrak{a}_M^\vee} = \int_{\mathrm{Im}\,\mathfrak{a}_G^*/[(\mathfrak{a}_M^{G*} + \mathfrak{a}_M^\vee) \cap \mathrm{Im}\,\mathfrak{a}_G^*]} \int_{\mathrm{Im}\,\mathfrak{a}_M^{G*}/(\mathfrak{a}_M^\vee \cap \mathrm{Im}\,\mathfrak{a}_M^{G*})}$$

and

$$\int_{\mathrm{Im}\,\mathfrak{a}_G^*/\mathfrak{a}_G^\vee} = \int_{\mathrm{Im}\,\mathfrak{a}_G^*/[(\mathfrak{a}_M^{G*}+\mathfrak{a}_M^\vee)\cap \mathrm{Im}\,\mathfrak{a}_G^*]} \int_{[(\mathfrak{a}_M^{G*}+\mathfrak{a}_M^\vee)\cap \mathrm{Im}\,\mathfrak{a}_G^*]/\mathfrak{a}_G^*} ;$$

by letting $X(\lambda)$ vary we can get rid of the first integrals; finally $[(\mathfrak{a}_M^{G*}+\mathfrak{a}_M^\vee)\cap \mathrm{Im}\,\mathfrak{a}_G^*]$ is equal to $(\frac{d_0}{d})\mathfrak{a}_G^\vee$.

Part (i) (resp. (ii), (iii)) of the theorem will be proved in sectins 6 to 9 (resp. 10, 11).

6. Proof of part (i) of theorem 5: reductions

We may assume that M is the diagonal Levi subgroup $GL_{d_1}(F) \times \ldots \times GL_{d_s}(F)$, that P_0 is the upper triangular element in $\mathcal{P}(M)$, i.e. $P_0 \supset B$, and that b corresponds to the first factor $GL_{d_1}(F)$. We then identify $\mathfrak{a}_M^*/\mathfrak{a}_M^\vee$ with

$$\bigoplus_{j=1}^{s}(\mathbb{C}/\frac{2\pi i}{d_j \log q}\mathbb{Z})$$

(map $z_j \in \mathbb{C}$ onto $\det \otimes (z_j/d_j) \in \mathfrak{a}_{GL_{d_j}(F)}^*$). For each $j = 2,\ldots,s$ let us denote by $\alpha_{1,j}$ the root $(1,0,\ldots,0,-1,0,\ldots,0) \in R(G,A_M)$ where -1 is the j-th entry.

For each $\alpha \in R(G,A_M)$ let M_α be the unique Levi subgroup of G containing M such that $R(M_\alpha,A_M) = \{\pm\alpha\}$, let R_α be the unique parabolic subgroup in $\mathcal{P}^{M_\alpha}(M)$ such that $\alpha >_{R_\alpha} 0$ and let $\rho_\alpha = \rho_{R_\alpha} \in \mathfrak{a}_M^{M_\alpha *}$.

PROPOSITION. — *In order to prove part* (i) *of theorem* 5 *it is sufficient to prove that there exists an integer n_1 such that, for every $n \geq n_1$, we have the equality*

(∗)

$$Z_M(\mathrm{St}^M, f; X(\lambda), b, n) = \sum_{\substack{\alpha=\alpha_{1,j}\\ j=2,\ldots,s}} Z_{M_\alpha}(\mathrm{St}^{M_\alpha}, f; X(\lambda+\rho_\alpha), b_\alpha, n) q^{-n\langle\rho_\alpha, b\rangle}$$

where b_α is the image of b under the natural map $\mathfrak{a}_M \to \mathfrak{a}_{M_\alpha}$.

Proof: We argue by induction on $\dim \mathfrak{a}_M^G$. If this dimension is equal to 0 part (i) of theorem 5 is trivial. If this dimension is at least 1 the right hand side of the equality (∗) may be computed by induction. We obtain that there exists an integer $n_0 \geq n_1$ such that, for every $n \geq n_0$, we have

$$Z_M(\mathrm{St}^M, f; X(\lambda), b, n)$$
$$= \sum_{\substack{\alpha=\alpha_{1,j}\\ j=2,\ldots,s}} \sum_{\substack{Q\in\mathcal{P}(M_\alpha)\\ b_\alpha\in\overline{\mathcal{C}}_Q}} Z_G(\mathrm{St}^G, f; X(\lambda+\rho_Q+\rho_\alpha), b_\alpha, n) q^{-n\langle\rho_Q, b_\alpha\rangle - n\langle\rho_\alpha, b\rangle}.$$

But the map $Q = M_\alpha N_Q \mapsto P_Q = (M_\alpha \cap P_0)N_Q$ is a bijection from

$$\coprod_{\substack{\alpha=\alpha_{1,j}\\ j=2,\dots,s}} \{Q \in \mathcal{P}(M_\alpha) \mid b_\alpha \in \overline{\mathcal{C}}_Q\}$$

onto

$$\{P \in \mathcal{P}(M) \mid b \in \overline{\mathcal{C}}_P\}.$$

Moreover this bijection has the property that $\rho_{P_Q} = \rho_Q + \rho_\alpha$ in $\mathfrak{a}_M^* = \mathfrak{a}_{M_\alpha}^* \oplus \mathfrak{a}_M^{M_\alpha *}$ and the equality in part (i) of the theorem follows. □

7. Proof of part (i) of theorem 5: moving the contour of integration

In this section we are going to prove the equality (∗) of proposition 6 by using the Cauchy formula to move the contour of integration of $Z_M(\mathrm{St}^M, f; X(\lambda), b, n)$ from the origin to "infinity". This will conclude the proof of part (i) of theorem 5.

We need a precise description of the poles (and their residues) of the rational function $\mathrm{tr}\big(\mathcal{R}_M(\mathrm{St}^M \otimes \lambda; P_0) \circ I_{P_0}^G(\mathrm{St}^M \otimes \lambda, f)\big)$.

Proposition. — *Let $M \in \mathcal{L}$, $P_0 \in \mathcal{P}(M)$ and let f be a very cuspidal function in $C_c^\infty(G)$.*

(i) *The singularities of the rational function $\mathrm{tr}\big(\mathcal{R}_M(\mathrm{St}^M \otimes \lambda; P_0) \circ I_{P_0}^G(\mathrm{St}^M \otimes \lambda, f)\big)$ of $\lambda \in \mathfrak{a}_M^*/\mathfrak{a}_M^\vee$ are contained in*

$$\bigcup_{\alpha \in R(G, A_M)} H_\alpha$$

where H_α is the "hyperplane" $\{\lambda \mid q^{\langle \lambda - \rho_\alpha, \alpha^\vee \rangle} = 1\}$.

(ii) *Let $\lambda \in H_\alpha$ be such that $\lambda \notin H_\beta$ for every $\beta \neq \alpha$. Then the limit*

$$\lim_{\lambda' \to \lambda} (1 - q^{\langle \lambda' - \rho_\alpha, \alpha^\vee \rangle})\, \mathrm{tr}\big(\mathcal{R}_M(\mathrm{St}^M \otimes \lambda'; P_0) \circ I_{P_0}^G(\mathrm{St}^M \otimes \lambda', f)\big)$$

exists and is equal to

$$\mathrm{tr}\big(\mathcal{R}_{M_\alpha}(\mathrm{St}^{M_\alpha} \otimes (\lambda - \rho_\alpha)_{M_\alpha}; Q_0) \circ I_{Q_0}^G(\mathrm{St}^{M_\alpha} \otimes (\lambda - \rho_\alpha)_{M_\alpha}, f)\big)$$

where Q_0 is an arbitrary element in $\mathcal{P}(M_\alpha)$.

This proposition will be proved in sections 8 and 9. Let us assume it for the moment and let us compute $Z_M(\mathrm{St}^M, f; X(\lambda), b, n)$ by the method of residues.

For simplicity let us set

$$R(\lambda) = \mathrm{tr}(\mathcal{R}_M(\mathrm{St}^M \otimes \lambda; P_0) \circ I_{P_0}^G(\mathrm{St}^M \otimes \lambda, f))$$

and for any $\Lambda \in \mathrm{Re}\,\mathfrak{a}_M^*$ for which it makes sense let us set

$$Z_\Lambda(n) = \int_{\Lambda + \mathrm{Im}\,\mathfrak{a}_M^*/\mathfrak{a}_M^\vee} R(\lambda)X(\lambda)q^{-n\langle\lambda,b\rangle}\,d\lambda.$$

Let us fix a point $\lambda(0)$ in general position but very near 0 in $\mathrm{Re}\,\mathfrak{a}_M^*$. As the function $R(\lambda)X(\lambda)q^{-n\langle\lambda,b\rangle}$ is holomorphic in a neighborhood of $\mathrm{Im}\,\mathfrak{a}_M^*$ we have

$$Z_M(\mathrm{St}^M, f; X(\lambda), b, n) = Z_0(n) = Z_{\lambda(0)}$$

by the residue theorem.

Under the assumptions of section 6 let us write $\lambda(0) = (\lambda_1(0), \ldots, \lambda_s(0))$ and let us consider the path

$$\mathbb{R}^+ \to \mathrm{Re}\,\mathfrak{a}_M^*,\ x \mapsto (\lambda_1(0) + x, \lambda_2(0) - \frac{xd_2}{d-d_1}, \ldots, \lambda_s(0) - \frac{xd_s}{d-d_1}) = \lambda(x).$$

Let us move the contour of integration $\Lambda + \mathfrak{a}_M^*/\mathfrak{a}_M^\vee$ of $Z_\Lambda(n)$ along this path. In this process each time we cross a singular hyperplane we pick a residue. But our path meets the hyperplane H_α if and only if α is one of the roots $\pm\alpha_{1,j}$ ($j = 2, \ldots, s$) and we have $\langle \rho_{\alpha_{1,j}}, \alpha_{1,j}^\vee\rangle = \frac{d_1+d_j}{2}$. Therefore it follows from the above proposition that, when we move the contour of integration along our path, we pick non-zero residues at most at the points $\lambda(x_2), \ldots, \lambda(x_s)$ where $x_j \in \mathbb{R}^+$ is defined by the relation

$$(\frac{\lambda_1(0)}{d_1} + \frac{x_j}{d_1}) - (\frac{\lambda_j(0)}{d_j} - \frac{x_j}{d-d_1}) - \frac{d_1+d_j}{2} = 0.$$

As $\lambda(0)$ is in general position the real numbers $x_2, \ldots, x_s$ are all distinct. For each $j = 2, \ldots, s$ let us fix two real numbers x_j^- and x_j^+ very near x_j such that $x_j^- < x_j < x_j^+$. Let us also fix a very large real number x_∞. Then by the residue theorem we have

$$(**)\qquad Z_{\lambda(0)}(n) = \sum_{j=2}^{s}(Z_{\lambda(x_j^-)}(n) - Z_{\lambda(x_j^+)}(n)) + Z_{\lambda(x_\infty)}(n).$$

Let us first consider the term $Z_{\lambda(x_\infty)}(n)$ of $(**)$. As $R(\lambda)$ (resp. $X(\lambda)$) is a rational (resp. polynomial) function in the variables $q^{\pm\lambda_j}$ there exists an integer n_1' (resp. n_1'') such that

$$R(\lambda(x) + \lambda') = \mathrm{O}(q^{n_1'x}) \qquad (\forall \lambda' \in \mathrm{Im}\,\mathfrak{a}_M^*)$$

(resp.

$$X(\lambda(x)+\lambda') = O(q^{n_1'' x}) \qquad (\forall \lambda' \in \operatorname{Im} \mathfrak{a}_M^*)).$$

when $x \to +\infty$. But we have $\langle \lambda(x), b\rangle = \lambda_1(0) + x$. Therefore, if $n \geq n_1 = n_1' + n_1'' + 1$ we have

$$\lim_{x_\infty \to +\infty} Z_{\lambda(x_\infty)}(n) = 0.$$

Then let us consider the term $Z_{\lambda(x_j^-)}(n) - Z_{\lambda(x_j^+)}(n)$ of (**) for some fixed $j = 2, \dots, s$ and let $\alpha = \alpha_{1,j}$ be the corresponding root. We have $\lambda(x_j) = \lambda(x_j)^{M_\alpha} \oplus \lambda(x_j)_{M_\alpha} \in \mathfrak{a}_M^{M_\alpha *} \oplus \mathfrak{a}_{M_\alpha}^*$ with $\lambda(x_j)^{M_\alpha} = \rho_\alpha$. Let us fix two real numbers y^-, y^+ very near 1 such that $y^- < 1 < y^+$ and let us set

$$\lambda^\pm = y^\pm \rho_\alpha \oplus \lambda(x_j)_{M_\alpha}.$$

As the segment between $\lambda(x_j^\pm)$ and $\lambda^\pm$ does not cross any singular hyperplane we have

$$Z_{\lambda(x_j^\pm)}(n) = Z_{\lambda^\pm}(n)$$

by the residue theorem. But, according to the splitting $\operatorname{Im} \mathfrak{a}_M^* = \operatorname{Im} \mathfrak{a}_M^{M_\alpha *} \oplus \operatorname{Im} \mathfrak{a}_{M_\alpha}^*$, we may decompose the integral $Z_{y\rho_\alpha + \lambda(x_j)_{M_\alpha}}(n)$ into

$$\int_{\lambda(x_j)_{M_\alpha} + \operatorname{Im} \mathfrak{a}_{M_\alpha}^* / \mathfrak{a}_{M_\alpha}^\vee} \frac{\log q}{2\pi i D} \int_y^{y + \frac{2\pi i D}{\log q}} \operatorname{tr}(\mathcal{R}_M(\mathrm{St}^M \otimes (z\rho_\alpha + \lambda_{M_\alpha}); P_0)$$
$$\circ I_{P_0}^G(\mathrm{St}^M \otimes (z\rho_\alpha + \lambda_{M_\alpha}), f))$$
$$\times X(z\rho_\alpha + \lambda_{M_\alpha}) q^{-n\langle z\rho_\alpha + \lambda_{M_\alpha}, b\rangle} dz d\lambda_{M_\alpha},$$

where D is a sufficiently divisible positive integer, so that the function

$$\operatorname{tr}(\mathcal{R}_M(\mathrm{St}^M \otimes (z\rho_\alpha + \lambda_{M_\alpha}); P_0) \circ I_{P_0}^G(\mathrm{St}^M \otimes (z\rho_\alpha + \lambda_{M_\alpha}), f))$$
$$\times X(z\rho_\alpha + \lambda_{M_\alpha}) q^{-n\langle z\rho_\alpha + \lambda_{M_\alpha}, b\rangle}$$

of the complex variable z is periodic of period $\frac{2\pi i D}{\log q}$, and where dz is the usual complex differential on the complex plane.

Now if we move the contour of integration of the inner integral from $y = y^-$ to $y = y^+$ we pick some residues when $y = 1$. In fact, by the above proposition we get a pole at each point z such that $q^{\langle (z-1)\rho_\alpha, \alpha^\vee\rangle} = 1$, i.e. at each point

$$z_u = 1 + \frac{2\pi i}{\log q} \frac{u}{\langle \rho_\alpha, \alpha^\vee\rangle} \qquad (u \in \mathbb{Z}/D\langle \rho_\alpha, \alpha^\vee\rangle \mathbb{Z}),$$

and the corresponding residue is

$$-\frac{1}{(\log q)\langle \rho_\alpha, \alpha^\vee\rangle} \operatorname{tr}(\mathcal{R}_{M_\alpha}(\mathrm{St}^{M_\alpha} \otimes (((z_u - 1)\rho_\alpha)_{M_\alpha} + \lambda_{M_\alpha}); P_0)$$
$$\circ I_{Q_0}^G(\mathrm{St}^{M_\alpha} \otimes (((z_u - 1)\rho_\alpha)_{M_\alpha} + \lambda_{M_\alpha}), f))$$
$$\times X(z_u \rho_\alpha + \lambda_{M_\alpha}) q^{-n\langle z_u \rho_\alpha + \lambda_{M_\alpha}, b\rangle}.$$

Therefore, by the Cauchy formula $Z_{\lambda^-}(n) - Z_{\lambda^+}(n)$ is equal to

$$\frac{1}{D\langle\rho_\alpha,\alpha^\vee\rangle}\sum_u \int_{\lambda(x_j)_{M_\alpha}+\mathrm{Im}\,\mathfrak{a}^*_{M_\alpha}/\mathfrak{a}^\vee_{M_\alpha}} \mathrm{tr}(\mathcal{R}_{M_\alpha}(\mathrm{St}^{M_\alpha}\otimes(((z_u-1)\rho_\alpha)_{M_\alpha}+\lambda_{M_\alpha});P_0)$$
$$\circ I^G_{Q_0}(\mathrm{St}^{M_\alpha}\otimes(((z_u-1)\rho_\alpha)_{M_\alpha}+\lambda_{M_\alpha}),f))$$
$$\times X(z_u\rho_\alpha+\lambda_{M_\alpha})q^{-n\langle z_u\rho_\alpha+\lambda_{M_\alpha},b\rangle}d\lambda_{M_\alpha}.$$

But we have $z_u\rho_\alpha + \lambda_{M_\alpha} = \rho_\alpha + ((z_u-1)\rho_\alpha)_{M_\alpha} + \lambda_{M_\alpha}$ in $\mathfrak{a}^*_{M_\alpha}/\mathfrak{a}^\vee_{M_\alpha}$ and, making the change of variable $\lambda'_{M_\alpha} := ((z_u-1)\rho_\alpha)_{M_\alpha}+\lambda_{M_\alpha}$, we see that the inner integral of this last formula does not depend on u. The equality (*) of proposition 6 follows. Part (i) of theorem 5 is thus proved.

8. Proof of proposition 7: preliminaries

Let us normalize the intertwining operators $M_{P'|P}(\mathrm{St}^M\otimes\lambda)$ for every $M\in\mathcal{L}$ and every $P,P'\in\mathcal{P}(M)$.

First let us consider the case $M=T$, so that St^T is simply the trivial character of T. The Hecke algebra $\mathcal{C}(\mathcal{B}\backslash K/\mathcal{B})$ acts on the space $(I^G_B\mathbb{C})^{\mathcal{B}}$ and the "Steinberg character" of this algebra occurs with multiplicity one in this space. Let us fix a vector v_B in the corresponding line in $(I^G_B\mathbb{C})^{\mathcal{B}}\subset I^G_B\mathbb{C}$. For any $P\in\mathcal{P}(T)$ there exists a unique $w\in W$ such that $P=wBw^{-1}$ and the map $\varphi\mapsto\varphi(w^{-1}\cdot)$ is an isomorphism of $I^G_B\mathbb{C}$ onto $I^G_P\mathbb{C}$. Let us denote by v_P the image of v_B under this isomorphism. Then for any $P,P'\in\mathcal{P}(T)$ we normalize the intertwining operator $M_{P'|P}(\mathrm{St}^T\otimes\lambda)$ by

$$R_{P'|P}(\mathrm{St}^T\otimes\lambda)v_P = v_{P'}.$$

If $G=GL_2(F)$ and $W=\{1,s\}$ we can take $v_B = 1_{\mathcal{B}} - \frac{1}{q+1}1_K$ and by a direct computation one checks that

$$M_{\overline{B}|B}(\mathrm{St}^T\otimes\lambda)1_K = \frac{q-q^{-(\lambda_1-\lambda_2)}}{q(1-q^{-(\lambda_1-\lambda_2)})}1_K,$$

$$M_{\overline{B}|B}(\mathrm{St}^T\otimes\lambda)1_{\mathcal{B}} = \frac{1}{q}1_{K-s\mathcal{B}} + \frac{(q-1)q^{-(\lambda_1-\lambda_2)}}{q(1-q^{-(\lambda_1-\lambda_2)})}1_{s\mathcal{B}},$$

and therefore that

$$M_{\overline{B}|B}(\mathrm{St}^T\otimes\lambda)v_B = -\frac{1-qq^{-(\lambda_1-\lambda_2)}}{q(1-q^{-(\lambda_1-\lambda_2)})}v_{\overline{B}} = -\frac{1-q^{\lambda_1-\lambda_2-1}}{1-q^{\lambda_1-\lambda_2}}v_{\overline{B}}.$$

By a reduction to the case where P and P' are adjacent, and thus to the case $G=GL_2(F)$, one easily sees that, in general, we have

$$r_{P'|P}(\mathrm{St}^T\otimes\lambda) = (-1)^{|\{\alpha\in R(G,T)|\alpha>_P 0,\ \alpha<_{P'}0\}|}\prod_{\substack{\alpha\in R(G,T)\\ \alpha>_P 0,\ \alpha<_{P'}0}}\frac{1-q^{\langle\lambda,\alpha^\vee\rangle-1}}{1-q^{\langle\lambda,\alpha^\vee\rangle}}.$$

For a general M let us fix $B^M \in \mathcal{P}^M(T)$. If $P = MN_P \in \mathcal{P}(M)$ we set $P(B^M) = B^M N_P \in \mathcal{P}(T)$. The representation St^M is a submodule of the induced representation $I_{B^M}^M(\delta_{B^M}^{1/2})$. Therefore we have an embedding of the space V^M of St^M into $I_{B^M}^M\mathbb{C}$ which is unique up to similarity. Let us fix such an embedding. For each $P \in \mathcal{P}(M)$ it induces an embedding $I_P^G V^M \to I_{P(B^M)}^G\mathbb{C}$. Then we normalize the intertwining operator $M_{P'|P}(\mathrm{St}^T \otimes \lambda)$ in such way that the diagram

$$\begin{array}{ccc} I_P^G V^M & \xrightarrow{R_{P'|P}(\mathrm{St}^M \otimes \lambda)} & I_{P'}^G V^M \\ \downarrow & & \downarrow \\ I_{P(B^M)}^G\mathbb{C} & \xrightarrow[R_{P'(B^M)|P(B^M)}(\delta_{B^M}\otimes\lambda)]{} & I_{P'(B^M)}^G\mathbb{C} \end{array}$$

commutes, the vertical arrows being the above embeddings.

In the case $M \cong GL_{d'}(F) \times GL_{d''}(F)$ one computes that, for each $P \in \mathcal{P}(M)$, we have

$$r_{\overline{P}|P}(\mathrm{St}^M \otimes \lambda) = (-1)^{d'd''} \prod_i \frac{1 - q^{\langle \lambda, \alpha^\vee \rangle - i}}{1 - q^{\langle \lambda, \alpha^\vee \rangle + i - 1}},$$

where i runs through the set of all the half integers in the interval $\left[\frac{|d'-d''|}{2} + 1, \frac{d'+d''}{2}\right]$ which are congruent to $\frac{d'+d''}{2}$ modulo $\mathbb{Z}$ and where α is the unique root in $R(G, A_M)$ which is positive for P.

For $M \cong GL_{d'}(F) \times GL_{d''}(F)$ and for any $P \in \mathcal{P}(M)$ and any $\lambda \in \mathfrak{a}_M^*$ let us set

$$r_{\overline{P}|P}^{\mathrm{num}}(\mathrm{St}^M \otimes \lambda) = \prod_i (1 - q^{\langle \lambda, \alpha^\vee \rangle - i})$$

with the same notations as before. This definition can be easily generalized for an arbitrary M and it gives $r_{P'|P}^{\mathrm{num}}(\mathrm{St}^M \otimes \lambda)$ if P and P' are adjacent parabolic subgroups in $\mathcal{P}(M)$. For P and P' arbitrary in $\mathcal{P}(M)$ we set

$$r_{P'|P}^{\mathrm{num}}(\mathrm{St}^M \otimes \lambda) = \prod_{j=1}^{\ell} r_{P_j|P_{j-1}}^{\mathrm{num}}(\mathrm{St}^M \otimes \lambda)$$

where $P = P_0, P_1, \ldots, P_\ell = P'$ is a sequence of minimal length such that, for each $j = 1, \ldots, \ell$, P_{i-1} and P_i are adjacent parabolic subgroups in $\mathcal{P}(M)$.

LEMMA A. — *Let $M \in \mathcal{L}$, $P, P' \in \mathcal{P}(M)$ and $\varphi \in I_P^G V^M$, where V^M is the space of the Steinberg representation* St^M *of M. Let us set*

$$S_{P'|P}(\mathrm{St}^M \otimes \lambda) = r_{P'|P}^{\mathrm{num}}(\mathrm{St}^M \otimes \lambda) R_{P'|P}(\mathrm{St}^M \otimes \lambda).$$

Then the function

$$\mathfrak{a}_M^* \to I_{P'}^G V^M, \ \lambda \mapsto S_{P'|P}(\mathrm{St}^M \otimes \lambda)\varphi,$$

is polynomial.

Proof: We may assume that $M = GL_{d'}(F) \times GL_{d''}(F)$ and that $P' = \overline{P}$.

In the case where $d' = d'' = 1$, so that $G = GL_2(F)$ and $M = T$, we have seen before that

$$R_{\overline{B}|B}(\mathrm{St}^T \otimes \lambda)1_K = -\frac{1 - q^{\lambda_1 - \lambda_2 + 1}}{q(1 - q^{\lambda_1 - \lambda_2 - 1})} 1_K$$

and we have

$$R_{\overline{B}|B}(\mathrm{St}^T \otimes \lambda)v_B = v_{\overline{B}}.$$

The lemma follows easily as $\{1_K, v_B\}$ is a basis of $(I_B^G \mathbb{C})^{\mathcal{B}}$.

In the case $d' \geq d'' = 1$ the above commutative square and the lemma for $GL_2(F)$ (which is already proved) show that

$$\widetilde{S}_{\overline{P}|P}(\mathrm{St}^M \otimes \lambda)\varphi := \Big(\prod_{i=0}^{d'-1} \big(1 - q^{\langle \lambda, \alpha^\vee \rangle - \frac{d'+1}{2} + i}\big)\Big) R_{\overline{P}|P}(\mathrm{St}^M \otimes \lambda)\varphi$$

is polynomial. The factor for $i = 0$ is nothing else than $r^{\mathrm{num}}_{\overline{P}|P}(\mathrm{St}^M \otimes \lambda)$. Therefore, to conclude the proof in the case $d' \geq d'' = 1$ it is sufficient to show that $\widetilde{S}_{\overline{P}|P}(\mathrm{St}^M \otimes \lambda)\varphi$ can be divided by the other factors, i.e. to show that the operator $\widetilde{S}_{\overline{P}|P}(\mathrm{St}^M \otimes \lambda)$ vanishes if $q^{\langle \lambda, \alpha^\vee \rangle - \frac{d'+1}{2} + i} = 1$ for some $i = 1, \ldots, d'-1$. Let us assume the contrary and let us search for a contradiction. We fix λ such that $q^{\langle \lambda, \alpha^\vee \rangle - \frac{d'+1}{2} + i} = 1$ for some $i \in \{1, \ldots, d'-1\}$ and such that the intertwining operator

$$\widetilde{S}_{\overline{P}|P}(\mathrm{St}^M \otimes \lambda) : I_P^G(\mathrm{St}^M \otimes \lambda) \to I_{\overline{P}}^G(\mathrm{St}^M \otimes \lambda)$$

is non-zero. Due to our normalization its kernel contains the line in $(I_P^G V^M)^{\mathcal{B}}$ on which the Hecke algebra $\mathcal{C}(\mathcal{B}\backslash G/\mathcal{B})$ acts by the Steinberg character and is thus non-zero. Therefore $I_P^G(\mathrm{St}^M \otimes \lambda)$ is reducible. But Zelevinsky has proved that this induced representation is reducible only if $q^{\langle \lambda, \alpha^\vee \rangle - \frac{d'+1}{2}} = 1$ or $q^{\langle \lambda, \alpha^\vee \rangle + \frac{d'+1}{2}} = 1$ (see [Ze]) and we get a contradiction.

Now if $d' \geq d'' \geq 1$ let us denote by T'' the maximal torus of diagonal matrices in $GL_{d''}(F)$ and by B'' the Borel subgroup of upper triangular matrices in $\mathcal{P}^{GL_{d''}(F)}(T'')$. Let us set $L = GL_{d'}(F) \times T'' \in \mathcal{L}$, $R = GL_{d'}(F) \times B'' \in \mathcal{P}^M(L)$, $Q = RN_P$ and $Q' = RN_{\overline{P}}$ (we have $Q, Q' \in \mathcal{P}(L)$). Then we have a commutative diagram

$$\begin{array}{ccc}
I_P^G V^M & \xrightarrow{R_{\overline{P}|P}(\mathrm{St}^M \otimes \lambda)} & I_{\overline{P}}^G V^M \\
\downarrow & & \downarrow \\
I_Q^G V^L & \xrightarrow[R_{Q'|Q}(\mathrm{St}^L \otimes (\lambda + \rho_R^M))]{} & I_{Q'}^G V^L
\end{array}$$

and it is sufficient to prove that, for every $\varphi \in I_Q^G V^L$,

$$r^{\mathrm{num}}_{\overline{P}|P}(\mathrm{St}^M \otimes \lambda) R_{Q'|Q}(\mathrm{St}^L \otimes (\lambda + \rho_R^M))\varphi$$

is polynomial. But from the hypothesis $d' \geq d'' \geq 1$ we get that

$$r^{\mathrm{num}}_{\overline{P}|P}(\mathrm{St}^M \otimes \lambda) = r^{\mathrm{num}}_{Q'|Q}(\mathrm{St}^L \otimes (\lambda + \rho_R^M))$$

and the operator $R_{Q'|Q}(\mathrm{St}^L \otimes (\lambda + \rho_R^M))$ (resp. the expression $r^{\mathrm{num}}_{Q'|Q}(\mathrm{St}^L \otimes (\lambda + \rho_R^M)))$ is a product of operators $R_{Q'_j|Q_j}(\mathrm{St}^{L_j} \otimes \lambda_j)$ (resp. expressions $r^{\mathrm{num}}_{Q'_j|Q_j}(\mathrm{St}^{L_j} \otimes \lambda_j))$, where $L_j = GL_{d'}(F) \times GL_1(F) \subset GL_{d'+1}(F)$ and $Q_j, Q'_j \in \mathcal{P}^{GL_{d'+1}(F)}(L_j)$, for $j = 1, \ldots, d''$. Therefore the case $d' \geq d'' \geq 1$ follows from the case where $G := GL_{d'+1}(F)$ and $M := GL_{d'}(F) \times GL_1(F)$ that we have already considered.

Finally, if $d'' \geq d' \geq 1$ we simply permute d' and d''. □

Let $M = GL_{d'}(F) \times GL_{d''}(F)$ be diagonally embedded in G and let $P \in \mathcal{P}(M)$ be the "upper triangular" parabolic subgroup ($P \supset B$). We denote by α the unique element of Δ_P. Let $i \in \frac{1}{2}\mathbb{Z}$ be such that $\frac{|d'-d''|}{2} < i \leq \frac{d'+d''}{2}$ and such that $i \equiv \frac{d'+d''}{2}$ modulo $\mathbb{Z}$. Then we set

$$M(i) = GL_{\frac{d'+d''}{2}+i}(F) \times GL_{\frac{d'+d''}{2}-i}(F),$$

diagonally embedded in G, and we denote by $P(i)$ the "upper triangular" parabolic subgroup in $\mathcal{P}(M(i))$. We also set

$$\lambda(i) = \frac{2i}{d}\rho_P, \quad \mu(i) = \frac{d'-d''}{d}\rho_{P(i)}.$$

Let us remark that, if $q^{\langle \lambda, \alpha^\vee \rangle - i} = 1$, we have $q^{\langle \lambda - \lambda(i), \alpha^\vee \rangle} = 1$.

LEMMA B. — *Let M and i be as above. Let us fix $\lambda \in \mathfrak{a}_M^*/\mathfrak{a}_M^\vee$ such that $q^{\langle \lambda, \alpha^\vee \rangle - i} = 1$. Then if we define $S_{\overline{P}|P}(\mathrm{St}^M \otimes \lambda)$ as in lemma A we have*

$$\mathrm{Ker}\, S_{\overline{P}|P}(\mathrm{St}^M \otimes \lambda) = \mathrm{Im}\, S_{P|\overline{P}}(\mathrm{St}^M \otimes \lambda)$$

and the restriction of the representation $I_P^G(\mathrm{St}^M \otimes \lambda)$ to this subspace of $I_P^G V^M$ is isomorphic to $I_{P(i)}^G(\mathrm{St}^{M(i)} \otimes ((\lambda - \lambda(i))_G + \mu(i)))$.

Proof: The operator $S_{\overline{P}|P}(\mathrm{St}^M \otimes \lambda)$ is non-zero. Indeed, on the one hand one checks that $S_{\overline{P}|P}(\mathrm{St}^M \otimes \lambda')$ is equal to $c(\lambda')M_{\overline{P}|P}(\mathrm{St}^M \otimes \lambda')$ where $c(\lambda')$ is a holomorphic function in a neighborhood of our fixed λ and where $c(\lambda) \neq 0$. On the other hand one checks that there exists $\varphi \in I_P^G V^M$ such that $(M_{\overline{P}|P}(\mathrm{St}^M \otimes \lambda')\varphi)(1)$ does not depend on λ' and is non-zero (to get the independence of λ' take φ which is supported by the parahoric subgroup $K_1(P \cap K)$).

From the equality

$$S_{\overline{P}|P}(\mathrm{St}^M \otimes \lambda) \circ S_{P|\overline{P}}(\mathrm{St}^M \otimes \lambda) = 0$$

we get that

$$\mathrm{Im}\, S_{P|\overline{P}}(\mathrm{St}^M \otimes \lambda) \subset \mathrm{Ker}\, S_{\overline{P}|P}(\mathrm{St}^M \otimes \lambda).$$

Due to our normalization $\mathrm{Im}\, S_{P|\overline{P}}(\mathrm{St}^M \otimes \lambda)$ contains the line ℓ in $(I_P^G V^M)^{\mathcal{B}}$ on which the Hecke algebra $\mathcal{C}(\mathcal{B}\backslash G/\mathcal{B})$ acts by the Steinberg character.

Therefore $\mathrm{Ker}\, S_{\overline{P}|P}(\mathrm{St}^M \otimes \lambda)$ and $\mathrm{Im}\, S_{P|\overline{P}}(\mathrm{St}^M \otimes \lambda)$ are two proper subspaces of $I_P^G V^M$ containing the line ℓ.

But Zelevinsky has proved in [Ze] that the length of $I_P^G(\mathrm{St}^M \otimes \lambda)$ is 2 and that $I_P^G(\mathrm{St}^M \otimes \lambda)$ admits $I_{P(i)}^G(\mathrm{St}^{M(i)} \otimes ((\lambda - \lambda(i))_G + \mu(i)))$ as a subquotient. As $I_{P(i)}^G V^{M(i)}$ also contains a line on which the Hecke algebra $\mathcal{C}(\mathcal{B}\backslash G/\mathcal{B})$ acts by the Steinberg character that subquotient must coincide with the subrepresentations $\mathrm{Im}\, S_{P|\overline{P}}(\mathrm{St}^M \otimes \lambda)$ and $\mathrm{Ker}\, S_{\overline{P}|P}(\mathrm{St}^M \otimes \lambda)$ and the proof of the lemma is completed. □

9. Proof of proposition 7: the end

Let $M \in \mathcal{L}$ and $P_0 \in \mathcal{P}(M)$. For any $\lambda \in \mathfrak{a}_M^*/\mathfrak{a}_M^\vee$ we define the (G, M)-families

$$c_P(\zeta) = (r_{P|P_0}^{\mathrm{num}}(\mathrm{St}^M \otimes \lambda))^{-1} r_{P|P_0}^{\mathrm{num}}(\mathrm{St}^M \otimes (\lambda + \zeta))$$

and

$$\begin{aligned} &\mathcal{S}_P(\mathrm{St}^M \otimes \lambda, \zeta; P_0) \\ &= (r_{P_0|P}^{\mathrm{num}}(\mathrm{St}^M \otimes \lambda) r_{P|P_0}^{\mathrm{num}}(\mathrm{St}^M \otimes \lambda))^{-1} S_{P_0|P}(\mathrm{St}^M \otimes \lambda) \circ S_{P|P_0}(\mathrm{St}^M \otimes (\lambda + \zeta)) \end{aligned}$$

$(P \in \mathcal{P}(M),\ \zeta \in \mathrm{Im}\, \mathfrak{a}_M^*)$. We have

$$\mathcal{S}_P(\mathrm{St}^M \otimes \lambda, \zeta; P_0) = c_P(\zeta) \mathcal{R}_P(\mathrm{St}^M \otimes \lambda, \zeta; P_0).$$

Therefore, if we apply the corollary of lemma 4 we get that

(1)

$$\mathrm{tr}\big(\mathcal{R}_M(\mathrm{St}^M \otimes \lambda; P_0) \circ I_{P_0}^G(\mathrm{St}^M \otimes \lambda, f)\big) = \mathrm{tr}\big(\mathcal{S}_M(\mathrm{St}^M \otimes \lambda; P_0) \circ I_{P_0}^G(\mathrm{St}^M \otimes \lambda, f)\big)$$

for each very cuspidal function $f \in C_c^\infty(G)$.

From lemma 8A we get that the singularities of $\mathcal{S}_M(\mathrm{St}^M \otimes \lambda; P_0)$ are contained in the zeros of the functions $r^{\mathrm{num}}_{P_0|P}(\mathrm{St}^M \otimes \lambda) r^{\mathrm{num}}_{P|P_0}(\mathrm{St}^M \otimes \lambda)$. Therefore the singularities of

$$\mathrm{tr}\big(\mathcal{S}_M(\mathrm{St}^M \otimes \lambda; P_0) \circ I^G_{P_0}(\mathrm{St}^M \otimes \lambda, f)\big)$$

are contained in

$$\bigcup_{\alpha \in R(G, A_M)} \bigcup_{i \in I_\alpha} H_{\alpha,i}$$

where I_α is a certain subset of non-negative half integers and where

$$H_{\alpha,i} = \{\lambda \in \mathfrak{a}^*_M \mid q^{\langle \lambda, \alpha^\vee \rangle - i} = 1\}$$

(in particular we have $H_{\alpha, \langle \rho_\alpha, \alpha^\vee \rangle} = H_\alpha$).

Thanks to (1), proposition 7 is a direct consequence of

LEMMA. — *Let f be a very cuspidal function in $C^\infty_c(G)$. Let $\lambda \in H_{\alpha,i}$ be such that $\lambda \notin H_{\beta,j}$ for every $(\beta, j) \neq (\alpha, i)$. Then the limit*

$$\lim_{\lambda' \to \lambda} (1 - q^{\langle \lambda', \alpha^\vee \rangle - i})\, \mathrm{tr}\big(\mathcal{S}_M(\mathrm{St}^M \otimes \lambda'; P_0) \circ I^G_{P_0}(\mathrm{St}^M \otimes \lambda', f)\big)$$

exists and is equal to

$$\mathrm{tr}\big(\mathcal{S}_{M_\alpha}(\mathrm{St}^{M_\alpha} \otimes (\lambda - \rho_\alpha)_{M_\alpha}; Q_0) \circ I^G_{Q_0}(\mathrm{St}^{M_\alpha} \otimes (\lambda - \rho_\alpha)_{M_\alpha}, f)\big),$$

if $i = \langle \rho_\alpha, \alpha^\vee \rangle$, Q_0 being an arbitrary element in $\mathcal{P}(M_\alpha)$, and is equal to 0 *otherwise.*

Proof: We may assume that $\alpha \in \Delta_P$. Then we have $P_0 \cap M_\alpha = R_\alpha$ where R_α is the unique parabolic subgroup in $\mathcal{P}^{M_\alpha}(M)$ which admits α as a positive root (see section 6). Let us denote by $Q_0 = M_\alpha N_{Q_0}$ the unique element of $\mathcal{P}(M_\alpha)$ containing P_0. The parabolic subgroup $P_1 = \overline{R}_\alpha N_{Q_0} \in \mathcal{P}(M)$ is adjacent to P_0, separated from P_0 by the root α. For $P \in \mathcal{P}(M)$ the factor $r^{\mathrm{num}}_{P_0|P}(\mathrm{St}^M \otimes \lambda') r^{\mathrm{num}}_{P|P_0}(\mathrm{St}^M \otimes \lambda')$ vanishes at $\lambda' = \lambda$ if and only if α separates P and P_0. If this is the case this factor is equal to

$$r^{\mathrm{num}}_{P_1|P}(\mathrm{St}^M \otimes \lambda') r^{\mathrm{num}}_{P|P_1}(\mathrm{St}^M \otimes \lambda') r^{\mathrm{num}}_{P_0|P_1}(\mathrm{St}^M \otimes \lambda') r^{\mathrm{num}}_{P_1|P_0}(\mathrm{St}^M \otimes \lambda').$$

Let us fix $\mu \in \mathrm{Im}\, \mathfrak{a}^*_M$ in general position. For each smooth function a on $\mathrm{Im}\, \mathfrak{a}^*_M$, each non-negative integer m and each $\nu \in \mathrm{Im}\, \mathfrak{a}^*_M$ it will be convenient to use the following notation:

$$\partial^m a(\nu) = \frac{d^m}{dx^m} a(\nu + x\mu)_{|x=0}.$$

Arthur has proved that, for every (G, M)-family $(c_P(\zeta) \mid P \in \mathcal{P}(M))$, we have the equality

$$c_M(0) = \frac{1}{(s-1)!} \sum_{P \in \mathcal{P}(M)} \frac{\partial^{s-1} c_P(0)}{\theta_P(\mu)}$$

where $s = \dim \mathfrak{a}_M$.

Let us set

$$\gamma_1 = \frac{1}{(s-1)!} \lim_{\lambda' \to \lambda} \frac{1 - q^{\langle \lambda', \alpha^\vee \rangle - i}}{r^{\mathrm{num}}_{P_0|P_1}(\mathrm{St}^M \otimes \lambda') r^{\mathrm{num}}_{P_1|P_0}(\mathrm{St}^M \otimes \lambda')}$$

(it is easy to see that this limit exists and does not depend on λ). Then it follows from the above considerations that the limit that we want to compute is equal to

(2)

$$\gamma_1 \sum_P \frac{\mathrm{tr}\big(S_{P_0|P}(\mathrm{St}^M \otimes \lambda) \circ \partial^{s-1} S_{P|P_0}(\mathrm{St}^M \otimes (\lambda + 0)) \circ I^G_{P_0}(\mathrm{St}^M \otimes \lambda, f)\big)}{r^{\mathrm{num}}_{P_1|P}(\mathrm{St}^M \otimes \lambda) r^{\mathrm{num}}_{P|P_1}(\mathrm{St}^M \otimes \lambda) \theta_P(\mu)}$$

where P runs through the set of parabolic subgroups in $\mathcal{P}(M)$ which are separated from P_0 by α.

Let $P \in \mathcal{P}(M)$ be such that α separates P from P_0. After having simplified the notations in an obvious way we have

$$\begin{aligned}
&\mathrm{tr}\big(S_{P_0|P} \circ \partial^{s-1} S_{P|P_0} \circ I^G_{P_0}\big) \\
&= \mathrm{tr}\big(S_{P_0|P_1} \circ S_{P_1|P} \circ \partial^{s-1}(S_{P|P_1} \circ S_{P_1|P_0}) \circ I^G_{P_0}\big) \\
&= \mathrm{tr}\big(S_{P_1|P} \circ \partial^{s-1}(S_{P|P_1} \circ S_{P_1|P_0}) \circ I^G_{P_0} \circ S_{P_0|P_1}\big) \\
&= \mathrm{tr}\big(S_{P_1|P} \circ \partial^{s-1}(S_{P|P_1} \circ S_{P_1|P_0}) \circ S_{P_0|P_1} \circ I^G_{P_1}\big)
\end{aligned}$$

(we have $d(P, P_0) = d(P, P_1) + d(P_1, P_0)$) and then by the Leibniz formula we get the equality

$$\begin{aligned}
&\mathrm{tr}\big(S_{P_0|P} \circ \partial^{s-1} S_{P|P_0} \circ I^G_{P_0}\big) \\
&\quad = \sum_{t=0}^{s-1} \binom{s-1}{t} \mathrm{tr}\big(S_{P_1|P} \circ \partial^{s-1-t} S_{P|P_1} \circ \partial^t S_{P_1|P_0} \circ S_{P_0|P_1} \circ I^G_{P_1}\big).
\end{aligned}$$

As

$$S_{P_1|P_0}(\mathrm{St}^M \otimes \lambda) \circ S_{P_0|P_1}(\mathrm{St}^M \otimes \lambda) = r^{\mathrm{num}}_{P_1|P_0}(\mathrm{St}^M \otimes \lambda) \circ r^{\mathrm{num}}_{P_0|P_1}(\mathrm{St}^M \otimes \lambda)\mathrm{id} = 0$$

the term for $t = 0$ in the above sum is zero.

Now let us fix μ^{M_α} and μ_{M_α} in general position in $\operatorname{Im}\mathfrak{a}_M^{M_\alpha *}$ and $\operatorname{Im}\mathfrak{a}_{M_\alpha}^*$ respectively. We assume that our fixed μ is equal to $y\mu^{M_\alpha} + \mu_{M_\alpha}$ and we denote by ∂_y^m the operator that we denoted by ∂^m before. For each smooth function a on $\operatorname{Im}\mathfrak{a}_M^*$ we have

$$\lim_{y\to 0} \partial_y^m a = \partial_0^m a$$

and, in the case where a is invariant under translation by $\operatorname{Im}\mathfrak{a}_{M_\alpha}^*$, we have

$$\partial_y^m a = y^m \partial_1^m a.$$

Therefore, as the function $\theta_P(y\mu^{M_\alpha} + \mu_{M_\alpha})$ has a zero of order 1 at $y = 0$ if $-\alpha \in \Delta_P$ and does not vanish at $y = 0$ otherwise and as the function $\zeta \mapsto S_{P_1|P_0}(\mathrm{St}^M \otimes (\lambda + \zeta))$ is invariant under translation by $\operatorname{Im}\mathfrak{a}_{M_\alpha}^*$, the limit

$$\lim_{y\to 0} \frac{\operatorname{tr}\big(S_{P_0|P} \circ \partial^{s-1} S_{P|P_0} \circ I_{P_0}^G\big)}{\theta_P(y\mu^{M_\alpha} + \mu_{M_\alpha})}$$

exists and is equal to

$$(3)\quad (s-1)\operatorname{tr}\big(S_{P_1|P} \circ \partial_0^{s-2} S_{P|P_1} \circ \partial_1^1 S_{P_1|P_0} \circ S_{P_0|P_1} \circ I_{P_1}^G\big) \lim_{y\to 0} \frac{y}{\theta_P(y\mu^{M_\alpha} + \mu_{M_\alpha})}$$

if $-\alpha \in \Delta_P$ and is equal to 0 otherwise (we have

$$\lim_{y\to 0} \frac{\operatorname{tr}\big(S_{P_1|P} \circ \partial_y^{s-1-t} S_{P|P_1} \circ \partial_y^t S_{P_1|P_0} \circ S_{P_0|P_1} \circ I_{P_1}^G\big)}{\theta_P(y\mu^{M_\alpha} + \mu_{M_\alpha})} = 0 \quad (\forall t = 2, \dots s-1)).$$

But the map $\mathcal{P}(M_\alpha) \to \mathcal{P}(M)$, $Q = M_\alpha N_Q \mapsto P = \overline{R}_\alpha N_Q$, is injective and its image is precisely $\{P \in \mathcal{P}(M) \mid -\alpha \in \Delta_P\}$. Moreover, for each $Q \in \mathcal{P}(M_\alpha)$ with image P under this map we have

$$\lim_{y\to 0} \frac{y}{\theta_P(y\mu^{M_\alpha} + \mu_{M_\alpha})} = -\frac{1}{(\log q)\langle \mu^{M_\alpha}, \alpha^\vee\rangle \theta_Q(\mu_{M_\alpha})}.$$

Therefore we have proved that

$$\lim_{\lambda'\to\lambda} (1 - q^{\langle \lambda', \alpha^\vee\rangle - i}) \operatorname{tr}\big(\mathcal{S}_M(\mathrm{St}^M \otimes \lambda'; P_0) \circ I_{P_0}^G(\mathrm{St}^M \otimes \lambda', f)\big)$$

is equal to

(4)

$$-\frac{(s-1)\gamma_1}{(\log q)\langle \mu^{M_\alpha}, \alpha^\vee\rangle} \sum_{Q\in\mathcal{P}(M_\alpha)} \frac{\operatorname{tr}\big(S_{P_1|P} \circ \partial_0^{s-2} S_{P|P_1} \circ \partial_1^1 S_{P_1|P_0} \circ S_{P_0|P_1} \circ I_{P_1}^G\big)}{r_{P_1|P}^{\mathrm{num}}(\mathrm{St}^M \otimes \lambda) r_{P|P_1}^{\mathrm{num}}(\mathrm{St}^M \otimes \lambda) \theta_Q(\mu_{M_\alpha})}$$

where P is the image of Q under the above map.

Let us fix some $Q = M_\alpha N_Q \in \mathcal{P}(M_\alpha)$ with image $P = \overline{R}_\alpha N_Q$ in $\{P \in \mathcal{P}(M) \mid -\alpha \in \Delta_P\}$ and let us compute the corresponding summand in (4).

It follows from lemma 8B that the intertwining map

$$S_{P_0|P_1} : I_{P_1}^G V^M \to I_{P_0}^G V^M$$

may be factorized into

$$I_{P_1}^G V^M \xrightarrow{e_0} I_{R(i)N_{Q_0}}^G V^{M(i)} \xrightarrow{j_0} I_{P_0}^G V^M$$

for some Levi subgroup $M(i) \in \mathcal{L}$ which is contained in M_α and which contains M, some parabolic subgroup $R(i) \in \mathcal{P}^{M_\alpha}(M(i))$, some $\lambda(i) \in \mathfrak{a}_M^{M_\alpha *}$ and some $\mu(i) \in \mathfrak{a}_{M(i)}^{M_\alpha *}$, where e_0 is surjective and intertwines $I_{P_1}^G(\mathrm{St}^M \otimes \lambda)$ and $I_{R(i)N_{Q_0}}^G\big(\mathrm{St}^{M(i)} \otimes ((\lambda - \lambda(i))_G + \mu(i))\big)$ and where j_0 is injective and intertwines $I_{R(i)N_{Q_0}}^G\big(\mathrm{St}^{M(i)} \otimes ((\lambda - \lambda(i))_G + \mu(i))\big)$ and $I_{P_0}^G(\mathrm{St}^M \otimes \lambda)$ (recall that $P_0 = R_\alpha N_{Q_0}$ and $P_1 = \overline{R}_\alpha N_{Q_0}$). Therefore we have

$$\begin{aligned}
(5) \qquad & \mathrm{tr}\big(S_{P_1|P} \circ \partial_0^{s-2} S_{P|P_1} \circ \partial_1^1 S_{P_1|P_0} \circ S_{P_0|P_1} \circ I_{P_1}^G\big) \\
&= \mathrm{tr}\big(S_{P_1|P} \circ \partial_0^{s-2} S_{P|P_1} \circ \partial_1^1 S_{P_1|P_0} \circ j_0 \circ e_0 \circ I_{P_1}^G\big) \\
&= \mathrm{tr}\big(S_{P_1|P} \circ \partial_0^{s-2} S_{P|P_1} \circ \partial_1^1 S_{P_1|P_0} \circ j_0 \circ I_{R(i)N_{Q_0}}^G \circ e_0\big) \\
&= \mathrm{tr}\big(e_0 \circ S_{P_1|P} \circ \partial_0^{s-2} S_{P|P_1} \circ \partial_1^1 S_{P_1|P_0} \circ j_0 \circ I_{R(i)N_{Q_0}}^G\big)
\end{aligned}$$

where $I_{R(i)N_{Q_0}}^G$ is a simplified notation for $I_{R(i)N_{Q_0}}^G\big(\mathrm{St}^{M(i)} \otimes ((\lambda - \lambda(i))_G + \mu(i)), f\big)$. Similarly, replacing Q_0 by $Q = M_\alpha N_Q$ we get a factorization

$$I_P^G V^M \xrightarrow{e} I_{R(i)N_Q}^G V^{M(i)} \xrightarrow{j} I_{R_\alpha N_Q}^G V^M$$

of

$$S_{R_\alpha N_Q|P} : I_P^G V^M \to I_{R_\alpha N_Q}^G V^M$$

with the same $M(i)$, $R(i)$, $\lambda(i)$ and $\mu(i)$ as before (recall that $P = \overline{R}_\alpha N_Q$). For suitable normalizations of e_0 and e the diagram

$$\begin{array}{ccc}
I_{P_1}^G V^M & \xrightarrow{\quad e_0 \quad} & I_{R(i)N_{Q_0}}^G V^{M(i)} \\
{\scriptstyle S_{P|P_1}} \Big\downarrow & & \Big\downarrow {\scriptstyle S_{R(i)N_Q|R(i)N_{Q_0}}} \\
I_P^G V^M & \xrightarrow{\quad e \quad} & I_{R(i)N_Q}^G V^{M(i)} \\
{\scriptstyle S_{P_1|P}} \Big\downarrow & & \Big\downarrow {\scriptstyle S_{R(i)N_{Q_0}|R(i)N_Q}} \\
I_{P_1}^G V^M & \xrightarrow{\quad e_0 \quad} & I_{R(i)N_{Q_0}}^G V^{M(i)}
\end{array}$$

is commutative. Indeed we have

$$(6)\quad \begin{cases} r^{\text{num}}_{P_1|P}(\text{St}^M \otimes \lambda) = r^{\text{num}}_{R(i)N_{Q_0}|R(i)N_Q}(\text{St}^{M(i)} \otimes ((\lambda - \lambda(i))_{M_\alpha} + \mu(i))), \\ r^{\text{num}}_{P|P_1}(\text{St}^M \otimes \lambda) = r^{\text{num}}_{R(i)N_Q|R(i)N_{Q_0}}(\text{St}^{M(i)} \otimes ((\lambda - \lambda(i))_{M_\alpha} + \mu(i))), \end{cases}$$

and the similar diagram, where we replace the operators S by the operators R and the operators e by the operators $j \circ e$, is obviously commutative. If we take the derivative of the upper square in the direction μ_{M_α} we get a commutative diagram

$$\begin{array}{ccc} I^G_{P_1} V^M & \xrightarrow{\quad e_0 \quad} & I^G_{R(i)N_{Q_0}} V^{M(i)} \\ \Big\downarrow \partial_0^{s-2} S_{P|P_1} & & \Big\downarrow \partial_0^{s-2} S_{R(i)N_Q|R(i)N_{Q_0}} \\ I^G_P V^M & \xrightarrow{\quad e \quad} & I^G_{R(i)N_Q} V^{M(i)} \\ \Big\downarrow S_{P_1|P} & & \Big\downarrow S_{R(i)N_{Q_0}|R(i)N_Q} \\ I^G_{P_1} V^M & \xrightarrow{\quad e_0 \quad} & I^G_{R(i)N_{Q_0}} V^{M(i)} \end{array}$$

and the equality

$$(7)\quad e_0 \circ S_{P_1|P} \circ \partial_0^{s-2} S_{P|P_1} = S_{R(i)N_{Q_0}|R(i)N_Q} \circ \partial_0^{s-2} S_{R(i)N_Q|R(i)N_{Q_0}} \circ e_0.$$

For any λ' we have the equality

$$S_{P_0|P_1}(\text{St}^M \otimes \lambda') \circ S_{P_1|P_0}(\text{St}^M \otimes \lambda') = r^{\text{num}}_{P_0|P_1}(\text{St}^M \otimes \lambda') \circ r^{\text{num}}_{P_1|P_0}(\text{St}^M \otimes \lambda')\text{id}.$$

Taking the derivative of this identity in the direction μ we obtain

$$S_{P_0|P_1} \circ \partial_1^1 S_{P_1|P_0} + \partial_1^1 S_{P_0|P_1} \circ S_{P_1|P_0} = -\frac{(\log q)\langle \mu^{M_\alpha}, \alpha^\vee \rangle}{(s-1)!\gamma_1}\text{id}$$

where γ_1 is the same constant as in (2). As $S_{P_1|P_0} \circ j_0 = 0$ it follows that

$$S_{P_0|P_1} \circ \partial_1^1 S_{P_1|P_0} \circ j_0 = -\frac{(\log q)\langle \mu^{M_\alpha}, \alpha^\vee \rangle}{(s-1)!\gamma_1} j_0,$$

i.e.

$$j_0 \circ e_0 \circ \partial_1^1 S_{P_1|P_0} \circ j_0 = -\frac{(\log q)\langle \mu^{M_\alpha}, \alpha^\vee \rangle}{(s-1)!\gamma_1} j_0.$$

By the injectivity of j_0 this implies that

$$e_0 \circ \partial_1^1 S_{P_1|P_0} \circ j_0 = -\frac{(\log q)\langle \mu^{M_\alpha}, \alpha^\vee \rangle}{(s-1)!\gamma_1}\text{id}.$$

Putting together the last equality with (5) and (7) we obtain the equality

$$\operatorname{tr}\big(S_{P_1|P}\circ\partial_0^{s-2}S_{P|P_1}\circ\partial_1^1 S_{P_1|P_0}\circ S_{P_0|P_1}\circ I_{P_1}^G\big)$$
$$=-\frac{(\log q)\langle\mu^{M_\alpha},\alpha^\vee\rangle}{(s-1)!\gamma_1}\operatorname{tr}\big(S_{R(i)N_{Q_0}|R(i)N_Q}\circ\partial_0^{s-2}S_{R(i)N_Q|R(i)N_{Q_0}}\circ I_{R(i)N_{Q_0}}^G\big)$$

and, putting together this equality with (4) and (5) we obtain that the limit

$$\lim_{\lambda'\to\lambda}(1-q^{\langle\lambda',\alpha^\vee\rangle-i})\operatorname{tr}\big(\mathcal{S}_M(\mathrm{St}^M\otimes\lambda';P_0)\circ I_{P_0}^G(\mathrm{St}^M\otimes\lambda',f)\big)$$

is equal to

$$\frac{1}{(s-2)!}\sum_{Q\in\mathcal{P}(M_\alpha)}\frac{\operatorname{tr}\big(S_{R(i)N_{Q_0}|R(i)N_Q}\circ\partial_0^{s-2}S_{R(i)N_Q|R(i)N_{Q_0}}\circ I_{R(i)N_{Q_0}}^G\big)}{r_{R(i)N_{Q_0}|R(i)N_Q}^{\mathrm{num}}r_{R(i)N_Q|R(i)N_{Q_0}}^{\mathrm{num}}\theta_Q(\mu_{M_\alpha})}$$

(the representation which is not written is $\mathrm{St}^{M(i)}\otimes((\lambda-\lambda(i))_G+\mu(i))$).

Let us introduce the (G,M_α)-family

$$d_Q(\xi)=\frac{\operatorname{tr}\big(S_{R(i)N_{Q_0}|R(i)N_Q}\circ S_{R(i)N_Q|R(i)N_{Q_0}}(\xi)\circ I_{R(i)N_{Q_0}}^G\big)}{r_{R(i)N_{Q_0}|R(i)N_Q}^{\mathrm{num}}r_{R(i)N_Q|R(i)N_{Q_0}}^{\mathrm{num}}}$$

$(Q\in\mathcal{P}(M_\alpha),\ \xi\in\operatorname{Im}\mathfrak{a}_{M_\alpha}^*)$ where, for simplicity, we have set

$$S_{R(i)N_Q|R(i)N_{Q_0}}(\xi)=S_{R(i)N_Q|R(i)N_{Q_0}}\big(\mathrm{St}^{M(i)}\otimes((\lambda-\lambda(i))_G+\mu(i)+\xi)\big).$$

On the one hand we have

$$d_{M_\alpha}(0)=\frac{1}{(s-2)!}\sum_{Q\in\mathcal{P}(M_\alpha)}\frac{\partial^{s-2}d_Q(0)}{\theta_Q(\mu_{M_\alpha})}$$

by the same formula of Arthur as we have already used, so that our limit

$$\lim_{\lambda'\to\lambda}(1-q^{\langle\lambda',\alpha^\vee\rangle-i})\operatorname{tr}\big(\mathcal{S}_M(\mathrm{St}^M\otimes\lambda';P_0)\circ I_{P_0}^G(\mathrm{St}^M\otimes\lambda',f)\big)$$

is nothing else than $d_{M_\alpha}(0)$. On the other hand d is the (G,M_α)-family deduced from the $(G,M(i))$-family

$$c_{P^{(i)}}^{(i)}(\zeta^{(i)})=\operatorname{tr}\big(\mathcal{S}_{P^{(i)}}(\mathrm{St}^{M(i)}\otimes((\lambda-\lambda(i))_{M_\alpha}+\mu(i)),\zeta^{(i)};R(i)N_{Q_0})$$
$$\circ I_{R(i)N_{Q_0}}^G(\mathrm{St}^{M(i)}\otimes((\lambda-\lambda(i))_{M_\alpha}+\mu(i)),f)\big)$$

$(P^{(i)}\in\mathcal{P}(M(i)),\ \zeta^{(i)}\in\operatorname{Im}\mathfrak{a}_{M(i)}^*)$. Therefore, by the very-cuspidality of f and by lemma 4 our limit is zero unless $M_\alpha=M(i)$, i.e. unless $i=\langle\rho_\alpha,\alpha^\vee\rangle$. Moreover, if the last condition is satisfied we have $R(i)=R_\alpha$, $\lambda(i)=\rho_\alpha$, $\mu(i)=0$ and

$$d_Q(\xi)=c_Q^{(i)}(\xi)=\operatorname{tr}\big(\mathcal{S}_Q(\mathrm{St}^{M_\alpha}\otimes(\lambda-\rho_\alpha)_{M_\alpha},\xi;Q_0)\circ I_{Q_0}^G(\mathrm{St}^{M_\alpha}\otimes(\lambda-\rho_\alpha)_{M_\alpha},f)\big).$$

This concludes the proof of the lemma. □

10. Proof of part (ii) of theorem 5

Let us fix $M \in \mathcal{L}$ and $P_0 \in \mathcal{P}(M)$. Let us denote by $\Pi(M)$ (resp. $\Sigma(M)$) the set of equivalence classes of irreducible (resp. standard) admissible representations of M. It follows from the Langlands classification of the irreducible admissible representations of M that there exist uniquely determined integers

$$\Delta(\pi, \sigma) \qquad (\forall \pi \in \Pi(M),\ \forall \sigma \in \Sigma(M))$$

such that $\Delta(\pi, \sigma) = 0$ unless $\chi_\pi = \chi_\sigma$, where $\mathcal{Z}(M)$ is the Bernstein center of $\mathcal{C}_c^\infty(M)$ and where $\chi_\pi : \mathcal{Z}(M) \to \mathbb{C}$ and $\chi_\sigma : \mathcal{Z}(M) \to \mathbb{C}$ are the infinitesimal characters of π and σ respectively, and such that, for each $\pi \in \Pi(M)$, we have

$$\mathrm{tr}(\pi) = \sum_{\sigma \in \Sigma(M)} \Delta(\pi, \sigma)\, \mathrm{tr}(\sigma)$$

in the space of traces on $\mathcal{C}_c^\infty(M)$ (the sum is finite by the above property: there exist only finitely many π with a given infinitesimal character).

In paragraphs 5 and 6 of [Ar] Arthur has explained how to normalize the intertwining operators $M_{P|P_0}(\sigma \otimes \lambda)$ ($\lambda \in \mathfrak{a}_M^*$) into

$$R_{P|P_0}(\sigma \otimes \lambda) = r_{P|P_0}(\sigma \otimes \lambda)^{-1} M_{P|P_0}(\sigma \otimes \lambda)$$

for any $\sigma \in \Sigma(M)$ and he has defined a (G, M)-family of scalar functions,

$$\begin{aligned} & r_P(\pi \otimes \lambda, \sigma \otimes \lambda, \zeta; P_0) \\ &= \Delta(\pi, \sigma) r_{P|P_0}(\pi \otimes \lambda) r_{P|P_0}(\sigma \otimes \lambda)^{-1} r_{P|P_0}(\pi \otimes (\lambda + \zeta))^{-1} r_{P|P_0}(\sigma \otimes (\lambda + \zeta)) \end{aligned}$$

($\zeta \in \mathrm{Im}\, \mathfrak{a}_M^*$) for each $\pi \in \Pi(M)$ and each $\sigma \in \Sigma(M)$. Then he has proved (see loc. cit. Proposition 6.1) that, for any $f \in \mathcal{C}_c^\infty(G)$, we have the equality of rational functions of λ,

$$\begin{aligned} (*) \quad & \mathrm{tr}\big(\mathcal{R}_M(\pi \otimes \lambda; P_0) \circ I_{P_0}^G(\pi \otimes \lambda, f)\big) \\ &= \sum_{\substack{L \in \mathcal{L} \\ L \supset M}} \sum_{\sigma \in \Sigma(M)} r_M^L(\pi \otimes \lambda, \sigma \otimes \lambda)\, \mathrm{tr}\big(\mathcal{R}_L((\sigma \otimes \lambda)^L; Q_0) \circ I_{Q_0}^G((\sigma \otimes \lambda)^L, f)\big), \end{aligned}$$

where Q_0 is an arbitrary element in $\mathcal{P}(L)$, where

$$r_M^L(\pi \otimes \lambda, \sigma \otimes \lambda) = \lim_{\zeta \to 0} \sum_{R \in \mathcal{P}^L(M)} r_{RN_{Q_0}}(\pi \otimes \lambda, \sigma \otimes \lambda, \zeta; P_0) \theta_R(\zeta)^{-1}$$

and where

$$(\sigma \otimes \lambda)^L = I_{R_0}^L(\sigma \otimes \lambda)$$

for some $R_0 \in \mathcal{P}^L(M)$.

Now let us assume that f is very cuspidal. Then it is easy to see that, for any σ, we have

$$\mathrm{tr}(\mathcal{R}_L((\sigma\otimes\lambda)^L;Q_0)\circ I_{Q_0}^G((\sigma\otimes\lambda)^L,f))=0$$

unless $L=M$. Moreover each $\sigma\in\Sigma(M)$ is the class of an induced representation $(\sigma'\otimes\lambda')^M$ for some $M'\in\mathcal{L}$ which is contained in M, some tempered representation σ' of M' and some regular point λ' in $\mathrm{Re}\,\mathfrak{a}_{M'}^*$ and it is easy to see that we also have

$$\mathrm{tr}\big(\mathcal{R}_M(\sigma\otimes\lambda;P_0)\circ I_{P_0}^G(\sigma\otimes\lambda,f)\big)=0$$

unless $M'=M$, i.e. unless σ is essentially tempered. Similarly each essentially tempered irreducible representation σ of M is the class of an induced representation σ'^M for some $M'\in\mathcal{L}$ which is contained in M and some essentially square-integrable representation σ' of M' and again we have

$$\mathrm{tr}\big(\mathcal{R}_M(\sigma\otimes\lambda;P_0)\circ I_{P_0}^G(\sigma\otimes\lambda,f)\big)=0$$

unless $M'=M$, i.e. unless σ is essentially square-integrable. Therefore, for a very cuspidal f the formula (*) simply reduces to

$$(**)\quad \mathrm{tr}\big(\mathcal{R}_M(\pi\otimes\lambda;P_0)\circ I_{P_0}^G(\pi\otimes\lambda,f)\big) = \sum_{\sigma\in\Pi_{\mathrm{ess}}^2(M)}\Delta(\pi,\sigma)\,\mathrm{tr}\big(\mathcal{R}_M(\sigma\otimes\lambda;P_0)\circ I_{P_0}^G(\sigma\otimes\lambda,f)\big).$$

where $\Pi_{\mathrm{ess}}^2(M)\subset\Sigma(M)$ is the set of classes of standard representations of M which are essentially square-integrable.

Part (ii) of theorem 5 immediately follows from the formula (**) (for every positive integer d' the class of the Steinberg representation $\mathrm{St}^{GL_{d'}(F)}$ is the unique class of essentially square-integrable irreducible representations of $GL_{d'}(F)$ having the same infinitesimal character as the trivial representation $1^{GL_{d'}(F)}$ and we have

$$\Delta(1^{GL_{d'}(F)},\mathrm{St}^{GL_{d'}(F)})=(-1)^{d'-1}).$$

11. Proof of part (iii) of theorem 5

Let us fix $M\in\mathcal{L}$, $P_0\in\mathcal{P}(M)$ and $\pi\in\Pi(F^\times\backslash M)$. Let f be a very cuspidal function in $C_c^\infty(F^\times\backslash G)$. By an obvious variant of the arguments in section 10 we obtain the equality

$$\mathrm{tr}\big(\mathcal{R}_M(\pi\otimes\lambda;P_0)\circ I_{P_0}^G(\pi\otimes\lambda,f)\big) = \sum_{\sigma\in\Pi_{\mathrm{ess}}^2(F^\times\backslash M)}\Delta(\pi,\sigma)\,\mathrm{tr}\big(\mathcal{R}_M(\sigma\otimes\lambda;P_0)\circ I_{P_0}^G(\sigma\otimes\lambda,f)\big),$$

where $\Pi^2_{\text{ess}}(F^\times\backslash M) \subset \Sigma(F^\times\backslash M)$ is the set of classes of standard representations of $F^\times\backslash M$ which are essentially square-integrable.

Now let us assume moreover that f is bi-invariant under the Iwahori subgroup $\mathcal{B}$. Obviously, in the above sum we can restrict ourself to the $\sigma \in \Pi^2_{\text{ess}}(F^\times\backslash M)$ which have non-zero fixed vectors under $M \cap \mathcal{B}$. There are very few such σ. In fact σ should be isomorphic to $\mathrm{St}^M \otimes (-\mu)$ for some $\mu \in \mathfrak{a}^*_M$. But then by considering the infinitesimal characters we see that

$$\Delta(\pi,\sigma) = \prod_{j=1}^{s} \Delta(\pi_j,\sigma_j) = 0$$

unless $\pi_j \otimes \mu_j$ is isomorphic to the Steinberg representation or to the trivial representation of $GL_{d_j}(F)$ for every $j = 1,\ldots,s$. Part (iii) of theorem 5 follows.

References

[Ar] J. Arthur. — Intertwining operators and residues I. Weighted characters, *J. Funct. Anal.* **84**, (1989), 19-84.

[Ze] A.V. Zelevinsky. — Induced representations of reductive p-adic groups II, *Ann. Scient. Ec. Norm. Sup.* **13**, (1980), 165-210.

Index